VITAMIN D

NUTRITION ◊ AND ◊ HEALTH
Adrianne Bendich, Series Editor

Vitamin D: *Molecular Biology, Physiology, and Clinical Applications,*
  edited by Michael F. Holick

Preventive Nutrition: *The Comprehensive Guide for Health Professionals,*
  edited by Adrianne Bendich and Richard J. Deckelbaum

VITAMIN D

MOLECULAR BIOLOGY, PHYSIOLOGY, AND CLINICAL APPLICATIONS

Edited by

MICHAEL F. HOLICK, PhD, MD

Boston University Medical Center, Boston, MA

HUMANA PRESS
TOTOWA, NEW JERSEY

© 1999 Humana Press Inc.
999 Riverview Drive, Suite 208
Totowa, New Jersey 07512

For additional copies, pricing for bulk purchases, and/or information about other Humana titles, contact Humana at the above address or at any of the following numbers: Tel.: 973-256-1699; Fax: 973-256-8341; E-mail: humana@humanapr.com

All rights reserved.

No part of this book may be reproduced, stored in a retrieval system, or transmitted in any form or by any means, electronic, mechanical, photocopying, microfilming, recording, or otherwise without written permission from the Publisher.

All articles, comments, opinions, conclusions, or recommendations are those of the author(s), and do not necessarily reflect the views of the publisher.

Cover design by Patricia F. Cleary.

This publication is printed on acid-free paper. ∞
ANSI Z39.48-1984 (American National Standards Institute) Permanence of Paper for Printed Library Materials.

Photocopy Authorization Policy:
Authorization to photocopy items for internal or personal use, or the internal or personal use of specific clients, is granted by Humana Press Inc., provided that the base fee of US $8.00 per copy, plus US $00.25 per page, is paid directly to the Copyright Clearance Center at 222 Rosewood Drive, Danvers, MA 01923. For those organizations that have been granted a photocopy license from the CCC, a separate system of payment has been arranged and is acceptable to Humana Press Inc. The fee code for users of the Transactional Reporting Service is: [0-89603-467-4/98 $8.00 + $00.25].

Printed in the United States of America. 10 9 8 7 6 5 4 3 2 1

Library of Congress Cataloging-in-Publication Data

Vitamin D: molecular biology, physiology, and clinical applications/edited by Michael F. Holick.
   p. cm.—(Nutrition and health)
  Includes bibliographical references and index.
  ISBN 0-89603-467-4 (alk. paper)
    1. Vitamin D—Physiological effect. 2. Vitamin D—Therapeutic use. I. Holick, M. F. (Michael F.)
II. Series: Nutrition and health (Totowa, NJ)
   [DNLM: 1. Vitamin D—physiology. 2. Vitamin D—pharmacology. 3. Vitamin D—therapeutic use.
QU 173 V8373 1999]
QP772.V53V585 1999
615'.328—dc21
DNLM/DLC
for Library of Congress   98-45813
                                                                                   CIP

IN MEMORIAM
John G. Haddad, Jr., MD (1937–1997)

John G. Haddad, Jr., MD, died unexpectedly at the age of 59 years on May 22, 1997, on his way to the vitamin D meeting in Strasbourg. He was elected the 10th President of the American Society of Bone and Mineral Research and was Associate Chairman for Research in the Department of Medicine at the University of Pennsylvania School of Medicine at the time of his untimely death. In order to review the impact this man had on all who knew and loved him, one has to recognize his character. He was honest, intelligent, energetic, ephemeral, compassionate, wise, industrious, clairvoyant, understanding, witty, unpretentious, and a damn good physician-scientist. Beloved by his wife, Julie, son John George, and daughter Margaret, he offered guidance and leadership to his students and peers alike with enviable grace and distinction. As a part time saxophonist, he could also jazz and jive with the best of them.

I had the good fortune to meet John for the first time in 1968 when he entered our fellowship program at Washington University as a first-year research fellow during a period when we were attempting to probe the mysteries of vitamin D metabolism. Although the vagaries of the intermediary metabolism of vitamin D were just being unraveled, the potent hormonal form of the vitamin, that is, $1,25(OH)_2D$, had not yet been identified. Moreover, available "vitamin D assays" were essentially limited to tedious bioassays and multistep chromatography–spectroscopy procedures. In this milieu, John became interested in the nature of the plasma transport of the major $25(OH)D$ metabolite of vitamin D. While analyzing the function of the human plasma binding of $25(OH)D$, he subsequently developed a simple but precise competitive radioimmunoassay for this sterol. The demand to measure circulating $25(OH)D$ in those days was certainly apparent from the flow of national and international scientists who visited his laboratory in order to become familiar with the technique he had developed. Once again, he gave of himself graciously to friends and potential collaborators. Having achieved the position of Associate Professor of Medicine at Washington University in 1980, John left us to become Professor of Medicine and Director of the Division of Endocrinology at the University of Pennsylvania. He subsequently developed a strong endocrine unit and continued his research on the multifunctional aspects of circulating vitamin D binding proteins with specific emphasis on the "immune system."

John often confided in me that during the pursuit of his many assigned duties, he loved to close his office door just to "have a smoke and drink a Coke!" Himself a man of wit, I believe that John would have concurred with Ogden Nash, often considered to be a causal person who dabbled in verse, but who, like John, had much of importance to say. John Haddad's contribution to this text on the "Clinical Significance of the Vitamin D-Binding Protein" reflects the depth of his knowledge and his enviable casual but directive style. John Haddad was a good husband, loving father, and devoted physician-scientist. As recently detailed by his close friend, Dr. Fred Kaplan, Professor of Orthopaedic Surgery at the University of Pennsylvania, "He loved the richness of existence, the fabric of life, the evolving drama, the deeply held feelings, and the texture of the day."

Louis V. Avioli, MD, FACE
Schoenberg Professor of Medicine
Professor of Orthopaedic Surgery
Director, Division of Bone and Mineral Diseases
Washington University School of Medicine
St. Louis, MO

SERIES INTRODUCTION

The *Nutrition and Health* series of books has as an overriding mission to provide health professionals with texts that are considered essential because each includes: a synthesis of the state of the science; timely, in-depth reviews by the leading researchers in their respective fields; extensive, up-to-date fully annotated reference lists; a detailed index; relevant tables and figures; identification of paradigm shifts and the consequences; virtually no overlap of information between chapters, but targeted, inter-chapter referrals, suggestions of areas for future research; and balanced, data-driven answers to patient questions that are based on the totality of evidence rather than the findings of any single study.

The series volumes are not the outcome of a symposium. Rather, each editor has the potential to examine a chosen area with a broad perspective, both in subject matter as well as in the choice of chapter authors. The international perspective, especially with regard to public health initiatives, is emphasized where appropriate. The editors, whose training is both research and practice oriented, have the opportunity to develop a primary objective for their book, define the scope and focus, and then invite the leading authorities from around the world to be part of their initiative. The authors are encouraged to provide an overview of the field, discuss their own research, and relate the research findings to potential human health consequences. Because each book is developed *de novo*, the chapters can be coordinated so that the resulting volume imparts greater knowledge than the sum of the information contained in the individual chapters.

Vitamin D: Physiology, Molecular Biology, and Clinical Applications, edited by Michael F. Holick, PhD, MD exemplifies the mission of the *Nutrition and Health* series. Dr. Holick provides the reader with the clear understanding that vitamin D is more than the vitamin added to milk to help our body absorb calcium. He includes the latest areas of research with critical significance to human health. Health care professionals not intimately involved in the fields of cancer, immune function, diabetes, dermatology, or numerous other specialized areas are provided with the most significant and timely information related to the action of vitamin D. The volume is true to its title and includes understandable chapters in highly technical areas of molecular biology such as genetics, enzymology, receptor biology, and genomics. Chapters dealing with physiology include the latest data on the "classic" mechanism of bone formation as well as the new information on the multifaceted role of vitamin D in other organ systems. The recent studies that examine the clinical effects of moderate vitamin D deficiency and the consequences of supplementation with natural as well as synthetic analogs are also reviewed. Thus, *Vitamin D: Physiology, Molecular Biology, and Clinical Applications* represents a comprehensive, essential resource for medical students, practicing physicians, and other health care providers, as well as clinical and laboratory researchers. Beyond its value as a resource, it is an important text for faculty and graduate students interested in the most up-to-date synthesis of information about vitamin D.

Adrianne Bendich, PhD

PREFACE

Vitamin D is truly a remarkable molecule. Naturally, vitamin D can only be made by the help of sunlight. It is the ultraviolet B portion of sunlight that provides the energy required for the transformation of provitamin D to previtamin D. Once formed, previtamin D, which is inherently unstable, undergoes a metamorphosis rearranging its double bonds to the more stable vitamin D structure. The photosynthesis of vitamin D has been occurring on earth almost from the initiation of life in the fertile oceans. As organisms evolved to utilize the energy from sunlight for photosynthesizing their food, i.e., carbohydrates, they also photosynthesized vitamin D. For reasons that are almost beyond comprehension, as life forms evolved, first in the oceans and then on land, it was the photosynthesis of vitamin D in the skin of aquatic and land vertebrates that permitted them efficiently to utilize their dietary calcium for the purpose of regulating a wide variety of metabolic functions and for the deposition of calcium into the skeleton to make it into a rigid structure. However, vitamin D, by itself, is inactive and must undergo two metabolic activations, first in the liver and then in the kidney, to form its biologically functioning form 1,25-dihydrovitamin D_3 [$1,25(OH)_2D_3$]. It is intuitive that if vitamin D were critically important for calcium homeostasis and skeletal health that specific receptors for $1,25(OH)_2D_3$ would be present in calcium-regulating tissues, the intestine, kidney, and bone. Since vitamin D is a sterol, it interacted with the nucleus to unlock genetic information to initiate its influence on calcium and bone metabolism. This functional form, $1,25(OH)_2D_3$, has been used clinically for treating a wide variety of disorders in calcium and bone metabolism, including the bone disease associated with kidney failure, renal osteodystrophy, as well as osteoporosis. As important as these functions are for vitamin D, in the late 1970s discovery of the localization of $1,25(OH)_2D_3$ in tissues not related to calcium metabolism, such as the skin, gonads, stomach, and thymus, the immune system, and some cancer cells, opened a new and exciting chapter in vitamin D that continues to evolve today. Remarkably, $1,25(OH)_2D_3$ is an extremely potent hormone/cytokine that can regulate the proliferation and alter the differentiation of a wide variety of cell types. These seminal observations have given rise to the use of $1,25(OH)_2D_3$ and its analogs for treating the proliferative skin disorder psoriasis and for investigating their anticancer potential. Indeed, active vitamin D treatment is now the first line of treatment for psoriasis. The recent revelations that $1,25(OH)_2D_3$ and its analogs can manipulate the immune system has given rise to exciting studies in animals that suggest that $1,25(OH)_2D_3$ could be useful for the prevention and/or treatment of insulin-dependent diabetes mellitus (type I) and multiple sclerosis.

Vitamin D, however, is taken for granted. It is assumed that there are sufficient amounts of vitamin D in our diet to satisfy our body's requirement. We have forgotten that it is sunlight that provides most humans, and in fact, most vertebrates with their vitamin D requirement. The change in lifestyle, especially for older adults, has given rise to an epidemic of vitamin D-deficiency bone disease. This is responsible for exacerbating osteoporosis in a large segment of the elder population. There is mounting evidence that vitamin D deficiency may also be associated with an increased risk of colon and breast cancer as well as osteoarthritis.

The inspiration for *Vitamin D: Physiology, Molecular Biology, and Clinical Aspects* came from discussions with medical students, house staff, health care professionals, internists, dermatologists, and basic scientists who have an incomplete understanding of the broad spectrum activities of vitamin D for human health. This endeavor was intended to bring together the world's leading experts in various aspects of vitamin D and organize an up-to-date review of the subject in a format that would be highly informative not only to medical students, graduate students, and health care professionals, but also provide a reference source for the most up-to-date information about the clinical applications of vitamin D for human health and disease. It is hoped that this book will stimulate new interest, ignite the imagination, and provide a new appreciation for this remarkable gift from the sun for the connoisseurs and novices in the field of vitamin D.

Michael F. Holick

DEDICATION

I dedicate this book to my best friend, colleague, business partner, and wife Sally who has been so supportive of my career and who has been the absolute love and joy of my life.

CONTENTS

In Memoriam: *John G. Haddad* .... *v*
Series Introduction .... *vi*
Preface .... *vii*
Contributors .... *xi*

1. Evolution, Biologic Functions, and Recommended Dietary Allowances for Vitamin D *1*
   Michael F. Holick

2. Photobiology of Vitamin D *17*
   Tai C. Chen

3. Functional Metabolism and Molecular Biology of Vitamin D Action *39*
   Laura C. McCary and Hector F. DeLuca

4. Metabolism and Catabolism of Vitamin D, Its Metabolites, and Clinically Relevant Analogs *57*
   Glenville Jones

5. The Enzymes Responsible for Metabolizing Vitamin D *85*
   Kyu-Ichiro Okuda and Yoshihiko Ohyama

6. The Vitamin D Binding Protein and Its Clinical Significance *101*
   John G. Haddad

7. Molecular Biology of the Vitamin D Receptor *109*
   Paul N. MacDonald

8. Vitamin D Receptor Translocation *129*
   Julia Barsony

9. Molecular Recognition and Structure–Activity Relations in Vitamin D-Binding Protein and Vitamin D Receptor *147*
   Rahul Ray

10. Mechanism of Action of 1,25-Dihydroxyvitamin D_3 on Intestinal Calcium Absorption and Renal Calcium Transport *163*
    Mihali Raval-Pandya, Angela R. Porta, and Sylvia Christakos

11. Biologic and Molecular Effects of Vitamin D on Bone *175*
    Jane B. Lian, Ada Staal, André van Wijnen, Janet L. Stein, and Gary S. Stein

12. Nongenomic Rapid Effects of Vitamin D *195*
    Daniel T. Baran

13. Noncalcemic Actions of 1,25-Dihydroxyvitamin D_3 and Clinical Implications *207*
    Michael F. Holick

| 14 | Regulation of Parathyroid Hormone Gene Expression and Secretion by Vitamin D 217
Tally Naveh-Many and Justin Silver

| 15 | Vitamin D Assays and Their Clinical Utility 239
Ronald L. Horst and Bruce W. Hollis

| 16 | Rickets and Vitamin D Deficiency 273
Michèle Garabédian and Hanifa Ben-Mekhbi

| 17 | Vitamin D and Bone Health in Adults and the Elderly 287
Clifford J. Rosen

| 18 | Inherited Defects of Vitamin D Metabolism 307
Marie Demay

| 19 | Molecular Defects in the Vitamin D Receptor Associated with Hereditary 1,25-Dihydroxyvitamin D-Resistant Rickets 317
Peter J. Malloy and David Feldman

| 20 | Extrarenal Production of 1,25-Dihydroxyvitamin D and Clinical Implications 337
John S. Adams

| 21 | Clinical Utility of 1,25-Dihydroxyvitamin D_3 and Its Analogs for the Treatment of Psoriasis and Other Skin Diseases 357
J. Reichrath and Michael F. Holick

| 22 | Epidemiology of Cancer Risk and Vitamin D 375
Cedric F. Garland, Frank C. Garland, and Edward D. Gorham

| 23 | Chemotherapeutic and Chemopreventive Actions of Vitamin D_3 Metabolites and Analogs 393
Thomas A. Brasitus and Marc Bissonnette

| 24 | Vitamin D and Breast Cancer 411
Johannes P. T. M. van Leeuwen, Trudy Vink-van Wijngaarden, and Huibert A. P. Pols

| 25 | Anticancer Activity of Vitamin D Analogs 431
Milan R. Uskoković, Candace S. Johnson, Donald L. Trump, and Robert H. Getzenberg

Index 447

CONTRIBUTORS

John S. Adams • *Cedars Sinai Medical Center, Los Angeles, CA*
Daniel T. Baran • *Department of Orthopedics, University of Massachusetts Medical Center, Worcester, MA*
Julia Barsony • *Laboratory of Cell Biochemistry and Biology, NIDDK, National Institutes of Health, Bethesda, MD*
Hanifa Ben-Mekhbi • *Department of Pediatry, Ben-Badis Hospital, Constantine, Algeria*
Marc Bissonnette • *Department of Medicine, The University of Chicago Hospitals and Clinics, Chicago, IL*
Thomas A. Brasitus • *Department of Medicine, The University of Chicago Hospitals and Clinics, Chicago, IL*
Tai C. Chen • *Boston University School of Medicine, Boston, MA*
Sylvia Christakos • *Department of Biochemistry and Molecular Biology, UMDNJ-New Jersey Medical School, Newark, NJ*
Hector F. DeLuca • *Department of Biochemistry, University of Wisconsin, Madison, WI*
Marie Demay • *Endocrine Unit, Massachusetts General Hospital, Boston, MA*
David Feldman • *Department of Endocrinology, Stanford Medical Center, Stanford, CA*
Michèle Garabédian • *Hopital St. Vincent De Paul, Paris, France*
Cedric F. Garland • *Department of Community Medicine, University of California, San Diego, CA*
Frank C. Garland • *Department of Community Medicine, University of California, San Diego, CA*
Robert H. Getzenberg • *Department of Medicine, Cancer Institute, University of Pittsburgh, PA*
Edward D. Gorham • *Department of Community Medicine, University of California, San Diego, CA*
John G. Haddad[†] • *School of Medicine, University of Pennsylvania, Philadelphia, PA*
Michael F. Holick • *Boston University School of Medicine, Boston, MA*
Bruce W. Hollis • *Metabolic Diseases and Immunology Research Unit, Agricultural Research Service, National Animal Disease Center, US Department of Agriculture, Ames, IA*
Ronald L. Horst • *Metabolic Diseases and Immunology Research Unit, Agricultural Research Service, National Animal Disease Center, US Department of Agriculture, Ames, IA*
Candace S. Johnson • *Department of Medicine, Cancer Institute, University of Pittsburgh, PA*
Glenville Jones • *Departments of Biochemistry and Medicine, Queen's University, Kingston, Ontario, Canada*
Jane B. Lian • *Department of Cell Biology, University of Massachusetts Medical Center, Worcester, MA*

PAUL N. MACDONALD • *Department of Pharmacological and Physiological Science, St. Louis University Health Science Center, St. Louis, MO*
PETER J. MALLOY • *Department of Medicine, Stanford University School of Medicine, Stanford, CA*
LAURA C. MCCARY • *Department of Biochemistry, University of Wisconsin, Madison, WI*
TALLY NAVEH-MANY • *Hadassah University Hospital, Jerusalem, Israel*
YOSHIHIKO OHYAMA • *Department of Surgery I, Miyazaki Medical College, Kiyotake Miyazaki, Japan*
KYU-ICHIRO OKUDA • *Department of Surgery I, Miyazaki Medical College, Kiyotake Miyazaki, Japan*
MIHALI RAVAL-PANDYA • *Department of Biochemistry and Molecular Biology, UMDNJ-New Jersey Medical School, Newark, NJ*
HUIBERT A. P. POLS • *Department of Internal Medicine, Erasmus University Medical School, Rotterdam, The Netherlands*
ANGELA R. PORTA • *Department of Biochemistry and Molecular Biology, UMDNJ-New Jersey Medical School, Newark, NJ*
RAHUL RAY • *Bioorganic Chemsitry and Structural Biology Group, Vitamin D Laboratory, Department of Medicine, Boston University School of Medicine, Boston, MA*
J. REICHRATH • *Boston University School of Medicine, Boston, MA*
CLIFFORD J. ROSEN • *Maine Center for Osteoporosis Research and Education, St. Joseph Hospital, Bangor, ME*
JUSTIN SILVER • *Hadassah University Hospital, Jerusalem, Israel*
ADA STAAL • *Department of Cell Biology, University of Massachusetts Medical Center, Worcester, MA*
GARY S. STEIN • *Department of Cell Biology, University of Massachusetts Medical Center, Worcester, MA*
JANET L. STEIN • *Department of Cell Biology, University of Massachusetts Medical Center, Worcester, MA*
DONALD L. TRUMP • *Department of Medicine, Cancer Institute, University of Pittsburgh, PA*
MILAN R. USKOKOVIĆ • *Hoffman-La Roche, Nutley, NJ*
JOHANNES P. T. M. VAN LEEUWEN • *Department of Internal Medicine, Erasmus University Medical School, Rotterdam, The Netherlands*
ANDRÉ VAN WIJNEN • *Department of Cell Biology, University of Massachusetts Medical Center, Worcester, MA*
TRUDY VINK-VAN WIJNGAARDEN • *Department of Internal Medicine, Erasmus University Medical School, Rotterdam, The Netherlands*

1 Evolution, Biologic Functions, and Recommended Dietary Allowances for Vitamin D

Michael F. Holick

1. EVOLUTIONARY PERSPECTIVE

1.1. The Calcium Connection

Approximately 400 million years ago, as vertebrates ventured from the ocean onto land, they were confronted with a significant crisis. As they had evolved in the calcium-rich ocean environment, they utilized this abundant cation for signal transduction and a wide variety of cellular and metabolic processes. In addition, calcium became a major component of the skeleton of marine animals and provided the "cement" for structural support. However, on land, the environment was deficient in calcium; as a result, early marine vertebrates that ventured onto land needed to develop a mechanism to utilize and process the scarce amounts of calcium in their environment in order to maintain their calcium-dependent cellular and metabolic activities and also satisfy the large requirement for calcium to mineralize their skeletons.

For most ocean-dwelling animals that were bathed in the high calcium bath (approx 400 mmol), they could easily extract this divalent cation from the ocean by specific calcium transport mechanisms in the gills or by simply absorbing it through their skin. However, once on land, a new strategy was developed whereby the intestine evolved to efficiently absorb the calcium present in the diet. For reasons that are unknown, an intimate relationship between sunlight and vitamin D evolved to play a critical role in regulating intestinal absorption of calcium from the diet to maintain a healthy mineralized skeleton and satisfy the body's requirement for this vital mineral.

Although vitamin D became essential for stimulating the intestine to absorb calcium from the diet, it could only do so by being activated first in the liver to 25-hydroxyvitamin D [25(OH)D] and then in the kidney to its active form 1,25-dihydroxyvitamin D [$1,25(OH)_2D$] *(1,2)* (Fig. 1). Once formed, $1,25(OH)_2D$ enters the circulation and travels to its principal calcium regulating target tissues, the small intestine and bone. The small intestinal absorptive cells contain specific receptors (known as vitamin D receptors) that specifically bind $1,25(OH)_2D_3$ and in turn activate vitamin D-responsive genes to enhance intestinal calcium absorption *(3)* (*see* Chapter 10 for details). However, when dietary calcium is insufficient to satisfy the body's requirement, $1,25(OH)_2D$ travels to

From: *Vitamin D: Physiology, Molecular Biology, and Clinical Applications*
Edited by: M. F. Holick © Humana Press Inc., Totowa, NJ

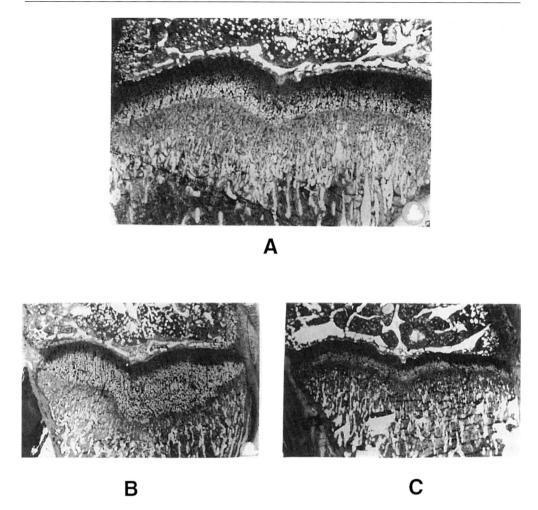

Fig. 2. Epiphyseal plates of tibias from rats that were fed (**A**) a vitamin D, deficient diet and supplemented with 125 ng (5 IU) of vitamin D_3 orally five times a week, (**B**) a vitamin D-deficient diet containing 3% calcium and 0.65% phosphorus, and (**C**) a vitamin D-deficient diet with 20% lactose, 4% calcium, and 1% phosphorus. Note the wide and disorganized hypertrophic zone in the vitamin D-deficient rat's tibial epiphyseal (B) fed high calcium and normal phosphorus diet compared with normal tibial epiphyseal plates from the rats that were either vitamin D repleted (A) or maintained on normal serum calcium and phosphorus by being on a high-calcium lactose, high-phosphorus diet (C). (Data from ref. 8.)

than bone formation, leading to gradual bone loss. Vitamin D deficiency in adults causes a decrease in the efficiency of intestinal calcium absorption and secondary hyperparathyroidism. This results in a mineralization defect of the skeleton that is similar to rickets. However, because the epiphyseal plates are closed in adults, there is no widening of the end of the long bones. Instead, there is a subtle but significant defect in bone mineralization of the collagen matrix (osteoid) that was laid down by the osteoblasts, leading to osteomalacia (unmineralized osteoid). In addition, since the body can no longer depend on the dietary calcium to satisfy its calcium requirements, it calls on the skeleton for its calcium stores. The increased PTH production mobilizes the precious calcium stores from

Fig. 3. Typical presentation of two children with rickets. The child in the middle is normal; the children on either side have severe muscle weakness and bony deformities including bowed legs (**right**) or knock knees (**left**). (Reproduced with permission from ref. *59*.)

the skeleton, thereby making the skeleton more porotic. Thus, vitamin D deficiency in adults can cause and exacerbate osteoporosis (*see* Chapter 17).

When dietary calcium intake, even in the presence of adequate vitamin D, is inadequate to satisfy the body's calcium needs, vitamin D [through $1,25(OH)_2D$], mobilizes osteoclast precursor stem cells and induces them to become mature osteoclasts. These osteoclasts, in turn, enhance bone calcium resorption, thereby maintaining the blood calcium in an acceptable physiologic range *(1,2)*. Vitamin D does not have a direct role in the mineralization of the skeleton *(8,9)*. It indirectly participates in skeletal mineralization by its effects on maintaining the serum calcium and phosphorus in the normal range so they are at supersaturating levels for deposition into the collagen matrix to form calcium hydroxyapatite crystals.

2.2. Sources of Vitamin D

A major source of vitamin D for most humans comes from exposure of the skin to sunlight *(5)*. A variety of factors limit the skin's production of vitamin D_3. An increase in skin pigmentation *(10)* or the topical application of a sunscreen will absorb solar ultraviolet B photons, thereby significantly reducing the production of vitamin D_3 in the skin *(11)*. Aging decreases the capacity of the skin to produce vitamin D_3 because of the decrease in the concentration of its precursor 7-dehydrocholesterol. Above the age of 65, there is a fourfold decrease in the capacity of the skin to produce vitamin D_3 when compared with a younger adult *(12)*. An alteration in the zenith angle of the sun caused by a change in latitude, season of the year, or time of day can dramatically influence the skin's production of vitamin D *(13)*. Above and below latitudes of approx 40° north and south, respectively, vitamin D synthesis in the skin is absent during most of the winter.

In nature, very few foods contain vitamin D. Oily fish and oils from the liver of some fish, including cod and tuna, are naturally occurring foods containing vitamin D. Although it is generally accepted that livers from meat animals such as cows, pigs, and chickens contain vitamin D, there is no evidence for this. However, the common practice of Eskimos eating a small amount of polar bear liver did provide them with their vitamin D and vitamin A requirements because this animal concentrates both of these vitamins in its liver.

In the United States and Canada, milk is routinely fortified with vitamin D, as are some bread products and cereals. However, three surveys of the vitamin D content in milk in the United States and Canada have revealed that upward of 70% of samples tested did not contain 80% of the amount of vitamin D stated on the label *(14–16)*. Approximately 10% of milk samples did not contain any vitamin D. In Europe, most countries do not fortify milk with vitamin D because in the 1950s there was an outbreak of vitamin D intoxication in young children, resulting in laws that forbade the fortification of foods with vitamin D *(15)*. However, with the recognition that vitamin D deficiency continues to be a great health concern for both children and older adults, many European countries are fortifying cereals, breads, and margarine with vitamin D. Indeed, in many European countries, margarine is the major dietary source of vitamin D *(17)*.

Multivitamin preparations often contain 400 IU of vitamin D. This is an excellent source of vitamin D, especially when there is inadequate exposure to sunlight. Pharmaceutical preparations of vitamin D include capsules and tablets that contain 50,000 IU of vitamin D_2 and liquids that usually contain 8000 IU/mL.

3. RECOMMENDED ADEQUATE DIETARY INTAKE OF VITAMIN D

3.1. Birth to 6 Months

In utero, the infant is wholly dependent on the mother for vitamin D. The 25(OH)D passes from the placenta into the fetus's blood stream. Since the half-life for 25(OH)D is approx 2–3 wk, after birth the infant can remain vitamin D sufficient for at least several weeks. The infant, however, depends on either sunlight exposure or dietary vitamin D for his/her vitamin D requirement *(17–19)*. Human breast milk and unfortified cow's milk have very little vitamin D *(18,20)*. Thus infants that are only fed human breast milk are prone to developing vitamin D deficiency, especially during the winter when they are not obtaining their vitamin D from exposure to sunlight *(21)*. Conservative estimates of the length of time a human milk-fed infant in the Midwest must be exposed to sunlight to maintain serum concentrations above the lower limit of normal are 2 h/wk with only the face exposed to sunlight or 30 min/wk with just a diaper on *(19)*. Human milk contains low amounts of vitamin D, and colostrum averages 15.9 ± 8.6 IU/L *(20)*. There is a direct

Table 1
Adequate Intake (AI), Previous Recommended Dietary Allowance (RDA),
Reasonable Daily Allowance, and Tolerable Upper Limit (UL) for Vitamin D

Age	AI [IU (µg)/d]	RDA [IU (µg)/d]	Reasonable daily allowance (IU)	UL [IU (µg)/d]
0–6 mo	200 (5)	300 (7.5)	200–400	1000 (25)
6 mo–12 yr	200 (5)	300 (7.5)	200–400	1000 (25)
1–18 yr	200 (5)	400 (10)	200–400	2000 (50)
19–50 yr	200 (5)	200 (5)	200–400	2000 (50)
51–70 yr	400 (10)	200 (5)	400–600	2000 (50)
71+ yr	600 (15)	200 (5)	600–800	2000 (50)
Pregnancy	200 (5)	400 (10)	200–400	2000 (50)
Lactation	200 (5)	400 (10)	200–400	2000 (50)

relationship between vitamin D intake and vitamin D content in human milk. However, even when women were consuming between 600 to 700 IU/d of vitamin D, the vitamin D content in their milk was only between 5 and 136 IU/L (18,20).

Vitamin D intakes between 340 and 600 IU/d have been reported to have the maximum effect on linear growth of infants (21–24). However, when Chinese infants were given either 100, 200, or 400 IU of vitamin D/d, none of the infants demonstrated any evidence of rickets (25). Markstad and Elzouki (17) reported that Norwegian infants fed infant formula containing 300 IU/d obtained blood levels of 25(OH)D of >11 ng/mL, which is considered the lower limit of normal for this assay.

Therefore, a minimum of 100 IU of vitamin D a day is satisfactory for preventing rickets. However, in the absence of any exposure to sunlight, it is likely that infants require a larger amount of vitamin D to optimize bone health. Thus infants may require as much as 300–400 IU of vitamin D to maximize bone health in the absence of sunlight exposure (21,22). Based on all the available evidence, the Institute of Medicine (IOM) of the U.S. National Academy of Sciences recommended in 1997 that an adequate dietary intake (AI) for infants be 200 IU (5 µg)/d (21). For infants not exposed to sunlight, the vitamin D intake should be at least 200 IU/d. However, twice this amount, i.e., 400 IU (10 µg)/d, the current amount included in 1 L of standard infant formula or 1 qt of commercial cows' milk, would not be excessive (Table 1).

3.2. Ages 6–12 Months

As for infants, it was recommended by the IOM (21) that the adequate dietary intake for ages 6–12 months be 200 IU (5 µg)/d. This was based on the observation that in the absence of sun-mediated vitamin D synthesis, approx 200 IU/d of vitamin D maintained circulating concentrations of 25(OH)D in the normal range of older infants, but below circulating concentrations attained in similar aged infants in the summer (17). It was also suggested that twice this amount, i.e., 400 IU/d, would not be excessive (21).

3.3. Ages 1–8 Years

Children aged 1–8 years of all races obtain most of their vitamin D from exposure to sunlight and therefore do not normally need to ingest vitamin D (21,26). However, for children who live in far northern and southern latitudes, such as northern Canada, Alaska, and Tierra del Fuego, vitamin D supplementation may be necessary. Although there were no data on how much vitamin D is required to prevent vitamin D deficiency in children aged 1–8 yr, extrapolating from available data in slightly older children who were not exposed to an adequate amount of sun-mediated vitamin D with a mean dietary intake between 75 and 100 IU/d showed no evidence of vitamin D deficiency, with normal circulating concentrations of 25(OH)D *(27)*. Therefore, the IOM *(21)* recommended 200 IU/d for this age group.

3.4. Ages 9–18 Years

Children aged 9–18 have a rapid growth spurt that requires a marked increase in their requirement for calcium and phosphorus to maximize skeletal mineralization. During puberty, the metabolism of 25(OH)D to 1,25(OH)$_2$D increases *(27)*. In turn, the increased blood levels of 1,25(OH)$_2$D enhance the efficiency of the intestine to absorb dietary calcium and phosphorus to satisfy the growing skeleton's requirement for these minerals during its rapid growth phase. However, despite the increased production of 1,25(OH)$_2$D, there is no scientific evidence demonstrating an increased requirement for vitamin D in this age group, probably because circulating concentrations of 1,25(OH)$_2$D are approx 500–1000 times lower than those of 25(OH)D (i.e., 15–60 pg/mL vs. 15–55 ng/mL, respectively).

A few studies conducted in children during the pubertal years show that children maintained a normal serum 25(OH)D with dietary vitamin D intakes of 2.5–10 µg/d *(27)*. When intakes were <2.5 µg/d, Turkish children aged 12–17 yr had 25(OH)D levels consistent with vitamin D deficiency, i.e., <11 ng/mL *(28)*. However, with regular sun exposure, there was no need for dietary vitamin D *(26,28–31)*. Children who live in the far northern and southern latitudes may not be able to synthesize enough vitamin D in their skin during the summer months to be stored for use during the winter and therefore may require a vitamin D supplement *(32)*. Therefore the IOM *(21)* recommended an AI for vitamin D of 200 IU (5 µg)/d.

3.5. Ages 19–50 Years

Most young and middle-aged adults obtain their vitamin D requirement from casual every day exposure to sunlight. Very few studies have evaluated the vitamin D requirement in this age group. An evaluation of 67 white and 70 black premenopausal women ingesting 138 ± 84 and 73 IU/d, respectively, revealed that serum 25(OH)D levels were in the low-normal range (circulating concentrations of 21.4 ± 12 and 18.3 ± 9 ng/mL, respectively) *(33)*.

A study conducted in submariners may provide an insight as to how much vitamin D is actually required for this age group in the absence of exposure to sunlight. Young groups of male submariners during a 3-mo voyage received either a placebo or 600 IU of vitamin D a day. Circulating concentrations of 25(OH)D decreased by 40% over the 3 mo in the submariners who did not receive vitamin D supplementation, whereas 25(OH)D levels were maintained in the group that received 600 IU/d. After the voyage, when the submariners were exposed to sunlight, there was a dramatic increase in 25(OH)D levels (Fig. 4) *(5)*.

In a cross-sectional study, 67 white (age 36.5 ± 6.4 yr) and 70 black (age 37 ± 6.7 yr) premenopausal women living in the New York City area who were active and exposed to sunlight had average daily vitamin D intakes of 139 ± 84 IU and 145 ± 73 IU/d,

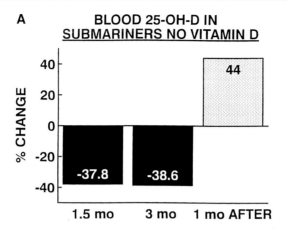

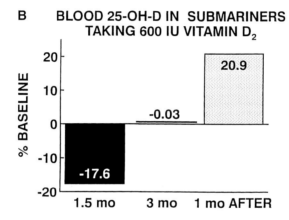

Fig. 4. (**A**) Percent change in circulating concentrations of 25-hydroxyvitamin D [25(OH)D] from baseline before the submariners entered the submarine compared with 1.5 and 3 mo aboard a submarine while receiving no vitamin D and 1 mo after leaving the submarine; (**B**) 25(OH)D concentrations in a similar a group of submariners who received a multivitamin that contained 15 μg (600 IU) vitamin D_2. (Data from ref. 5.)

respectively. Both groups of women had normal circulating concentrations of 25(OH)D and PTH concentrations (33). During the winter months (November through May) in Omaha, Nebraska, 6% of a group of young women aged 25–35 yr ($n = 52$) maintained serum concentrations of 25(OH)D >20 ng/mL when the daily vitamin D intake was estimated to be between 131 and 135 IU/d (34).

Therefore, based on all of the available literature, both sunlight and diet play an essential role in providing vitamin D to this age group. Thus, the IOM (21) recommended an AI for this age group of 200 IU (5 μg)/d.

3.6. Ages 51–70 Years

Men and women aged 51–70 are very dependent on sunlight for most of their vitamin D requirement. However, this age group is much more health conscious and is especially concerned about their skin health as it relates to skin cancer and wrinkles. Therefore, there is increased use of clothing and sunscreen over sun-exposed areas, which decreases the amount of ultraviolet B radiation penetrating the skin and thereby decreasing the

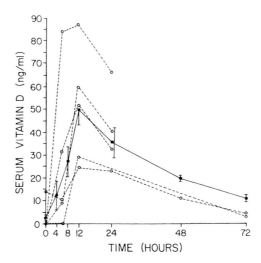

Fig. 5. Vitamin D absorption in young (closed circles) and elderly (open circles) adults. Each subject received an oral dose of 50,000 IU of vitamin D_2, and at various times blood determinations were made for circulating concentrations of vitamin D. (Data from ref. *35*.)

cutaneous production of vitamin D_3 by 95–100% *(5,11)*. In addition, age decreases the capacity of the skin to produce vitamin D_3 *(12)*. Although it has been suggested that aging may decrease the ability of the intestine to absorb dietary vitamin D, two studies have revealed that aging does not alter the absorption of pharmacologic doses of vitamin D *(35,36)* (Fig. 5).

It is now recognized that men and women in this age group require twice the amount of vitamin D compared with younger adults. This is in part based on the data of Krall and Dawson-Hughes *(37)*, who observed in 66 healthy postmenopausal women (mean age 60 ± 5 yr) a seasonal variation in calcium retention that positively correlated with serum 25(OH)D levels in women on low-calcium diets. Furthermore, an evaluation of bone loss in 247 postmenopausal women (mean age 64 ± 5 yr), who consumed an average of 100 IU/d, found that women who received an additional supplement of 700 IU of vitamin D/d lost less bone than women who were on 200 IU/d *(38)*. A similar observation was made in women who were placed on placebo or 400 IU/d *(39)*. It was also observed that when the vitamin D intake was up to 220 IU/d, women had higher PTH values between March and May than women studied between August and October *(40)*. This suggested that, in the absence of sunlight-mediated vitamin D synthesis, 220 IU/d may be inadequate to maximize bone health. Vitamin D deficiency in this age group should not be minimized. Recently, a study of 169 men and women ages 49–83 yr suggested that 37% were vitamin D deficient *(41)*. Therefore, because this age group is more dependent on dietary and supplementary sources of vitamin D to satisfy the body's requirement, it has now been recommended by the IOM *(21)* that the AI be 400 IU(10 µg)/d. It is, however, reasonable that this age group may require at least 400 and probably as much as 600 IU of vitamin D to maximize skeletal health *(21)* (Table 1).

3.7. Age 71 Years and Older

Those who are 71 yr and older are at high risk of developing vitamin D deficiency *(37–50)*, owing to a decrease in outdoor activity, which reduces cutaneous production of

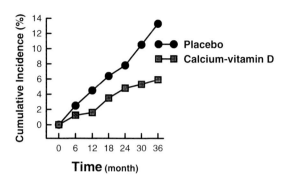

Fig. 6. Cumulative percentage of 389 men and women older than 65 years with a first nonvertebral fracture, according to study group. By 36 mo, 26 of 202 subjects in the placebo group and 11 of 187 subjects in the calcium–vitamin D group had had a fracture ($p = 0.02$). (Data from ref. 56.)

vitamin D. Many studies have looked at dietary supplementation of vitamin D in older men and women and its influence on serum 25(OH)D, PTH, and bone health as measured by bone mineral density and fracture risks. Several randomized, double-blind clinical trials of elderly men and women who had an intake of 400 IU/d showed normal 25(OH)D and serum PTH levels *(38,45)*. When men and women were supplemented with 400–1000 IU/d, they had a significant reduction in bone resorption *(51–56)*. When a group of elderly French women were supplemented with calcium and 800 IU of vitamin D, there was a significant increase in the bone mineral content, as well as a significant increase in vertebral and nonvertebral fractures *(52)*. A similar observation was made in free-living men and women 65 years and older who received 500 mg of calcium and 700 IU of vitamin D *(56)* (Fig. 6).

A comparison of the vitamin D status of elderly people in Europe and North American revealed that the elderly are prone to vitamin D deficiency and associated abnormalities in blood chemistries and bone density *(57)*. Therefore, based on the available data and on the fact that this age group is less likely to take advantage of the sun-mediated cutaneous synthesis of vitamin D, the IOM *(21)* recommended an AI for this age group of 600 IU (15 µg)/d, which is three times the previous recommended daily allowance for vitamin D (Table 1). Those not exposed to any sunlight may need as much as 800 IU/d, and therefore a reasonable allowance of 600–800 IU/d is appropriate to maximize bone health in this age group.

3.8. Pregnancy

It has always been assumed that during pregnancy the vitamin D requirement is increased because of fetal utilization of vitamin D and the requirement of vitamin D to increase the maternal intestinal calcium absorption. During the first and second trimesters, the fetus is developing most of its organ systems and laying down the collagen matrix for its skeleton. During the last trimester, the fetus begins to calcify the skeleton, thereby increasing its maternal demand for calcium. This demand is met by increased production of $1,25(OH)_2D$ by the mother's kidneys and placental production of $1,25(OH)_2D$. Circulating concentrations of $1,25(OH)_2D$ gradually increase during the first and second trimesters owing to an increase in vitamin D binding protein concentrations in the maternal circulation. However, the free levels of $1,25(OH)_2D$, which are responsible for enhancing intestinal calcium absorption, are only increased during the third trimester. Although there is ample evidence that the placenta converts 25(OH)D to 1,25(OH)$_2$D and transfers 1,25(OH)$_2$D to the fetus, these quantities are relatively small and appear not to affect the overall vitamin D status of pregnant women. Therefore, women (whether pregnant or not) who receive regular exposure to sunlight do not need additional vitamin D supplementation above what is recommended for their age group. However, if vitamin D intakes are <150 IU/d during the winter months in pregnant women who live at high latitudes above and below 40° north and south, respectively, they are prone to vitamin deficiency at the time of delivery. Therefore, although the IOM *(21)* recommended an AI for pregnant women of 200 IU (5 µg)/d, an intake of twice this amount, i.e., 400 IU (10 µg)/d, which is supplied by prenatal vitamin D supplements, is very reasonable and will ensure that the pregnant woman satisfies her vitamin D requirement (Table 1).

3.9. Lactation

During lactation, the mother needs to increase the efficiency of dietary absorption of calcium to ensure an adequate calcium content in her milk. The metabolism of 25(OH)D to 1,25(OH)$_2$D is enhanced in response to this new demand. However, because circulating concentrations of 1,25(OH)$_2$D$_3$ are 500–100 times less than 25(OH)D, the increased metabolism probably does not significantly alter the daily requirement for vitamin D. Therefore the IOM recommended *(21)* AI for lactating women is the same as that for nonlactating women, i.e., 200 IU (5 µg)/d. However, an intake of 400 IU (10 µg)/d, which is supplied in a postnatal vitamin D supplement, is reasonable to ensure vitamin D adequacy (Table 1).

3.10. Tolerable Upper Intake Levels

In the 1950s, when vitamin D supplementation of milk in Great Britain was not monitored with great care, there was an outbreak of hypervitaminosis D. This condition has also been associated with prolonged intakes (months to years) of pharmacologic amounts of vitamin D. It is associated with hypercalcemia, hyperphosphatemia, and markedly elevated levels of 25(OH)D. This can lead to nephrocalcinosis with renal failure, calcification of soft tissues including large and small blood vessels, and kidney stones. Infants aged 0–12 mo may be more sensitive to pharmacologic amounts of vitamin D, and therefore the IOM *(21)* to be recommended an upper safe level of 1000 IU (40 µg)/d. For children and adults, the recommended upper safe level was 2000 IU (80 µg)/d (Table 1).

3.11. Redefining Vitamin D Deficiency

It has long been suspected that the lower limit of the normal range for 25(OH)D may be too low and therefore falsely provides reassurance to clinicians that a person is vitamin D sufficient. The normal range is usually determined by obtaining blood samples from a large group of healthy individuals with no known disease or other factors that could influence the subject's vitamin D status. The mean ±2 standard deviations from the mean is considered to be the normal range. To evaluate further whether the lower 20% of the normal range for 25(OH)D was adequate to satisfy the body's requirement, patients aged 49–83 yr attending a Boston bone health care clinic had their 25(OH)D determined. The normal range for the assay was 10–55 ng/mL. All individuals with 25(OH)D of between

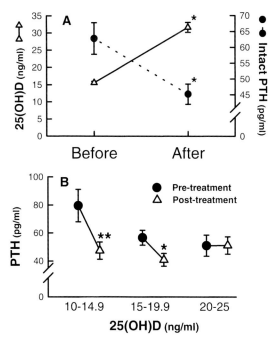

Fig. 7. (**A**) Serum levels of 25(OH)D (△) and parathyroid hormone (PTH; ●) before and after therapy with 50,000 IU of vitamin D_2 and calcium supplementation once a week for 8 wk. (**B**) Serum levels of PTH levels in patients who had serum 25(OH)D levels of between 10 and 25 ng/mL and who were stratified in increments of 5 ng/mL before and after receiving 50,000 IU of vitamin D_2 and calcium supplementation for 8 wk, $*p < 0.001$; $**p < 0.02$. (Data from ref. *41*.)

10 and 25 ng/mL received 50,000 IU of vitamin D once a week for 8 wk. Their 25(OH)D, intact PTH, and serum calcium levels were determined before and at the end of the study. As can be seen in Fig. 7, there was an average 109% increase in serum 25(OH)D and an average 22% decrease in serum PTH values in the volunteers. Forty percent of the patients had secondary hyperparathyroidism. Patients with serum 25(OH)D levels between 10 and 14.9 ng/mL and 15 and 20 ng/mL and who received the 50,000 IU vitamin D treatment had remarkable 35 and 26% decreases in PTH levels, respectively. There was no significant change in PTH values in individuals whose 25(OH)D was between 20 and 25 ng/mL. These data suggest that the lower limit of 25(OH)D, at least for adults aged 49 and above, should be 20 ng/mL rather than 10 ng/mL.

4. CONCLUSIONS

Vitamin D is taken for granted and is not usually considered an important nutrient. Because most humans obtain their vitamin D passively from exposure to sunlight, it would normally not be necessary to recommend a dietary allowance for vitamin D. Everyday casual exposure to sunlight provides most active children with their vitamin D requirement. Children who live in far northern and southern latitudes and who make vitamin D during the summer months store the excess in their body fat and utilize these stores during the winter when they are unable to make vitamin D in the skin. Young adults who are active outdoors obtain their vitamin D requirement from sunlight. However, middle-aged and older adults who are not outdoors as often or who wear sun protection before going outdoors are at risk of vitamin D deficiency. Vitamin D deficiency is associated with osteoporosis and osteomalacia, and also with increased risk of colon cancer and osteoarthritis *(58)*.

ACKNOWLEDGMENTS

This work was supported in part by NIH grants RO1 AR36963, MO1RR 00533, and NAGW 4936.

REFERENCES

1. Holick MF. Vitamin D: photobiology, metabolism, mechanism of action, and clinical application. In: Primer on the Metabolic Bone Diseases and Disorders of Mineral Metabolism, 3rd ed. Favus MJ, ed. Philadelphia: Lippincott-Raven, 1996; 74–81.
2. Holick MF. Vitamin D: photobiology, metabolism, and clinical applications. In: Endocrinology, 3rd ed. DeGroot L, et al., eds. Philadelphia: WB Saunders, 1995; 990–1013.
3. Strugnell SA, DeLuca HF. The vitamin D receptor—structure and transcriptional activation. Proc Soc Exp Biol Med 1997; 215:223–228.
4. Holick MF. Phylogenetic and evolutionary aspects of vitamin D from phytoplankton to humans. In:. Vertebrate Endocrinology: Fundamentals and Biomedical Implications, Vol 3. Pang PKT, Schreibman MP, eds. Orlando, FL: Academic Press, 1989; 7-43.
5. Holick MF. McCollum Award Lecture, 1994: Vitamin D: new horizons for the 21st century. Am J Clin Nutr 1994; 60:619–630.
6. Holick MF. The evolution of vitamin D from phytoplankton to man. In: Vitamin D: Chemistry, Biology and Clinical Applications of the Steroid Hormone. (Proceedings of the Tenth Workshop on Vitamin D). Norman AW, Bouillon R, Thomasset M, eds. Riverside: University of California Press, 1997; 771–776.
7. Fraser D, Scriver CR. Disorders associated with hereditary or acquired abnormalities of vitamin D function: hereditary disorders associated with vitamin D resistance or defective phosphate metabolism. In: Endocrinology. De Groot et al., eds. New York: Grune & Stratton, 1979; 797–808.
8. Holtrop ME, Cox KA, Carnes DL, Holick MF. Skeletal mineralization in vitamin D-deficient rats. Am J Physiol 1986; 251:E20.
9. Underwood JL, DeLuca HF. Vitamin D is not directly necessary for bone growth and mineralization. Am J Physiol 1984; 246:E493–E498.
10. Clemens,TL, Henderson SL, Adams JS, Holick MF. Increased skin pigment reduces the capacity of skin to synthesize vitamin D_3. Lancet 1982; 1:74–76.
11. Matsuoka LY, Ide L, Wortsman J, MacLaughlin J, Holick MF. Sunscreens suppress cutaneous vitamin D_3 synthesis. J Clin Endocrinol Metab 1987; 64:1165–1168.
12. Holick MF, Matsuoka LY, Wortsman J. Age, vitamin D, and solar ultraviolet radiation. Lancet 1989; 2:1104–1105.
13. Webb AR, Kline L, Holick MF. Influence of season and latitude on the cutaneous synthesis of vitamin D_3: exposure to winter sunlight in Boston and Edmonton will not promote vitamin D_3 synthesis in human skin. J Clin Endocrinol Metab 1988; 67:373–378.
14. Tanner JT, Smith J, Defibaugh P, Angyal G, Villalobos M, Bueno M, McGarrahan E. Survey of vitamin content of fortified milk. J Assoc Off Anal Chem 1988; 71:607-610.
15. Holick MF, Shao Q, Liu WW, Chen TC. The vitamin D content of fortified milk and infant formula. N Engl J Med 1992; 326:1178–1181.
16. Chen TC, Heath H, Holick MF. An update on the vitamin D content of fortified milk from the United States and Canada. N Engl J Med 1993; 329:1507.
17. Markestad T, Elzouki AY. Vitamin D deficiency rickets in northern Europe and Libya. In: Rickets Nestle Nutrition Workshop Series. Glorieux FH, ed. New York: Raven, 1991; 203–213.
18. Specker BL, Valanis B, Hertzberg V, Edwards N, Tsang RC. Sunshine exposure and serum 25-hydroxyvitamin D. J Pediatr 1985; 107:372–376.
19. Specker BL, Tsang RC. Cyclical serum 25-hydroxyvitamin D concentrations paralleling sunshine exposure in exclusively breast-fed infants. J Pediatr 1987; 110:744–747.
20. Food and Nutrition Board Institute of Medicine. Dietary Reference Intakes for Calcium, Phosphorus, Magnesium, Vitamin D, and Fluoride. Washington, DC: IOM National Academy Press, 1997; 7–30.

21. Nakao H. Nutritional significance of human milk vitamin D in neonatal period. Kobe J Med Sci 1988; 34:21–128.
22. Holick MF. Reasonable vitamin D daily allowance. ARES-Serono Symposium, 1997, in press.
23. Feliciano ES, Ho ML, Specker BL, Falciglia G, et al. Seasonal and geographical variations in the growth rate of infants in China receiving increasing dosages of vitamin D supplements. J Trop Pediatr 1994; 40:162–165.
24. Foman SJ, Younoszai K, Thomas L. Influence of vitamin D on linear growth of normal full-term infants. J Nutr 1966; 88:345–350.
25. Specker B, Ho M, Oestreich A, Yin T, et al. Prospective study of vitamin D supplementation and rickets in China. J Pediatr 1992; 120:733–739.
26. Pettifor JM, Ross FP, Moodley G, Wang J, et al. Serum calcium, magnesium, phosphorus, alkaline, phosphatase and 25-hydroxyvitamin D concentrations in children. A Afr Med J 1978; 53:751–754.
27. Aksnes L, Aarskog D. Plasma concentrations of vitamin D metabolites in puberty: effect of sexual maturation and implications for growth. J Clin Endocrinol Metab 1982; 55:94–101.
28. Gultekin A, Ozalp I, Hasanoglu A, Unal A. Serum-25-hydroxycholecalciferol levels in children and adolescents. Turkish J Pediatr 1987; 29:155–162.
29. Ala-Houhala M, Koskinen T, Terho A, Koivula T, Visakorpi J. Maternal compared with infant vitamin D supplementation. Arch Dis Child 1986; 61:1159–1163.
30. Riancho JA, del Arco C, Arteaga R, Herranz JL, Albajar M, Macias JG. Influence of solar irradiation on vitamin D levels in children on anticonvulsant drugs. Acta Neurol Scand 1989; 79:296–299.
31. Taylor AF, Norman ME. Vitamin D metabolite levels in normal children. Pediatr Res 1984; 18:886–890.
32. Oliveri MB, Ladizesky M, Mautalen CA, Alonso A, Martinez L. Seasonal variations of 25-hydroxyvitamin D and parathyroid hormone in Ushuaia (Argentina), the southernmost city of the world. Bone Miner 1993; 20:99–108.
33. Meier DE, Luckey MM, Wallenstein S, Clemens TL, et al. Calcium, vitamin D, and parathyroid hormone status in young white and black women: association with racial differences in bone mass. J Clin Endocrinol Metab 1991; 72:703–710.
34. Kinyamu HK, Gallagher JC, Galhorn KE, Petranick KM, Rafferty KA. Serum vitamin D metabolites and calcium absorption in normal young and elderly free-living women and in women living in nursing homes. Am J Clin Nutr 1997; 65:790–797.
35. Holick MF. Vitamin D requirements for the elderly. Clin Nutr 1986; 5:121–129.
36. Clemens TL, Zhou X, Myles M, Endres D, Lindsay R. Serum vitamin D_2 and vitamin D_3 metabolite concentrations and absorption of vitamin D_2 in elderly subjects. J Clin Endocrinol Metab 1986; 63:656–660.
37. Krall EA, Dawson-Hughes B. Relation of fractional ^{47}Ca retention to season and rates of bone loss in healthy postmenopausal women. J Bone Miner Res 1991; 6:1323–1329.
38. Dawson-Hughes B, Harris SS, Krall EA, Dallal GE, Falconer G, Green CL. Rates of bone loss in postmenopausal women randomly assigned to one of two dosages of vitamin D. Am J Clin Nutr 1995; 61:1140–1145.
39. Dawson-Hughes B, Dallal GE, Krall EA, Harris S, Sokoll LJ, Falconer G. Effect of vitamin D supplementation on wintertime and overall bone loss in healthy postmenopausal women. Ann Intern Med 1991; 115:505–512.
40. Krall E, Sahyoun N, Tannenbaum S, Dallal G, Dawson-Hughes B. Effect of vitamin D intake on seasonal variations in parathyroid hormone secretion in postmenopausal women. N Engl J Med 1989; 321:1777–1783.
41. Malabanan A, Veronikis IE, Holick MF. Redefining vitamin D deficiency. Lancet 1998; 351:805,806.
42. Chevalley T, Rizzoli R, Nydegger V, Slosman D, Rapin CH, Michel JP, Vasey H, Bonjour JP. Effects of calcium supplements on femoral bone mineral density and vertebral fracture rate in vitamin-D-replete elderly patients. Osteopor Int 1994; 4:245–252.
43. Hordon LD, Peacock M. Vitamin D metabolism in women with femoral neck fracture. Bone Miner 1987; 2:413–426.
44. Lamberg-Allardt C, Karkkainen M, Seppanen R, Bistrom H. Low serum 25-hydroxyvitamin D concentrations and secondary hyperparathyroidism in middle-aged white strict vegetarians. Am J Clin Nutr 1993; 58:684–689.
45. Lips XX, Wiersinga A, van Ginkel FC, Jongen MJM, Netelenbos C, Hackeng WHL, et al. The effect of vitamin D supplementation on vitamin D status and parathyroid function in elderly subjects. J Clin Endocrinol Metab 1988; 67:644–650.

46. McGrath N, Singh V, Cundy T. Severe vitamin D deficiency in Auckland. NZ Med J 1993; 106: 525,526.
47. Ng K, St. John A, Bruce DG. Secondary hyperparathyroidism, vitamin D deficiency and hip fracture: importance of sampling times after fracture. Bone Miner 1994; 25:103–109.
48. Ooms ME, Lips P, Roos JC, Van Der Hijgh WJF, Popp-Snijders C, Bezemer PD, Bouter LM. Vitamin D status and sex hormone binding globulin: determinants of bone turnover and bone mineral density in elderly women. J Bone Miner Res 1995; 10:1177–1184.
49. Villareal DT, Civitelli R, Chines A, Avioli LV. Subclinical vitamin D deficiency in postmenopausal women with low vertebral bone mass. J Clin Endocrinol Metab 1991; 72:628–634.
50. Webb AR, Pilbeam C, Hanafin N, Holick MF. An evaluation of the relative contributions of exposure to sunlight and diet on the circulating concentrations of 25-hydroxyvitamin D in an elderly nursing home population in Boston. Am J Clin Nutr 1990; 51:1075–1081.
51. Brazier M, Kamel S, Maamer M, Agbomson F, Elesper I, Garabedian M, Desmet G, Sebert JL. Markers of bone remodeling in the elderly subject: effects of vitamin D insufficiency and its correction. J Bone Miner Res 1996; 10:1753–1761.
52. Chapuy MC, Arlot M, Duboeuf F, Brun J, Crouzet B, Arnaud S, Delmas P, Meuner P. Vitamin D_3 and calcium to prevent hip fractures in elderly women. N Engl J Med 1992; 327:1637–1642.
53. Egsmose C, Lund B, McNair P, Lund B, Storm T, Sorensen OH. Low serum levels of 25-hydroxyvitamin D and 1,25-dihydrovitamin D in institutionalized old people: influence of solar exposure and vitamin D supplementation. Age Ageing 1987; 16:35–40.
54. Fardellone P, Sebert JL, Garabedian M, Bellony R, Maamer M, Agbomson F, Brazier M. Prevalence and biological consequences of vitamin D deficiency in elderly institutionalized subjects. Rev Rhum 1995; 62:576–581.
55. Kamel S, Brazier M, Rogez JC, Vincent O, Maamer M, Desmet G, Sebert JL. Different responses of free and peptide-bound cross-links to vitamin D and calcium supplementation in elderly women with vitamin D insufficiency. J Clin Endocrinol Metab 1996; 81:3717–3721.
56. Dawson-Hughes B, Harris SS, Krall EA, Dallal GE. Effect of calcium and vitamin D supplementation on bone density in men and women 65 years of age or older. N Engl J Med 1997; 337:670–676.
57. McKenna MJ. Differences in vitamin D status between countries in young adults and the elderly. Am J Med 1992; 93:69–77.
58. McAlindon TE, Felson DT, Zhang Y, Hannan MT, Aliabadi P, Weissman B, Rush D, Wilson PWF, Jacques P. Relation of dietary intake and serum levels of vitamin D to progression of osteoarthritis of the knee among participants in the Framingham Study. Ann Intern Med 1996; 125:353–359.
59. Bicknell F, Prescott F. Vitamin D. In: The Vitamins in Medicine, 2nd ed. Heineman W, ed. London: Random House UK, 1948; 630–708.

2 Photobiology of Vitamin D

Tai C. Chen

1. INTRODUCTION

Vitamin D is one of the four fat-soluble vitamins that have been recognized to possess important biologic functions. The major physiologic effect of vitamin D is on calcium and bone metabolism, by maintaining extracellular concentrations of calcium and phosphorus within the normal range *(1–3)*. During the past three decades, intensive research on vitamin D has revealed that it is a hormone and not a vitamin. Once vitamin D is formed in the skin, it requires two sequential hydroxylation reactions, first in the liver to form 25-hydroxyvitamin D (25-OH-D), and then in the kidneys to form 1,25-dihydroxyvitamin D [1,25(OH)$_2$D]. It is 1,25(OH)$_2$D that is responsible for enhancing the efficiency of intestinal absorption of dietary calcium and phosphorus, as well as the mobilization of calcium and phosphorus stores from bone *(1–3)*. In addition, 1,25(OH)$_2$D has other biologic actions in many tissues or cells that possess the 1,25(OH)$_2$D receptor, including enhancement of cellular differentiation and/or inhibition of cellular proliferation in cultured fibroblasts and keratinocytes *(2)*.

2. HISTORICAL PERSPECTIVE

Rickets is a disease identified by deformities of the skeleton, including enlargement of the head, and the joints of the long bones and rib cage, curvature of the spine and thighs, and generalized muscular weakness. It was recognized in the mid-17th century in Northern Europe as a major health problem for young children as people began to migrate to city centers and live in an environment that was devoid of direct exposure to sunlight (Fig. 1). The incidence of the disease continued to climb during the industrial revolution. By the turn of the 20th century rickets had become epidemic, not only in Northern Europe but also in the industrialized northeastern region of the United States.

The first published observation into the potential cause of this devastating bone disease and the implication of exposure to the sun in prevention and cure was that of Sniadecki in 1822 *(4)*. He observed that children who lived in Warsaw had a high incidence of rickets, whereas children who lived on the farms surrounding Warsaw were free of the disease. He concluded that lack of sunlight was the likely cause of rickets. In 1890 Palm *(5)* conducted an epidemiologic survey regarding factors that might cause rickets. He concluded that the common denominator was lack of sunlight exposure, and he encouraged systematic sunbathing for prevention and cure. These insightful observations went unnoticed, like Sniadecki's earlier finding.

From: *Vitamin D: Physiology, Molecular Biology, and Clinical Applications*
Edited by: M. F. Holick © Humana Press Inc., Totowa, NJ

Fig. 1. A typical photograph of Glasgow in the mid-1800s, taken by Thomas Annan. (Reproduced with permission from ref. *86*).

In the mid-1800s, Bretonneau administered cod liver oil to a 15-mo-old child with acute rickets and noted an incredibly speedy recovery *(6)*. Later, Trousseau, a student of Bretonneau, used liver oils from a variety of fish and marine animals for the treatment of rickets and osteomalacia. In his monograph on therapeutics he advocated the use of fish liver oil, preferably accompanied by exposure to sunlight, for rapid cure of both rickets and osteomalacia *(6)*. These clinical observations led many to believe that rickets was caused by some type of nutritional deficiency. Unfortunately, these important observations like many previous ones, were also disregarded. It was not until 1918, when Mellanby *(7)* gave cod liver oil to rachitic beagle puppies and found that their rickets was cured, that the scientific community began to consider rickets as a nutritional deficiency disease. Mellanby concluded that cod liver oil possessed a fat-soluble nutritional factor that he called the antirachitic factor, originally thought to be the newly discovered vitamin A. However, McCollum et al. *(8)* were able to dissociate vitamin A activity from antirachitic activity by heating cod liver oil and aerating it with oxygen to destroy the vitamin A activity without eliminating the antirachitic activity. This finding therefore established a new fat-soluble principle in cod liver oil, which was subsequently named vitamin D. About the same time as Mellanby's observations *(7)*, Huldschinsky *(9)* exposed four rachitic children to radiation from a mercury vapor arc lamp and demonstrated by X-ray analysis that the rickets was cured after 4 mo of therapy. He further demonstrated that the effect was not localized at the site of irradiation because exposing one arm to the radiation resulted in healing of both arms. In 1921, Hess and Unger *(10)* exposed seven rachitic children in New York City to sunlight and, based on X-ray examination, reported marked improvement in all of them. This finding prompted Goldblatt and Soames *(11,12)*, Hess and Weinstock *(13)*, and Steenbock and Black *(14)* to expose a variety of foods (such as wheat, lettuce, olive, linseed oils, and rat chow) and other substances (such as human and rat plasma) to ultraviolet (UV) radiation. UV irradiation imparted antirachitic activity, a finding that led Steenbock *(15)* to recommend that the irradiation of milk might be an excellent method to provide vitamin D to children and to prevent rickets. This suggestion was followed first by the addition of ergosterol (provitamin D_2) to milk, then irradiation to impart antirachitic activity, and ultimately the addition of synthetic vitamin D directly. This simple concept led to the eradication of rickets as a significant health problem in the United States and other countries that used this practice. Thus, nearly a century after Sniadecki had first suggested the importance of sunlight exposure for prevention of rickets, it was finally demonstrated that exposure to sunlight alone or ingesting UV irradiated foods or substances could prevent and cure this crippling bone disease.

2.1. Vitamin D Fortification in Milk

In the 1940s, the amount of vitamin D added to milk was not well regulated in Europe, and excessive fortification of the vitamin caused intoxication in infants, leading to hypercalcemia and irreversible brain damage. As a result, vitamin D fortification was severely restricted in Europe. Consequently, the incidence of rickets became more frequent, and this childhood disease continues to be a health problem in Europe *(16,17)*. Realizing this problem, many Europian countries have begun adding vitamin D to various foods, including cereals and margarine. In the United States, the American Medical Association's Council on Foods and Nutrition recommended in 1957 that milk should contain 400 IU (10 µg) a quart and that the vitamin D content should be measured at least twice yearly by an independent laboratory. The Food and Drug Administration has modified its guidelines for safe levels of vitamin D in milk and has stipulated that 1 qt should contain 400 IU and no more than 600 IU of vitamin D. Fortification above 800 IU/qt may create the potential for a public health threat and should be prohibited. A survey of vitamin D content in fortified milk within the last decade revealed that 80% and 73% of the milk samples from the United States and Canada, respectively, did not contain 80–120% of that claimed on the label *(18–20)*. Some samples even had an undetectable amount of vitamin D, and others had a two to threefold excess.

The fact that exposure of skin or foods to UV radiation produced antirachitic activity prompted scientists to identify the precursor of vitamin D. The first one was isolated from yeast and identified as ergosterol. Ergosterol is a major sterol in the fungal and plant kingdoms and contains a four-member ring system with double bonds between carbon 5 and 6, and carbon 7 and 8 in ring B, and an additional double bond between carbon 22 and 23 of its side chain (Fig. 2). When ergosterol was first irradiated, it was believed to form a single product, which was named vitamin D_1. However, the term vitamin D_1 was dropped after it was found out that the preparation was a mixture of several compounds. Later, the vitamin D that was isolated in pure form from the irradiated ergosterol was named ergocalciferol, or vitamin D_2. Vitamin D_2 was later chemically synthesized and

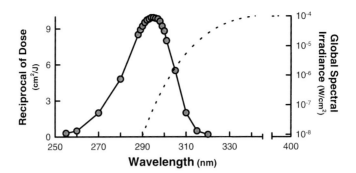

Fig. 4. The action spectrum of previtamin D_3 formation from 7-DHC in human epidermis (○) and the spectral irradiance curve for sunlight (------). The action spectrum was obtained by plotting the reciprocal of the dose as a function of the wavelength. At any wavelength, no more than 5% of product was made. The overlay of the curve of the action spectrum with that of the solar spectrum demonstrates the small portion of the solar UV spectrum that is involved with the production of previtamin D_3 from 7-DHC. (Reproduced with permission from ref. *38*).

ergosterol solution with 295 nm of radiation gave the maximum yield of previtamin D_2. MacLaughlin et al. *(38)* examined the photosynthesis of previtamin D_3 from 7-DHC in human skin after exposing the tissue to narrow-band radiation or simulated solar radiation. They reported that the optimum wavelengths for the production of previtamin D_3 were between 295 and 300 nm (Fig. 4) *(38)*. In Caucasians with skin type II, 20–30% of the radiation of 295 nm is transmitted through the epidermis; most of the UVB photons (290–320 nm) are absorbed by the stratum spinosum of the epidermis. In blacks with skin type V, only about 2–5% of the UVB photons penetrate the epidermis *(39,40)*. As the UV radiation penetrates the epidermis, it is absorbed by a variety of molecules including DNA, RNA, proteins, and 7-DHC. The 5,7-diene of 7-DHC absorbs solar radiation between 290 and 315 nm, causing it to isomerize, resulting in a bond cleavage between carbon 9 and 10 to form a 9,10-seco-sterol previtamin D_3 *(25,35,38)*. Because most of the radiation responsible for producing previtamin D_3 is absorbed in the epidermis, >95% of the previtamin D_3 produced is in the epidermis *(35)*. Once previtamin D_3 is synthesized in the skin, it can undergo either photoconversion to lumisterol, tachysterol, and 7-DHC or a heat-induced isomerization to vitamin D_3 (Fig. 5).

4. CONVERSION OF PREVITAMIN D_3 TO VITAMIN D_3 IN THE SKIN

The isomerization of previtamin D_3 to vitamin D_3 is the last step in the synthesis of vitamin D_3 in human skin. The reaction rate of this isomerization is temperature dependent and is enhanced by raising the temperature. Earlier studies found that this process was not affected by acids, bases, catalysts, or inhibitors of radical chain processes *(41)*. Furthermore, no intermediate was detected during the isomerization. This led to the conclusion that the reaction was an intramolecular concerted process involving a [1,7]-sigmatropic hydrogen rearrangement *(41)*, which is an antarafacially (oppositive side of a plane) allowed and suprafacially (same side of a plane) forbidden process *(42)*. Much of the information about the previtamin D_3 isomerization was obtained from experiments using organic solvents for the conversion *(43,44)* and assumed to be the same in human skin. There was no evidence for the the existence of an enzymatic process in the skin that

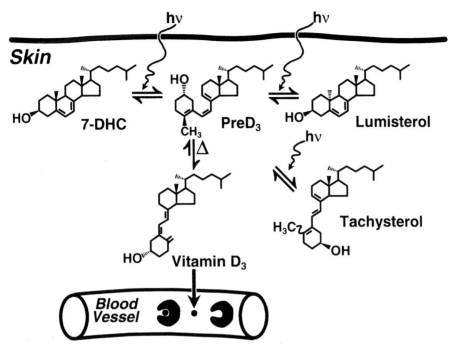

Fig. 5. Schematic representation of the formation of previtamin D_3 in the skin during exposure to the sun and the thermal isomerization of previtamin D_3 to vitamin D_3, which is specifically translocated by the vitamin-D-binding protein into the circulation. During continual exposure to the sun, previtamin D_3 also photoisomerizes to lumisterol$_3$ and tachysterol$_3$. (Reproduced with permission from ref. 68).

could convert previtamin D_3 to vitamin D_3. In an organic solvent such as hexane, it takes about 6–10 d for the isomerization reaction between previtamin D_3 and vitamin D_3 to reach equilibrium at 37°C (Fig. 6). This slow isomerization rate could not explain the relatively rapid increases in serum vitamin D_3 concentrations 12–24 h after whole-body UVB irradiation (Fig. 7). One explanation for the relatively rapid rise in serum vitamin D_3 concentration was that the specific translocation of vitamin D_3 from skin into the circulation by the serum vitamin D binding protein that would result in the removal of vitamin D_3 from the skin as it was being produced, thereby, changing the isomerization reaction from a reversible process to an irreversible one *(35,45)*. However, this change would have little effect on the kinetics of previtamin D_3 isomerization to vitamin D_3 because of the much slower reverse reaction rate at 37°C. As would be expected, the vitamin D_3 synthesis in the biologic system should be more efficient, with a faster reaction rate and higher equilibrium constant. A comparative study of the kinetic and thermodynamic properties of the isomerization reaction in human skin and in an organic solvent revealed that not only was the equilibrium of the reaction shifted in favor of vitamin D_3 synthesis in human skin [equilibrium constant (K) at 37°C = 11.44] compared with hexane (K = 6.15), but the rate of the reaction was also increased by more than 10-fold in human skin ($T_{1/2}$ at 37°C = 2.5 h) compared with hexane ($T_{1/2}$ = 30 h) (Fig. 6) *(44)*. This accelerated rate of isomerization was also observed in chicken, frog, and lizard skin *(45,46)*. The enthalpy ($H°$) changes for the reaction were –21.58 and –15.60 kJ/mol in human skin and in hexane, respectively. The activation energies for both the forward and

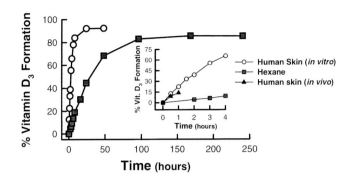

Fig. 6. Thermoconversion of previtamin D_3 to vitamin D_3 as a function of time in human skin and in *n*-hexane at 37°C. The insert depicts the thermoconversion of previtamin D_3 to vitamin D_3 in human skin in vivo (s) compared with levels in *n*-hexane (n) and in human skin in vitro (O) at 37°C. (Reproduced with permission from ref. *44*).

the reverse reactions were lower in human skin than in hexane. Thus human skin profoundly changed the rate constant and equilibrium constant in favor of vitamin D_3 formation.

Determination of the subcellular localization of 7-DHC and previtamin D_3 in human epidermal tissue revealed that most 7-DHC and previtamin D_3 were in the membrane fraction; only 20% was in the cytosol. Thus, it has been postulated that most 7-DHC is entrapped in membrane. It is likely that the 3β-hydroxyl group of the 7-DHC molecule is near the polar head group of the membrane phospholipids and interacts with it through hydrophilic forces, whereas the nonpolar rings and side chain are associated with the nonpolar tail by hydrophobic van der Waals interactions. As was discussed in the previous section on the photosynthesis of previtamin D_3 in the skin, there are two previtamin D_3 conformers; one is s-*cis*, s-*cis*-previtamin D_3, and the other is s-*trans*, s-*cis*-previtamin D_3. When 7-DHC in the skin's plasma membrane is exposed to UVB radiation, the conformation of the s-*cis*, s-*cis*-previtamin D_3 is preserved through hydrophobic and hydrophilic interactions with different moieties of the membrane phospholipid bilayer, whereas a rotation around carbon 5 and carbon 6 to form the thermodynamically more stable s-*trans*, c-*cis* conformer is only permissible in hexane. The maintenance of the previtamin D_3 in the s-*cis*, s-*cis* conformation would greatly facilitate its conversion to vitamin D_3, whereas the more stable s-*trans*, s-*cis* conformation would not be easily converted to vitamin D_3. Thus, instead of taking 30 h for 50% of previtamin D_3 to convert to vitamin D_3 at 37°C in hexane, it took only 2.5 h in the human skin at the same temperature, suggesting that the interaction of previtamin D_3 with membrane lipid and/or protein in skin may be responsible for the increased vitamin D_3 formation rate in the skin. During the formation of vitamin D_3, the hydrophilic and hydrophobic interactions of the s-*cis*, s-*cis*-previtamin D_3 with the membrane phospholipids are disrupted, thereby facilitating the specific translocation of vitamin D_3 from the skin cells into the extracellular space.

The importance of membrane microenvironments on previtamin D_3 ⇌ vitamin D_3 isomerization received further support from a kinetic study in an aqueous solution of β-cyclodextrin *(47)*. Cyclodextrins, a group of naturally occurring, truncated cone-shaped oligosaccharides, have an unique ability to complex a variety of foreign molecules including steroids into their hydrophobic cavities in aqueous solution *(48)* and catalyze reactions of a wide variety of guest molecules *(49)*. Among the various cyclodextrins,

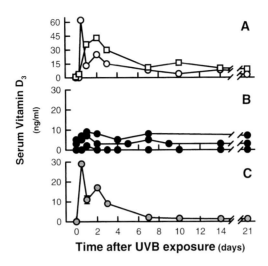

Fig. 7. Change in serum concentrations of vitamin D in two lightly pigmented white (**A**) and three heavily pigmented black subjects (**B**) after total-body exposure to 0.054 J/cm² of UVB. (**C**) Serial change in circulating vitamin D after reexposure of one black subject in panel B to a 0.32 J/cm² dose of UV radiation. (Reproduced with permission from ref. *52*).

β-cyclodextran has been shown to form 2:1 (host/guest) inclusion complexes with vitamin D_3 *(50)*. Using this model, Tian and Holick *(47)* demonstrated that, at 5°C, the forward (κ_1) and reversed (κ_2) rate constants for previtamin $D_3 \rightleftarrows$ vitamin D_3 isomerization were increased by more than 40 and 600 times, respectively, compared with those in *n*-hexane *(47)*. The equilibrium constant of the reaction was significantly reduced by more than 12-fold when compared with that in *n*-hexane at 5°C, and the percentage of vitamin D_3 at equilibrium was increased as the temperature was increased. When complexed with β-cyclodextrin, the previtamin $D_3 \rightleftarrows$ vitamin D_3 isomerization became endothermic ($H° = 13.05$ kJ mol), whereas it was exothermic in *n*-hexane, suggesting that the thermodynamically unfavorable s-*cis*, s-*cis*-previtamin D_3 conformers are stabilized by β-cyclodextrin, and therefore the rate of the isomerization is increased.

5. TRANSLOCATION OF VITAMIN D_3 FROM THE SKIN INTO THE CIRCULATION

After vitamin D_3 is formed from the thermally induced isomerization of previtamin D_3 in the epidermis, it is transported into the dermal capillary bed beneath the dermoepidermal junction. Little is known about the mechanism of this translocation process. In an attempt to understand this event, Tian el al. *(45)* studied the kinetics of vitamin D_3 formation and the time course of appearance of vitamin D_3 in the circulation after exposure of chickens to UVB radiation. Their data indicate a much faster rate of formation of vitamin D_3 from previtamin D_3 than the reverse reaction (return back to previtamin D_3 from vitamin D_3) and a relatively fast rate of translocation from skin to circulation. By examining the time course of the appearance of vitamin D_3 in the circulation, they found a rapid phase of vitamin D_3 appearance from 8 h to about 30 h after irradiation and a relatively slower phase of its disappearance in the circulation after the concentration of circulating vitamin D_3 reached its peak value (Fig. 8). No previtamin D_3

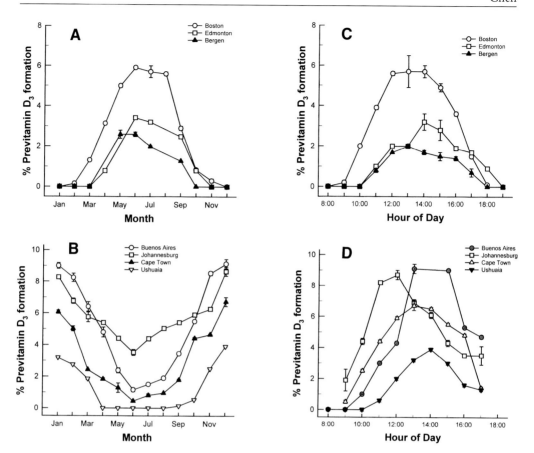

Fig. 9. Influence of season, time of day, and latitude on the synthesis of previtamin D_3 in the Northern (**A** and **C**) and Southern hemispheres (**B** and **D**). The hour indicated in **C** and **D** is the end of the 1-h exposure time. The data represent the means ± SEM of duplicate determinations.

basis throughout the day once a month in seven cities located in both the Northern and Southern hemispheres. Because of the absorption properties of human skin and the effect of skin pigmentation on the absorption of solar UVB radiation, the results obtained from the ampules represented the maximal conversion of 7-DHC to previtamin D_3. To correlate the synthesis of previtamin D_3 in ampules with human skin, 1 cm^2 of white skin samples of skin type III *(63)*, which sometimes burns and always tans, were exposed to simulated sunlight along with the ampules containing 50 μg 7-DHC in 1 mL of absolute alcohol. It was found that 0.8% of 7-DHC was converted to previtamin D_3 in the ampules before any previtamin D_3 appeared in the skin samples.

Based on this correlation, we have estimated the capacity of type III skin to synthesize vitamin D_3 from natural sunlight each month at seven locations with different latitudes in both the Northern and Southern Hemispheres. It was found that following exposure to noon sunlight for 1 h, previtamin D_3 formation was negligible during the months of December through February in Boston, MA (42°N); from November through March in Edmonton, Canada (52°N); and from October through March in Bergen, Norway (61°N) (Fig. 9A). Similarly, people with the same skin type cannot synthesize previtamin D_3 from April through September in Ushuaia, Argentina (55°S) (Fig. 9B). However, individuals with this skin type can synthesize previtamin D_3 year round in Cape Town, South Africa (35°S); Johannesburg, South Africa (26°S); and Buenos Aires, Argentina (34°S). The figures show that previtamin D_3 synthesis increased from the spring to summer months and decreased thereafter. These changes are reflected in the seasonal variation of serum 25-hydroxyvitamin D_3 concentrations in children and adults *(64–67)*.

The effects of sunlight on previtamin D_3 synthesis within a day were also studied (Fig. 9C). The maximal previtamin D_3 formation after 1 h of exposure was $5.7 \pm 0.1\%$ in Boston (42°N) in July, whereas it reached only $2.0 \pm 0.1\%$ in Bergen (61°N). Figure 9C also demonstrates that in the Northern Hemisphere, individuals with skin type III exposed to July sunlight for 1 h can synthesize previtamin D_3 as early as between 9:00 and 10:00 AM in Boston (42°N), 10:00 and 11:00 AM in Edmonton (52°N), and 10:00 and 11:00 AM in Bergen (61°N). Similarly, in the Southern Hemisphere, people with type III skin exposed to summer sunlight in December for 1 h are capable of forming previtamin D_3 as early as between 9:00 and 10:00 AM in Buenos Aires (34°S), 10:00 and 11:00 AM in Ushuaia (55°S), 8:00 and 9:00 AM in Cape Town (35°S), and 9:00 and 10:00 AM in Johannesburg (26°S) (Fig. 9D). Individuals living in these places can synthesize previtamin D_3 at least as late as between 4:00 and 5:00 PM.

Thus the periods when humans cannot synthesize vitamin D_3 by natural sunlight increase with increasing latitude and zenith angle of the sun during the shift of the season and time of day. The length of these periods will increase in people with higher degrees of skin pigmentation *(68)*. We have correlated the synthesis of previtamin D_3 in ampules with skin type V black skin samples and found that 1.8% of 7-DHC was converted to previtamin D_3 in ampules before any previtamin D_3 was detected in the skin samples. Therefore, individuals with skin type V would require a considerably longer period of sun exposure to make the same amount of vitamin D in their skin compared with those with skin type III.

Several studies have shown significant changes in spine and hip bone mineral density during the winter among postmenopausal women with low calcium intake *(69–71)*. In the study performed in a remote area of northern Maine, Rosen et al. *(72)* observed drops of 3.6% and 3.0% in spine and hip bone mineral density, respectively, from August to February. During that same period, serum 25(OH)D fell $13 \pm 6\%$ and parathyroid hormone rose $27 \pm 11\%$. By contrast, during spring and summer (February to August), no changes in spine and hip bone mineral density mass were evident.

6.3. Effect of Aging on the Cutaneous Production of Previtamin D_3

Several studies have suggested that osteomalacia caused by vitamin D deficiency is becoming an epidemic in Asia, Europe, and the United States *(69,73–78)*. It has been estimated that about 30–40% of elderly people with hip fracture in the United States and Great Britain are vitamin D deficient. Lamberg-Allard *(78)* reported that circulating concentrations of 25(OH)D were low in long-stay geriatric patients and in residents at the senior citizens' home due to lack of exposure to sunlight and low vitamin D intake. Similarly, a study conducted at a nursing home in the Boston area demonstrated that approximately 60% of the nursing home residents were vitamin D deficient during the winter months *(79)*. Even near the end of the summer, we found that in the Boston area 30, 43, and 80% of the free-living elderly white, Hispanic, and black subjects, respectively, had 25(OH)D levels <20 ng/mL.

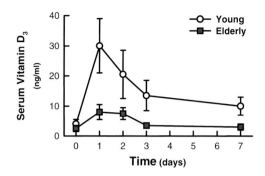

Fig. 10. Circulating concentrations of vitamin D in healthy young (○) and elderly (■) volunteers exposed to ultraviolet radiation. (Reproduced with permission from ref. 82).

Furthermore, Lester et al. *(67)* reported that elderly subjects had lower circulating levels of 25(OH)D than healthy young adults. These data suggested that aging decreased the capacity of human skin to produce vitamin D_3. It has been known that skin thickness decreases linearly with age after the age of 20 yr *(80)*. This is correlated with an age-dependent decrease in epidermal concentrations of 7-DHC *(81)*. Exposure of human skin samples from various age groups to simulated sunlight showed that the skin from 8- and 18-yr-old subjects produced two- to threefold greater amounts of previtamin D_3 than the skin from 77- and 82-yr-old subjects. Whole-body exposure of healthy young and older adults to the same amount of UV radiation confirmed the in vitro finding. When the circulating concentrations of vitamin D were determined before and at various times after the exposure, the young volunteers (age range 22–30 yr) raised their circulating concentrations of vitamin D_3 to a maximum of 30 ng/mL within 24 h, whereas the elderly subjects (62–80 yr) were able to reach a maximum concentration of only 8 ng/mL (Fig. 10) *(82)*. In a recent study, we determined the circulating concentrations of 25(OH)D in 15 healthy young adults after most of the body was exposed to a suntanning lamp called MedSun (Wolff Systems Technology, Atlanta, GA). Three times a week for 7 wk (total radiation was approx 4 MED for the period). After 1 wk, there was a 50% increase in 25(OH)D that continued to increase for 5 wk before reaching a plateau of about 150% above baseline values (Fig. 11).

6.4. Effect of Sunscreen Use and Clothing on Previtamin D_3 Formation

There is great concern about the damaging effects of chronic exposure to sunlight on the skin. Long-term exposure to sunlight can cause dry, wrinkled skin as well as an increased risk of skin cancers such as squamous cell and basal cell carcinoma. As a result of this concern, there has been a major effort by dermatologists to encourage people to apply sunscreens on their body before going outdoors. There is no question of the benefit of sunscreen use for the protection of the skin from the damaging effects of sunlight. However, the radiation that is responsible for causing sunburn, skin wrinkling, and skin cancer is the same radiation that is responsible for producing previtamin D_3. Thus sunscreen use can also prevent the beneficial effect of sunlight, the photoconversion of 7-DHC to previtamin D_3 *(83,84)*. Matsuoka et al. *(84)* studied the effects of paraaminobenzoic acid (PABA) on previtamin D_3 formation in skin samples in vitro and on cutaneous

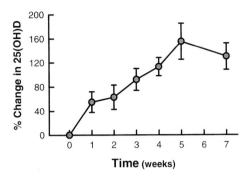

Fig. 11. Circulating concentrations of 25-hydroxyvitamin D in healthy young volunteers exposed to MedSun ultraviolet radiation.

vitamin D synthesis in vivo. They found that 5% PABA solution totally blocked the photoisomerization of 7-DHC to previtamin D_3 in human skin specimens. The volunteers who received an application of 5% PABA, which had a sun protection factor of 8, did not raise their circulating concentrations of vitamin D_3 after exposure to 1 MED UV radiation. The same dose of UV radiation increased the serum level of vitamin D_3 10-fold in control subjects (without PABA) 24 h after exposure. Therefore, any UVB blocking agent that prevents the damaging effects of sunlight will also prevent the cutaneous production of previtamin D_3. The impact of chronic sunscreen use on the vitamin D status of the elderly has been evaluated *(84)*. It was found that the 25-OH-D concentrations in serum of long-term PABA users were significantly lower than those of the nonusers. Among the 19 long-term PABA users, 4 had borderline to overt vitamin D deficiency based on low circulating concentrations of 25(OH)D.

Clothing also blocks transmission of nonvisible UV radiation *(59,85)*. Robson and Diffey *(85)* tested 60 different fabrics and observed that the nature and type of the textures and the structure of the fabric affect the sun protection factor (SPF). For example, a polyester blouse would have an SPF of 2; cotton twill jeans would have an SPF of 1000, or complete protection against extreme sunlight exposures. Matsuoka and her colleagues *(59)* examined the effect of commonly used fabrics (including cotton, wool, and polyester in black and white colors) on UVB transmission properties, on the photoproduction of previtamin D_3 from 7-DHC, and on the elevation of serum vitamin D_3 after irradiation with 1 MED of UVB in volunteers wearing jogging garments made of these fabrics. They found that direct transmission of UVB was attenuated the most by black wool and the least by white cotton. None of the fabrics allowed conversion of 7-DHC to previtamin D_3 or increase in serum vitamin D_3 in volunteers after irradiation with up to 40 min of simulated sunlight. Increasing the whole-body irradiation dose to 6 MEDs still failed to elevate serum vitamin D_3 levels in garment-clad subjects. It was concluded that clothing significantly impairs the formation of vitamin D_3 after 6 MEDs of UVB photostimulation. They also studied the effect of regular (seasonal) street clothing on serum vitamin D_3 response to whole-body irradiation with a suberythematous dose of UVB (27 mJ/cm^2) radiation in seven healthy subjects. They found that summer clothing partially and autumn clothing totally prevented an elevation of vitamin D_3 in response to UVB radiation 24 h later *(59)*.

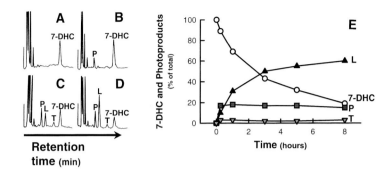

Fig. 12. High-performance liquid-chromatographic profiles of a lipid extract from the basal cells of surgically obtained hypopigmented skin that was previously shielded from (**A**) or exposed to equatorial simulated solar ultraviolet radiation that reaches the Earth at sea level at noon for (**B**) 10 min, (**C**) 1 h, and (**D**) 8 h. (**E**) Analysis of the photolysis of 7-DHC (○) in the basal cell layer and the appearance of the photoproducts previtamin D_3 (P)(■), lumisterol$_3$ (L)(▲), and tachysterol$_3$(T)(▼) with increasing exposure time. (Reproduced with permission from ref. 68).

6.5. Photodegradation of Vitamin D_3

Once vitamin D is formed in the skin from previtamin D_3, it is transported into the circulation by a diffusion process from the epidermis into the dermal capillary bed. It is believed that this process is prompted by its attraction to the DBP (35). If vitamin D_3 in the skin is exposed to sunlight prior to its transfer into the circulation, the triene system of vitamin D_3 structure will absorb solar UV radiation and photolyze to three major photoproducts, 5,6-*trans*-vitamin D_3, suprasterol 1, and suprasterol 2 (55). Exposure to as little as 10 min of sunlight in Boston in the summer resulted in the photodegradation of 30% of [^{3}H]vitamin D_3 in a test tube model system (55). Likewise, exposure to 0.5, 1, and 3 h of the summer sunlight caused 50, 75, and 95% degradation of the original [^{3}H]vitamin D_3. Although winter sunlight in Boston does not promote vitamin D_3 synthesis, the longer wavelength UV radiation such as UVA present in the winter sunlight could potentially photodegrade vitamin D stores in the skin and in the circulation.

6.6. Photoisomers of Vitamin D_3

Previtamin D_3 photosynthesized in the skin can either isomerize to vitamin D_3 or to a variety of products, including tachysterol$_3$ and lumisterol$_3$. For example, during the first 10 min of simulated equatorial solar radiation, about 10–15% of the epidermal 7-DHC in white skin was converted to previtamin D_3 without any detectable amounts of tachysterol$_3$ or lumisterol$_3$ (68). After 1 h of exposure, 5 and 30% of the original 7-DHC was converted to tachysterol$_3$ and lumisterol$_3$, respectively, whereas the amount of previtamin D_3 remained at about 15%. The concentrations of lumisterol increased with increasing exposure times, reaching 60% by 8 h (Fig. 12). When Black skin was irradiated under the same conditions, longer exposure times were required to reach the maximal previtamin D_3 formation (68). Thus it is the photochemical degradation of previtamin D_3 rather than melanin pigmentation that is most responsible for limiting the production of previtamin D_3 in human skin.

7. SUMMARY

Vitamin D was discovered after a long search for the causes and treatments of the bone-deforming disease rickets (a product of the industrial revolution), beginning with an early observation by Sniadecki in 1822 that lack of sunlight exposure was a likely cause. Ensuing efforts by numerous clinicians and scientists, including Bretonneau, Trousseau, Mellanby, McCollum, Huldschinsky, Hess, Steenbock, Windaus, and their colleagues, finally led to the demonstration a century later that exposure to sunlight alone or ingesting UV-irradiated foods could prevent and cure this bone disease.

It is well established that during exposure to sunlight, the UVB portion of the solar spectrum (290–315 nm) that penetrates to the earth's surface is responsible for the photolysis of epidermal stores of 7-DHC to previtamin D_3, which is then thermoisomerized to vitamin D_3. During the past 20 years significant progress has been made in our understanding of the basic photobiologic process that results in the production of previtamin D_3 and the regulation of this process in the skin by (1) skin pigmentation, (2) season, latitude, and time of day, (3) aging, (4) sunscreen use and clothing and (5) prolonged sunlight exposure.

A major advance in the field of vitamin D photobiology in recent years is the demonstration that the thermoisomerization reaction from previtamin D_3 to vitamin D_3 is greatly enhanced by the lipid-membrane environment in which it is made. For many years, it had been assumed that the isomerization rate of previtamin D_3 to vitamin D_3 in the skin was a relatively slow process, similar to that occurring in organic solvents such as hexane, which took approximately 30 h for 50% of the previtamin D_3 to convert to vitamin D_3 at body temperature. Recent evidence indicates that 50% of the previtamin D_3 can convert to vitamin D_3 within 2.5 h in the skin. This observation explains the rapid rise in blood levels of vitamin D_3 after exposure to simulated solar ultraviolet radiation. Within 12–24 h after exposure to simulated sunlight, the circulating concentrations of vitamin D_3 are at their maximum levels. However, little is known about the mechanism by which vitamin D_3 is translocated from the epidermis into the circulation. It has been suggested that DBP may be involved in this transport process.

It is recognized not only that exposure to sunlight produces previtamin D_3, which ultimately is thermally isomerized to vitamin D_3, but also that a variety of photoisomers of both previtamin D_3 and vitamin D_3 are generated. Little is known as to whether any of these photoisomers have unique biologic properties in the skin or systemically. It is intriguing to speculate that sunlight exposure produces some of these photoisomers for an important biologic purpose. If this is true, then the simple supplementation of foods with vitamin D, although satisfying the body's vitamin D requirement, does not provide the other potentially beneficial photoproducts of previtamin D_3 and vitamin D_3.

ACKNOWLEDGMENTS

The authors thank Dr. David Jackson for the graphics. This work was supported in part by grants RO1AR36963, RO1DK43690, and MO1RR 00533 from the National Institutes of Health.

REFERENCES

1. DeLuca HF. The metabolism, physiology, and function of vitamin D. In: Vitamin D, Basic and Clinical Aspects. Kumar R, ed. Boston: Nijhoff Publishing, 1984; 259–302.

2. Holick MF. Photobiology, metabolism, and clinical applications. In: Endocrinology, 3rd ed. DeGroot LJ, ed. Philadelphia: WB Saunders, 1995; 990–1013.
3. Reichel H, Koeffler HP, Norman AW. The role of the vitamin D endocrine system in health and disease. N Engl J Med 1989; 320:981–991.
4. Mozolowski W. Jedrzej Sniadecki (1768–1883) on the cure of rickets. Nature 1939; 143:121.
5. Palm TA. The geographic distribution and etiology of rickets. Practitioner 1890; 45:270–279, 321–342.
6. Holick MF. Vitamin D and the skin: photobiology, physiology and therapeutic efficacy for psoriasis. In: Bone and Mineral Research, vol 7. Heersche JNM, Kanis JA, eds. Amsterdam: Elsevier, 1990: 313–366.
7. Mellanby T. The part played by an 'accessory factor' in the production of experimental rickets. J Physiol 1918; 52:11–14.
8. McCollum EF, Simmonds N, Becker JE, Shipley PG. Studies on experimental rickets and experimental demonstration of the existence of a vitamin which promotes calcium deposition. J Biol Chem 1922; 53:293–312.
9. Huldschinsky K. Heilung von Rachitis durch kunstliche Honensonne. Dtsch Med Wochenschr 1919; 45:712,713.
10. Hess AF, Unger LJ. Cure of infantile rickets by sunlight. JAMA 1921; 77:39–43.
11. Goldblatt H, Soames KN. A study of rats on a normal diet irradiated daily by the mercury vapor quartz lamp or kept in darkness. Biochem J 1924; 17:294–297.
12. Goldblatt H. A study of the relation of the quantity of fat-soluble organic factor in the diet to the degree of calcification of the bones and the development of experimental rickets in rats. Biochem J 1924; 17:298–326.
13. Hess AF, Weinstock M. Antirachitic properties imparted to inert fluids and green vegetables by ultraviolet irradiation. J Biol Chem 1924; 62:301–313.
14. Steenbock H, Black A. The induction of growth-promoting and calcifying properties in a ration by exposure to ultraviolet light. J Biol Chem 1924; 61:408–422.
15. Steenbock H. The induction of growth-promoting and calcifying properties in a ration by exposure to light. Science 1924; 60:224–225.
16. Stamp TC, Walker PG, Perry W, Jenkins, MV. Nutritional osteomalacia and late rickets in greater London, 1974–1979: clinical and metabolic studies in 45 patients. Clin Endocrinol Metab 1980; 9: 81–105.
17. Marksted T, Halvorsen S, Halvorsen KS et al. Plasma concentrations of vitamin D metabolites before and during treatment of vitamin D deficiency rickets in children. Acta Paediatr Scand 1984; 73: 225–231.
18. Tanner JT, Smith J, Defibaugh P et al. Survey of vitamin content of fortified milk. J Assoc Off Anal Chem 1988; 71:607–610.
19. Holick MF, Shao Q, Liu WW, Chen TC. The vitamin D content of fortified milk and infant formula. N Engl J Med 1992; 326:1178–1181.
20. Chen TC, Shao Q, Heath H, Holick MF. An update on the vitamin D content of fortified milk from the United states and Canada. N Engl J Med 1993; 329:1507.
21. Fieser LD, Fieser M. Vitamin D. In: Steroids. New York: Reinhold, 1959; 90–168.
22. Windaus A, Bock F. Uber das provitamin aus dem sterin der schweineschwarte. Hoppe-Seylers Z Physiol Chem 1937; 245:168–170.
23. Rauschkolb EW, Winston D, Fenimore DC, Black HS, Fabre LF. Identification of vitamin D_3 in human skin. J Invest Dermatol 1969; 53:289–293.
24. Okano T, Yasumura M, Mizuno K, Kobayashi T. Photochemical conversion of 7-dehydrocholesterol into vitamin D_3 in rat skins. J Nutr Sci Vitaminol (Tokyo) 1977; 23:165–168.
25. Esvelt RR, Schnoes HK, DeLuca HF. Vitamin D_3 from rat skins irradiated in vitro with ultraviolet light. Arch Biochem Biophys 1978; 188:282–286.
26. Pask-Hughes PA, Calam DH. Determination of vitamin D_3 in cod-liver oil by high performance liquid chromatography. J Chromatogr 1982; 246:95–104.
27. Muller-Mulot VW, Rohrer G, Schwarzbauer K. Zur auffindung naturilich vorkommender vitamin D_3-ester im lebertran chemische bestimmung des freien, verester-ten und gesamt-vitamin D_3. Fette Seifen Anstrichm 1979; 81:38–40.
28. St Lezin MA. Phylogenetic occurence of vitamin D and provitamin D sterols. MS dissertation. Cambridge, MA: MIT Press, 1983.
29. Koch EM, Koch FC. The provitamin D of the covering tissues of chickens. J Poult Sci 1941; 20:33–35.

30. Wheatley RH, Sher DW. Studies of the lipids of dog skin. J Invest Dermatol 1969; 36:169–170.
31. Kenny DE, Irlbeck NA, Chen TC, Lu Z, Holick MF. Determination of vitamins D, A and E in sera and vitamin D in milk from captive and free-ranging polar bears (*Ursus maritimus*) and 7-dehydrocholesterol levels in skin from captive polar bears. 1998, submitted.
32. Morris JG. Ineffective vitamin D synthesis of vitamin D in kittens exposed to sun and ultraviolet light is reversed by an inhibitor of 7-dehydrocholesterol-Δ^7-reductase. In: Proceedings of the Tenth Workshop on Vitamin D, Strasbourg, France, May 24–29, 1997. Norman AW, Bouillon R, Thomasset M, eds. University of California Press, Riverside, CA. 1997; 721–722.
33. Velluz L, Petit A, Amiard G. Sur un stage non photochimique dans la formation des calciferols: essais d'interpretation. Bull Soc Chim Fr 1948; 15:1115–1120.
34. Velluz L, Amiard G, Petit A. Le precalciferol—ses relations d'equilibre avec le calciferol. Bull Soc Chim Fr 1949; 16:501–508.
35. Holick MF, MacLaughlin JA, Clark MB, Holick SA, Potts JT Jr, Anderson RR, Blank IH, Parrish JA, Elias P. Photosynthesis of vitamin D_3 in human skin and its physiologic consequences. Science 1980; 210:203–205.
36. Bunker JWM, Harris RS, Mosher ML. Relative efficiency of active wavelengths of ultraviolet light in activation of 7-dehydrocholesterol. J Am Chem Soc 1940; 62:508–511.
37. Kobayashi T, Yasumara M. Studies on the ultraviolet irradiation of previtamin D and its related compounds. Effect of wavelength on the formation of potential vitamin D_2 in the irradiation of ergosterol by monochromatic ultraviolet rays. J Nutr Sci Vitam 1973; 119:123–128.
38. MacLaughlin JA, Anderson RR, Holick MF. Spectral character of sunlight modulates photosynthesis of previtamin D_3 and its photoisomers in human skin. Science 1982; 216:1001–1003.
39. Anderson RR, Parrish JA. Optical properties of human skin. In:. The Science of Photomedicine. Regan JD, Parrish JA, eds. New York: Plenum, 1982; 147–194.
40. MacLaughlin JA, Holick MF. Photobiology of vitamin D in the skin. In: Biochemistry and Physiology of the Skin. Goldsmith LA, ed. New York: Oxford University Press, 1983; 734–754.
41. Havinga E. Vitamin D, example and challenge. Experientia 1973; 29:1181–1193.
42. Woodward RB, Hoffmann R. Selection rules for sigmatropic reactions. J Am Chem Soc 1965; 87:2511–2513.
43. Hanewald KH, Rappoldt MP, Roborgh X Jr. Antirachitic activity of previtamin D_3. Recl Trav Chim Pays-Bas Belg 1961; 80:1003–1014.
44. Tian XQ, Chen TC, Matsuoka LY, Wortsman J, Holick MF. Kinetic and thermodynamic studies of the conversion of previtamin D_3 to vitamin D_3 in human skin. J Biol Chem 1993; 268:14,888–14,892.
45. Tian XQ, Chen TC, Lu Z, Shao Q, Holick MF. Characterization of the translocation process of vitamin D_3 from the skin into the circulation. Endocrinology 1994;135:655–661.
46. Holick MF, Tian XQ, Allen M. Evolutionary importance for the membrane enhancement of the production of vitamin D_3 in the skin of poikilothermic animals. Proc Natl Acad Sci USA 1995; 92:3124–3126.
47. Tian X, Holick MF. Catalyzed thermal isomerization between previtamin D_3 and vitamin D_3 via β-cyclodextrin complexation. J Biol Chem 1995; 270:8706–8711.
48. Albers E, Muller BW. Complexation of steroid hormones with cyclodextrin derivatives: substituent effects of the guest molecule on solubility and stability in aqueous solution. Pharm Sci 1992; 81:756–761.
49. Chen ET, Pardue HL. Analytical applications of catalytic properties of modified cyclodextrins. Anal Chem 1993; 65:2563–2567.
50. Bogoslovsky NA, Kurganov BI, Samochvalova NG, Isaeva TA, Sugrobova NP, Gurevich VM, Valashek IE, Samochvalov GI. Vitamin D: Molecular, Cellular and Clinical Endocrinology. Berlin: Walter de Gruyter, 1988; 1021–1023.
51. Haddad JG, Matsuoko LY, Hollis BW, Hu YZ, Wortsman, J. Human plasma transport of vitamin D after its endogenous synthesis. J Clin Invest 1993; 91:2552–2555.
52. Clemens TL, Henderson SL, Adams JS Holick, MF. Increased skin pigment reduces the capacity of skin to synthesize vitamin D_3. Lancet 1982; January 9:74–76.
53. Lo C, Paris PW, Holick MF. Indian and Pakistani immigrants have the same capacity as Caucasians to produce vitamin D in response to ultraviolet irradiation. Am J Clin Nutr 1986; 44:683–685.
54. Loomis, F. Skin-pigment regulation of vitamin D biosynthesis in man. Science 1967; 157:501–506.
55. Webb AR, de Costa B, Holick MF. Sunlight regulates the cutaneous production of vitamin D_3 by causing its photodegradation. J Clin Endocrinol Metab 1989; 68:882–887.

56. Kassowitz M. Tetanie and autointoxication in kindersalter. Wien Med Presse 1897; 97:139–141.
57. Schmorl G. Die Pathologigische Anatomie de Rachitischen Knochenerkrankung mit Besonderer ber Ucksichtigung imer Histologie und Pathogenese. Ergeb Inn Med Kinderheilkd 1909; IV:403.
58. Hansemann D. Veber den Einfluss der Domestikation auf die Entstehung der Krankheiten. Berl Klin Wochenschr 1906; 629:670.
59. Matsuoko LY, Wortsman J, Dannenberg MJ, Hollis B, Lu Z, Holick MF. Clothing prevents ultraviolet-B radiation-dependent photosynthesis of vitamin D_3. J Clin Endocrinol Metab 1992; 75: 1099–1103.
60. Webb AR, Kline L, Holick MF. Influence of season and latitude on the cutaneous synthesis of vitamin D_3: exposure to winter sunlight in Boston and Edmonton will not promote vitamin D_3 synthesis in human skin. J Clin Endocrinol Metab 1988; 67:373–378.
61. Lu Z, Chen TC, Holick MF. Influence of season and time of day on the synthesis of vitamin D_3. In: Biological Effects of Light. Holick MF, Kligman AM, eds. Berlin: Walter de Gruyter, 1992; 57–61.
62. Ladizesky M, Lu Z, Oliver B, Roman NS, Diaz S, Holick MF, Mautalen C. Solar ultraviolet B radiation and photoproduction of vitamin D_3 in Central and Southern areas of Argentina. J Bone Miner Res 1995; 10:545–549.
63. Fitzpatrick TB. The validity and practicality of sun-reactive skin types I through VI. Arch Dermatol 1988; 124:869–871.
64. Ala-Houhala M, Parviainen MT, Pyykko K, Visakorpi JK. Serum 25-hydroxyvitamin D levels in Finnish children aged 2–17 years. Acta Paediatr Scand 1984; 73:232–236.
65. Oliveri MB, Ladizesky M, Mautalen CA, Alonso A, Martinez L. Seasonal variations of 25-hydroxyvitamin D and parathyroid hormone in Ushuaia (Argentina), the southernmost city of the world. Bone Miner 1993; 20:99–108.
66. Sherman SS, Hollis BW, Tobin JD. Vitamin D status and related parameters in a healthy population: the effects of age, sex and season. J Clin Endocrinol Metab 1990; 71:405–413.
67. Lester E, Skinner RK, Wills MR. Seasonal variation in serum 25-hydroxyvitamin D in the elderly in Britain. Lancet 1977; 1:979–980.
68. Holick MF, MacLaughlin JA, Dopplet SH. Regulation of cutaneous previtamin D_3 photosynthesis in man: skin pigment is not an essential regulator. Science 1981; 211:590–593.
69. Chalmers J, Conacher DH, Gardner DL, Scott PJ. Osteomalacia- a common disease in elderly women. J Bone Joint Surg [Br] 1967; 49B:403–423.
70. Chapuy MC, Arlot ME, Duboeuf F, Brun J, Crouzet B, Arnaud S, Delmas PD, Meunier PJ. Vitamin D_3 and calcium to prevent hip fractures in elderly women. N Engl J Med 1992; 327:1637–1642.
71. Dawson-Hughes B, Dallal GE, Krall EA, Harris S, Sokoll LJ, Falconer G. Effect of vitamin D supplementation on wintertime and overall bone loss in healthy postmenopausal women. Ann Intern Med 1991; 115:505–512.
72. Rosen CJ, Morrison A, Zhou H, Storm D, Hunter SJ, Musgrave K, Chen TC, Liu WW, Holick MF. Elderly women in northern New England exhibit seasonal changes in bone mineral density and calciotropic hormones. Bone Miner 1994; 25:83–92.
73. Jenkins DH, Roberts JG, Webster D, Williams EO. Osteomalacia in elderly patients with fracture of the femoral neck. J Bone Joint Surg [Br] 1973; 55B:575–580.
74. Doppelt SH, Neer RM, Daly M, Bourret L, Schiller A, Holick MF. Vitamin D deficiency and osteomalacia in patients with hip fractures. Orthop Trans 1983; 7:512–513.
75. Sokoloff L. Occult osteomalacia in American patients with fracture of the hip. Am J Surg Pathol 1978; 2:21–30.
76. Whitelaw GP, Abramowitz AJ, Kavookjian H, Holick MF. Fractures and vitamin D deficiency in the elderly. Complications Ortho 1991; 6:70–80.
77. Omdahl JL, Garry PJ, Hunsaker LA, Junt WC, Goodwin JS. Nutritional status in a healthy elderly population: vitamin D. Am J Clin Nutr 1982; 36:1225–1233.
78. Lamberg-Allard T. Vitamin D intake, sunlight exposure, and 25-hydroxyvitamin D levels in elderly during one year. Ann Nutr Met 1984; 28:144–150.
79. Webb AR, Pilbeam C, Hanafin N, Holick MF. An evaluation of the relative contributions of exposure to sunlight and diet on the circulating concentrations of 25-hydroxyvitamin D in an elderly nursing home population in Boston. Am J Clin Nutr 1990; 51:1075–1081.
80. Tan CY, Strathum B, Marks R. Skin thickness measurement by pulsed ultrasound: its reproducibility, validation and variability. Br J Dermatol 1982; 106:657–667.

81. MacLaughlin JA, Holick MF. Aging decreases the capacity of human skin to produce vitamin D_3. J Clin Invest 1985; 76:1536–1538.
82. Holick MF, Matsuoka LY, Wortsman J. Age, vitamin D, and solar ultraviolet radiation. Lancet 1989; November 4:1104,1105.
83. Matsuoko L, Ide L, Wortsman J, MacLaughlin JA, Holick MF. Sunscreens suppress cutaneous vitamin D_3 synthesis. J Clin Endocrinol Metab 1987; 64:1165–1168.
84. Matsuoko LY, Wortsman J, Hanifan N, Holick MF. Chronic sunscreen use decreases circulating concentrations of 25-hydroxyvitamin D: a preliminary study. Arch Dermatol 1988; 124:1802–1804.
85. Robson J, Diffey BL. Textiles and sun protection. Photodermatol Photoimmunol Photomed 1990; 7:32–34.
86. Annan T. Thomas Annan's Photographs of the Old Closes and Streets of Glasgow 1868/1877. New York: Dover, 1977.

3 Functional Metabolism and Molecular Biology of Vitamin D Action

Laura C. McCary and Hector F. DeLuca

1. INTRODUCTION

The dietary requirement for vitamin D and/or the necessity for exposure to the sun's ultraviolet (UV) rays were first recognized in the prevention or cure of the disease rickets *(1–4)*. This disease is the result of a failure to mineralize the organic matrix of bone. The resultant decrease in bone strength causes skeletal malformations and may result in death *(5)*. In adults, the disease osteomalacia occurs, in which undermineralized osteoid seams appear and bones are easily fractured *(6)*.

Rickets in children and osteomalacia in adults became prevalent at the time of the industrial revolution because smoke and urbanization deprived the industrialized population of sunlight *(5)*. Steenbock and Black *(7)* found that (UV) light could be used to induce vitamin D activity in the lipid portions of food. This process was then used to fortify food, which eliminated rickets as a major medical problem. Nutritional rickets then became rare in the developed countries. However, forms of rickets and osteomalacia not cured by nutritional levels of vitamin D remained. Of these, two result from genetic defects in the vitamin D system. Vitamin D-dependent rickets type I occurs when the kidneys are unable to metabolize 25-hydroxyvitamin D_3 [$25(OH)D_3$] to the more polar and the hormonally active 1,25-dihydroxyvitamin D_3 [$1,25(OH)_2D_3$] *(8)*. The genetic defect is thought to occur in the $25(OH)D_3$-1-hydroxylase enzyme in the mammalian kidney *(9)*. The inability to produce $1,25(OH)_2D_3$ greatly impairs skeletal mineralization since $1,25(OH)_2D_3$ is the functional or hormonal form of vitamin D *(10)*. A more recently discovered form of rickets, vitamin D-dependent rickets type II, occurs when the vitamin D receptor (VDR) protein is nonfunctional; thus, a tissue resistance to $1,25(OH)_2D_3$ occurs *(11–13)*. Several genetic mutations in the VDR gene have been identified that render the protein nonfunctional or dramatically less functional *(14–16)*. The absence of functional VDR precludes the transcriptional regulation of vitamin D-regulated genes involved in calcium and phosphorus homeostasis. Therefore, calcium and phosphorus are not maintained at high levels in the serum to support bone mineralization. These two types of rickets illustrate both the functional metabolism and molecular biology of vitamin D action.

From: *Vitamin D: Physiology, Molecular Biology, and Clinical Applications*
Edited by: M. F. Holick © Humana Press Inc., Totowa, NJ

2. OVERALL ROLE OF THE VITAMIN D HORMONE IN CALCIUM AND PHOSPHORUS HOMEOSTASIS

Ultimately, the function of the vitamin D hormone is to increase serum calcium and serum phosphorus levels required for skeletal mineralization and therefore the prevention of rickets *(17–19)*. The hormone accomplishes this through its actions on the intestine, bone, and kidney. In the intestine, $1,25(OH)_2D_3$, stimulates the active transport of calcium and phosphorus from the lumen of the intestine to the blood *(20–28)*. Through this action, the vitamin D hormone has indirect anabolic effects on bone by increasing serum calcium and phosphorus to levels that can allow proper and complete skeletal mineralization *(17–19)*.

The role of the vitamin D hormone in skeletal mineralization is entirely separate from the actions of $1,25(OH)_2D_3$ and parathyroid hormone (PTH) on kidney and bone. At times of extreme hypocalcemia, $1,25(OH)_2D_3$ and PTH will act upon bone to mobilize both calcium and phosphorus from this tissue *(29–32)*. Obviously, mobilization of mineral from bone counters the effect of vitamin D on skeletal mineralization. Furthermore, $1,25(OH)_2D_3$ and PTH will act on the kidney to increase the reabsorption of calcium, but not phosphorus, from the urine to the serum *(33)*. Indirectly, $1,25(OH)_2D_3$ activates the reabsorption of phosphorus from the urine to the serum by decreasing PTH levels because PTH causes the kidneys to excrete phosphorus *(34,35)*. However, if both $1,25(OH)_2D_3$ and PTH are present, there is no net increase in serum phosphorus through the action of $1,25(OH)_2D_3$ and PTH on the kidney, but there is a net increase in serum calcium levels. The purpose of $1,25(OH)_2D_3$ and PTH under dietary calcium restriction and hypocalcemia is to provide calcium in the serum for neuromuscular function. Extreme hypocalcemia places one in grave danger of death resulting from hypocalcemic tetany. To avoid this situation, $1,25(OH)_2D_3$ and PTH increase serum calcium levels, without increasing serum phosphorus levels, through actions on kidney and bone. If serum phosphorus is not increased in conjunction with serum calcium, then calcium will not be used for skeletal mineralization but, instead, will be employed to maintain proper neuromuscular function. Figure 1 depicts the actions of $1,25(OH)_2D_3$ and PTH on their classical target tissues in order to maintain serum calcium.

3. FUNCTIONAL METABOLISM

The ability of researchers to synthesize [^{3}H] vitamin D chemically was the major breakthrough in elucidating the metabolism of vitamin D *(36)*. Upon administration of [^{3}H] vitamin D to a rat, many polar tritiated metabolites were observed in the serum, intestine, kidney, and bone *(36)*. Metabolite identification, in vitro synthesis, and biological activity assays were vital to our present knowledge of the vitamin D endocrine system.

Vitamin D may be obtained by dietary means from only a few sources, i.e., fish liver oils or fortified foods. Otherwise, the principal source of vitamin D is skin. The epidermis contains a pool of 7-dehydrocholesterol *(37)*. This compound contains a chromophore because of the conjugated double bond system, and it absorbs 280–310-nm UV light. Upon absorption of the UV light, 7-dehydrocholesterol undergoes a photochemical isomerization to form previtamin D_3 *(38,39)*. Previtamin D_3 exists in equilibrium with its isomer, vitamin D_3 *(40)*. Vitamin D_3 may be transported in the blood by the vitamin D binding protein to the liver *(41)*. In the liver, the vitamin D_3 molecule is enzymatically hydroxylated on carbon 25 to form $25(OH)D_3$ *(42,43)*. This metabolite is transported to

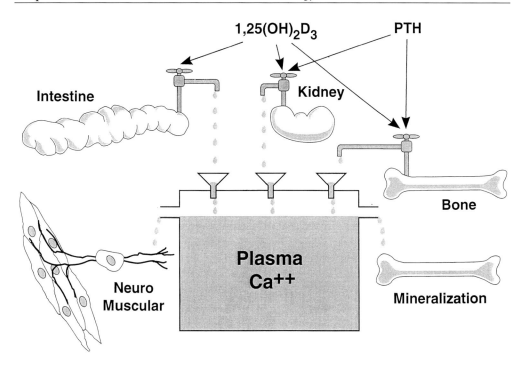

Fig. 1. A diagrammatic representation of the classic physiologic actions of vitamin D in healing the disease rickets and in preventing the convulsive disease hypocalcemic tetany. Note that the vitamin D hormone functions in the intestine, in the skeleton, and in the distal renal tubule to mobilize calcium and, in the case of intestine, phosphorus to the plasma. The saturating levels of plasma calcium and phosphorus support bone mineralization and neuromuscular junction activity.

the kidney, where it is again enzymatically hydroxylated, but this time on carbon 1, to form $1,25(OH)_2D_3$ *(44,45)*. $1,25(OH)_2D_3$ is the biologically active form of vitamin D *(46–48)*, classified as a hormone because its production is at a site distant from the targets of hormone action, and it is carried to its target tissues by the blood *(49)*.

The catabolism of $1,25(OH)_2D_3$ also proceeds by an enzymatic pathway. The first enzyme in this pathway is the $25(OH)D_3$-24-hydroxylase. The 24-hydroxylase enzyme is present in highest concentrations in kidney mitochondria *(50)*. Its substrates are $25(OH)D_3$ and $1,25(OH)_2D_3$, but the 24-hydroxylase has a 40-fold higher affinity for $1,25(OH)_2D_3$ *(51)*. There is increasing evidence that the 24-hydroxylase enzyme is able to execute all enzymatic reactions in the catabolism of $1,25(OH)_2D_3$ by the C23/C24 pathway *(52,53)*. Purified 24-hydroxylase enzyme has been shown to possess 23-hydroxylase activity in vitro *(52,53)*. Calcitroic acid has been found in the bile; therefore, it is thought to be the eventual excretory product of vitamin D *(54–55)*. Figure 2 demonstrates the functional metabolism of vitamin D. Greater details are given in Chapter 4, by Dr. Glenville Jones.

4. REGULATION OF PRODUCTION AND CATABOLISM OF $1,25(OH)_2D_3$

Calcium and phosphorus levels in the plasma are tightly regulated by the vitamin D/PTH/calcitonin endocrine system. As a result, in vivo, three situations result in differ-

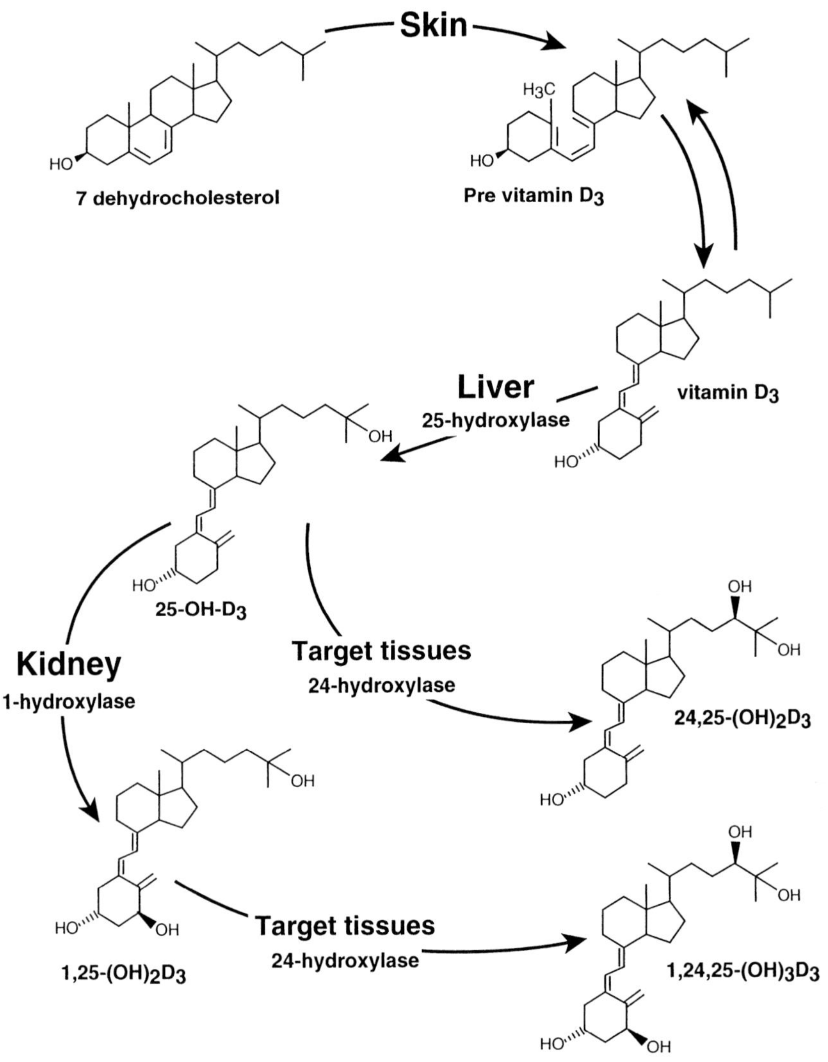

Fig. 2. The functional metabolism of vitamin D. Note that the 24-hydroxylase enzyme can use both 25(OH)D$_3$ and 1,25(OH)$_2$D$_3$ as substrates, although 24-hydroxylation of 25(OH)D$_3$ occurs predominantly in the kidney because other vitamin D target tissues are not exposed to large amounts of 25(OH)D$_3$.

ential regulation of 1,25(OH)$_2$D$_3$ production. First, serum calcium levels may be low and serum phosphorus levels normal. Second, serum phosphorus levels may be low and serum calcium levels normal. Third, serum calcium and serum phosphorus levels may both be low. In each situation, 1,25(OH)$_2$D$_3$, either alone or in conjunction with PTH, normalizes serum calcium and serum phosphorus levels.

Under conditions of low serum calcium, the calcium receptor on parathyroid glands activates the secretion of PTH *(56)*, which acts on the kidney to increase the activity of the 25(OH)D$_3$-1-hydroxylase enzyme *(47,48,57–59)*. Such increased activity increases the circulating levels of 1,25(OH)$_2$D$_3$. 1,25(OH)$_2$D$_3$ acts on the intestine to stimulate the active transport of calcium and phosphorus from the intestinal lumen to the serum *(20–28)*. In the bone, PTH and 1,25(OH)$_2$D$_3$ activate bone calcium and phosphorus mobilization *(29–32)*. In addition, 1,25(OH)$_2$D$_3$ and PTH act on the kidney to stimulate the active reabsorption of calcium from the urine to the blood *(33)*. PTH also acts on the kidney to reduce phosphorus reabsorption *(35)*. Therefore, in this situation, serum calcium levels will increase to normal levels and serum phosphorus levels will remain unchanged. The function of $1,25(OH)_2D_3$ and PTH in this situation is to increase serum calcium levels to avoid hypocalcemic tetany. Once serum calcium levels have returned to normal, PTH secretion is no longer stimulated, and, thus, the production of $1,25(OH)_2D_3$ is no longer activated.

Under conditions of low serum phosphorus and normal serum calcium, PTH secretion is not stimulated since the serum calcium levels are normal. However, low serum phosphorus levels can trigger an increase in the activity of the $25(OH)D_3$-1-hydroxylase enzyme by an unknown mechanism *(60–62)*. Growth hormone (GH) and insulin-like growth factor-1 (IGF-1) have been proposed as mediators of 1-hydroxylase activity in response to hypophosphatemia *(63–65)*. Although IGF-1 and GH do appear to stimulate 1-hydroxylase activity at low serum phosphorus levels in hypophysectomized rats, they are not likely to mediate the 1-hydroxylase response to serum phosphorus because IGF-1 and GH do not increase 1-hydroxylase activity when administered to hypophysectomized rats with normal serum phosphorus levels *(64)*. If IGF-1 and GH were mediators of serum phosphorus effects on the 1-hydroxylase, then administration of these compounds to rats should be able to increase 1-hydroxylase activity independent of serum phosphorus. Regardless, the 1-hydroxylase response to serum phosphorus seems to be mediated by some pituitary gland hormone since hypophysectomy eliminates the 1-hydroxylase response to serum phosphorus and since hypophysectomy decreases serum $1,25(OH)_2D_3$ levels *(66–68)*. In hypophosphatemia, increased activity of the 1-hydroxylase enzyme increases circulating levels of $1,25(OH)_2D_3$. Without PTH present, $1,25(OH)_2D_3$ will act solely on the intestine to stimulate the active absorption of calcium and phosphorus from the lumen of the intestine to the blood *(31)*. This action will increase both serum calcium and serum phosphorus levels. Once serum phosphorus is increased, $25(OH)D_3$-1-hydroxylase activity is no longer stimulated. To date, it is not explained how the body avoids hypercalcemia under these conditions, but there has been some speculation that high serum calcium levels can signal the calcium receptor in the parathyroid glands to inhibit the secretion of PTH. Therefore, the kidneys would enhance calcium excretion and decrease phosphorus excretion.

When both serum calcium and serum phosphorus levels are low, the $25(OH)D_3$-1-hydroxylase is superstimulated. Low serum phosphorus increases 1-hydroxylase activity *(60–62)*. In addition, low serum calcium stimulates PTH secretion, which also activates 1-hydroxylase activity *(47–48,57–59)*. The result is extremely high circulating levels of $1,25(OH)_2D_3$. $1,25(OH)_2D_3$ will act alone in the intestine to increase calcium and phosphorus absorption in the intestine. The high PTH levels in combination with $1,25(OH)_2D_3$ will increase bone calcium and phosphorus mobilization and calcium reabsorption from the urine; PTH itself will increase renal phosphate excretion. The result will be a net increase in serum calcium and phosphorus *(31)*.

The catabolism of $1,25(OH)_2D_3$ must also be strictly regulated to prevent hypercalcemia and hyperphosphatemia. Thus PTH, $1,25(OH)_2D_3$, and serum phosphorus regulate the principal catabolic enzyme in the vitamin D endocrine system, the $25(OH)D_3$-24-hydroxylase *(69,70)*. PTH is secreted under conditions of hypocalcemia and not when calcium levels are adequate. Therefore, PTH decreases $25(OH)D_3$-24-hydroxylase activity and mRNA levels during hypocalcemia *(69,71)*. As long as serum calcium remains low, PTH will continue to stimulate the production of $1,25(OH)_2D_3$, and repress the

REGULATION OF 1-HYDROXYLASE AND 24-HYDROXYLASE IN KIDNEY

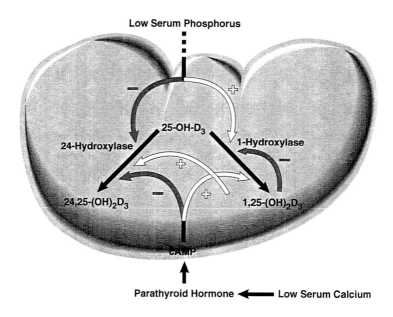

Fig. 3. The reciprocal regulation of the 1- and 24-hydroxylase enzymes by serum calcium, serum phosphorus, PTH, and $1,25(OH)_2D_3$.

production of $1,24,25(OH)_3D_3$ and $24,25(OH)_2D_3$. Thus, while low serum calcium persists, $1,25(OH)_2D_3$ catabolism to biologically inactive metabolites will be low. In addition, $1,25(OH)_2D_3$ activates its own breakdown by stimulating transcription of the 24-hydroxylase gene, thereby increasing 24-hydroxylase activity so that, once serum calcium is normalized, $1,25(OH)_2D_3$ will stimulate its own breakdown *(70,72,73)*. Low serum phosphorus decreases the mRNA levels of the 24-hydroxylase to prevent catabolism of $1,25(OH)_2D_3$ under hypophosphatemic conditions *(74)*. Since the synthesis of PTH is directly responsive to serum calcium, this hormone has the major control of vitamin D catabolism and 24-hydroxylase activity, providing for increased synthesis and decreased destruction of $1,25(OH)_2D_3$. The mechanism(s) whereby PTH controls 1α- and 24-hydroxylase activity is unclear at the present time, but it must partially involve transcriptional regulation *(71,75)*. Figure 3 shows the regulation of $1,25(OH)_2D_3$ production and catabolism by calcium and phosphorus.

There have been some suggestions that 24-hydroxylated vitamin D metabolites have some biologic function *(76–80)*. It is instructive to note that this laboratory has maintained vitamin D-deficient rats on $25(OH)D_3$ fluorinated at the 24 position for two consecutive generations *(10,81)*. The fluoro-derivative at the 24 position precludes metabolism of $1,25(OH)_2D_3$ to 24-hydroxylated metabolites *(82)*. When vitamin D-deficient rats were maintained on 24,24 difluoro-$25(OH)D_3$ as their sole source of vitamin D_3, growth, reproduction, and skeletal mineralization remained normal compared with rats maintained on $25(OH)D_3$ or $1,25(OH)_2D_3$ *(10,81,83)*. Thus, it is clear that 24-hydroxylation of vitamin D plays no significant role in the functions of vitamin D. An important experiment will be the production of a 24-hydroxylase transgenic null mouse.

This has been reported, but the results are not yet clear *(84)*. The authors found some death of the homozygotes prior to weaning and accumulation of osteoid at sites for intramembranous ossification in second-generation homozygotes, but the phenotype of the homozygotes is unexpected due to normal serum calcium, serum phosphorus, and $1,25(OH)_2D_3$ levels *(84)*.

5. MOLECULAR MECHANISM OF VITAMIN D ACTION

By using radiolabeled $1,25(OH)_2D_3$, Stumpf and colleagues *(85)* demonstrated that this compound localized in the nucleus of target cells. Brumbaugh and Haussler *(86)* and Kream et al. *(87)* demonstrated a nuclear VDR. Subsequent isolation of the VDR cDNA revealed that it was homologous to other steroid hormone receptors *(88,89)*. The VDR is a member of the steroid hormone receptor superfamily and, more specifically, the thyroid/vitamin D/retinoic acid receptor subfamily. Therefore, the VDR possesses both a DNA binding domain (C) and a ligand binding domain (E) in addition to other regions of the protein denoted A/B, D, and F *(90)*. As stated above, the autosomal recessive disease, vitamin D-dependency rickets type II, results from mutations in the VDR gene. The different VDR mutations that result in either a truncated VDR protein or the inability of the VDR protein to bind DNA demonstrate that the VDR is a necessary requirement in the vitamin D signal transduction cascade *(14–16)*.

Because of the ability of the VDR to influence gene transcription directly in the presence of its ligand, the correct term for the VDR is ligand-induced transcription factor. The sequences in DNA of the target genes to which the VDR binds are called VDREs or vitamin D response elements (VDREs). The sequences usually consist of two hexameric repeats of the consensus sequence AGGTCA separated by three nucleotides. The shorthand notation for the VDRE is direct repeat 3 (DR3) *(91)*. Specificity for responsiveness to $1,25(OH)_2D_3$ is determined by the three-nucleotide spacer *(91)*. When the spacer is modified to four or five nucleotides, the response element becomes responsive to thyroid hormone and retinoic acid, respectively *(91)*. VDREs are found in the promoter regions of genes that regulate calcium and phosphorus homeostasis as well as those genes involved in vitamin D metabolism and other functions.

When purified VDR was incubated in the presence of a VDRE sequence, it was noted that the receptor did not bind the DNA *(92,93)*. However, upon addition of mammalian nuclear extract stripped of VDR with anti-VDR monoclonal antibodies, binding occurred *(93)*. Therefore, there appeared to be a requirement for an accessory factor(s) for the VDR to bind DNA *(92)*. Subsequent to this observation, Yu et al. *(94)* isolated the cDNA for retinoid X receptor β (RXRβ) by using two criteria. First, they found that RXRβ heterodimerized with retinoic acid receptor (RAR) and second, they noted that RXRβ and RAR bound retinoic acid response elements (RAREs) in DNA. Because of the recent functional relationship reported for RAREs, VDREs, and thyroid hormone response elements (TREs), Yu et al. *(94)* further reported that RXRβ enhanced VDR–VDRE binding, but coexpression of RXRβ and VDR had little effect on VDR-dependent transcription. Furthermore, RXRβ is not expressed at high levels in mammalian intestine *(94)*. Kliewer and colleagues *(95)* reported that RXRα coprecipitated with VDR and enhanced VDR–VDRE DNA binding. These authors did not test the ability of RXRα to influence VDR-mediated gene transcription *(95)*. Meanwhile, Munder et al. *(96)* purified the nuclear accessory factor from porcine intestinal nuclear extract and showed that it was an RXR. It is now widely accepted that VDR heterodimerizes with one or more isoforms of RXR to influence gene transcription.

Interaction between VDR and RXR is thought to occur at a C-terminal dimerization interface in both proteins. Crosslinking studies showed that a mutated VDR, which did not contain a DNA binding domain, could still interact with the nuclear accessory factor *(97)*. Therefore, binding of the VDR to the accessory factor did not require DNA *(97)*. Also, the addition of ligand enhanced heterodimerization of the VDR with the accessory factor *(97)*. Rosen et al. *(98)* demonstrated that the human VDR possessed a dimerization domain between amino acids 244 and 263, which, when deleted, reduced VDR–VDRE binding. A single amino acid change in this region slightly reduced VDR-mediated transcription *(98)*. Furthermore, Nakajima and colleagues *(99)* showed evidence for another dimerization domain from amino acid 382–402 of the human VDR. Deletion of this region reduced VDR–VDRE binding and VDR-mediated transcription *(99)*. It is thought that the VDR–RXR heterodimer bound to DNA then interacts with basal eukaryotic transcription factors including RNA polymerase II to influence gene transcription.

Recently, a major area of research has emerged in the area of steroid hormone receptor molecular biology. Researchers have focused on the interactions of steroid hormone receptors with other proteins in the transcription machinery *(100)*. Originally, Blanco et al. *(101)* reported that the VDR bound transcription factor IIB (TFIIB) and that TFIIB was able to activate $1,25(OH)_2D_3$-dependent transcription when TFIIB and VDR were cotransfected into P19 embryonal carcinoma cells. Furthermore, they analyzed deletion mutants of both the VDR and TFIIB and found that the N terminus of the VDR interacted with the C terminus of TFIIB *(101)*. MacDonald et al. *(102)*, using the yeast two-hybrid system to study protein/protein interactions between the VDR and other eukaryotic proteins, also found that TFIIB interacted with the VDR. By contrast, this study showed that the C terminus of the VDR interacts with the N terminus of TFIIB in a ligand-independent manner *(102)*. Although these two studies appear to be contradictory, Baniahmad et al. *(103)* have proposed that the presence or absence of thyroid hormone can regulate which regions of the thyroid hormone receptor and TFIIB interact. More recently, Zierold and DeLuca *(104)* demonstrated the necessity for TFIIB to form the hetereodimer complex on the VDRE.

Several laboratories have isolated corepressor and coactivator proteins that influence transcription of steroid hormone-regulated genes. Steroid receptor coactivator-1 (SRC-1) activates ligand-dependent transcription by many steroid hormone receptors including progesterone receptor, glucocorticoid receptor, estrogen receptor, thyroid receptor, and RXR through a direct interaction with the receptor *(105)*. Recently, Gill et al. *(106)* found that SRC-1 also interacts with the VDR in a ligand-dependent manner. Suppressor of gal1 (SUG1), receptor interacting protein (RIP140), and glucocorticoid receptor interacting protein 1(GRIP1), all showed interaction with the ligand binding domain of the VDR in a ligand-dependent manner *(107,108)*. The SMRT protein acts as a corepressor to silence transcription in the absence of ligand of thyroid hormone and retinoic acid-regulated genes *(109)*. The VDR was not tested *(109)*. Burris et al. *(110)* isolated thyroid hormone receptor uncoupling protein (TRUP), which interferes with DNA binding of thyroid receptor and retinoic acid receptor to their response elements in the presence and absence of ligand. The VDR was not mentioned *(110)*. Lastly, a protein termed nuclear receptor co-repressor (N-CoR) also mediates ligand-independent transcriptional repression for thyroid receptor and retinoic acid receptor by interacting with the steroid receptor hinge region. No interaction was detected between N-CoR and the VDR *(111)*. A region exists in the VDR termed the AF-2 domain (activation function). This region is at the C terminus of the VDR (amino acids 408–427) in the ligand binding domain. Mutations here abolish VDR-mediated transactivation and interaction between VDR and SUG1, SRC-1, and RIP140 but maintain ligand binding, heterodimerization with RXR, and DNA binding; therefore, this region is thought to be crucial for coactivation of transcription *(106,107,112)*.

The role of $1,25(OH)_2D_3$ in the molecular mechanism of vitamin D action remains ambiguous. It is widely accepted that the binding of ligand to the VDR results in a conformational change in the protein, which can then regulate transcription. From transfection assays, it is known that $1,25(OH)_2D_3$ is required for transcriptional regulation of a reporter gene downstream of a promoter sequence containing a VDRE *(113–117)*. Pike and Sleator *(118)* were the first to identify a functional role for ligand in vitamin D-mediated signal transduction when they found that mouse fibroblast VDR was phosphorylated on addition of ligand. Brown and DeLuca *(119)* extended this work to demonstrate that VDR phosphorylation occurred in vivo 15 min after the addition of ligand to embryonic chick intestine. This phosphorylation event was independent of protein synthesis since addition of cycloheximide did not affect phosphorylation *(119)*. When the VDR was incubated in the presence of [^{32}P]-orthophosphate and ligand and then subjected to protease digestion, the radioactivity localized to a 23-kDa fragment that did not include the N terminus of the protein *(120)*. Photoaffinity labeling of the VDR with [^{3}H] $1,25(OH)_2D_3$ followed by protease digestion demonstrated that most of the ligand also localized to this 23-kDa fragment; thus, the site(s) of VDR phosphorylation were thought to be in the ligand binding domain *(120)*. Phosphoamino acid analysis determined that the VDR was phosphorylated on serine residues *(120)*. The enzyme responsible for phosphorylating the VDR after ligand addition may be a DNA-dependent protein kinase, as has been reported for the progesterone receptor *(121,122)*. Presumably, VDR phosphorylation results in a conformational change that exposes domains of the VDR that may interact with the basal transcription machinery of eukaryotic cells.

Recently, several laboratories have attempted to localize the serine residues in the VDR that are phosphorylated. To be an actual in vivo phosphorylated serine, phosphorylation of a specific serine must be shown to be dependent on the presence of $1,25(OH)_2D_3$. For the human VDR, Hilliard et al. *(123)* have demonstrated ligand-dependent phosphorylation of serine 205, but when serine 205 is mutated to alanine, the mutated VDR is still able to function in $1,25(OH)_2D_3$-dependent transcription assays. However, adjacent serines became phosphorylated instead *(123)*. Other proposed phosphorylation sites of the human VDR include serine 208 and serine 51, but phosphorylation of these sites was not shown to be ligand dependent *(124,125)*. The elucidation of the phosphorylated serines in the VDR is the first step in demonstrating the functional role of phosphorylation. Figure 4 depicts $1,25(OH)_2D_3$-regulated gene transcription.

6. VITAMIN D-REGULATED GENES

Several genes have been identified that possess VDREs in their gene promoters *(113,115–117)*. These promoter regions are typically identified by using deletion analysis. Portions of the gene promoter are deleted, binding to DNA is monitored by gel-shift assay, and transactivation is monitored by transfection assay *(115–117)*. If the VDRE sequence in a gene promoter is deleted, the VDR will no longer bind that promoter in a

Model for Vitamin D–influenced Target Gene Expression

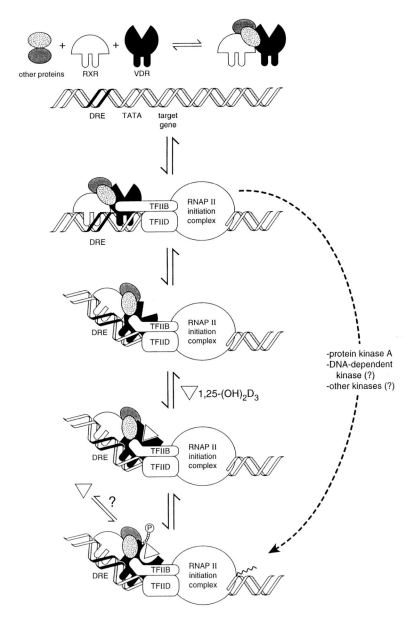

Fig. 4. A proposed model for vitamin D-influenced target gene expression. This model shows the ability of the VDR to interact with the RXR and the VDRE in the absence of $1,25(OH)_2D_3$, whereas the presence of ligand increases affinity of the VDR–RXR complex for DNA, presumably through some conformational change. Also, following ligand and DNA binding, this model predicts phosphorylation of the VDR. Transcriptional activation proceeds thereafter through the interaction of the VDR with other transcriptional regulatory factors and the eukaryotic basal transcription machinery.

gel-shift binding assay. In addition, when the VDRE sequence in a gene promoter is deleted, 1,25(OH)$_2$D$_3$ will no longer influence transcription of a reporter gene under the influence of the deleted promoter. Conversely, the putative VDRE sequence in a gene promoter must be able to bind the VDR in a gel-shift assay and regulate ligand-dependent transcription of a reporter gene. By these methods, several vitamin D-regulated genes have been identified *(113,115,117)*.

By identifying genes regulated by vitamin D, one can obtain clues about the mechanisms of calcium and phosphorus homeostasis, other functions of vitamin D, and control of vitamin D metabolism. Vitamin D-regulated genes in intestine and bone are discussed thoroughly in later chapters. The end of this chapter is dedicated to vitamin D-regulated genes involved in the functional metabolism of vitamin D.

Although the parathyroid glands are not a classical target organ for 1,25(OH)$_2$D$_3$, a VDRE has been identified in the promoter of the PTH gene *(126)*. In competition experiments, it was demonstrated that a nucleotide sequence in the PTH gene promoter from −125 to −101 successfully competed with the osteocalcin DRE for binding of VDR *(126)*. Within this nucleotide sequence, one motif of the consensus sequence AGGTCA was identified *(126)*. The identification of one motif instead of the traditional direct repeat is disturbing because specificity to 1,25(OH)$_2$D$_3$ is determined by the spacing between a direct repeat. Demay and colleagues *(126)* have suggested that flanking sequences to this half site are involved in the response to 1,25(OH)$_2$D$_3$. When this oligonucleotide was placed upstream of a heterologous viral promoter, it mediated transcriptional repression in response to the vitamin D hormone in GH4C1 rat pituitary cells, but not in ROS 17/2.8 bone cells *(126)*. Furthermore, Mackey and colleagues *(127)* have recently proposed that the VDR heterodimerizes with a non-RXR nuclear protein present in GH4C1 and bovine parathyroid nuclear extracts but not in ROS 17/2.8 nuclear extracts. Although parathyroid cells express both RXR and VDR, the authors propose that VDR preferentially heterodimerizes with a novel nuclear protein(s) in order to mediate transcriptional repression *(127)*. The authors' argument is weakened by the lack of binding constant data for each VDR-containing complex seen in parathyroid nuclear extract. Thus, the authors propose the existence of cellular factors other than the VDR that mediate transcriptional repression *(127)*. Recently, a second arm has been found for the PTH DRE and this response element binds RXR, although the RXR may occupy the 3' position instead of the '5 position as it does for an activation DRE *(128)*. Physiologically, vitamin D represses PTH gene transcription once vitamin D levels are sufficient to maintain serum calcium levels and PTH is no longer required to activate the synthesis of 1,25(OH)$_2$D$_3$.

In addition, two VDREs have been identified and characterized in the 25(OH)D$_3$-24-hydroxylase gene promoter *(72,73)*. Both VDREs consist of the traditional direct repeat separated by a three-nucleotide spacer. Transcriptional regulation of the 24-hydroxylase is unique since it is a gene that is completely transcriptionally inactive in the absence of 1,25(OH)$_2$D$_3$ and highly induced in its presence *(129)*. The mRNA for the 24-hydroxylase is upregulated by 1,25(OH)$_2$D$_3$; therefore, 1,25(OH)$_2$D$_3$ can catalyze its own breakdown, once serum calcium and serum phosphorus levels are normalized, through a mechanism involving transcriptional regulation of the key catabolic enzyme in vitamin D metabolism.

Lastly, it has been proposed that 1,25(OH)$_2$D$_3$ can feedback inhibit its own synthesis by regulating transcription of the 25(OH)D$_3$-1-hydroxylase gene through a VDRE sequence in the promoter of the 1-hydroxylase gene. In preliminary analysis of the murine

1-hydroxylase promoter, thus far, a consensus VDRE sequence has not been found *(75)*. Presently, it is still assumed that $1,25(OH)_2D_3$ will negatively regulate transcription of the 1-hydroxylase gene similarly to transcription of the PTH gene.

7. FUTURE RESEARCH

In the past 30 years, our understanding of the vitamin D endocrine system has grown at an exponential rate, but many questions remain unanswered. In the realm of the functional metabolism of vitamin D, three important areas of research remain. An important area is understanding regulation of the 1α-hydroxylase and the 24-hydroxylase. The recent cloning of the 1-hydroxylase will provide information on the regulation of that enzyme by phosphorus, PTH, and $1,25(OH)_2D_3$ itself *(130–132)*. Lastly, the mechanism of 24-hydroxylase regulation by PTH must be clarified.

The molecular biology of vitamin D is a newer field, and therefore has many more unanswered questions. First, the role of ligand, $1,25(OH)_2D_3$, in influencing gene transcription must be explained. To approach this problem, the VDR must be crystallized in the absence and presence of ligand to examine the differences in VDR conformation upon ligand addition. Second, the role of VDR phosphorylation in gene transcription must be determined. Possibly, phosphorylation is required for liganded VDR to bind DNA or other eukaryotic transcription factors. To investigate the role of phosphorylation in vitamin D-mediated gene transcription, site-directed mutagenesis can be performed on specific amino acids of the VDR that may be targets for phosphorylation. If, when a phosphorylation site is eliminated, the mutated VDR can no longer bind DNA and/or influence gene transcription in a transfection assay, the results may indicate that phosphorylation of the VDR is required for gene transcription. Next, it is well known that liganded VDR stimulates transcription of some vitamin D-regulated genes and represses transcription of other vitamin D-regulated genes. Whereas transcription of genes, such as osteocalcin and calbindin 9K, is stimulated by $1,25(OH)_2D_3$, transcription of the PTH gene is repressed by $1,25(OH)_2D_3$. Determining how $1,25(OH)_2D_3$ can both positively and negatively regulate transcription is an interesting problem. Lastly, the interaction of the ligand VDR–RXR heterodimer with other transcription factors and nuclear proteins must be examined. Techniques, such as the yeast two-hybrid system, will elucidate proteins bound in the VDR–RXR transcription complex. By these means, one can determine to which proteins the VDR binds to influence gene transcription. Ultimately, a model in vitro transcription system that is fully responsive in a specific way to $1,25(OH)_2D_3$ is needed to identify which transcriptional components are required and when each is needed in the activation or repression of target gene transcription.

ACKNOWLEDGMENTS

This work was supported in part by a program project grant from the National Institutes of Health (DK14881), a grant from the National Foundation for Cancer Research, and a grant from the Wisconsin Alumni Research Foundation.

REFERENCES

1. Mellanby E. An experimental investigation on rickets. Lancet 1919; 1:407–412.
2. McCollum EV, Simmonds N, Becker JE, Shipley PG. Studies on experimental rickets. XXI. An experimental demonstration of the existence of a vitamin which promotes calcium deposition. J Biol Chem 1922; 53:293–312.

3. Huldshinsky K. Heilung von rachitis durch kunstlickhe hohensonne. Dtsch Med Wochschr 1919; 45:712–713.
4. Chick H, Palzell EJ, Hume EM. Studies of rickets in Vienna 1919–1922. Medical Research Council, 1923: Special Report No. 77.
5. Hess A, ed. Rickets, Including Osteomalacia and Tetany. Philadelphia: Lea & Febiger, 1929.
6. Sebrell WH, Harris RS, eds. The Vitamins. New York: Academic Press, 1954.
7. Steenbock H, Black A. Fat-soluble vitamins. XVII. The induction of growth-promoting and calcifying properties in a ration by exposure to ultraviolet light. J Biol Chem 1924; 61:405–422.
8. Scriver CR, Reade TM, DeLuca HF, Hamstra AJ. Serum 1,25-$(OH)_2D_3$ levels in normal subjects and in patients with hereditary rickets or bone disease. N Engl J Med 1978; 299:976–979.
9. Fraser D, Kooh SW, Kind HP, Holick MF, Tanaka Y, DeLuca HF. Pathogenesis of hereditary vitamin D-dependent rickets: an inborn error of vitamin D metabolism involving defective conversion of 25-hydroxyvitamin D to 1,25-dihydroxyvitamin D. N Engl J Med 1973; 289:817–822.
10. Brommage R, Jarnagin K, DeLuca HF, Yamada S, Takayama H. 1- But no 24- hydroxylation of vitamin D is required for skeletal mineralization in rats. Am J Physiol 1983; 244:E298–E304.
11. Eil C, Lieberman UA, Rosen JF, Marx SJ. A cellular defect in hereditary vitamin D-dependent rickets type II: defective nuclear uptake of 1,25-dihydroxyvitamin D in cultured skin fibroblasts. N Engl J Med 1981; 304:1588–1591.
12. Bell NH, Hamstra AJ, DeLuca HF. Vitamin D-dependent rickets type II: resistance of target organs to 1,25-dihydroxyvitamin D. N Engl J Med 1978; 298:996–999.
13. Rosen JF, Fleischman AR, Finberg L, Hamstra A, DeLuca HF. Rickets with alopecia: An inborn error of vitamin D metabolism. J Pediatr 1979; 94:729–735.
14. Marx SJ, Liberman UA, Eil C, Gamblin GT, DeGrange DA, Balsan S. Hereditary resistance to 1,25-dihydroxyvitamin D. Rec Prog Horm Res 1984; 40:589–620.
15. Wiese RJ, Goto H, Prahl JM, Marx SJ, Thomas M, Al-Aqeel A, DeLuca HF. Vitamin D-dependency rickets type II: truncated vitamin D receptor in three kindreds. Mol Cell Endocrinol 1993; 90:197–201.
16. Liberman UA, Marx SJ. Vitamin D dependent rickets. In: Primer on the Metabolic Bone Diseases and Disorders of Mineral Metabolism, 1st ed. Favus MJ, ed. Richmond: William Byrd Press, 1990; 178–182.
17. Underwood JL, DeLuca HF. Vitamin D is not directly necessary for bone growth and mineralization. Am J Physiol 1984; 246:E493–E498.
18. DeLuca HF. Mechanism of action and metabolic fate of vitamin D. Vitam Horm 1967; 25:315–367.
19. DeLuca HF, Schnoes HK. Vitamin D: recent advances. Annu Rev Biochem 1983; 52:411–439.
20. Schachter D, Rosen SM. Active transport of Ca^{45} by the small intestine and its dependence on vitamin D. Am J Physiol 1959; 196:357–362.
21. Higaki M, Takahashi M, Suzuki T, Sahashi Y. Metabolic activities of vitamin D in animals. III. Biogenesis of vitamin D sulfate in animal tissues. J Vitaminol 1965; 11:261–265.
22. Martin DL, DeLuca HF. Calcium transport and the role of vitamin D. Arch Biochem Biophys 1969; 134:139–148.
23. Walling MW, Rothman SS. Phosphate-independent, carrier-mediated active transport of calcium by rat intestine. Am J Physiol 1969; 217:1144–1148.
24. Wasserman RH, Kallfelz FA, Comar CL. Active transport of calcium by rat duodenum *in vivo*. Science 1961; 133:883–884.
25. Schachter D. Vitamin D and the active transport of calcium by the small intestine. In: The Transfer of Calcium and Strontium Across Biological Membranes. Wasserman RH, ed. New York: Academic Press, 1963; 197–210.
26. Chen TC, Castillo L, Korycka-Dahl M, DeLuca HF. Role of vitamin D metabolites in phosphate transport of rat intestine. J Nutr 1974; 104:1056–1060.
27. Walling MW. Effects of 1,25-dihydroxyvitamin D_3 on active intestinal inorganic phosphate absorption. In: Vitamin D: Biochemical, Chemical, and Clinical Aspects Related to Calcium Metabolism. Norman AW, Schaefer K, Coburn JW, eds. Berlin: Walter de Gruyter, 1977; 321–330.
28. Harrison HE, Harrison HC. Intestinal transport of phosphate: action of vitamin D, calcium, and potassium. Am J Physiol 1961; 201:1007–1012.
29. Carlsson A. Tracer experiments on the effect of vitamin D on the skeletal metabolism of calcium and phosphorus. Acta Physiol Scand 1952; 26:212–220.
30. Nicolaysen R, Eeg-Larsen N. The mode of action of vitamin D. In: Ciba Foundation Symposium on Bone Structure and Metabolism. Wolstenholme GWE, O'Connor CM, eds. Boston: Little, Brown, 1956; 175–186.

31. Garabedian M, Tanaka Y, Holick MF, DeLuca HF. Response of intestinal calcium transport and bone calcium mobilization to 1,25-dihyoxyvitamin D_3 in thyroparathyroidectomized rats. Endocrinology 1974; 94:1022–1027.
32. Rasmussen H, DeLuca H, Arnaud C, Hawker C, Stedingk MV. The relationship between vitamin D and parathyroid hormone. J Clin Invest 1963; 42:1940–1946.
33. Yamamoto M, Kawanobe Y, Takahashi H, Shimazawa E, Kimura S, Ogata E. Vitamin D deficiency and renal calcium transport in the rat. J Clin Invest 1984; 74:507–513.
34. Bonjour JP, Preston C, Fleisch H. Effect of 1,25-dihydroxyvitamin D_3 on renal handling of Pi in thyroparathyroidectomized rats. J Clin Invest 1977; 60:1419–1428.
35. Forte LR, Nickols GA, Anast CS. Renal adenylate cyclase and the interrelationship between parathyroid hormone and vitamin D in the regulation of urinary phosphate and adenosine cyclin 3' 5'-monophosphate excretion. J Clin Invest 1976; 57:559–568.
36. Neville PF, DeLuca HF. The synthesis of [1,2-^{3}H]vitamin D_3 and the tissue localization of a 0.25 μg (10 IU) dose per rat. Biochemistry 1966; 5:2201–2207.
37. Windus A, Bock F. Uber das Provitamin aus dem Sterin der Schweineschwarte. Z Physiol Chem 1937; 245:168–170.
38. Esvelt RP, Schnoes HK, DeLuca HF. Vitamin D_3 from rat skins irradiated *in vitro* with ultraviolet light. Arch Biochem Biophys 1978; 188:282–286.
39. Windus A, Schenck F, Weder Fv. Uber das Antirachitisch Wirksame Bestrahlungs-produkt aus 7-Dehydro-Cholesterin. Hoppe-Seylers Z Physiol Chem 1936; 241:100–103.
40. Velluz L, Amiard G. Chimie organique—le precalciferol. Compt Rend 1949; 228:692–694.
41. Holick MF, Clark MB. The photobiogenesis and metabolism of vitamin D. Fed Proc 1978; 37:2567–2574.
42. Ponchon G, DeLuca HF, Suda T. Metabolism of [1,2]-^{3}H-vitamin D_3 and [26,27]-^{3}H-25-hydroxyvitamin D_3 in rachitic chicks. Arch Biochem Biophys 1970; 141:397–408.
43. Horsting M, DeLuca HF. *In vitro* production of 25-hydroxycholecalciferol. Biochem Biophys Commun 1969; 36:251–256.
44. Fraser DR, Kodicek E. Unique biosynthesis by kidney of a biologically active vitamin D metabolite. Nature 1970; 228:764–766.
45. Gray R, Boyle I, DeLuca HF. Vitamin D matabolism: the role of kidney tissue. Science 1971; 172:1232–1234.
46. Boyle IT, Miravet L, Gray RW, Holick MF, DeLuca HF. The response of intestinal calcium transport to 25-hydroxy and 1,25-dihydroxyvitamin D in nephrectomized rats. Endocrinology 1972; 90:605–608.
47. Holick MF, Garabedian M, DeLuca HF. 1,25-Dihydroxycholecalciferol: metabolite of vitamin D_3 active on bone in anephric rats. Science 1972; 176:1146–1147.
48. Wong RG, Norman AW, Reddy CR, Coburn JW. Biologic effects of 1,25-dihydroxycholecalciferol (a highly active vitamin D metabolite) in acutely uremic rats. J Clin Invest 1972; 51:1287–1291.
49. DeLuca HF. Vitamin D: the vitamin and the hormone. Fed Proc 1974; 33:2211–2219.
50. Pedersen JI, Shobaki HH, Holmberg I, Bergseth S, Bjorkhem I. 25-Hydroxyvitamin D_3-24-hydroxylase in rat kidney mitochondria. J Biol Chem 1983; 258:742–746.
51. Burgos-Trinidad M, DeLuca HF. Kinetic properties of 25-hydroxyvitamin D- and 1,25-dihydroxyvitamin D-24-hydroxylase from chick kidney. Biochim Biophys Acta 1991; 1078:226–230.
52. Akiyoshi-Shibata M, Sakaki T, Ohyama Y, Noshiro M, Okuda K, Yabusaki Y. Further oxidation of hydroxycalcidiol by calcidiol 24-hydroxylase. Eur J Biochem 1994; 224:335–343.
53. Beckman MJ, Tadikonda P, Werner E, Prahl J, Yamada S, DeLuca HF. Human 25-hydroxyvitamin D_3-24-hydroxylase, a multicatalytic enzyme. Biochemistry 1996; 35:8465–8472.
54. Esvelt RP, Rivizzani MA, Paaren HE, Schnoes HK, DeLuca HF. Synthesis of calcitroic acid, a metabolite of 1,25-dihydroxycholecalciferol. J Org Chem 1981; 46:456–458.
55. Onisko BL, Esvelt RP, Schnoes HK, DeLuca HF. Metabolites of 1,25-dihydroxyvitamin D_3 in rat bile. Biochemistry 1980; 19:4124–4130.
56. Brown EM, Gamba G, Riccardi D, Lombardi M, Butters R, Kifor O, Sun A, Hediger MA, Lutton J, Hebert SC. Cloning and characterization of an extracellular Ca^{+2}-sensing receptor from bovine parathyroid. Nature 1993; 366:575–580.
57. Omdahl JL, Gray RW, Boyle IT, Knutson J, DeLuca HF. Regulation of metabolism of 25-hydroxycholecalciferol metabolism by kidney tissue *in vitro* by dietary calcium. Nature New Biol 1972; 237:63,64.

58. Garabedian M, Holick MF, DeLuca HF, Boyle IT. Control of 25-hydroxycholecalciferol metabolism by the parathyroid glands. Proc Natl Acad Sci USA 1972; 69:1673–1676.
59. Fraser DR, Kodicek E. Regulation of 25-hydroxycholecalciferol-1-hydroxylase activity in kidney by parathyroid hormone. Nature New Biol 1973; 241:163–166.
60. Tanaka Y, DeLuca HF. The control of 25-hydroxyvitamin D metabolism by inorganic phosphorus. Arch Biochem Biophys 1973; 154:566–574.
61. Baxter LA, DeLuca HF. Stimulation of 25-hydroxyvitamin D_3-1-hydroxylase by phosphate depletion. J Biol Chem 1976; 251:3158–3161.
62. Hughes MR, Brumbaugh PF, Haussler MR, Wergedal JE, Baylink DJ. Regulation of serum 1,25-dihydroxyvitamin D_3 by calcium and phosphate in the rat. Science 1975; 190:578–580.
63. Gray RW. Evidence that somatomedins mediate the effect of hypophosphatemia to increase serum 1,25-dihydroxyvitamin D_3 levels in rats. Endocrinology 1987; 121:504–512.
64. Halloran BP, Spencer EM. Dietary phosphorus and 1,25-dihydroxyvitamin D metabolism: influence of insulin-like growth factor-1. Endocrinology 1988; 123:1225–1229.
65. Spencer EM, Tobiassen O. The mechanism of the action of growth hormone on vitamin D metabolism in the rat. Endocrinology 1981; 108:1064–1070.
66. Gray RW. Control of plasma 1,25-$(OH)_2$-vitamin D concentrations by calcium and phosphorus in the rat: effects of hypophysectomy. Calcif Tiss Int 1981; 33:485–488.
67. Pahuja DN, DeLuca HF. Role of the hypophysis in the regulation of vitamin D metabolism. Mol Cell Endocrinol 1981; 23:345–350.
68. Brown DJ, Spanos E, MacIntyre I. Role of pituitary hormones in regulating renal vitamin D metabolism in man. BMJ 1980; 280:277.
69. Tanaka Y, Lorenc RS, DeLuca HF. The role of 1,25-dihydroxyvitamin D_3 and parathyroid hormone in the regulation of chick renal 25-hydroxyvitamin D_3-24-hydroxylase. Arch Biochem Biophys 1975; 171:521–526.
70. Tanaka Y, DeLuca HF. Stimulation of 24,25-dihydroxyvitamin D_3 production by 1,25-dihydroxyvitamin D_3. Science 1974; 183:1198–1200.
71. Shinki T, Jin CH, Nishimura A, Nagai Y, Ohyama Y, Noshiro M, Okuda K, Suda T. Parathyroid hormone inhibits 25-hydroxyvitamin D_3-24-hydroxylase mRNA expression stimulated by 1,25-dihydroxyvitamin D_3 in rat kidney but not in intestine. J Biol Chem 1992; 267:13,757–13,762.
72. Zierold C, Darwish HM, DeLuca HF. Identification of a vitamin D-response element in the rat calcidiol (25-hydroxyvitamin D_3) 24-hydroxylase gene. Proc Natl Acad Sci USA 1994; 91:900–902.
73. Ohyama Y, Ozono K, Uchida M, Shinki T, Klato S, Suda T, Yamamoto O, Noshiro M, Kato Y. Identification of a vitamin D-responsive element in the 5'-flanking region of the rat 25-hydroxyvitamin D_3 24-hydroxylase gene. J Biol Chem 1994; 269:10,545–10,550.
74. Wu SX, Finch J, Zhong M, Slatopolsky E, Grieff M, Brown AJ. Expression of the renal 25-hydroxyvitamin D-24-hydroxylase gene-regulation by dietary phosphate. Am J Phys 1996; 40:F203–F208.
75. Brenza HL, Kimmel-Jehan C, Jehan F, Shinki T, Wakino S, Anazawa H, Suda T, DeLuca HF. Parathyroid hormone activation of the 25-hydroxyvitamin D_3-1α-hydroxylase gene promoter. Proc Natl Acad Sci USA 1998; 95:1387–1391.
76. Rasmussen H, Bordier P. Vitamin D and bone. Metab Bone Dis Rel Res 1978; 1:7–13.
77. Ornoy A, Goodwin D, Noff D, Edelstein S. 24,25-dihydroxyvitamin D is a metabolite of vitamin D essential for bone formation. Nature 1978; 276:517–519.
78. Henry HL, Taylor AN, Norman AW. Response of chick parathyroid glands to the vitamin D metabolites 1,25-dihydroxyvitamin D_3 and 24,25-dihydroxyvitamin D_3. J Nutr 1977; 107:1918–1926.
79. Garabedian M, Lieberherr M, Nguyen TM, Corvol MT, DuBois MB, Balsan S. *In vitro* production and activity of 24,25-dihydroxycholecalciferol in cartilage and calvarium. Clin Orthop 1978; 135:241–248.
80. Henry HL, Norman AW. Vitamin D: two dihydroxylated metabolites are required for normal chicken egg hatchability. Science 1978; 201:835–837.
81. Jarnagin K, Brommage R, DeLuca HF, Yamada S, Takayama H. 1-But not 24-hydroxylation of vitamin D is required for growth and reproduction in rats. Am J Physiol 1983; 244:E290–E297.
82. Halloran BP, DeLuca HF, Barthell E, Yamada S, Ohmori M, Takayama H. An examination of the importance of 24-hydroxylation to the function of vitamin D during early development. Endocrinology 1981; 108:2067–2071.
83. Miller SC, Halloran BP, DeLuca HF, Yamada S, Takayama H, Jee WSS. Studies on the role of 24-hydroxylation of vitamin D in the mineralization of cartilage and bone of vitamin D-deficient rats. Calcif Tiss Int 1981; 33:489–497.

127. Mackey SL, Heymont JL, Kronenberg HM, Demay MB. Vitamin D receptor binding to the negative human parathyroid hormone vitamin D response element does not require the retinoid X receptor. Mol Endocrinol 1996; 10:298–305.
128. Darwish HM, DeLuca HF. Analysis of binding of the 1,25-dihydroxyvitamin D_3 receptor to positive and negative vitamin D response elements. Arch Biochem Biophys 1996; 334:223–234.
129. Zierold C, Darwish HM, DeLuca HF. Two vitamin D response elements function in the rat 1,25-dihydroxyvitamin D 24-hydroxylase promoter. J Biol Chem 1995; 270:1675–1678.
130. Takeyama KI, Khanaka D, Sato T, Kobori M, Yanagisawa J, Kato S. 25-Hydroxyvitamin D-3 1α-hydroxylase and vitamin D synthesis. Science 1997; 277:1827–1830.
131. St-Arnaud R, Messerlain S, Moir JM, Omdahl JL, Glorieux FH. The 25-hydroxyvitamin D 1α-hydroxylase gene maps to the pseudovitamin D deficiency rickets (PDDR) disease locus. J Bone Miner Res 1997; 12:1552–1559.
132. Shinki T, Shimada H, Wakino S, Anazawa H, Hayashi M, Saruta T, DeLuca HF, Suda T. Cloning and expression of rat 25-hydroxyvitamin D-3-1-α-hydroxylase cDNA. Proc Natl Acad Sci USA 1997; 94:12,920–12,925.

4
Metabolism and Catabolism of Vitamin D, Its Metabolites, and Clinically Relevant Analogs

Glenville Jones

1. METABOLISM OF VITAMIN D_3 AND 25(OH)D_3

The elucidation of the metabolism of vitamin D_3 is arguably one of the most important developments in nutritional sciences over the latter half of the 20th century. An appreciation that vitamin D_3 represents a precursor to the functionally active form and that two steps of activation are necessary to produce the hormone 1α,25-dihydroxyvitamin D_3 [1α,25(OH)$_2D_3$] constitute historical landmarks in modern vitamin research *(1)*. These developments spawned not only detailed studies of the biologic properties of vitamin D metabolites produced and the regulation of cytochrome P450-containing enzymes involved in their production but also provided the stimulus for the chemical synthesis of a plethora of vitamin D analogs (approx 300 at last count). Furthermore, it appears that susceptibility to vitamin D catabolic pathways and other important parameters such as binding to the vitamin D receptor (VDR) functional complex and binding to the vitamin D binding protein (DBP) are probably key elements in dictating the differences in the actions of so-called calcemic and noncalcemic vitamin D analogs. Therefore, from the perspective of its historical significance and relevance, it seems entirely logical to consider the metabolism of vitamin D and its analogs together at this stage of a general text on vitamin D.

1.1. 25- and 1α-Hydroxylation

Vitamin D_3 can be synthesized in the skin (*see* Chapter 1) or be derived from dietary sources. This precursor does not circulate for long in the bloodstream, but instead is immediately taken up by adipose tissue or liver for storage. In humans, tissue storage of vitamin D can last for months or even years. Ultimately, the vitamin D_3 undergoes its first step of activation, namely 25-hydroxylation in the liver (Fig. 1). Over the years, there has been some controversy over whether 25-hydroxylation is carried out by one enzyme or two and whether this cytochrome P450-based enzyme is found in the mitochondrial or microsomal fractions of liver *(2)*. Currently, only one of these enzymes, the mitochondrial form, has been purified to homogeneity, subsequently cloned from several species, and studied in any detail *(3–5)*. The cytochrome P450 involved is known as CYP27 or

From: *Vitamin D: Physiology, Molecular Biology, and Clinical Applications*
Edited by: M. F. Holick © Humana Press Inc., Totowa, NJ

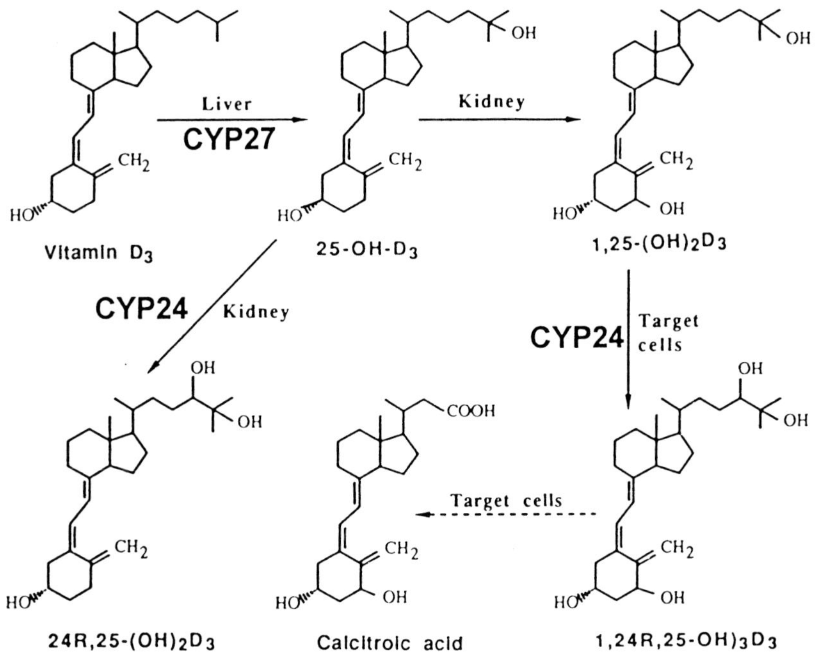

Fig. 1. Metabolism of vitamin D_3.

P450 c27 because it is a bifunctional cytochrome P450, which, in addition to 25-hydroxylating vitamin D_3, also carries out the side-chain hydroxylation of intermediates involved in bile acid biosynthesis. Even though 25-hydroxylation of vitamin D_3 has been clearly demonstrated in cells transfected with CYP27, there is still some skepticism in the vitamin D field over whether a single cytochrome P450 can explain all the metabolic findings observed over the past 2 decades of research. These unexplained observations include the following:

1. Using the perfused rat liver, Fukushima et al. *(6)* demonstrated *two* 25-hydroxylase enzyme activities: a high-affinity, low-capacity form (presumably microsomal) and a low-affinity, high-capacity form (presumably mitochondrial; CYP27).
2. Regulation, albeit weak, of the liver 25-hydroxylase in animals given normal dietary intakes of vitamin D after a period of vitamin D deficiency *(7)* is not explained by a transcriptional mechanism since the gene promoter of CYP27 lacks a vitamin D response element (VDRE).
3. No obvious $25(OH)D_3$ or $1\alpha,25(OH)_2D_3$ deficiency in patients suffering from the genetically inherited disease cerebrotendinous xanthomatosis, in which CYP27 is mutated. [Although a subset of these patients can suffer from osteoporosis, this is more likely owing to biliary defects, leading to altered enterohepatic circulation of $25(OH)D_3$ *(8)*].
4. CYP27 does not appear to 25-hydroxylate vitamin D_2.

Very recently, a pig liver *microsomal* 25-hydroxylase was purified to homogeneity and subsequently cloned *(8a)*. However, it remains to be seen whether this cytochrome P450, which belongs to the CYP2D family, has a human counterpart, represents a physiologically important enzyme in vivo, and therefore explains the list of anomalous findings listed above. One thing that the existence of CYP27 does explain is the occasional reports of extrahepatic 25-hydroxylation of vitamin D_3 *(9)*. CYP27 mRNA has been detected in a number of extrahepatic tissues, including kidney and bone (osteoblast) *(10,11)*.

The product of the 25-hydroxylation step, 25-hydroxyvitamin D_3 [25(OH)D_3], is the major circulating form of vitamin D_3 and in humans is present in plasma at concentrations in the range of 10 to 40 ng/mL (25–125 nM). The main reason for the stability of this metabolite is its strong affinity for the DBP (globulin) of blood.

The second step of activation, 1α-hydroxylation, occurs in the kidney *(12)*, and its synthesis in the normal mammal appears to be the *exclusive* domain of that organ. The main evidence for this stems from clinical medicine: patients with chronic renal failure exhibit frank rickets or osteomalacia because of a deficiency of 1α,25(OH)$_2$$D_3$ caused by lack of 1α-hydroxylase, a situation that is reversible by 1α,25(OH)$_2$$D_3$ administration. As the result of a tremendous amount of attention over the past two decades, the cytochrome P450 CYP1α, representing the 1α-hydroxylase enzyme, was finally cloned from a rat renal cDNA library by St. Arnaud's group in Montreal *(12a)*. This was rapidly followed by cloning of cDNAs representing mouse and human CYP1α *(12a–12c)* as well as the human and mouse genes *(12a,12c–12e)*. It had been known for some time that the mitochondria-located 1α-hydroxylase enzyme is composed of three proteins (a cytochrome P450, ferredoxin, and ferredoxin reductase for activity) and is strongly downregulated by 1α,25(OH)$_2$$D_3$ and upregulated by parathyroid hormone *(13,14)*. It will be interesting to see whether the promoter for the CYP1α gene contains the necessary transcriptional regulatory elements necessary to explain the observed physiologic regulation or whether the 1α-hydroxylase regulation is achieved at some other level. There had been claims that CYP27 can catalyze 1α-hydroxylation *(10)*, but the physiologic relevance of these observations must now be in question following the finding of the new cytochrome P450 by St. Arnaud *(12a)*. There is little doubt that CYP1α is important since its human gene colocalizes to the chromosomal location of vitamin D-dependent rickets type 1, a human disease state first proposed to be due to a mutation of 1α-hydroxylase 25 years ago *(12a,12d,14a,14b)*.

Over the past 20 years, it has been suggested that extrarenal 1α-hydroxylase activity may exist in several physiologic or pharmacologic situations. Placental 1α-hydroxylase was reported *(15)* but has proved to be difficult to purify, and there have been suggestions that the activity may be artifactual. Adams and Gacad *(16)* have demonstrated the existence of a 25(OH)D_3 1α-hydroxylase in sarcoid tissue, and this poorly regulated enzyme results in elevated plasma 1α,25(OH)$_2$$D_3$ levels, which in turn cause hypercalciuria and hypercalcemia in sarcoidosis patients. The induction of this 1α-hydroxylase in macrophages by cytokines and other growth factors has been postulated, but its role, if any, in normal macrophages is unknown. The availability of molecular probes for the renal CYP1α should now allow for exact characterization of these various extrarenal 1α-hydroxylase activities and clarification of their physiologic roles.

1.2. 24-Hydroxylation

24-Hydroxylation of both 25(OH)D_3 and 1α,25(OH)$_2$$D_3$ has been shown to occur in vivo *(17,18)*. The importance of this step has been immersed in controversy since it has been claimed that 24-hydroxylated metabolites might play a role in (1) bone mineralization *(19,20)*; and (2) egg hatchability *(21)*. Experimental evidence favors a different function for 24-hydroxylation: *inactivation* of the vitamin D molecule. This concept comes from four main lines of evidence:

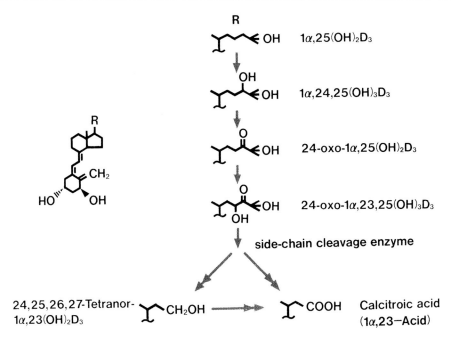

Fig. 2. C-24 oxidation pathway. (Reproduced with permission from ref. 23.)

1. The levels of 24,25(OH)$_2$D$_3$ do not appear to be regulated, reaching >100 ng/mL in hypervitaminotic animals (22).
2. There is no apparent 24,25(OH)$_2$D$_3$ receptor similar to VDR within the steroid receptor superfamily.
3. Synthesis of vitamin D analogs blocked with fluorine atoms at the various carbons of the side chain [e.g., 24F$_2$-1α,25(OH)$_2$D$_3$] results in molecules with the *full* biologic activity of vitamin D in vivo.
4. 24-Hydroxylation appears to be the first step in a degradatory pathway demonstrable in vitro (23,24) (Fig. 2), which culminates in a biliary excretory form, calcitroic acid observed in vivo (25).

The 25(OH)D$_3$-24-hydroxylase was originally characterized as a P450-based enzyme over 20 years ago (26) and more recently the cytochrome P450 species purified and cloned by Okuda's group (27). The enzyme appears to 24-hydroxylate both 25(OH)D$_3$ and 1α,25(OH)$_2$D$_3$, the latter with a 10-fold higher efficiency (28,29). However, since the circulating level of 25(OH)D$_3$ is approx 1000 times higher than 1α,25(OH)$_2$D$_3$, the role of the enzyme in vivo is not clear. The enzyme, particularly the renal form, which appears to be expressed at high constitutive levels in the normal animal, may be involved in the inactivation and clearance of excess 25(OH)D$_3$ in the circulation (22). On the other hand, the 24-hydroxylase may be involved in target cell destruction of 1α,25(OH)$_2$D$_3$. This topic is discussed further in Section 2.1.

1.3. 26-Hydroxylation and 26,23-Lactone Formation

25,26(OH)$_2$D$_3$ was the first dihydroxylated metabolite to be identified back in the late 1960s (30), and yet it is still the most poorly understood. The metabolite is readily

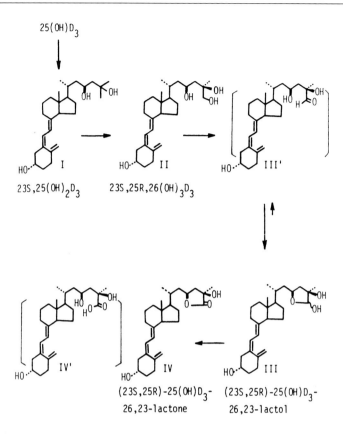

Fig. 3. 26,23-Lactone pathway. The pathway shown is for 25(OH)D$_3$. An analogous pathway exists for 1α,25(OH)$_2$D$_3$. (Reproduced with permission from ref. *33*.)

detectable in the plasma of animals given large doses of vitamin D$_3$, and it retains strong affinity for DBP *(31)*. However, its biologic activity is inferior to that of other endogenous vitamin D compounds, and it is presumed to be a minor catabolite. The knowledge that CYP27 is involved in vitamin D$_3$ activation and that 26(27)(OH)D$_3$ and 1α,27(OH)$_2$D$_3$ are formed from vitamin D$_3$ and 1α (OH)D$_3$, respectively, in CYP27 transfection systems *(5)* suggests that 26(27)-hydroxylation may be a consequence of errant side-chain hydroxylation.

The most abundant 26-hydroxylated analog appearing in vivo is the 26,23-lactone derivative of 25(OH)D$_3$. 25(OH)D$_3$-26,23-lactone accumulates in hypervitaminotic animals in vivo because of its extremely strong affinity for DBP *(32)*. The route of synthesis of this metabolite is depicted in Fig. 3; research indicates that 26-hydroxylation follows 23-hydroxylation in this process *(33)*. The site of this synthesis is probably extrahepatic, suggesting the existence of extrahepatic 26-hydroxylation. The recent evidence *(34,35)* that CYP24 carries out 23- as well as 24-hydroxylation raises the possibility that CYP24 may also be involved in 26,23-lactone formation. The roles of 25(OH)D$_3$-26,23-lactone or its 1α,25(OH)$_2$D$_3$ counterpart, which has also been reported *(36)*, are currently unknown. It is presumed that 26,23-lactone formation represents a backup degradatory pathway for 25(OH)D$_3$ and/or 1α,25(OH)$_2$D$_3$.

2. CATABOLISM OF 1α,25(OH)$_2$D$_3$

2.1. C-24 Oxidation Pathway to Calcitroic Acid

As described in Section 1.2. above, 1α,25(OH)$_2$D$_3$ is a very good substrate for the 24-hydroxylase. Using a variety of cell lines representing specific vitamin D target organs (intestine: CaCo2 cells; osteosarcoma: UMR-106 cells; kidney: LLC-PK1 cells; keratinocyte: HPK1A and HPK1A-ras) a number of researchers have shown that 24-hydroxylation is the first step in the C-24 oxidation pathway, a five-step, vitamin D-inducible, ketoconazole-sensitive pathway, which changes the vitamin D molecule to water-soluble truncated products (Fig. 2) *(24,25)*. In most biologic assays, the intermediates and truncated products of this pathway possess lower or negligible activity. Furthermore, many of these compounds have little or no affinity for DBP, making their survival in plasma tenuous at best. The recent cloning of the cytochrome P450 component (CYP24) of the 24-hydroxylase enzyme has led to detection of CYP24mRNA in a wide range of tissues, corroborating the earlier studies reporting widespread 24-hydroxylase enzyme activity in most, if not all, vitamin D target cells. Additional studies have shown that mRNA transcripts for CYP24 are virtually undetectable in naive target cells not exposed to 1α,25(OH)$_2$D$_3$ but increase dramatically by a VDR-mediated mechanism within hours of exposure to 1α,25(OH)$_2$D$_3$ *(37)*. It is therefore attractive to propose that not only is 24-hydroxylation an important step in inactivation of excess 25(OH)D$_3$ in the circulation but it is also involved in the inactivation of 1α,25(OH)$_2$D$_3$ inside target cells. One can thus hypothesize that C-24 oxidation is a target cell attenuation or desensitization process that constitutes a molecular switch to turn off vitamin D responses inside target cells *(38)*. The recent development of CYP24 knockout animals *(38a)*, resulting in hypercalcemia, hypercalciuria, nephrocalcinosis, and premature death in 50% of null animals, seems to support this hypothesis. On the other hand, surviving animals have unexplained changes in bone morphology, which could suggest an alternative role for 24-hydroxylase in bone mineralization *(38a,38b)*. These surviving CYP24 null animals might also be useful in demonstrating the importance of proposed backup systems, such as 26,23-lactone formation to vitamin D catabolism (*see* previous section).

Calcitroic acid, the final product of 1α,25(OH)$_2$D$_3$ catabolism, is probably not synthesized in the liver because C-24 oxidation does not occur in hepatoma cells and therefore must presumably be transferred from target cells to the liver via some plasma carrier. Although calcitroic acid has been found in various tissues in vivo *(39)*, the nature of this transfer mechanism has not been elucidated.

3. METABOLISM AND CATABOLISM OF THE ANALOGS OF VITAMIN D

3.1. Activation of Prodrugs

Prodrugs are synthetic analogs of vitamin D$_3$ requiring *one or more step(s)* of activation by endogenous enzyme systems before they are biologically active (e.g., *one step*: 1α(OH)D$_2$, 1α(OH)D$_3$, or 25(OH)D$_3$; *multiple steps*: vitamin D$_2$ and dihydrotachysterol).

3.1.1. Vitamin D$_2$

Although vitamin D$_2$ (for structure, *see* Table 1) can be synthesized naturally by irradiation of ergosterol, little finds its way into the human diet unless it is provided as

Table 1
Vitamin D Prodrugs

Vitamin D prodrugs [ring structure][a]	Side chain structure (R)	Company	Possible target diseases	Mode of delivery	Reference
1α-OH-D_3 [3]	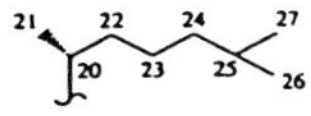 (21, 22, 24, 27, 20, 23, 25, 26)	Leo	Osteoporosis	Systemic	Barton et al. (1973)
1α-OH-D_2 [3]	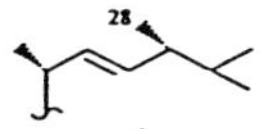 (28)	Lunar	Osteoporosis	Systemic	Paaren et al. (1978)
Dihydrotachysterol [2]	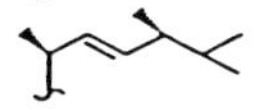	Duphar	Renal failure	Systemic	Jones et al. (1988)
Vitamin D_2 [1]	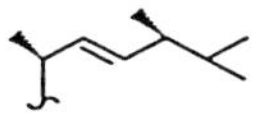	Various	Rickets osteomalacia	Systemic Systemic	Fraser et al. (1973)

[a] Structure of the vitamin D nucleus (seco-sterol ring structure).

Vitamin D Nucleus

[1] [2] [3]

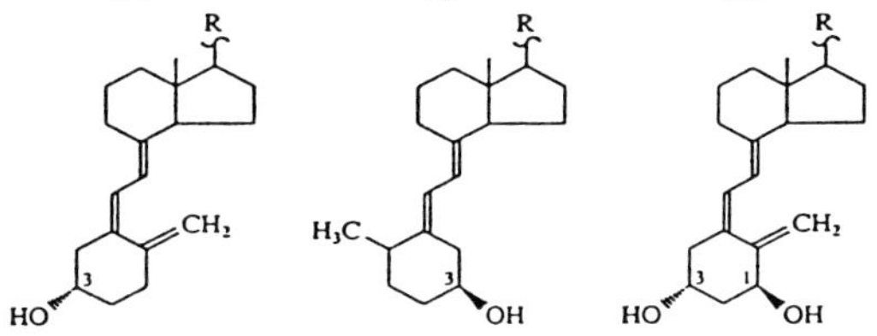

a dietary supplement. Indeed, vitamin D_2 has been used as a dietary supplement in lieu of vitamin D_3 since the 1930s. Since vitamin D_2 is an artificial form of the vitamin and it is dependent on the same activation steps as vitamin D_3 to become biologically active, one could make a strong case for considering it as a prodrug. In many ways the metabolism of vitamin D_2 is analogous to that of vitamin D_3. For instance, it has been established that vitamin D_2 gives rise to a similar series of metabolites in the form of 25(OH)D_2 (40), 1α,25(OH)$_2D_2$ (41), and 24,25(OH)$_2D_2$ (42). The formation of these metabolites suggests that the enzymes involved in side-chain metabolism, namely, 25-, 1α-, and 24-hydroxylases, do not discriminate against compounds bearing the vitamin D_2 side chain. However, metabolic studies have also revealed the formation of several additional metabolites including 24(OH)D_2 (43), 1α,24S(OH)$_2D_2$ (44), 24,26(OH)$_2D_2$ (45), and 1α,25,28(OH)$_3D_2$ (46). Pharmaceutical companies have exploited these subtle differences in the metabolism of vitamin D_2 by synthesizing molecules incorporating the features of the vitamin D_2 side chain, namely, the C-22–23 double bond or the C-24 methyl group (Table 1), into the structure of other analogs (e.g., calcipotriol).

Biologically active vitamin D_2 compounds, such as 1α,25(OH)$_2D_2$ and 1α,24S(OH)$_2D_2$, are also subject to further metabolism although it differs from that of 1α,25(OH)$_2D_3$, essentially because the modifications in the vitamin D_2 side chain prevent the C-23–24 cleavage observed during calcitroic acid production. Instead, the principal products are more polar tri- and tetrahydroxylated metabolites such as 1α,24,25(OH)$_3D_2$, 1α,25,28(OH)$_3D_2$, and 1α,25,26(OH)$_3D_2$ from 1α,25(OH)$_2D_2$ (47–48) and 1α,24,26(OH)$_3D_2$ from 1α,24S(OH)$_2D_2$ (49) produced by as yet undefined enzymes. In this latter case, the rate of 1α,24S(OH)$_2D_2$ metabolism appears to be slower than that of 1α,25(OH)$_2D_3$ (49). Some catabolites retain considerable biologic activity and at least one, 1α,25,28(OH)$_3D_2$, is patented for use as a drug.

3.1.2. DIHYDROTACHYSTEROL

This example of a vitamin D prodrug represents the oldest vitamin D analog and was developed in the 1930s as a method of stabilizing the triene structure of one of the photoisomers of vitamin D. The structure of dihydrotachysterol$_2$ (DHT$_2$) shown in Table 1 contains an A-ring rotated through 180°, a reduced C-10–19 double bond, and the side-chain structure of ergosterol/vitamin D$_2$. This side chain is depicted because the clinically approved drug form of DHT is DHT$_2$. However, it should be noted that DHT$_3$ can also be chemically synthesized with the side chain of vitamin D$_3$. The metabolism of both DHT$_2$ and DHT$_3$ has been extensively studied over the past three decades *(50–53)*. Initial studies performed in the early 1970s showed that DHT is efficiently converted to its 25-hydroxylated metabolite *(54)*.

The effectiveness of DHT for relieving hypocalcemia of chronic renal failure in the absence of functional renal 1α-hydroxylase led to the hypothesis *(55)* that *25(OH)DHT might represent the biologically active form of DHT, by virtue of its 3β-hydroxy group being rotated 180° into a pseudo-1α-hydroxyl position.*

It was thus believed that 1α-hydroxylation of 25(OH)DHT was unnecessary. This viewpoint prevailed for at least a decade, but debate was renewed when Bosch et al. *(51)* were able to provide evidence for the existence of a mixture of 1α- and 1β-hydroxylated products of 25(OH)DHT$_2$ in the blood of rats dosed with DHT$_2$. Studies involving the perfusion of kidneys from vitamin D-deficient rats with an incubation medium containing 25(OH)DHT$_3$ and using diode-array spectrophotometry to analyze the extracts, showed this molecule to be subject to extensive metabolism by renal enzymes but failed to give the expected 1-hydroxylated metabolites (Fig. 4), opening up the possibility that the 1α- and 1β-hydroxylated metabolites observed by Bosch et al. *(51)* might be formed by an extrarenal 1-hydroxylase *(52)*. Following the synthesis of appropriate authentic standards, subsequent research *(53)* has rigorously confirmed the in vivo formation and identity of 1α,25(OH)$_2$DHT and 1β,25(OH)$_2$DHT in both rat and human. The ability of these 1α- and 1β-hydroxylated forms of both DHT$_2$ and DHT$_3$ to stimulate a VDRE-inducible growth hormone reporter system exceeded that of 25(OH)DHT and in the process established 1α,25(OH)$_2$DHT and 1β,25(OH)$_2$DHT as the most potent derivatives of DHT identified to date *(53)*. The importance of the pseudo-1α-hydroxyl group hypothesis is now in question, although current findings do not rule out that the biologic activity of DHT might be owing to the collective action of a group of metabolites including 25(OH)DHT, 1α,25(OH)$_2$DHT, and 1β,25(OH)$_2$DHT. Other data suggest that 25(OH)DHT might be a substrate for an extrarenal hydroxylase of bone marrow origin in vivo *(56,57)*.

Although the enzymes involved in the activation of DHT, especially the 1-hydroxylation step, have an altered specificity toward this molecule, the enzymes involved in the catabolism of DHT$_3$ appear to treat the molecule as they would 25(OH)D$_3$ or 1α,25(OH)$_2$D$_3$. Side-chain hydroxylated derivatives of both 25(OH)DHT$_3$ and 1,25(OH)$_2$DHT$_3$ have been identified and appear to be analogous to intermediates of the C-24 oxidation and 26,23-lactone pathways of vitamin D$_3$ metabolism *(58,59)*.

3.1.3. 1α(OH)D$_2$ AND 1α(OH)D$_3$

The prodrug 1α(OH)D$_3$ was developed in the early 1970s following the discovery of the hormone 1α,25(OH)$_2$D$_3$ and the realization that the kidney was the site of its synthesis *(60,61)*. The rationale behind its use was to circumvent the 1α-hydroxylation step

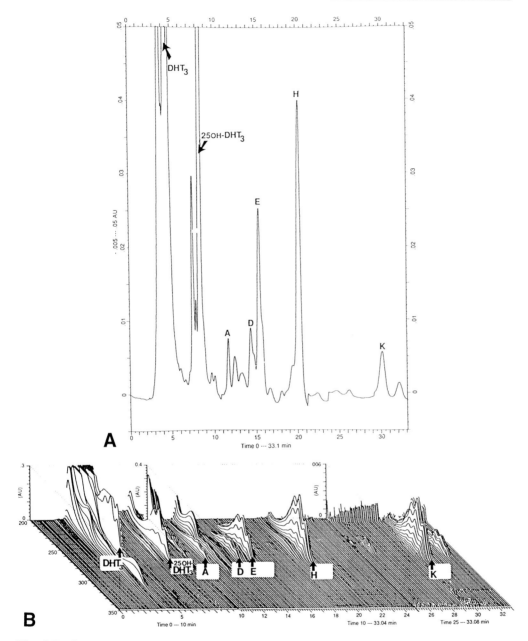

Fig. 4. In vivo metabolism of dihydrotachysterol₃ in the rat. Diode-array high-performance liquid chromatography of the plasma extract of a rat administered 1 mg dihydrotachysterol₃ (DHT₃) 18 h prior to sacrifice. Metabolites are labeled 25(OH)DHT₃ and peaks A–L. All possess the distinctive tricusped UV spectrum (λ_{max}, 242.5, 251, and 260.5 nm). Metabolites A–L were subsequently identified as side chain modified compounds analogous to vitamin D metabolites of the C-24 oxidation and 26,23-lactone pathways depicted in Figures 1 and 2. (Reproduced with permission from ref. 52.)

involved in vitamin D activation thereby providing a molecule that could still be activated even in the absence of a functional kidney. It soon became an alternative drug therapy to synthetic 1α,25(OH)₂D₃ in renal osteodystrophy and other hypercalcemic conditions. Aside from the advantage of a reduced cost of synthesis, 1α(OH)D₃ offers

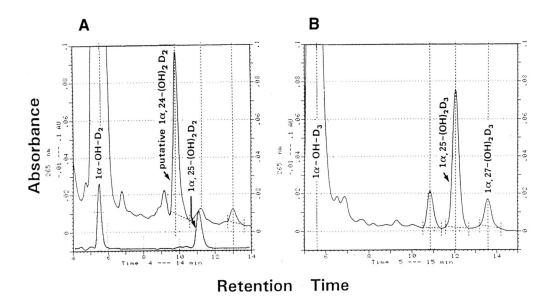

Fig. 5. In vitro metabolism of 1α(OH)D$_2$ and 1α(OH)D$_3$ by the hepatoma Hep3B. (**A**) High-performance liquid chromatography (HPLC) trace using diode-array detector at 265 nm of an extract of Hep3B cells incubated with 10 μM 1α(OH)D$_2$. The peak at 9.79 min was later conclusively identified by gas chromatography-mass spectrometry and comigration with authentic standard as 1α,24S(OH)$_2$D$_2$. Note that inset is a trace of standards: 1α(OH)D$_2$, 5.6 min; 1α,25(OH)$_2$D$_2$, 11.1 min. (**B**) HPLC trace using diode-array detector at 265 nm of an extract of Hep3B cells incubated with 10 μM 1α(OH)D$_3$. Note that the peak at 12.04 min comigrated with authentic 1α,25(OH)$_2$D$_3$. Subsequent work with standard 1α,26(27)(OH)$_2$D$_3$ (synthesized by Martin Calverley, Leo Pharmaceuticals) has confirmed its identity. (Reproduced with permission from ref. *63*.)

the potential biologic edge of requiring a step of activation in the form of 25-hydroxylation to produce an active molecule. It was believed that the requirement for an activation step might alter the pharmacokinetics of the drug compared with 1α,25(OH)$_2$D$_3$, delaying slightly its initial effects and extending its duration of action, thereby making the drug less likely to cause acute hypercalcemia. 25-Hydroxylation of 1α(OH)D$_3$ was first investigated using the isolated perfused rat liver *(6)* and confirmed that the liver was the main site of activation. Little subsequent work has been carried out to characterize this enzyme, although it is widely assumed that the enzyme is only loosely regulated and therefore constitutes an insignificant barrier to drug activation. The theoretical advantages of 1α(OH)D$_3$ over 1α,25(OH)$_2$D$_3$ have not materialized in clinical practice *(62)*.

A prodrug based on vitamin D$_2$ has also been synthesized in the form of 1α(OH)D$_2$ *(63)*. Although developed recently as a potential antiosteoporosis drug, this molecule has proved to be a valuable tool in studying hydroxylation reactions in the liver. At low substrate concentrations, 1α(OH)D$_2$, like 1α(OH)D$_3$, is 25-hydroxylated by liver hepatomas, Hep3B and HepG2 producing the well-established, biologically active compound 1α,25(OH)$_2$D$_2$. However, when the substrate concentration is increased to micromolar values the principal site of hydroxylation of 1α (OH)D$_2$ becomes the C-24 position, the product being 1α,24S(OH)$_2$D$_2$ (Fig. 5), another compound with significant biologic activity in several calcemia and cell proliferation assay systems *(49,63)*. This metabolite has been previously reported in cows receiving massive doses of vitamin D_2 *(44)*. Transfection studies using the liver cytochrome P450 CYP27 expressed in COS-1 cells suggest that $1\alpha,24S(OH)_2D_2$ is a product of this cytochrome *(6)*. Whether the formation of this unique metabolic product of $1\alpha(OH)D_2$ is the reason for the relative lower toxicity of $1\alpha(OH)D_2$ compared with $1\alpha(OH)D_3$ *(64)* has not been established definitively. $1\alpha(OH)D_2$ recently underwent clinical trials for the treatment of secondary hyperparathyroidism associated with end-stage renal disease *(65)*.

3.2. Metabolism-Sensitive Analogs

These synthetic analogs of $1\alpha,25(OH)_2D_3$ require no activation in vivo but are susceptible to attack by catabolic enzyme systems, in most cases rendering them biologically inactive (e.g., calcipotriol, OCT, KH1060).

3.2.1. CYCLOPROPANE RING-CONTAINING ANALOGS OF VITAMIN D

These analogs are modified in their side chains such that C-26 is joined to C-27 to give a cyclopropane ring consisting of C-25, C-26, and C-27. The best known member of this group of compounds is MC 903 or calcipotriol *(66)*, the structure of which is shown in Table 2. In addition to the cyclopropane ring, calcipotriol features a C-22=C-23 double bond and a 24S-hydroxyl group, which has been proposed to act as a surrogate C-25 hydroxyl in interactions of the molecule with the VDR. As is presented in Chapter 24, calcipotriol was the first vitamin D analog to be approved for topical use in psoriasis and is currently used worldwide for the successful control of this skin lesion *(67,68)*.

Pharmacokinetic data acquired for calcipotriol showed that it had a very short $T_{1/2}$, in the order of minutes, results that are consistent with the lack of a hypercalciuric /hypercalcemic effect when it is administered in vivo *(69)*. The first metabolic studies *(70)* revealed that calcipotriol was rapidly metabolized by a variety of different liver preparations from rat, minipig, and human to two novel products. These workers *(70)* were able to isolate and identify the two principal products as a C-22=C-23 unsaturated, 24-ketone (MC1046) and a C-22=C-23 reduced, 24-ketone (MC1080). These results were confirmed and extended by others *(71)* who showed that calcipotriol metabolism was not confined to liver tissue, but could be carried out by a variety of cells including those cells exposed to topically administered calcipotriol in vivo, namely, keratinocytes. Furthermore, these workers *(71)* proposed further metabolism of the 24-ketone in these vitamin D target cells to side-chain cleaved molecules including calcitroic acid (Fig. 6). The main implications of this work are that calcipotriol is subject to rapid metabolism, initially by non-vitamin-D-related enzymes, and then by vitamin D-related pathways to a side-chain cleaved molecule. Catabolites are produced in a variety of tissues and appear to have lower biologic activity than the parent molecule. Since calcipotriol is administered topically, the work suggests that it acts and is broken down locally and may never reach detectable levels in the bloodstream. Should calcipotriol enter the circulation, the ability of the liver and target cells to break down calcipotriol provides a backup system to prevent hypercalcemia.

The reduction of the C-22=C-23 double bond during the earliest phase of calcipotriol catabolism was an unexpected event given that this bond in vitamin D_2 compounds is extraordinarily stable to metabolism. It thus appears that metabolism of calcipotriol provides evidence that the C-24 methyl group in the vitamin D_2 side chain must play a stabilizing role, preventing the formation of the 24-ketone that facilitates the reduction

Table 2
Analogs of 1α,25(OH)$_2$D$_3$

Vitamin D analog [ring structure][a]	Side chain structure (R)	Company	Possible target diseases	Mode of delivery	Reference
1α,25-(OH)$_2$D$_3$ [3]	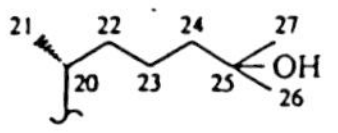 (21, 22, 24, 27, 20, 23, 25, 26) OH	Roche Duphar	Hypocalcemia psoriasis	Systemic Topical	Baggiolini et al. (1982)
26,27-F$_6$-1α,25-(OH)$_2$D$_3$ [3]	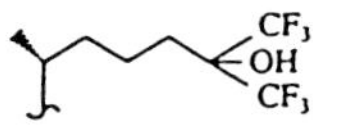 CF$_3$, OH, CF$_3$	Sumitomo-Taisho	Osteoporosis hypoparathyroidism	Systemic Systemic	Kobayashi et al. (1982)
19-nor-1α,25-(OH)$_2$D$_2$ [5]	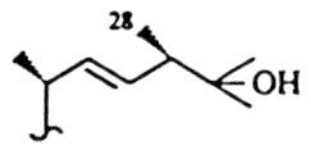 28, OH	Abbott	Hyperparathyroidism	Systemic	Perlman et al. (1990)
22-oxacalcitriol (OCT) [3]	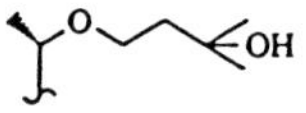 O, OH	Chugai	Hyperparathyroidism psoriasis	Systemic Topical	Murayama et al. (1986)
Calcipotriol (MC903) [3]	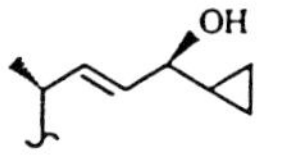 OH	Leo	Psoriasis cancer	Topical Topical	Calverley (1987)
1α,25-(OH)$_2$-16-ene-23-yne-D$_3$ (Ro 23-7553) [6]	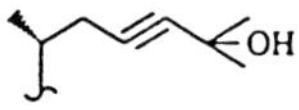 OH	Roche	Leukemia	Systemic	Baggiolini et al. (1989)
EB1089 [3]	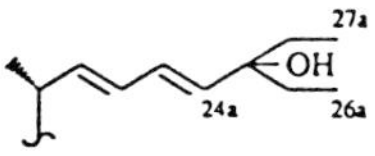 27a, OH, 24a, 26a	Leo	Breast cancer	Systemic	Binderup et al. (1991)
20-epi-1α,25-(OH)$_2$D$_3$ [3]	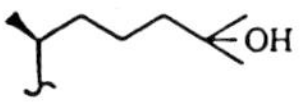 OH	Leo	?	Systemic	Calverley et al. (1991)
KH1060 [3]	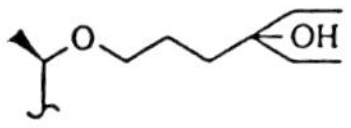 O, OH	Leo	Immune diseases	Systemic	Hansen et al. (1991)
ED71 [4]	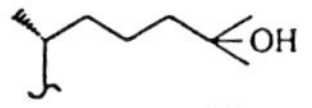 OH	Chugai	Osteoporosis	Systemic	Nishii et al. (1993)
1α,24(S)-(OH)$_2$D$_2$ [3]	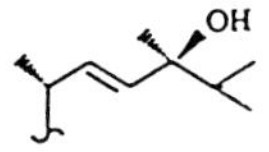 OH	Lunar	Psoriasis	Topical	Strugnell et al. (1995a)
1α,24(R)-(OH)$_2$D$_3$ (TV-02) [3]	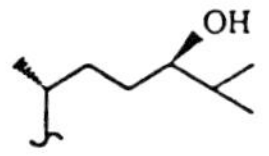 OH	Teijin	Psoriasis	Topical	Morisaki et al. (1975)

[a] Structure of the vitamin D nucleus (seco-sterol ring structure).

Vitamin D Nucleus

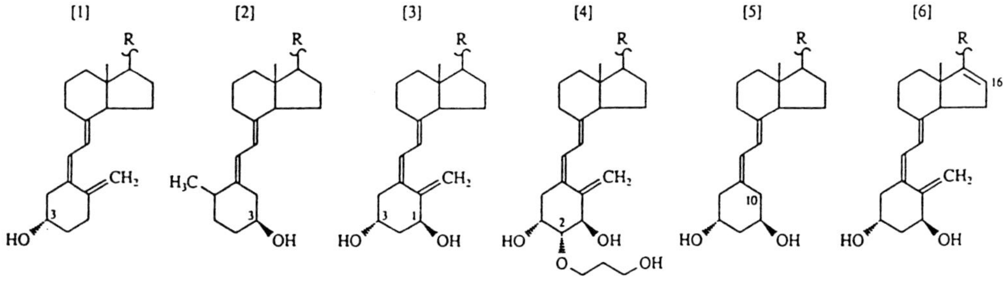

[1] [2] [3] [4] [5] [6]

of the C-22=C-23 double bond. However, it is still unknown which enzyme is responsible for this reduction in the side chain of calcipotriol.

3.2.2. OXA-GROUP-CONTAINING ANALOGS

These compounds involve the replacement of a carbon atom (usually in the side chain) with an oxygen atom. The best known are the 22-oxa analogs, including 22-oxa-calcitriol

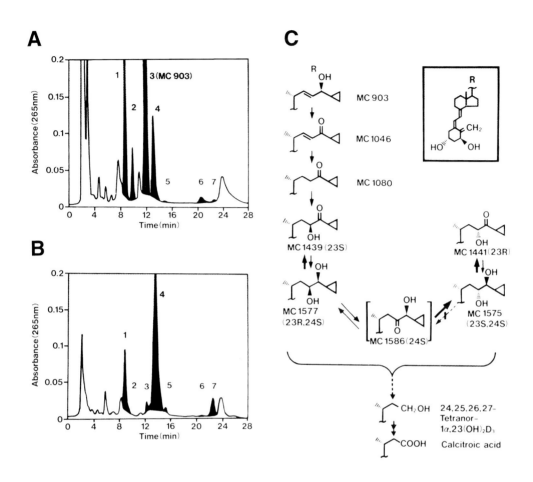

Fig. 6. (A and B) In vitro metabolism of calcipotriol (MC903) by HPK1A-ras cells. High-performance liquid chromagraphy of lipid extracts following incubation of MC903 with (A) HPK1A human keratinocytes and (B) HPK1A-ras human keratinocytes. Peak 1, MC1080; peak 2, MC1046; peak 3, MC903 (calcipotriol); peak 4, mixture of MC1439 and MC1441; peak 5, tetranor-1α,23(OH)$_2$D$_3$; peak 6, MC1577; peak 7, MC1575. (C) Proposed pathway of calcipotriol metabolism in cultured keratinocytes. (Reproduced with permission from ref. *71*.)

(OCT) and KH1060 *(72,73)*. Both of these molecules are metabolically fascinating to study because *the oxa-atom makes the molecule inherently unstable should it be hydroxylated at the adjacent carbon atom.* The hydroxylation at an adjacent carbon generates a unstable hemi-acetal, which spontaneously breaks down to eliminate the carbons distal to the oxa-group. In the case of the 22-oxa compounds, the expected product(s) would be C-20 alcohol/ketone.

The metabolism of OCT has been extensively studied in a number of different biologic systems including primary parathyroid *(74)* and keratinocyte cells *(75)* as well as cultured osteosarcoma, hepatoma, and keratinocyte cell lines *(76)*. In all these systems, OCT is rapidly broken down. Judicious use of two different radioactive labels in the form of (26-^{3}H)OCT and (2β-^{3}H)OCT enabled Brown et al. *(74)* to suggest that the side chain was truncated, although definitive proof of the the identity of the products was not

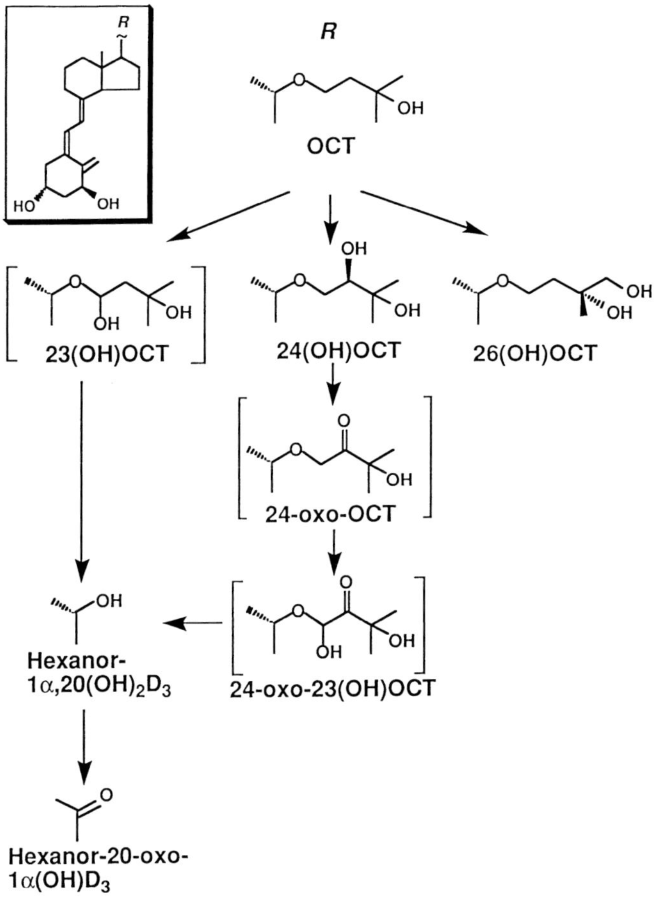

Fig. 7. Proposed pathways of 22-oxa-calcitriol (OCT) metabolism in cultured vitamin D target cells in vitro. Metabolic pathways were worked out using cultured cell lines representing hepatoma, osteosarcoma, and keratinocyte. (Reproduced with permission from ref. 76.)

immediately forthcoming. It was not until later work (76) that the principal metabolites were unequivocably identified by gas chromatography-mass spectrometry (GC-MS) as 24(OH)OCT, 26(OH)OCT, and hexanor-1α,20-dihydroxyvitamin D_3 (Fig. 7). In the case of the keratinocyte HPK1A-ras, an additional product, hexanor-20-oxo-1α-hydroxyvitamin D_3, is formed. These latter two truncated products are suggestive of hydroxylation of OCT at the C-23 position to give the theoretical unstable intermediate. Although all these products were isolated from in vitro systems, there is evidence that the processes also occur in vivo because Kobayashi et al. (77) have generated data suggesting that the biliary excretory form of OCT in the rat is a glucuronide ester of the truncated 20-alcohol.

The above example of a simple oxa-analog provides useful knowledge that can help in predicting the metabolic fate of a complex oxa-analog, such as KH1060. This highly potent compound, which possesses in vitro cell-differentiating activity exceeding that of any other analog synthesized to date, has four different modifications to the side chain of 1α,25(OH)$_2$D$_3$, namely, (1) the 22-oxa group, (2) the 20-epi side chain stereochemistry, (3) 24a-homologation, and (4) 26- and 27-dimethyl homologation (see Table 2 for structure).

Since all these changes are known to affect biologic activity in vitro and in vivo as well as side-chain metabolism (78,79), it comes as no surprise that the metabolism of KH1060

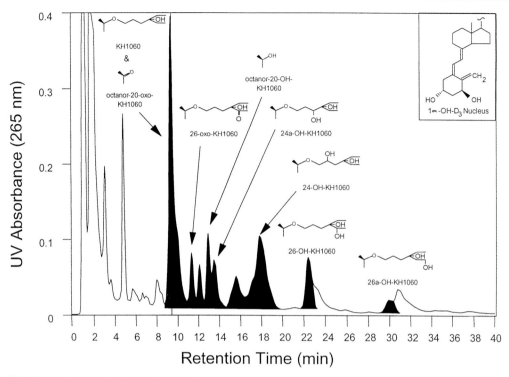

Fig. 8. In vitro metabolism of KH1060 by HPK1A-ras cells. High-performance liquid chromatography (HPLC) of lipid extracts following incubation of KH1060 (10 µM) with the human keratinocyte, HPK1A-ras for 72 h. Nine peaks (darkened) possessing the characteristic UV chromophore of vitamin D (λ_{max}, 265 nm; λ_{min}, 228 nm) are visible in the HPLC profile reproduced here. Rechromatography of these peaks on a second HPLC system resulted in the further resolution of these nine peaks into 22 separate metabolites. Many of these metabolites were identified *(81)* by comparison with synthetic standards on HPLC gas chromatography-mass spectrometry. Some examples of the types of structures corresponding to each peak are provided. These include peak at 13.39 min, 24a(OH)KH1060; peak at 22.14 min, 26(OH)KH1060. (F. J. Dilworth, A.-M. Kissmeyer, and G. Jones, unpublished results.)

is extremely complex. KH1060 has a very short $T_{1/2}$ in pharmacokinetic studies in vivo, giving a metabolic profile with at least 16 unknown metabolites *(80)*. Very recently, Dilworth et al. *(81)* reported the first in vitro study using micromolar concentrations of KH1060 incubated with the keratinocyte cell line HPK1A-ras. These workers were able to discern 22 different metabolites after multiple high performance liquid chromatography steps and assigned structures to 12 of these metabolites (Fig. 8). As would be expected from consideration of the studies of other oxa-compounds, two of these were truncated products identical to the molecules formed from another 22-oxa compound, OCT. As would be expected from consideration of the studies of other homologated compounds, other products are hydroxylated at specific carbons of the side chain including C-26 and C-26a. As with the metabolism-resistant analog EB1089 (*see* following Section 3.3.2.) and 26,27-dimethyl-1α,25(OH)$_2$D$_3$ (Leo code: MC1548), the presence of dimethyl groups in the terminus of the side chain appears to attract hydroxylation to these sites in KH1060. One novel metabolite found only for KH1060 is 24a(OH)KH1060, observed both in broken cell and in intact cell models *(81,82)*.

An important facet of this complex metabolic profile is that rather than simplifying our understanding of the mechanism of action of KH1060, these data complicate it. This is because biologic assays performed on each of the metabolic products of KH1060 have shown that several of the principal and long-lived metabolites retain significant vitamin D-dependent gene-inducing activity in reporter gene expression systems *(81)*. Although current published assays have demonstrated a high biologic activity for KH1060, these assays are performed in whole cell assay systems in which metabolism is known to occur (cell culture, organ culture, transfected cell systems) over extended periods (usually 24–72 h). However, assayists do not employ inhibitors of metabolism and often assume that the biologic effects observed are due to the parent compound, not to its metabolic products. In the case of KH1060, for which metabolism is rapid, it would seem to be prudent to assess the rate of metabolism in the bioassay model or else attempt to block metabolism by the use of appropriate inhibitors (e.g., ketoconazole).

3.3. Metabolism-Resistant Analogs

These synthetic analogs of $1\alpha,25(OH)_2D_3$ require no activation in vivo and are resistant to attack by catabolic enzyme systems because of blocking groups in metabolically sensitive regions (e.g., F_6-$1\alpha,25(OH)_2D_3$; $1\alpha,25(OH)_2D_3$-16-ene,23-yne; EB1089).

3.3.1. F_6-$1\alpha,25(OH)_2D_3$

This analog was first synthesized in the early 1980s *(83)*, along with a number of other side-chain fluorinated analogs, to test the importance of certain key hydroxylation sites (e.g., C-23, C-24, C-25, C-26(27), C-1) to biologic activity. It was noted immediately that $26,27$-F_6-$1\alpha,25(OH)_2D_3$ was extremely potent (10-fold higher than $1\alpha,25(OH)_2D_3$) in calcemia assays both in vitro and in vivo *(84–86)*. Using a bone cell line, UMR106, Lohnes and Jones *(38)* presented evidence that $26,27$-F_6-$1\alpha,25(OH)_2D_3$ had a longer $T_{1/2}$ inside target cells due to the apparent lack of 24-hydroxylation of $26,27$-F_6-$1\alpha,25(OH)_2D_3$. At around the same time, Morii's group *(87)* noted the appearance of a metabolite of $26,27$-F_6-$1\alpha,25(OH)_2D_3$ that they have identified as $26,27$-F_6-$1\alpha,23,25(OH)_3D_3$. This compound possesses excellent calcemic activity in its own right *(88)*, but whether this derivative is in part responsible for the biologic activity of $26,27$-F_6-$1\alpha,25(OH)_2D_3$ is not conclusively proved. Nonetheless, $26,27$-F_6-$1\alpha,25(OH)_2D_3$ has undergone clinical trials for hypocalcemia associated with hypoparathyroidism and uremia *(89,90)*.

3.3.2. UNSATURATED ANALOGS

The idea of introducing double bond(s) into the side chain of vitamin D analogs arose from experience with vitamin D_2. Vitamin D_2 metabolites have a biologic activity similar to that of vitamin D_3, so the introduction of the double bond is not deleterious. As mentioned earlier, the metabolism of the side chain is significantly altered by this relatively minor change.

The modification has not been confined to the introduction of a C-22=C-23 double bond. Roche has developed molecules with two novel modifications (introduction of a C-16=C-17 double bond, and introduction of a C-23=C-24 triple bond), which when combined produce the well-studied 16-ene, 23-yne analog of $1\alpha,25(OH)_2D_3$ *(91)* (*see* Table 2 for structure). Leo Pharmaceuticals has introduced the unsaturated analog EB1089, which contains a conjugated double-bond system at C-22=C-23 and C-24=C-24a, in addition to both main side chain and terminal dimethyl types of homologation *(92)* (*see* Table 2 for structure). These two series of Roche and Leo compounds have shown strong antiproliferative activity both in vitro and in vivo *(91,93,94)*.

The metabolism of the 16-ene compound by the perfused rat kidney has been studied *(95)*. These workers *(95)* found that the introduction of the C-16=C-17 double bond reduces 23-hydroxylation of the molecule, and the implication is that the D-ring modification must alter the conformation of the side chain sufficiently to change subtly the site of hydroxylation by CYP24, the cytochrome P450 thought to be responsible for 23- and 24-hydroxylation. Dilworth et al. *(79)* also noted the absence of measurable 23-hydroxylation of the analog 20-epi-1α,25(OH)$_2$D$_3$ in their studies, reinforcing the view that modifications around the C-17–20 bond profoundly influence the rate of 23-hydroxylation.

The metabolism of the 16-ene, 23-yne analog of 1α,25(OH)$_2$D$_3$ by WEHI-3 myeloid leukemic cells has recently been reported *(96)*. Although one might predict that because this molecule is blocked in the C-23 and C-24 positions it must be stable to a C-24 oxidation pathway enzyme(s), it was found experimentally that the 16-ene,23-yne analog has the same $T_{1/2}$ as 1α,25(OH)$_2$D$_3$ when incubated with this cell line (approx 6.8 h). The main product of [25-^{14}C]1α,25(OH)$_2$-16-ene-23-yne-D$_3$ was not identified by these workers but appeared to be more polar than the starting material. Similar work *(97)* used the perfused rat kidney and GC-MS to identify the main metabolite as 1α,25,26(OH)$_3$-16-ene-23-yne-D$_3$. Again it appears that the C-26 becomes vulnerable to attack when C-23 and C-24 are blocked. As in the case of calcipotriol, not all vitamin D analogs containing unsaturation are resistant to metabolism, suggesting that the type of unsaturation, exact position, and context (neighboring groups) must also be considered. Another unsaturated analog that one might predict would be relatively metabolically stable is EB1089, with its conjugated double-bond system. However, as pointed out earlier, EB1089 contains three structural modifications: the conjugated double-bond system is accompanied by two types of side chain homologation. Nevertheless, as expected, the conjugated double-bond system dominates the metabolic fate of EB1089, there being no C-24 oxidation activity because of the blocking action of the conjugated diene system. When metabolism is studied with either in vitro liver cell systems or the cultured keratinocyte cell line HPK1A-ras, the disappearance of EB1089 is much slower than that of 1α,25(OH)$_2$D$_3$ *(98,99)*. Such data are consistent with the fairly long $T_{1/2}$ observed in pharmacokinetic studies in vivo *(100)*. Since the conjugated system of EB1089 blocks C-24 oxidation reactions, it is not surprising that a different site in the molecule becomes the target for hydroxylation, albeit at a much reduced rate. Diode-array spectrophotometry has allowed for identification of the principal metabolic products of EB1089 as 26- and 26a-hydroxylated metabolites *(99,100)* (Fig. 9). These metabolites have been chemically synthesized and, as in the case of KH1060, shown to retain significant biologic activity in cell differentiation and antiproliferative assays *(100)*.

Again, it is interesting to note that with EB1089 and other molecules blocked in the C-23 and C-24 positions such as 1α,24S(OH)$_2$D$_2$ *(49)*, the terminal carbons C-26 and C-26a become the alternative sites of further hydroxylation. However, it should also be considered that even in molecules not blocked in the C-23 and C-24 positions but containing the terminal 26- and 27-dimethyl homologation such as 26,27-dimethyl-1α,25(OH)$_2$D$_3$ (MC1548) *(101)*, there seems to be significant terminal 26a-hydroxylation occurring. Therefore, the hydroxylation of EB1089 at C-26 and C-26a may be in part a consequence of the introduction of the conjugated double-bond system and in part a consequence of the introduction of the terminal homologation.

When the C-22=C-23 double bond is present in the side chain in the absence of a C-24 methyl group, as in calcipotriol, the double bond appears vulnerable to reduction.

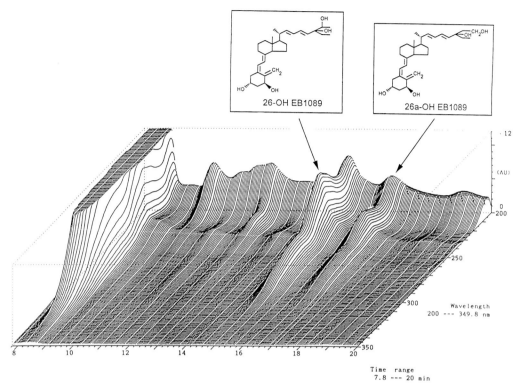

Fig. 9. In vitro metabolism of EB1089 by HPK1A-ras cells. Diode-array high-performance liquid chromatography (HPLC) of a lipid extract following incubation of EB1089 (10 μM) with the human keratinocyte HPK1A-ras for 72 h. In addition to the substrate at 8.5 min, two metabolites showing the distictive UV chromophore of EB1089 (λ_{max}, 235 nm; shoulder, 265 nm) are visible in the part of the HPLC profile reproduced here (8–20 min). Metabolite peaks at 15.03 and 16.55 min were isolated by extensive HPLC and identified (*98–100*) by comparison with synthetic standards on HPLC, gas chromatography-mass spectrometry, and nuclear magnetic resonance. The identifications are as follows: peak A at 15.03 min, 26(OH)EB1089; peak B at 16.55 min, 26a(OH)EB1089. The structures of each of the metabolite peaks are depicted in the insets. (V.N. Shankar, A-M. Kissmeyer, and G. Jones, unpublished results.)

As pointed out earlier, the *principal metabolites of calcipotriol are reduced in the C-22=C-23 bond* except for one, the C-22=C-23 unsaturated, 24-ketone (MC1046) *(71,72)*. This suggests that a C-24 ketone must be present to allow for this reduction to occur. Work using the Roche compound, Δ^{22}-1α,25(OH)$_2$D$_3$, an analog that contains the C-22=C-23 double bond but lacks a C-24 subsituent, tends to support this theory indirectly *(102)*. When incubated with the chronic myelogenous leukemic cell line RWLeu-4, this molecule, like 1α,25(OH)$_2$D$_3$, is converted, presumably via metabolites analogous to intermediates in the C-24 oxidation pathway, to the side-chain truncated product 24,25,26,27-tetranor-1,23(OH)$_2$D$_3$, a molecule that lacks the C-22=C-23 double bond.

4. IMPORTANT IMPLICATIONS DERIVED FROM METABOLISM STUDIES

4.1. Implications for Mechanism of Action of Vitamin D Analogs

There is currently tremendous interest in defining the mechanism of action of vitamin D analogs, particularly for clarifying the difference between *calcemic* or *noncalcemic* analogs. The susceptibility of a vitamin D analog to metabolism and excretion undoubtedly

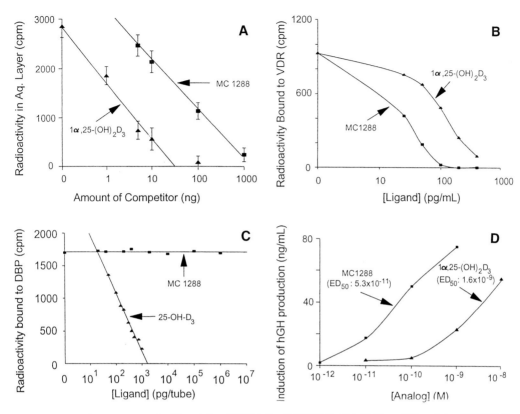

Fig. 10. Biologic parameters for 20-epi-1α,25(OH)$_2$D$_3$ (MC1288). (**A**) Ability of 20-epi-1α,25(OH)$_2$D$_3$ to compete for C-24 oxidation pathway enzymes. (**B**) Vitamin D receptor binding affinity of 20-epi-1α,25(OH)$_2$D$_3$ compared with 1α,25(OH)$_2$D$_3$. (**C**) Vitamin D binding protein binding affinity of 20-epi-1α,25(OH)$_2$D$_3$ compared with 1α,25(OH)$_2$D$_3$. (**D**) Gene transactivation (vitamin D response element placed upstream of GH reporter gene) by 20-epi-1α,25(OH)$_2$D$_3$ compared with 1α,25(OH)$_2$D$_3$ in the COS-1 cell line. (Reproduced with permission from ref. *78*.)

plays a significant role in determining the biologic activity of that analog in vivo. However, the susceptibility to metabolism is not the only important parameter dictating whether a vitamin D analog will be calcemic or noncalcemic. The main factors include:

1. Affinity of the analog for plasma DBP, which dictates cell entry and plasma clearance.
2. Affinity of the analog for target cell VDR–retinoid X receptor (RXR) heterodimeric complex and the resultant affinity of this complex for the VDRE upstream of target genes.
3. Rate of metabolism of the analog by both liver and target cell enzymes.

Data for each of these parameters are being collected in various ways. Binding assays for DBP and VDR have been available for some time now, and such data are easy to acquire. The rate of metabolism of an analog can be studied in cultured cells in vitro, and several cell lines from liver or target cell sources are available to act as valuable tools to reflect this process occurring in vivo *(103)*. In general, these parameters *when considered alone* do not fully explain why some analogs have improved biologic activity when compared with 1α,25(OH)$_2$D$_3$. However, when these metabolic, VDR, and DBP parameters *are considered together* for a given analog, such as 20-epi-1α,25(OH)$_2$D$_3$ *(79)*, as in Fig. 10, then one begins to understand the complexity and the fact that all components contribute to the overall superiority of 20-epi-1α,25(OH)$_2$D$_3$ over 1α,25(OH)$_2$D$_3$ observed in gene transactivation models in vitro (Fig. 10D). One might anticipate this complexity to be even greater in vivo.

Of late, refinements to VDR binding assays are now emerging that measure not only the affinity of the analog for the VDR but the strength of the liganded VDR–RXR heterodimeric complex and the stability of the liganded VDR–RXR–VDRE complex *(104–106)*. The data from these more sophisticated assays suggests that analogs that form stable complexes are more active than 1α,25(OH)$_2$D$_3$ in vitro. Although these data are creating much excitement in vitamin D circles, it is our view and that of many of the molecular biologists performing the in vitro gene expression work that vitamin D analog action in vivo is even more complex than is portrayed in Fig. 10 and cannot be explained by one parameter alone.

In the whole animal in vivo, for each analog pharmacokinetic data can be acquired that probably reflect more than one of these parameters to different degrees. These data reflect the following important parameters:

1. The affinity of the vitamin D analog for DBP in the bloodstream.
2. The rate of target cell uptake and metabolism by target cell enzymes.
3. The rate of liver cell uptake, hepatic metabolism, and biliary clearance.
4. The rate of storage depot uptake and release.

In the case of some of the analogs shown in Tables 1 and 2, pharmacokinetic data *(81, 107,108)* are available and can be compared with the data provided by in vitro metabolic studies. It is apparent from perusal of pharmacokinetic and metabolic data that the analogs we have defined as *metabolically resistant are usually calcemic* and those termed *metabolically sensitive are noncalcemic*. In fact, this classification can be refined along the lines suggested by Kissmeyer et al. *(80)*, with all their compounds segregating into at least two groups (perhaps more) on the basis of their pharmacokinetic parameters:

1. Calcemic analogs (strong or weak): Those analogs with a long $T_{1/2}$, which is a function of either strong DBP binding *or* a reduced rate of metabolism (or both). There appear to be a group of analogs in which a long $T_{1/2}$ is correlated with a slower rate of metabolism (e.g., F$_6$-1α,25(OH)$_2$D$_3$, EB1089, 1α,24S(OH)$_2$D$_2$, and ED-71). With the exception of ED-71, which has a strong affinity for DBP, most of these active analogs bind DBP poorly.
2. Noncalcemic analogs: Those analogs with a short $T_{1/2}$, which is a function of either poor DBP binding *or* a rapid rate of metabolism (or both) (e.g., calcipotriol, KH1060 and OCT).

Classifications used in the vitamin D literature are somewhat artificial since analogs that are *purely* noncalcemic have not been created. All noncalcemic analogs will stimulate in vitro calcemic gene expression and will eventually cause hypercalcemia in vivo if their concentration is raised sufficiently. The crucial question is whether systemically administered weakly calcemic or noncalcemic analogs can produce their anti-cell proliferation/ pro-cell-differentiation effects in vivo at concentrations lower than that required to produce calcemia. Various in vivo clinical trials currently in progress will be the acid test for this question.

When considering molecular mechanisms of action at the target cell level, metabolism is often disregarded or given too little emphasis. Furthermore, certain invalid metabolic assumptions made during biologic activity testing include the following: (1) the analog is biologically active as administered, and (2) the analog is stable in the in vitro target cell model used, whether organ culture, cultured cell, or transfected cell. The validity of this approach is made more tenuous *when data acquired with different in vitro models in which metabolic considerations may or may not apply are compared with data acquired in vivo, for which metabolic considerations definitely apply.* Invalid comparisons of in vivo and in vitro data abound in this field.

In summary, from studies performed thus far, metabolism appears to be one of a number of key parameters dictating the survival and hence biologic activity of the analog when administered topically or systemically in vivo. At this point in time, a pattern is emerging suggesting that noncalcemic or calcemic analogs are metabolically sensitive and metabolically resistant, respectively. However, it seems unlikely that permutations of these biochemical parameters will translate into pure noncalcemic and calcemic analogs, but rather into therapeutic agents with improved cell-differentiating or calcemic activities suitable for different applications.

4.2. Future Directions

4.2.1. VITAMIN D-RELATED CYTOCHROME P450S

The study of vitamin D metabolism will remain a productive focus over the next decade of research. This will be mainly due to the continuing renaissance in the study of the vitamin D-related enzymes stimulated by the elucidation of the cytochrome P450 isoforms involved. The study of the structure, properties, and regulation of these cytochrome P450s has reached new levels of sophistication and will continue to develop. For instance, the cytochromes CYP27 and CYP24 thought to be responsible for 25- and 24-hydroxylation, respectively, have now been overexpressed and studied in a variety of transfection systems (*Escherichia coli*, baculovirus, yeast, COS-1 monkey kidney cells), and this has led to new knowledge of substrate preferences, site of hydroxylation, catalytic properties, and mechanisms of regulation. No X-ray crystal model yet exists for a mammalian cytochrome P450, but molecular modeling has begun based on structures of prokaryotic cytochrome P450s and the high sequence conservation observed for cytochromes P450 throughout the phylogenetic tree. One can predict that the cytochromes P450 representing the renal and extrarenal 1α-hydroxylase(s) will be recognized in the near future; if they are novel, these will be cloned. The complex regulation of the 1α-hydroxylase is of central interest to physiologists. Similarly, the minor hydroxylases involved in 26-hydroxylation and 26,23-lactone formation of 25(OH)D_3 will be characterized in molecular terms and their specificity for vitamin D analogs [e.g., EB1089, 1α,24S(OH)$_2D_2$] fully defined.

4.2.2. HYDROXYLASE, VDR, AND DBP GENE KNOCKOUTS

Equally exciting is the application of new genetic technologies allowing for creation of mice overexpressing or lacking the genes coding for vitamin D-related hydroxylase enzymes, VDR, or DBP. These mutations will allow a new look at complex interactions between vitamin D metabolic machinery and the other elements of the vitamin D signal transduction system. Of particular importance will be testing of the hypothesis that certain pathways are catabolic in nature (e.g., 24-hydroxylation and 26,23-lactone formation), a concept that has emerged from other approaches over the past 10 years. Gene knockout experiments in the related retinoid field *(109)* have emphasized the redundancy built into these systems, and it will be interesting to observe whether this also applies in the vitamin D signaling pathways.

4.2.3. FUTURE VITAMIN D ANALOG DESIGN

The success of vitamin D analogs in a variety of clinical applications, particularly hypocalcemic conditions and psoriasis, will continue to fuel interest in the development of more effective vitamin D analogs for use in these conditions as well as in osteoporosis, cancer, and immunosuppression. Over the immediate future we can anticipate (1) a continuing search for novel synthetic modifications to the vitamin D molecule, (2) a combination of "useful" modifications to fine-tune the best analogs, and (3) synthesis of "smart" molecules based on emerging metabolic and structure–activity information gained from earlier generations of molecules.

In particular, one can envision that the VDR ligand-binding pocket studies and cytochrome P450 substrate-binding pocket studies just beginning to emerge will provide valuable information for the design of further generations of vitamin D analogs. With the first application of molecular modeling techniques to the study of vitamin D-related proteins, we will be in a position to view the fit of vitamin D analogs to active site topology and fine-tune this fit. The future of vitamin D metabolism research remains bright, and the pursuit of novel vitamin D analogs continues to be attractive to the pharmaceutical industry.

ACKNOWLEDGMENTS

The author thanks F. Jeffrey Dilworth and David Prosser for compilation of the references, tables, and figures. Some of the work cited here was supported through grants to the author from the Medical Research Council of Canada. The collaborative research reviewed here required the input of some excellent research trainees (David Lohnes, Fuad Qaw, Stephen Strugnell, and Sonoko Masuda) as well as the interdisciplinary involvement of talented scientists from around the world. I gratefully acknowledge the essential contributions to our analog work of Hugh L.J. Makin, Martin Calverley, Joyce Knutson, Charles Bishop, Noboru Kubodera, Anne-Marie Kissmeyer, Richard Kremer, Mark R. Haussler, and Hector F. DeLuca.

REFERENCES

1. Holick MF. Vitamin D: photobiology, metabolism and clinical applications. In: Endocrinology, 3rd ed., vol. 2. Degroot L, ed. Philadelphia: WB Saunders, 1995; 990–1014.
2. Bhattacharyya MH, DeLuca HF. The regulation of the rat liver calciferol-25-hydroxylase. J Biol Chem 1973; 248:2969–2973.
3. Andersson S, Davis DL, Dahlback H, Jornvall H, Russell DW. Cloning, structure and expression of the mitochondrial cytochrome P450 sterol 26-hydroxylase, a bile acid biosynthetic enzyme. J Biol Chem 1989; 246:8222–8229.
4. Okuda KI, Usui E, Ohyama Y. Recent progress in enzymology and molecular biology of enzymes involved in vitamin D metabolism. J Lipid Res 1995; 36:1641–1652.
5. Guo Y-D, Strugnell S, Back DW, Jones G. Transfected human liver cytochrome P-450 hydroxylates vitamin D analogs at different side-chain positions. Proc Natl Acad Sci USA 1993; 90:8668–8672.
6. Fukushima M, Suzuki Y, Tohira Y, Nishii Y, Suzuki M, Sasaki S, Suda T. 25-Hydroxylation of 1α-hydroxyvitamin D_3 in vivo and in the perfused rat liver. FEBS Lett 1976; 65:211–214.
7. Baran DT, Milne ML. 1,25-Dihydroxyvitamin D increases heapatocyte cytosolic calcium levels: a potential regulator of vitamin D-25-hydroxylase. J Clin Invest 1986; 77:1622–1626.
8. Berginer VM, Shany S, Alkalay D, Berginer J, Dekel S, Salen G, Tint GS, Gazit D. Osteoporosis and increased bone fractures in cerebrotendinous xanthomatosis. Metabolism 1993; 42:69–74.
8a. Postlind H, Axen E, Bergman T, Wikvall K. Cloning, structure and expression of a cDNA encoding vitamin D_3 25-hydroxylase. Biochem Biophys Res Commun 1997; 241:491–497.

9. Tucker G, Gagnon RE, Haussler MR. Vitamin D_3-25-hydroxylase: tissue occurrence and lack of regulation. Arch Biochem Biophys 1973; 155:47–57.
10. Axen E, Postlind H, Wikvall K. Effects of CYP27 mRNA expression in rat kidney and liver by $1\alpha,25$-dihydroxyvitamin D_3, a suppressor of renal 25-hydroxyvitamin D_3-1α-hydroxylase activity. Biochem Biophys Res Commun 1995; 215:136–141.
11. Ichikawa F, Sato K, Nanjo M, Nishii Y, Shinki T, Takahashi N, Suda T. Mouse primary osteoblasts express vitamin D_3 25-hydroxylase mRNA and convert 1α-hydroxyvitamin D_3 into $1\alpha,25$-dihydroxyvitamin D_3. Bone 1995; 16:129–135.
12. Fraser DR, Kodicek E. Unique biosynthesis by kidney of a biologically active vitamin D metabolite. Nature 1970; 228:764–766.
12a. St-Arnaud R., Messerlian S, Moir JM, Omdahl JL, Glorieux FH. The 25-hydroxyvitamin D 1-alpha-hydroxylase gene maps to the pseudovitamin D-deficiency rickets (PDDR) disease locus. J. Bone Miner Res 1997; 12:1552–1559.
12b. Takeyama K-I, Kitanaka S, Sato T, Kobori, M, Yanagisawa J, Kato S. 25-Hydroxyvitamin D_3 1α-hydroxylase and vitamin D synthesis. *Science* 1997; 277:1827–1830.
12c. Monkawa T, Yoshida T, Wakino S, Shinki T, Anazawa H, DeLuca HF, Suda T, Hayashi M, Saruta T. Molecular cloning of cDNA and genomic DNA for human 25-hydroxyvitamin D_3 1-alpha hydroxylase. Biochem Biophys Res Commun 1997; 239:527–533.
12d. Fu GK, Lin D, Zhang MY, Bikle DD, Shackleton CH, Miller WL, Portale AA. Cloning of human 25-hydroxyvitamin D-1-alpha-hydroxylase and mutations causing vitamin D dependent rickets type 1. Mol Endocrinol 1997; 11:1961–1970.
12e. Fu GK, Portale AA, Miller WL. Complete structure of the human gene for the vitamin D 1α-hydroxylase, P450c1α. DNA Cell Biol 1997; 16:1499–1507.
13. Gray RW, Omdahl JL, Ghazarian JG, DeLuca HF. 25-Hydroxycholecalciferol-1-hydroxylase: subcellular location and properties. J Biol Chem 1972; 247:7528–7532.
14. Henry HL. Regulation of the hydroxylation of 25-hydroxyvitamin D_3 in vivo and in primary cultures of chick kidney cells. J Biol Chem 1979; 254:2722–2729.
14a. Kitanaka S, Takeyama K, Murayama A, Sato T, Okumura K, Nogami M, Hasegawa Y, Niimi H, Yanigisawa J, Tanaka T, Sato K. Inactivating mutations in the 25-hydroxyvitamin D3-1α-hydroxylase gene in patients with pseudovitamin D deficiency rickets. N Engl J Med 1998; 338: 653–661.
14b. Fraser D, Kooh SW, Kind P, Holick MF, Tanaka Y, DeLuca HF. Pathogenesis of hereditary vitamin D dependency rickets. N Engl J Med 1973; 289:817–822.
15. Lester GE, Gray TK, Williams ME. In vitro 1α-hydroxylation of ^{3}H-25-hydroxyvitamin D_3 by isolated cells from rat kidneys and placentae In: Hormonal Control of Calcium Metabolism. Cohn DV, Talmage RV, Matthews JL, eds. Amsterdam: Excerpta Medica, 1981; 376.
16. Adams JS, Gacad MA. Characterization of 1α-hydroxylation of vitamin D_3 sterols by cultured alveolar macrophages from patients with sarcoidosis. J Exp Med 1985; 161:755–765.
17. Holick MF, Schnoes HK, DeLuca HF, Gray RW, Boyle IT, Suda T. Isolation and identification of 24,25-dihydroxycholecalciferol: a metabolite of vitamin D_3 made in the kidney. Biochemistry 1972; 11:4251–4255.
18. Holick MF, Kleiner-Bossaller A, Schnoes HK, Kasten PM, Boyle IT, DeLuca HF. 1,24,25-Trihydroxyvitamin D_3. A metabolite of vitamin D_3 effective on intestine. J Biol Chem 1973; 248:6691–6696.
19. Ornoy A, Goodwin D, Noff D, Edelstein S. 24,25-Dihydroxyvitamin D is a metabolite of vitamin D essential for bone formation. Nature 1978; 276:517–519.
20. Rasmussen H, Bordier P. Vitamin D and bone. Metab Bone Dis Rel Res 1978; 1:7–13.
21. Henry HL, Norman AW. Vitamin D: two dihydroxylated metabolites are required for normal chicken egg hatchability. Science 1978; 201:835–837.
22. Jones G, Vriezen D, Lohnes D, Palda V, Edwards NS. Side chain hydroxylation of vitamin D_3 and its physiological implications. Steroids 1987; 49:29–55.
23. Makin G, Lohnes D, Byford V, Ray R, Jones G. Target cell metabolism of 1,25-dihydroxyvitamin D_3 to calcitroic acid. Evidence for a pathway in kidney and bone involving 24-oxidation. Biochem J 1989; 262:173–180.
24. Reddy GS, Tserng K-Y. Calcitroic acid, end product of renal metabolism of 1,25-dihydroxyvitamin D_3 through C-24 oxidation pathway. Biochemistry 1989; 28:1763–1769.
25. Esvelt RP, Schnoes HK, DeLuca HF. Isolation and characterization of 1α-hydroxy-23-carboxytetranorvitamin D: a major metabolite of 1,25-dihydroxyvitamin D_3. Biochemistry 1979; 18:3977–3983.
26. Knutson JC, DeLuca HF. 25-Hydroxyvitamin D3-24-hydroxylase. Subcellular location and properties. Biochemistry 1974; 13:1543–1548.

27. Ohyama Y, Noshiro M, Okuda K. Cloning and expression of cDNA encoding 25-hydroxyvitamin D_3 24-hydroxylase. FEBS Lett 1991; 278:195–198.
28. Ohyama Y, Okuda K. Isolation and characterization of a cytochrome P450 from rat kidney mitochondria that catalyzes the 24-hydroxylation of 25-hydroxyvitamin D3. J Biol Chem 1991; 266: 8690–8695.
29. Tomon M, Tenenhouse HS, Jones G. Expression of 25-hydroxyvitamin D_3-24-hydroxylase activity in CaCo-2 cells. An in vitro model of intestinal vitamin D catabolism. Endocrinology 1990; 126: 2868–2875.
30. Bouillon R, Okamura WH, Norman AW. Structure-function relationships in the vitamin D endocrine system. Endocr Rev 1995; 16:200–257.
31. Suda T, DeLuca HF, Schnoes HK, Tanaka Y, Holick MF. 25,26-dihydroxyvitamin D_3, a metabolite of vitamin D_3 with intestinal transport activity. Biochemistry 1970; 9:4776–4780.
32. Horst RL. 25-OH-D_3-26,23-Lactone: a metabolite of vitamin D_3 that is 5 times more potent than 25-OH-D_3 in the rat plasma competitive protein binding radioassay. Biochem Biophys Res Commun 1979; 89:286–293.
33. Yamada S, Nakayama K, Takayama H, Shinki T, Takasaki Y, Suda T. Isolation, identification and metabolism of (23S,25R)-25-hydroxyvitamin D_3-26,23-lactol: a biosynthetic precursor of (23S,25R)-25-hydroxyvitamin D_3-26,23-lactone. J Biol Chem 1984; 259:884–889.
34. Akiyoshi-Shibata M, Sakaki T, Ohyama Y, Noshiro M, Okuda K, Yabusaki Y. Further oxidation of hydroxycalcidiol by calcidiol 24-hydroxylase—a study with the mature enzyme expressed in *Escherichia coli*. Eur J Biochem 1994; 224:335–343.
35. Beckman M, Tadikonda P, Werner E, Prahl JM, Yamada S, DeLuca HF. Human 25-hydroxyvitamin D_3-24-hydroxylase, a multicatalytic enzyme. Biochemistry 1996; 35:8465–8472.
36. Ishizuka S, Ishimoto S, Norman AW. Isolation and identification of 1α,25-dihydroxy-24-oxo-vitamin D_3, 1α,25-dihydroxyvitamin D_3-26,23-lactone, 1α,24(S),25-trihydroxyvitamin D_3: in vivo metabolites of 1α,25-dihydroxyvitamin D_3. Biochemistry 1984; 23:1473–1478.
37. Shinki T, Jin CH, Nishimura A, Nagai Y, Ohyama Y, Noshiro M, Okuda K, Suda T. Parathyroid hormone inhibits 25-hydroxyvitamin D_3-24-hydroxylase mRNA expression stimulated by 1α,25-dihydroxyvitamin D_3 in rat kidney but not in intestine. J Biol Chem 1992; 267:13,757–13,762.
38. Lohnes D, Jones G. Further metabolism of 1α,25-dihydroxyvitamin D_3 in target cells. J Nutr Sci Vitam Special Issue. 1992; 75–78.
38a. St-Arnaud R, Arabian A, Travers R, Glorieux FH. Abnormal intramembranous ossification in mice deficient for the vitamin D 24-hydroxylase gene. In: Vitamin D. Chemistry, Biology and Clinical Applications of the Steroid Hormone. Norman AW, Bouillon, R, Thomasset M, eds. Vitamin D Workshop, Inc., University of California Press, Riverside, CA. 1997; 635–639.
38b. St-Arnaud R, Arabian A, Travers R, Glorieux FH. Partial rescue of abnormal bone formation in 24-hydroxylase knock-out mice supports a role for 24,25$(OH)_2D_3$ in intramembranous ossification. J Bone Miner Res 1997; 12:33 (abstract S111).
39. Esvelt RP, DeLuca HF. Calcitroic acid: biological activity and tissue distribution studies. Arch Biochem Biophys 1980; 206:404–413.
40. Suda T, DeLuca HF, Schnoes HK, Blunt JW. Isolation and identification of 25-hydroxyergocalciferol. Biochemistry 1969; 8:3515–3520.
41. Jones G, Schnoes HK, DeLuca, HF. Isolation and identification of 1,25-dihydroxyvitamin D_2. Biochemistry 1975; 14:1250–1256.
42. Jones G, Rosenthal A, Segev D, Mazur Y, Frolow F, Halfon Y, Rabinovich D, Shakked Z. Isolation and identification of 24,25-dihydroxyvitamin D_2 using the perfused rat kidney. Biochemistry 1979; 18:1094–1101.
43. Jones G, Schnoes HK, Levan L, DeLuca HF. Isolation and identification of 24-hydroxyvitamin D_2 and 24,25-dihydroxyvitamin D_2. Arch Biochem Biophys 1980; 202:450–457.
44. Horst RL, Koszewski NJ, Reinhardt TA. 1α-Hydroxylation of 24-hydroxyvitamin D_2 represents a minor physiological pathway for the activation of vitamin D_2 in mammals. Biochemistry 1990; 29:578–582.
45. Koszewski NJ, Reinhardt TA, Napoli JL, Beitz DC, Horst RL. 24,26-Dihydroxyvitamin D_2: a unique physiological metabolite of vitamin D_2. Biochemistry 1988; 27:5785–5790.
46. Reddy GS, Tserng K-Y. Isolation and identification of 1,24,25-trihydroxyvitamin D_2, 1,24,25,28-tetrahydroxyvitamin D_2, 1,24,25,26-tetrahydroxyvitamin D_2: new metabolites of 1,25-dihydroxyvitamin D_2 produced in the rat kidney. Biochemistry 1986; 25:5328–5336.

47. Horst RL, Koszewski NJ, Reinhardt TA. Species variation of vitamin D metabolism and action: lessons to be learned from farm animals. In: Vitamin D. Molecular, Cellular and Clinical Endocrinology. Norman AW, Schaefer K, Grigoleit H-G, von Herrath D, eds. Berlin: de Gruyter, 1988; 93–101.
48. Clark JW, Reddy GS, Santos-Moore A, Wankadiya KF, Reddy GP, Lasky S, Tserng K-Y, Uskokovic MR. Metabolism and biological activity of 1,25-dihydroxyvitamin D_2 and its metabolites in a chronic myelogenous leukemia cell line, RWLEU-4. Bioorg Med Lett 1993; 3:1873–1878.
49. Jones G, Byford V, Kremer R, Makin HLJ, Rice RH, deGraffenreid LA, Knutson JC, Bishop CA. Anti-proliferative activity and target cell catabolism of the vitamin D analog, $1\alpha,24(S)$-dihydroxyvitamin D_2 in normal and immortalized human epidermal cells. Biochem Pharmacol 1996; 52:133–140.
50. Suda T, Hallick RB, DeLuca HF, Schnoes HK. 25-hydroxydihydrotachysterol$_3$ Synthesis and biological activity. Biochemistry 1970; 9:1651–1657.
51. Bosch R, Versluis C, Terlouw JK, Thijssen JHH, Duursma SA. Isolation and identification of 25-hydroxydihydrotachysterol$_2$, $1\alpha,25$-dihydroxydihydrotachysterol$_2$ and $1\beta,25$-dihydroxydihydrotachysterol$_2$. J Steroid Biochem 1985; 23:223–229.
52. Jones G, Edwards N, Vriezen D, Porteous C, Trafford DJH, Cunningham J, Makin HLJ. Isolation and identification of seven metabolites of 25-hydroxy-dihydrotachysterol$_3$ formed in the isolated perfused rat kidney: a model for the study of side-chain metabolism of vitamin D. Biochemistry 1988; 27:7070–7079.
53. Qaw F, Calverley MJ, Schroeder NJ, Trafford DJH, Makin HLJ, Jones G. *In vivo* metabolism of the vitamin D analog, dihydrotachysterol. Evidence for formation of $1\alpha,25$- and $1\beta,25$-dihydroxydihydrotachysterol metabolites and studies of their biological activity. J Biol Chem 1993; 268:282–292.
54. Bhattacharyya MH, DeLuca HF. Comparative studies on the 25-hydroxylation of vitamin D_3 and dihydrotachysterol$_3$. J Biol Chem 1973; 248:2974–2977.
55. Wing RM, Okamura WH, Pirio MP, Sine SM, Norman AW. Vitamin D in solution: conformations of vitamin D_3, 1,25-dihydroxyvitamin D_3 and dihydrotachysterol$_3$. Science 1974; 186:939–941.
56. Shany S, Ren S-Y, Arbelle JE, Clemens TL, Adams JS. Subcellular localization and partial purification of the 25-hydroxyvitamin D-1-hydroxylation reaction in the avian myelomonocytic cell line HD-11. J Bone Miner 1993; 8:269–276.
57. Qaw F, Schroeder NJ, Calverley MJ, Maestro M, Mourino A, Trafford DJH, Makin, HLJ, Jones G. *In vitro* synthesis of 1,25-dihydroxydihydrotachysterol in the myelomonocytic cell line, HD-11. J Bone Miner Res 1992; 7:S161 (abstract 274).
58. Qaw FS, Makin HLJ, Jones G. Metabolism of 25-hydroxy-dihydrotachysterol$_3$ in bone cells *in vitro*. Steroids 1992; 57:236–243.
59. Schroeder NJ, Qaw F, Calverley MJ, Trafford DJH, Jones G, Makin HLJ. Polar metabolites of dihydrotachysterol$_3$ in the rat: comparison with *in vitro* metabolites of $1\alpha,25$-dihydroxydihydrotachysterol$_3$. Biochem Pharm 1992; 43:1893–1905.
60. Holick MF, Semmler E, Schnoes HK, DeLuca HF. 1α-Hydroxy derivative of vitamin D_3: a highly potent analog of $1\alpha,25$-dihydroxyvitamin D_3. Science 1973; 180:190,191.
61. Barton DH, Hesse RH, Pechet MM, Rizzardo E. A convenient synthesis of 1α-hydroxy-vitamin D_3. J Am Chem Soc 1973; 95:2748–2749.
62. Gallagher JC, Goldgar D. Treatment of postmenopausal osteoporosis with high doses of synthetic calcitriol. A randomized control study. Ann Intern Med 1990; 113:649–655.
63. Strugnell S, Byford V, Makin HLJ, Moriarty RM, Gilardi R, LeVan LW, Knutson JC, Bishop CW, Jones G. $1\alpha,24(S)$-dihydroxyvitamin D_2: a biologically active product of 1α-hydroxyvitamin D_2 made in the human hepatoma, Hep3B. Biochem J 1995; 310:233–241.
64. Sjoden G, Smith C, Lindgren V, DeLuca HF. 1-Alpha-hydroxyvitamin D_2 is less toxic than 1-alpha-hydroxyvitamin D_3 in the rat. Proc Soc Exp Biol Med 1985; 178:432–436.
65. Frazao JM, Chesney RW, Coburn JW, the 1α-OH-D2 Study Group. Intermittent oral 1α-hydroxyvitamin D2 is effective and safe for the suppression of secondary hyperparathyroidism in haemedialysis patients. Nephrol Dial Transplant 1998; 13(Suppl 3):68–72.
66. Calverley MJ. Synthesis of MC-903, a biologically active vitamin D metabolite analog. Tetrahedron 1987; 43:4609–4619.
67. Kragballe K, Gjertsen BT, De Hoop D, Karlsmark T, van de Kerkhof PC, Larko O, Nieboer C, Roed-Petersen J, Strand A, Tikjob G. Double-blind, right/left comparison of calcipotriol and betamethasone valerate in treatment of psoriasis vulgaris. Lancet 1991; 337:193–196.
68. Jones G, Calverley MJ. A dialogue on analogues: newer vitamin-D drugs for use in bone disease, psoriasis, and cancer. Trends Endocrinol Metab 1993; 4:297–303.

69. Binderup L. MC903—A novel vitamin D analogue with potent effects on cell proliferation and cell differentiation. In: Vitamin D. Molecular, Cellular and Clinical Endocrinology. Norman AW, Schaefer K, Grigoleit H-G, von Herrath D, eds. Berlin: de Gruyter, 1988; 300–309.
70. Sorensen H, Binderup L, Calverley MJ, Hoffmeyer L, Rastrup Anderson N. *In vitro* metabolism of calcipotriol (MC 903), a vitamin D analogue. Biochem Pharmacol 1990; 39:391–393.
71. Masuda S, Strugnell S, Calverley MJ, Makin HLJ, Kremer R, Jones, G. In vitro metabolism of the anti-psoriatic vitamin D analog, calcipotriol, in two cultured human keratinocyte models. J Biol Chem 1994; 269:4794–4803.
72. Murayama E, Miyamoto K, Kubodera N, Mori T, Matsunaga I. Synthetic studies of vitamin D analogues. VIII. Synthesis of 22-oxavitamin D_3 analogues. Chem Pharm Bull (Tokyo) 1986; 34: 4410–4413.
73. Hansen K, Calverley MJ, Binderup L. Synthesis and biological activity of 22-oxa vitamin D analogues. In: Vitamin D: Gene Regulation, Structure-Function Analysis and Clinical Application. Norman AW, Bouillon R, Thomasset M, eds. Berlin: de Gruyter, 1991; 161–162.
74. Brown AJ, Berkoben M, Ritter C, Kubodera N, Nishii Y, Slatopolsky E. Metabolism of 22-oxacalcitriol by a vitamin D-inducible pathway in cultured parathyroid cells. Biochem Biophys Res Commun 1992; 189:759–764.
75. Bikle DD, Abe-Hashimoto J, Su MJ, Felt S, Gibson DFC, Pillai S. 22-Oxa-calcitriol is a less potent regulator of keratinocyte proliferation and differentiation due to decreased cellular uptake and enhanced catabolism. J Invest Dermatol 1995; 105:693–698.
76. Masuda S, Byford V, Kremer R, Makin HLJ, Kubodera N, Nishii Y, Okazaki A, Okano T, Kobayashi T, Jones G. *In vitro* metabolism of the vitamin D analog, 22-oxacalcitriol, using cultured osteosarcoma, hepatoma and keratinocyte cell lines. J Biol Chem 1996; 271:8700–8708.
77. Kobayashi T, Tsugawa N, Okano T, Masuda S, Takeuchi A, Kubodera N, Nishii Y. The binding properties with blood proteins and tissue distribution of 22-oxa-1α,25-dihydroxyvitamin D_3, a noncalcemic analogue of 1α,25-dihydroxyvitamin D_3 in rats. J Biochem 1994; 115:373–380.
78. Dilworth FJ, Calverley MJ, Makin HLJ, Jones G. Increased biological activity of 20-epi-1,25-dihydroxyvitamin D_3 is due to reduced catabolism and altered protein binding. Biochem Pharmacol 1994; 47:987–993.
79. Dilworth FJ, Scott I, Green A, Strugnell S, Guo Y-D, Roberts EA, Kremer R, Calverley MJ, Makin HLJ, Jones G. Different mechanisms of hydroxylation site selection by liver and kidney cytochrome P450 species (CYP27 and CYP24) involved in vitamin D metabolism. J Biol Chem 1995; 270: 16,766–16,774.
80. Kissmeyer A-M, Mathiasen IS, Latini S, Binderup L. Pharmacokinetic studies of vitamin D analogues: relationship to vitamin D binding protein (DBP). Endocrine 1995; 3:263–266.
81. Dilworth FJ, Williams GR, Kissmeyer A-M, Løgsted-Nielsen J, Binderup E, Calverley MJ, Makin HLJ, Jones G. The vitamin D analog, KH1060 is rapidly degraded both in vivo and in vitro via several pathways: principal metabolites generated retain significant biological activity. Endocrinology 1997; 138:5485–5496.
82. Rastrup-Anderson N, Buchwald FA, Grue-Sorensen G. Identification and synthesis of a metabolite of KH1060, a new potent 1α,25-dihydroxyvitamin D_3 analogue. Bioorg Med Chem Lett 1992; 2: 1713–1716.
83. Kobayashi Y, Taguchi T, Mitsuhashi S, Eguchi T, Ohshima E, Ikekawa N. Studies on organic fluorine compounds. XXXIX. Studies on steroids. LXXIX. Synthesis of 1α,25-dihydroxy-26,26,26,27,27,27-hexaflurovitamin D_3. Chem Pharm Bull (Tokyo) 1982; 30:4297–4303.
84. Koeffler HP, Armatruda T, Ikekawa N, Kobayashi Y, DeLuca HF. Induction of macrophage differentiation of human normal and leukemic myeloid stem cells by 1α,25-dihydroxyvitamin D_3 and its fluorinated analogs. Cancer Res 1984; 44:6524–6528.
85. Inaba M, Okuno S, Nishizawa Y, Yukioka K, Otani S, Matsui-Yuasa I, Morisawa S, DeLuca HF, Morii H. Biological activity of fluorinated vitamin D analogs at C-26 and C-27 on human promyelocytic leukemia cells, HL-60. Arch Biochem Biophys 1987; 258:421–425.
86. Kistler A, Galli B, Horst R, Truitt GA, Uskokovic MR. Effects of vitamin D derivatives on soft tissue calcification in neonatal and calcium mobilization in adult rats. Arch Toxicol 1989; 63:394–400.
87. Inaba M, Okuno S, Nishizawa Y, Imanishi Y, Katsumata T, Sugata I, Morii H. Effect of substituting fluorine for hydrogen at C-26 and C-27 on the side chain of 1α,25-dihydroxyvitamin D_3. Biochem Pharmacol 1993; 45:2331–2336.
88. Sasaki H, Harada H, Hanada Y, Morino H, Suzawa M, Shimpo E, Katsumata T, Masuhiro Y, Matsuda K, Ebihara K, Ono T, Matsushige S, Kato S. Transcriptional activity of a fluorinated vitamin D analog on VDR-RXR-mediated gene suppression. Biochemistry 1995; 34:370–377.

89. Nakatsuka K, Imanishi Y, Morishima Y, Sekiya K, Sasao K, Miki T, Nishizawa Y, Katsumata T, Nagata A, Murakawa S. Biological potency of a fluorinated vitamin D analogue in hypoparathyroidism. Bone Miner 1992; 16:73–81.
90. Nishizawa Y, Morii H, Ogura Y, DeLuca HF. Clinical trial of 26,26,26,27,27,27-hexafluro-1α,25-dihydroxyvitamin D_3 in uremic patients on hemodialysis: preliminary report. Contrib Nephrol 1991; 90:196–203.
91. Zhou J-Y, Norman AW, Chen D-L, Sun G, Uskokovic M, Koeffler HP. 1,25-Dihydroxy-16-ene-23-yne-vitamin D_3 prolongs survival time of leukemic mice. Proc Natl Acad Sci USA 1990; 87: 3929–3932.
92. Binderup E, Calverley MJ, Binderup L. Synthesis and biological activity of 1α-hydroxylated vitamin D analogues with poly-unsaturated side chains. In: Vitamin D: Gene Regulation, Structure-Function Analysis and Clinical Application. Norman AW, Bouillon R, Thomasset M, eds. Berlin: de Gruyter, 1991; 192,193.
93. Colston KW, Mackay AG, James SY, Binderup L, Chandler S, Coombes RC. EB1089: a new vitamin D analogue that inhibits the growth of breast cancer cells *in vivo* and *in vitro*. Biochem Pharmacol 1992; 44:2273–2280.
94. James SY, Mackay AG, Binderup L, Colston KW. Effects of a new synthetic analogue, EB1089, on the oestrogen-responsive growth of human breast cancer cells. J Endocrinol 1994; 141:555–563.
95. Reddy GS, Clark JW, Tserng K-Y, Uskokovic MR, McLane JA. Metabolism of 1,25(OH)$_2$-16-ene D_3 in kidney: influence of structural modification of D-ring on side chain metabolism. Bioorg Med Lett 1993; 3:1879–1884.
96. Satchell DP, Norman AW. Metabolism of the cell differentiating agent 1,25(OH)$_2$-16-ene-23-yne vitamin D_3 by leukemic cells. J Steroid Biochem Mol Biol 1996; 57:117–124.
97. Dantuluri PK, Haning C, Uskokovic MR, Tserng K-Y, Reddy GS. Isolation and identification of 1,25,26(OH)$_3$-16-ene-23-yne D_3, a metabolite of 1,25(OH)$_2$-16-ene-23-yne D_3 produced in the kidney. In: Ninth Workshop on Vitamin D Abstract Book, Orlando, May 28–June 2 1994, abstract #43, p. 32.
98. Shankar VN, Makin HLJ, Schroeder NJ, Trafford DJH, Kissmeyer A-M, Calverley MJ, Binderup E, Jones G. Metabolism of the antiproliferative vitamin D analogue, EB1089, in a cultured human keratinocyte model. Bone 1995; 17:326 (abstract).
99. Shankar VN, Dilworth FJ, Makin HLJ, Schroeder NJ, Trafford DAJ, Kissmeyer A-M, Calverley MJ, Binderup E, Jones G. Metabolism of the vitamin D analog EB1089 by cultured human cells: redirection of hydroxylation site to distal carbons of the side chain. Biochem Pharmacol 1997; 53:783–793.
100. Kissmeyer A-M, Binderup E, Binderup L, Hansen CM, Andersen NR, Schroeder NJ, Makin HLJ, Shankar VN, Jones G. The metabolism of the vitamin D analog EB 1089: identification of *in vivo* and *in vitro* metabolites and their biological activities. Biochem Pharmacol 1977; 53:1087–1097.
101. Dilworth FJ, Scott I, Calverley MJ, Makin HLJ, Jones G. Enzymes of side chain oxidation pathway not affected by addition of methyl groups to end of the vitamin D_3 side chain. J Bone Miner Res 1995; 10:S388 (abstract M546).
102. Wandkadiya KF, Uskokovic MR, Clark J, Tserng K-Y, Reddy GS. Novel evidence for the reduction of the double bond in Δ^{22}-1,25-dihydroxyvitamin D_3. J Bone Miner Res 1992; 7:S171 (abstract 315).
103. Jones G, Lohnes D, Strugnell S, Guo Y-D, Masuda S, Byford V, Makin HLJ, Calverley MJ. Target cell metabolism of vitamin D and its analogs. In: Vitamin D. A Pluripotent Steroid Hormone: Structural Studies, Molecular Endocrinology and Clinical Applications. Norman AW, Bouillon R, Thomasset M, eds. Berlin: de Gruyter, 1994; 161–169.
104. Cheskis B, Lemon BD, Uskokovic MR, Lomedico PT, Freedman LP. Vitamin D_3-retinoid X receptor dimerization, DNA binding, and transactivation are differentially affected by analogs of 1,25-dihydroxyvitamin D_3. Mol Endocrinol 1995; 9:1814–1824.
105. Peleg S, Sastry M, Collins ED, Bishop JE, Norman AW. Distinct conformational changes induced by 20-epi analogues of 1α,25-dihydroxyvitamin D_3 are associated with enhanced activation of the vitamin D receptor. J Biol Chem 1995; 270:10,551–10,558.
106. Nayeri S, Danielsson C, Kahlen J, Schräder M, Mathiasen IS, Binderup L, Carlberg C. The anti-proliferative effect of vitamin D_3 analogues is not mediated by inhibition of the AP-1 pathway, but may be related to promoter selectivity. Oncogene 1995; 11:1853–1858.
107. Bouillon R, Allewaert K, Xiang DZ, Tan BK, Van Baelen H. Vitamin D analogs with low affinity for the vitamin D binding protein: Enhanced *in vitro* and decreased *in vivo* activity. J Bone Miner Res 1991; 6:1051–1057.

108. Dusso AS, Negrea L, Gunawardhana S, Lopez-Hilker S, Finch J, Mori T, Nishii Y, Slatopolsky E, Brown AJ. On the mechanisms for the selective action of vitamin D analogs. Endocrinology 1991; 128:1687–1692.
109. Lohnes D, Mark M, Mendelsohn C, Dolle P, Dierich A, Gorry P, Gansmuller A, Chambon P. Function of the retinoic acid receptors (RARs) during development. (I) Craniofacial and skeletal abnormalities in RAR double mutants. Development 1994; 120:2723–2748.
110. Paaren HE, Hamer DE, Schnoes HK, DeLuca HF. Direct C-1 hydroxylation of vitamin D compounds: convenient preparation of 1α-hydroxyvitamin D_3, 1α,25-dihydroxyvitamin D_3 and 1α-hydroxyvitamin D_2. Proc Natl Acad Sci USA 1978; 75:2080,2081.
111. Baggiolini EG, Wovkulich PM, Iacobelli JA, Hennessy BM, Uskokovic MR. Preparation of 1-alpha hydroxylated vitamin D metabolites by total synthesis. In: Vitamin D: Chemical, Biochemical and Clinical Endocrinology of Calcium Metabolism. Norman AW, Schaefer K, von Herrath D, Grigoleit H-G, eds. Berlin: de Gruyter, 1982; 1089–1100.
112. Perlman KL, Sicinski RR, Schnoes HK, DeLuca HF 1α,25-Dihydroxy-19-nor-vitamin D_3, a novel vitamin D-related compound with potential therapeutic activity. Tetrahedron Lett 1990; 31: 1823,1824.
113. Baggiolini EG, Partridge JJ, Shiuey S-J, Truitt GA, Uskokovic MR. Cholecalciferol 23-yne derivatives, their pharmaceutical compositions, their use in the treatment of calcium-related diseases, and their antitumor activity, US 4,804,502. Chem Abstr 1989; 111:58160d (abstract).
114. Calverley MJ, Binderup E, Binderup L. The 20-epi modification in the vitamin D series: selective enhancement of "non-classical" receptor-mediated effects. In: Vitamin D: Gene Regulation, Structure-Function Analysis and Clinical Application. Norman AW, Bouillon R, Thomasset M, eds. Berlin: de Gruyter, 1991; 163,164.
115. Nishii Y, Sato K, Kobayashi T. The development of vitamin D analogues for the treatment of osteoporosis. Osteoporos Int 1993; 1(Suppl):S190–193.
116. Morisaki M, Koizumi N, Ikekawa N, Takeshita T, Ishimoto S. Synthesis of active forms of vitamin D. Part IX. Synthesis of 1α,24-dihydroxycholecalciferol. J Chem Soc Perkin Trans 1975; 1:1421–1424.

5
The Enzymes Responsible for Metabolizing Vitamin D

Kyu-Ichiro Okuda and Yoshihiko Ohyama

1. INTRODUCTION

Vitamin D is a generic name for the antirachitic agents capable of curing rickets of which vitamin D_3 is the major constituent. Vitamin D_3 is formed in the skin from 7-dehydrocholesterol, a precursor of cholesterol, by the action of ultraviolet light *(1,2)*. It is then transported to the liver and is hydroxylated at position 25 by vitamin D_3 25-hydroxylase (25-hydroxylase) to 25-hydroxyvitamin D_3 [25(OH)D_3]. The 25(OH)D_3 formed is again transported to the kidney through the blood stream. In the kidney 25(OH)D_3 is further hydroxylated at position 1α by 25-hydroxyvitamin D_3 1α-hydroxylase (1α-hydroxylase) to 1α,25-dihydroxyvitamin D_3 [1,25(OH)$_2D_3$] which is now considered the hormonal form of the vitamin. It plays a central role in the physiology of calcium homeostasis *(2,3)*. In some calcium statuses, another metabolite of vitamin D_3, 24R,25-dihydroxyvitamin D_3 [24,25(OH)$_2D_3$], is formed from 25(OH)D_3 by 25-hydroxyvitamin D_3 24-hydroxylase (24-hydroxylase), existing in proximal convoluted tubules of the kidney *(4)*. The two kidney hydroxylases are induced in the reciprocal calcium statuses. 1α-Hydroxylase activity is enhanced in a calcium- and/or 1,25(OH)$_2D_3$-depleted status, whereas that of 24-hydroxylase is enhanced in the replete status *(4,5)*.

25- and 24-Hydroxylases were purified to homogeneity *(6,7)* and their cDNA clones isolated *(8,9)*. The primary structures of the enzymes were deduced from the nucleotide sequence of the cDNAs. These studies clearly established that 25- and 24-hydroxylases are P450 enzymes and belong to the new families in the P450 superfamily, CYP27 and CYP24, respectively *(10)*. P450 is a generic name for the enzymes existing in mitochondria and microsomes catalyzing monooxygenation that involves the uptake of two electrons from reduced nicotinamide adenosine diphosphate (NADPH) with the reduction of one atom of O_2 to water and insertion of the other into the substrate. In mitochondria, both flavoprotein containing FAD as a prosthetic group (NADPH-ferredoxin reductase) and iron sulfur protein (ferredoxin) participate in the electron transfer; in microsomes, flavoprotein containing both FAD and FMN (NADPH-cytochrome P450 reductase) is involved in transfer of electrons *(11,12)*.

From: *Vitamin D: Physiology, Molecular Biology, and Clinical Applications*
Edited by: M. F. Holick © Humana Press Inc., Totowa, NJ

Subsequent studies on the promoter region of the gene of 24-hydroxylase showed that two vitamin D-responsive elements (VDREs) exist *(13,14)*. They resemble those reported for vitamin D-inducible proteins such as osteocalcin, calbindin, and osteopontin. They bind the vitamin D receptor (VDR) that is known to be a member of the superfamily of steroid/thyroid/retinoid receptors that function as ligand-dependent transcription factors *(15)*. In this chapter, properties and regulations of 25- and 24-hydroxylases are described.

2. VITAMIN D 25-HYDROXYLASE

2.1. Liver Microsomal Enzyme

Vitamin D_3 is hydroxylated at 25 in liver microsomes by a P450 that may be identical with P4502C11 *(16,17)*. P4502C11 is a sex-specific P450 existing only in liver microsomes of the male rats. Although in female rat liver microsomes a 25-hydroxylation activity exists that is about fivefold less than that of the male, the enzyme is not identical with CYP2C11 and is less well characterized *(18)*. The significance of these microsomal enzymes in calcium homeostasis remains to be clarified.

2.2. Liver Mitochondrial Enzyme

The mitochondrial 25-hydroxylase exists in liver mitochondria of both female and male animals. The enzyme was purified from rat liver mitochondria to homogeneity based on the enzyme activity *(6)*. The molecular weight of the purified enzyme was 52,500. The enzyme shows the absorption spectra characteristic of P450. To reconstitute the enzyme activity, the mitochondrial-type electron-transferring proteins were needed. The specific activity per mole of P450 was 0.36 min^{-1}. The microsomal-type electron transfer protein (NADPH-cytochrome P450 reductase) is completely inactive. The enzyme works toward 1α-hydroxyvitamin D_3 more strongly than toward vitamin D_3, but did not work toward xenobiotics such as benzphetamine, 7-ethoxycoumarine, and benzo[a]pyrene.

The cDNA clone of rat liver enzyme was isolated and the nucleotide sequence determined *(8)*. The amino acid sequence deduced from the nucleotide sequence of the cDNA was 73% identical to that of CYP27 of rabbit liver mitochondria *(19)*. The deduced primary structure contained a presequence characteristic of the mitochondrial proteins in the NH_2-terminal region. From these results it was established that the enzyme is a form of P450 belonging to a family of the P450 superfamily (CYP 27) *(10)*.

2.3. Unique Function of 25-Hydroxylase

Curiously, the purified enzyme revealed a high activity toward 5β-cholestane-3α, 7α,12α-triol, an intermediate in the conversion of cholesterol to cholic acid, and hydroxylated it at position 27 *(20,21)* (Fig. 1). This was verified later on the basis of enzymology *(22)*. 27-Hydroxylation of cholestanetriol is the first step of the side chain degradation of cholesterol (C27) to cholic acid (C24), a bile acid. Expression of the cDNA in COS- and yeast cells clearly demonstrated that the expressed protein showed the two enzyme activities *(23,24)*. Identity of 25-hydroxylase with 27-hydroxylase is thus confirmed on the basis of molecular biology. This fact was also confirmed with the human 25-hydroxylase expressed in COS cells *(25)*.

Fig. 1. Substrate specificity and regioselectivity of vitamin D_3 25-hydroxylase (CYP27). a: Turnover number is not known, as experiment was done with COS cells transfected with human CYP27 cDNA.

Chapter 5 / Metabolizing Enzymes

Turnover number
(nmol × min^{-1} × nmol P450^{-1})

Substrate	Turnover number
5β-Cholestane-3α,7α,12α-triol → (product)	36
Vitamin D$_3$ → (product)	0.36
1α-Hydroxyvitamin D$_3$ → (product)	1.6
Vitamin D$_2$ → (product)	— a
Dihydrotachysterol$_3$ → (product)	— a

2.4. Anomaly in the Gene Structure of Vitamin D_3 25-Hydroxylase

Cerebrotendinous xanthomatosis (CTX) *(26)* is caused by an anomaly of the gene structure of 27-hydroxylase *(27,28)*. Berginer et al. *(29)* found that extensive osteoporosis and increased risk of bone fractures occurred in some CTX patients. The serum level of some vitamin D_3 metabolites such as $25(OH)D_3$, and $24,25(OH)_2D_3$ was lower than normal. In particular, the serum level of $24,25(OH)_2D_3$ was more than twofold lower than the normal; that of $1,25(OH)_2D_3$ was within normal limits. It was considered that extensive osteoporosis and increased risk of bone fracture are the components of the inherited disease. They considered that these symptoms may be explained by low or null activity of abnormal 25-hydroxylase formed from the defective gene. The reason why some of vitamin D derivatives exist in CYP27-deficient patients may be as follows. In liver microsomes many P450s exist that act on xenobiotics. Generally, these P450s are inducible. Some of these P450 species may be induced by increased concentration of vitamin D in the pathologic condition and may work as 25-hydroxylase, although the activity may be insufficient to maintain calcium homeostasis perfectly.

2.5. Miscellaneous Factors

Previously, cholestanetriol 27-hydroxylase activity was observed in cultured skin fibroblasts from healthy individuals, and also in the liver, which is known as the sole organ to catabolize cholesterol *(30)*. Subsequent studies employing Northern blot analysis of the mRNA of CYP27 demonstrated that it exists not only in the liver but also in extrahepatic rat tissues, such as duodenum, adrenal, and lung, in much smaller amounts in the kidney, and little if any in the spleen *(19)*. By the same token, 25-hydroxylase is known to exist in liver and in extrahepatic tissues of cockerels such as kidney and intestine *(31)*. The enzyme activities in these tissues are comparable to that in the liver. Although these reports suggest that both enzyme activities are distributed similarly, no report has been published in which both were measured at the same time in tissues other than liver. Therefore, whether both enzyme activities observed in extrahepatic tissues are caused by a common enyzme remains to be resolved. Since hepatectomy almost eliminates the ability of rats to produce $25(OH)D_3$, the physiologic significance of the extrahepatic 25-hydroxylase is unclear *(32)*.

It has been shown shown that 25-hydroxylation of chick liver homogenate is not strongly inhibited in the presence of excess $25(OH)D_3$ *(32)*. The enzyme activity was also unaffected by the vitamin D status of the chick. Accordingly, the enzyme does not seem to be under the strict control of vitamin D and calcium status.

Saarem and Pedersen *(33)* reported that 25- and 27-hydroxylase activities were 4.6- and 2.7-fold higher, respectively, in female rat liver mitochondria than in the male. Injection of testosterone into female rats decreased both enzyme activities, but not to a statistically significant extent. By contrast, injection of estradiol valerate into males increased both enzyme activities to the same levels as those of control females. Masumoto (personal communication) has confirmed the difference of 25-hydroxylase activity between both sexes when enzyme assay was carried out with mitochondria.

In pig liver mitochondria, this enzyme is also capable of hydroxylating $25(OH)D_3$ at the 1α-position in addition to position 25 of vitamin D_3 and position 27 of cholestanetriol. Since this 1α-hydroxylation activity is extremely low compared with that of other enzymes *(34)*, the physiologic meaning of this activity is questionable.

Human 25-hydroxylase expressed in COS cells showed a broad substrate specificity and a varied regioselectivity toward vitamin D derivatives, i.e., it hydroxylated vitamin D_3, 1α-hydroxyvitamin D_3, and dihydrotachysterol at position 25, 5β-cholestane-3α,7α,12α-triol at position 27, and vitamin D_2 at positions 24 and 27 (Fig. 1). Regioselectivity of the enzyme may therefore depend on the stereospecific structure of the substrates.

3. 25-HYDROXYVITAMIN D_3 1α-HYDROXYLASE

3.1. Kidney Mitochondrial Enzyme

25-Hydroxyvitamin D_3 transported from the liver to kidney is hydroxylated at position 1α by 1α-hydroxylase existing in kidney mitochondria. The product, 1α,25-dihydroxyvitamin D_3, is transported again to the target organs, intestine, bone, and kidney. It increases calcium absorption in the intestine, mobilizes calcium in the bone, and stimulates reabsorption of calcium in the kidney, playing a central role in calcium homeostasis.

It was established by the photochemical action spectrum that 1α-hydroxylase is a P450 enzyme *(35)*. Since then numerous reports of the solubilization and partial purification of the enzyme have been published. However, the enzyme has not been purified to homogeneity so far and its cDNA is not isolated. At the present time no Nebert classification number *(10)* is assigned to this enzyme. Owing to the lability and low abundance of the enzyme, its purification is considered to be a largely intractable problem.

3.2. Liver Microsomal and Mitochondrial 1α-Hydroxylase

In fishes, 1α-hydroxylase activity exists not only in the kidney, but also in the liver *(36)*. Recently, it was reported that in fetal rats the enzyme activity was also observable in both the kidney and liver *(37)*. Since no 1α-hydroxlase activity exists in the liver of born rats, the enzyme may play some role in the fetus. In pigs, 1α-hydroxlase was observed in both liver microsomes *(38)* and mitochondria *(34,38)*. However, how these enzymes are involved in calcium homeostasis remains to be elucidated.

3.3. Kidney Microsomal 1α-Hydroxylase

It was also recently, reported that 1α-hydroxylase activity exists in pig kidney microsomes, and the enzyme was purified to apparent homogeneity *(39)*. The enzyme copurified with the 25-hydroxylase and vitamin D_3 26(27)-hydroxylase activities. The purified enzyme showed the same catalytic activity and apparent molecular weight as the liver microsomal 25-hydroxylase. The amino acid sequence of this enzyme is presently unknown, and thus, no Nebert's classification number has yet been assigned. Although it was shown that it has the catalytic potential to play a role in production of $1,25(OH)_2D_3$, the physiologic significance of this enzyme is unclear because the relationship between enzyme activity and calcium status is unknown.

4. 25-HYDROXYVITAMIN D_3 24-HYDROXYLASE

24R,25-Dihydroxyvitamin D_3 was isolated by Suda et al. *(40)*, and the structure was determined by Holick et al. *(4,41)*. The configuration of carbon 24 was determined by Tanaka et al. *(42)*. Subsequently, it was found that its synthesis occurred in kidney mitochondria of rats fed calcium-replete diets *(5,43)*.

The physiologic function of 24,25(OH)$_2$D$_3$ remains unsolved. Some authors consider that it is important for mineralization of bone matrix, as massive doses of 24,25(OH)$_2$D$_3$ stimulate bone formation without inducing hypercalcemia *(44)*. Others consider that the true substrate of 24-hydroxylase is 1,25(OH)$_2$D$_3$ because the enzyme shows a much higher affinity for 1,25(OH)$_2$D$_3$ than for 25(OH)D$_3$ *(45)*. According to this hypothesis, 1,24,25-trihydroxyvitamin D$_3$ is the end metabolic product formed whenever 1,25(OH)$_2$D$_3$, the active form of vitamin D$_3$, is present in excess in the target tissue. If this is true, 24-hydroxylase works only for catabolism of the active form of the vitamin. From that point of view it may be difficult to explain the fact that CTX patients, who maintain the normal levels of serum 1,25(OH)$_2$D$_3$ but lower levels of 24,25(OH)$_2$D$_3$, reveal extensive osteoporosis and frequent bone fracture. Studies on the knockout mouse that lacks the 24-hydroxylase gene could solve this problem.

4.1. Purification of 25-Hydroxyvitamin D$_3$ 24-Hydroxylase

The kidney is the major organ for producing 24,25(OH)$_2$D$_3$, although extrarenal production of 24,25(OH)$_2$D$_3$ has been reported in some organs. In this section, purification of 24-hydroxylase from kidney mitochondria is described *(7)*. Rats were injected daily with 50,000 IU/animal vitamin D$_3$ dissolved in 0.1 mL of corn oil for 1 wk. Enzyme activity increased about 10-fold. The inductive effect of vitamin D$_3$ may be caused by 1,25(OH)$_2$D$_3$ that is formed from vitamin D$_3$ in vivo *(45,46)*. 24-Hydroxylase purified from kidney mitochondria had a molecular weight of 53,000. The purified enzyme showed the absolute and CO-difference spectra characteristic of P450. The enzyme activity was reconstituted using adrenodoxin and NADPH-adrenodoxin reductase as the electron-transferring proteins. Purified enzyme hydroxylated 25(OH)D$_3$ to 24,25(OH)$_2$D$_3$ with a specific activity of 54.6 nmol/min/mg of protein and a turnover number of 22 min^{-1}. Since all P450s reveal the characteristic absorption- and CO-difference spectra, the amount of P450 that is spectrally measured is the sum of all P450s present; therefore, this is not the real turnover number but the quasiturnover number. The quasiturnover number increases when purification proceeds, whereas the true turnover number does not. Therefore, the value of the quasiturnover number is as an index of purification. The purified enzyme also hydroxylates 1,25(OH)$_2$D$_3$ with a turnover number of 6.9 min^{-1}; it did not hydroxylate 1α(OH)D$_3$. It was also inactive toward xenobiotics such as benzphetamine, 7-ethoxycoumarine, and benzo[a]pyrene. Since the purified enzyme did not hydroxylate 25(OH)D$_3$ at position 1α, it is a different enzyme from 1α-hydroxylase. The reaction followed Michaelis-Menten kinetics, and the K_m value was 2.8 μM. The enzyme activity was inhibited significantly by ketoconazole and 7,8-benzoflavone, whereas it was only slightly inhibited by aminogluthetimide, metyrapone, and SKF 525A, known inhibitors for adrenal P450s.

Both polyclonal and monoclonal antibodies against the purified enzyme were prepared from mice. The polyclonal antibodies were specific since they inhibited the 24-hydroxylase about 70%, whereas they did not affect other species of P450s at all.

4.2. Cloning of cDNA for 25-Hydroxyvitamin D$_3$ 24-Hydroxylase

A cDNA encoding 24-hydroxylase was isolated from a rat kidney cDNA library using the specific antibodies *(9)*. The isolated cDNA was 3.2 kbp in length and contained a 1542-bp open reading frame encoding 514 amino acids. The deduced amino acid

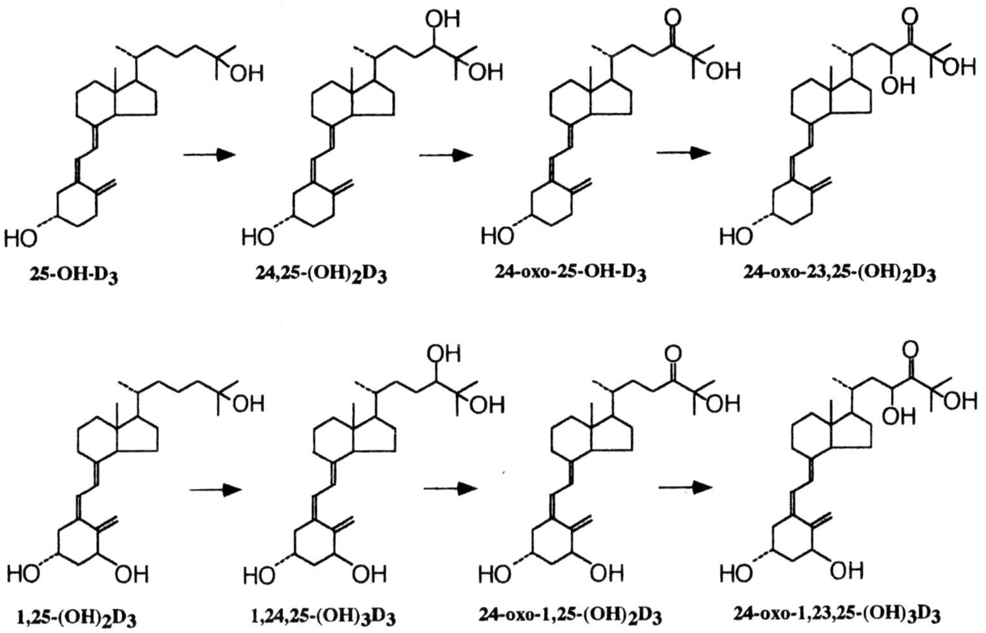

Fig. 2. Further oxidation of 24,25-dihydroxyvitamin D_3 and 1,24,25-trihydroxy vitamin D_3 by 25(OH)D_3 24-hydroxylase.

sequence contained a presequence typical of mitochondrial enzymes in the NH_2-terminal region. The amino acid sequence showed <40% similarity to those of other P450s reported to that date. The enzyme therefore constitutes a novel family in the P450 superfamily and was classified as CYP24 *(10)*.

4.3. Expression of the 24-Hydroxylase cDNA in COS Cells

Expression assay of CYP24 cDNA was performed in COS cells. The specific enzyme activity of the enzyme, expressed in COS cells that were transfected with the expression vector containing the coding region of CYP24 cDNA, was able to hydroxylate 25(OH)D_3 at position 24 in the presence of adrenodoxin and NADPH-adrenodoxin reductase, both purified from bovine adrenals. The specific activity of the enzyme was 14.2 pmol/min/mg of mitochondrial protein and was 10-fold higher than that of the control *(9)*. A cDNA encoding for human CYP24 was recently isolated *(47)*.

4.4. Expression of CYP24 cDNA in Bacterial Cells

Expression assay of the enzyme in *Escherichia coli* was performed in a specific manner without deletion or substitution of any parts of the coding region of the cDNA *(48)*. When 25(OH)D_3 or 1,25(OH)$_2D_3$ was incubated with the expressed enzyme, not only 24-hydroxylated compounds but also further oxidation products were produced. The authors concluded that 24-hydroxylase itself catalyzes a series of reactions, as shown in Figure 2. The enzyme thus catalyzes 24-hydroxylation, 24-dehydrogenation, and subsequent 23-hydroxylation of the 24-oxo compounds. The multiple oxidation is not specific to CYP24 but also is known for other species of P450 such as CYP19, CYP51, and CYP27.

4.5. Gene Structure of 25-Hydroxyvitamin D_3 24-Hydroxylase

The structural gene encoding 24-hydroxylase was isolated from the rat genomic DNA *(49)*. It spans 15 kbp and is composed of 12 exons. From Southern blot analysis it was found that the gene exists as a single copy. A putative TATA box (ATTAAATA) was observed at position nt −30 and a putative CCAAT box was at nt -58. Some VDREs that may be involved in regulation of the enzyme expression were observed in the 5'-flanking region. Alignment with mitochondrial P450 proteins showed that 7 of 11 intron insertion sites of CYP24 occupied positions identical to those in the CYP11 family (CYP11A1 and CYP11B1).

5. REGULATION OF 25-HYDROXYVITAMIN D_3 24-HYDROXYLASE

5.1. Site of Regulation of 24-Hydroxylase in Kidney

Recently, Iwata et al. *(50)* studied induction of 24-hydroxylase occuring in rat kidney using an immunogold technique. They found that the enzyme was distributed in mitochondria along the renal tubules of rat kidney in normal rats. However, when vitamin D_3 was administered, about a 12-fold increase in the amount of the enzyme was observed only in the S1 and S2 segments of kidney, a much smaller increase being observed in other parts of the tubules (Fig. 3). This indicates that induction of CYP24 occurs at the S1 and S2 segments.

5.2. Northern Blot Analysis of 24-Hydroxylase mRNA in Kidney and Intestine

Ohyama et al. *(9)* injected 50,000 IU 25(OH)D_3 to rats daily for 5 days and then carried out Northern blot analysis using CYP24 cDNA as a probe. They found that mRNA for 24-hydroxylase markedly increased in the rat given vitamin D_3. From this result they concluded that the increase of 24-hydroxylase by vitamin D is pretranslational. Since other authors observed that even a single injection of 1,25(OH)$_2$$D_3$ clearly induced 24-hydroxylase mRNA time dependently (Fig. 4) *(45, 46)*, the true inducer might have been 1,25(OH)$_2$$D_3$ that formed in vivo from vitamin D_3. That the mRNA was also induced in the intestine was shown by Armbrecht and Boltz *(46)* and by Shinki et al. *(45)*. The latter authors also observed the following facts: induction of intestinal 24-hydroxylase was far more rapid than that of the renal CYP24 mRNA. Thyroparathyroidectomy shortened the time required to induce expression of renal, but not intestinal 24-hydroxylase. The stimulative effect of 1,25(OH)$_2$$D_3$ on expression of 24-hydroxylase mRNA in the kidney of vitamin D-deficient rats is reduced by administration of parathyroid hormone (PTH) or cyclic adenosine monophosphate, although the effect of the latter was more striking. When rats were fed a vitamin D-replete diet containing 0.7% (adequate) or 0.03% (low) calcium for 2 wk, intestinal expression of 24-hydroxylase mRNA could be induced only in the low calcium group. By contrast, renal 24-hydroxylase mRNA was preferentially stimulated in the adequate calcium group. They thus concluded that expression of 24-hydroxylase mRNA is downregulated by PTH in the kidney but not in the intestine.

5.3. Mechanism of the Regulation of 24-Hydroxylase

Zierold et al. *(14)* studied transcriptional regulation using a gel retardation assay of polymerase chain reaction products for the promoter region of CYP24. They identified

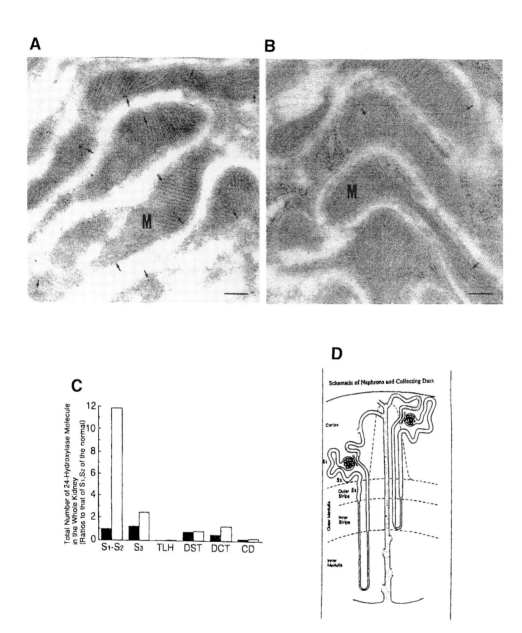

Fig. 3. Immunoelectron microscopy of mitochondria in S1 and S2 segments of the proximal tubules of rat kidney. Ultrathin LR White sections prepared from vitamin D_3-treated (**A**) and normal rats (**B**) were incubated with an anti-24-hydroxylase antibody. Some of gold particles are indicated by arrows. Original magnification, ×50,000. Scale bars ≈0.2 μm. (Reproduced with permission from ref. *50*.) (**C**) Ratio of the total number of enzyme molecules in the various segments in normal rats (solid line bars) and vitamin D_3-treated rats (open bars). Calculated from the data given in A and B. (K. Iwata, personal communication). S1–S2, S1 and S2 segments of proximal tubules; S3, S3 segment; TLH, thin limb of Henle's loops; DST, distal straight tubules; DCT, distal convoluted tubules; CD, collecting ducts. (**D**) Schematic of nephrons and collecting ducts. (Reproduced with permission from ref. *58*.)

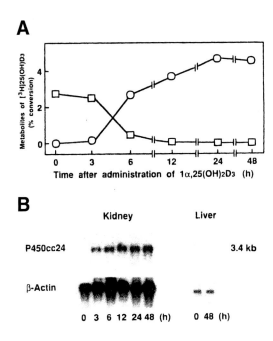

Fig. 4. Time course of changes in the in vivo metabolism of $[^3H]25(OH)D_3$ and in renal 24-hydroxylase mRNA expression after a single injection of $1,25(OH)_2D_3$. (**A**) In vivo conversion of $[^3H]25(OH)D^3$ into $[^3H]1,25(OH)_2D_3$ (□) and $[^3H]24,25(OH)_2D_3$ (○). (**B**) mRNA expression of 24-hydroxylase (CYP24) and β-actin in the kidney.

a VDRE between nt −262 and nt −238 in the 5' flanking region. The DNA sequence of the identified VDRE was 5'-GGTTCAgcg GGTGCG-3', residing on the antisense strand between nt −259 and nt −245.

On the other hand, Ohyama et al. *(13)* identified a VDRE between nt −167 and nt −102 by unidirectional deletion analysis of the 5'-flanking region as well as gel retardation study. By closer analysis they identified a direct repeat motif, 5'-AGGTGAgtgAGGGCG-3', between nt −151 and nt −137 on the antisense strand as a VDRE. Both VDREs resemble those reported for vitamin D-inducible proteins such as osteocalcin, calbindin, and osteopontin (Fig. 5). The promoter regions of the genes for these proteins contain specific DNA sequence elements that bind the VDR, a known member of the superfamily of steroid/thyroid/retinoid receptors that function as ligand-dependent transcription factors.

Zierold et al. *(51)* recently reported that the two VDREs found between nt −154 to nt −125 and nt −262 to nt −238 (a total of five half-sites) apparently account for most if not all the transcription activation of the rat 24-hydroxylase by $1,25(OH)_2D_3$, that are the most powerful of the VDREs reported to date.

5.4. Miscellaneous Factors

Chen et al. *(52)* observed the induction of 24-hydroxylase in cultured human keratinocytes. Intracellular concentration of $1,25(OH)_2D_3$ is regulated by 24-hydroxylase, as 24-

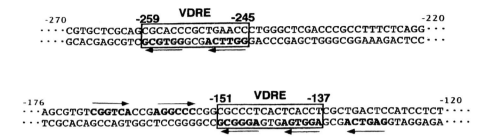

Fig. 5. Structures of vitamin D-responsive elements (VDREs) and their locations in the rat 25(OH)D$_3$ 24-hydroxylase gene.

hydroxylation is the first step of catabolism of 1,25(OH)$_2$D$_3$. Indeed, 1,25(OH)$_2$D$_3$ caused dose- and time-dependent increase in CYP24 mRNA. They surmised that 1,25(OH)$_2$D$_3$ upregulates CYP24 mRNA as an important first step in the initiation of catabolism of 1,25(OH)$_2$D$_3$.

Roy et al. *(53)* studied the effect of phosphate deprivation of Hyp mice that exhibit rachitic bone disease, hypophosphatamia, impaired renal phosphate reabsorption, and abnormal regulation of renal 1,25(OH)$_2$D$_3$ metabolism. They found that phosphate deprivation of Hyp mice resulted in a threefold increase in the maximum velocity of 24-hydroxylase activity, in the amount of 24-hydroxylase-immunoreactive protein, and in CYP24 mRNA in kidney. Hyp mice also exhibited an appropriate increase in CYP24 mRNA and catalytic activity in response to increasing doses of 1,25(OH)$_2$D$_3$.

Induction of 24-hydroxylase was also observed in bone. Nishimura et al. *(54)* have shown that both activity and mRNA level of 24-hydroxylase increased when osteoblast cells were incubated with 1,25(OH)$_2$D$_3$.

6. MECHANISM OF CALCIUM HOMEOSTASIS

Because 1α-hydroxylase, the key enzyme in the maintenance of calcium homeostasis, has not been purified and its cDNA so far not isolated, we cannot understand the mechanism of calcium homeostasis perfectly. Recent purification of 24-hydroxylase, isolation of its cDNA, and studies of the promoter region of the 24-hydroxylase gene provide us some clues.

It is now known how expression of 24-hydroxylase is controlled by serum 1,25(OH)$_2$D$_3$ status, as shown in the previous section. In the promoter region of 24-hydroxylase gene two VDREs exist that control expression of 24-hydroxylase depending on VDR, which binds 1,25(OH)$_2$D$_3$. Iida et al. *(55)* performed Northern blot analysis of mRNAs of VDR and CYP24 along the microdissected rat nephron segments and found that CYP24 mRNA was mainly expressed in proximal convoluted tubules, where the 24-hydroxylase activity was observed. By contrast, VDR mRNA was expressed ubiquitously along the nephron segments. They found that expression of VDRs in proximal convoluted tubular cells is downregulated when renal production of 1,25(OH)$_2$D$_3$ is stimulated in a low calcium status, suppressing the expression of 24-hydroxylase at proximal convoluted tubular cells *(56)*. PTH seems to be involved in this control.

It is well known that PTH is released from the parathyroid gland when serum calcium levels decrease. This hormone is considered to downregulate 24-hydroxylase, as described before *(45)*. By contrast, when serum calcium increases, the serum level of PTH decreases, which seems to allow enhancement of 24-hydroxylase expression by $1,25(OH)_2D_3$.

The situation may be reversed with 1α-hydroxylase. When serum calcium level increases, serum PTH decreases, leading to decrease in expression of 1α-hydroxylase. Inversely, when serum calcium decreases, serum PTH increases, enhancing 1α-hydroxylase expression. To understand calcium homeostasis fully on the molecular level, analysis of the promoter region of 1α-hydroxylase is a must.

7. CONCLUSIONS

P450s are a group of proteins belonging to a superfamily called the P450 superfamily. They were originally found in rat liver microsomes as a pigment with a peculiar carbon monoxide-binding spectrum. Subsequently, Omura and Sato *(57)* determined that they are a b-type cytochrome group, showing a prominent peak at 450 nm in the CO-difference spectrum, and named them cytochrome P-450. Now they are renamed P450 and classified according to the amino acid sequence by the Nebert's group *(10)*. Extensive studies carried out in the following three decades established that P450 catalyzes monooxygenation of xenobiotics as well as endogenous substrates, such as steroids, and some P450s are inducible by xenobiotics and endogenous substrates. Recent studies of the mechanism of regulation of P450 at the molecular level clearly demonstrated that expression is controlled at the promoter region by the trans-active factor(s). The structure of some effector, as well as that of DNA responsive to the effectors, was elucidated. Now two of three hydroxylases that are involved in vitamin D metabolism have been purified to homogeneity and their cDNA clones isolated. These studies definitely established that the two hydroxylases are the P450 enzyme. Analysis of the promoter region of the gene of one of these enzymes (24-hydroxylase) demonstrated that its expression is indeed regulated at the promoter region through VDREs. It is well known that 1α-hydroxylase is under the strict control of calcium and vitamin D status. It has also been established that 1α-hydroxylase is a P450 enzyme. The next problem is to purify the enzyme to homogeneity, isolate its cDNA, and elucidate at the molecular level the mechanism by which the expression of 1α-hydroxylase is controlled by calcium and/or vitamin D status.

REFERENCES

1. Omdahl JL, DeLuca HF. Regulation of vitamin D metabolism and function. Physiol Rev 1973; 53: 327–372.
2. Holick MF, MacLaughlin JA, Clark BM, Holick SA, Potts JT Jr. Photosynthesis of previtamin D in human skin and the physiologic consequences. Science 1980; 210:203–205.
3. Kodicek E. The story of vitamin D. Lancet 1974; 325–329.
4. Holick MF, Schnoes HK, DeLuca HF, Gray RW, Boyle IT, Suda T. Isolation and identification of 24,25-dihydroxycholecalciferol, a metabolite of vitamin D_3 made in kidney. Biochemistry 1972; 11: 4251–4255.
5. Omdahl JL, Gray RW, Boyle IT, Knutson J, DeLuca HF. Regulation of metabolism of 25-hydroxycholecalciferol by kidney tissue in vitro by dietary calcium. Nature New Biol 1972; 237:63,64.
6. Masumoto O, Ohyama Y, Okuda K-I. Purification and characterization of a cytochrome P450 from female rat liver mitochondria. J Biol Chem 1988; 263:14,256–14,260.

7. Ohyama Y, Okuda K-I. Isolation and characterization of a cytochrome P450 from rat liver kidney mitochondria that catalyzes the 24-hydroxylation of 25-hydroxyvitamin D_3. J Biol Chem 1991; 266: 8690–8695.
8. Usui E, Noshiro M, Okuda K-I. Molecular cloning of cDNA for vitamin D_3 25-hydroxylase from rat liver mitochondria. FEBS Lett 1990; 262:135–138.
9. Ohyama Y, Noshiro M, Okuda K-I. Cloning and expression of cDNA encoding 25-hydroxyvitamin D_3 24-hydroxylase. FEBS Lett 1991; 278:195–198.
10. Nelson DR, Kamataki T, Waxman DJ, Guengerich FP, Estabrook RW, Feyereisen R, Gonzalez FJ, Coon MJ, Gunsalus IC, Gotoh O, Okuda K-I, Nebert DW. The P450 superfamily. DNA Cell Biol 1993; 12:1–51.
11. Omura T, Ishimura Y, Fujii-Kuriyama Y, eds. Cytochrome P450, 2nd ed. Weinheim:VCH, 1993; 1–292.
12. Schenkman JB, Greim H. eds. Handbook of Experimental Pharmacology (Cytochrome P450), vol 105. Berlin: Springer-Verlag, 1993; 1–739.
13. Ohyama Y, Ozono K, Uchida M, Shinki T, Kato S, Suda T, Yamamoto O, Noshiro M, Kato Y. Identification of a vitamin D-responsive element in the 5'-flanking region of the rat 25-hydroxyvitamin D_3 24-hydroxylase gene. J Biol Chem 1994; 269:10,545–10,550.
14. Zierold C, Darwish HM, DeLuca HF. Identification of a vitamin D-responsive element in the rat calcidiol (25-hydroxyvitamin D_3) 24-hydroxylase gene. Proc Natl Acad Sci USA 1994; 91:900–902.
15. Baker AR, McDonnel D P, Hughes M, Crisp TM, Mangelsdorf DJ, Haussler MR, Pike JW, Shine J, O'Malley BW. Cloning and expression of full-length cDNA encoding vitamin D receptor. Proc Natl Acad Sci USA 1988; 85:3294–3298.
16. Hayashi S, Noshiro M, Okuda K-I. Purification of cytochrome P-450 catalyzing 25-hydroxylation of vitamin D_3 from rat liver microsomes. Biochem Biophys Res Commun 1984; 121:994–1000.
17. Hayashi S, Noshiro M, Okuda K-I. Isolation of a cytochrome P-450 that catalyzes the 25-hydroxylation of vitamin D_3 from rat liver microsomes. J Biochem (Tokyo) 1986; 99:1753–1763.
18. Hayashi S, Usui E, Okuda K-I. Sex-related differences in vitamin D_3 25-hydroxylase of rat liver microsomes. J Biochem (Tokyo) 1988; 103:863–866.
19. Andersson S, Davis D, Dahlbäck H, Jörnval H, Russell D. Cloning, structure and expression of the mitochondrial cytochrome P-450 sterol 27-hydroxylase, a bile acid biosynthetic enzyme. J Biol Chem 1989; 264:8222–8229.
20. Okuda K-I, Usui E, Ohyama,Y. Recent progress in enzymology and molecular biology of enzymes involved in vitamin D metabolism. J Lipid Res 1995; 36:1641–1652.
21. Okuda K-I, Masumoto O, OhyamaY. Purification and characterization of 5β-cholestane-3α,7α,12α-triol 27-hydroxylase from female rat liver mitochondria. J Biol Chem 1988; 263:18,138–18,142.
22. Ohyama Y, Masumoto O, Usui E, Okuda K-I. Multi-functional property of rat liver mitochondrial cytochrome P450. J Biochem (Tokyo) 1991; 109:389–393.
23. Usui E, Noshiro M, Ohyama Y, Okuda K-I. Unique property of liver mitochondrial P450 to catalyze the two physiologically important reactions involved in both cholesterol catabolism and vitamin D activation. FEBS Lett 1990; 274:175–177.
24. Akiyoshi-Shibata M, Usui E, Sakaki T, Yabusaki Y, Noshiro M, Okuda K-I, and Ohkawa H. Expression of rat liver vitamin D_3 25-hydroxylase cDNA in *Saccharomyces cerevisiae*. FEBS Lett 1991; 280: 367–370.
25. Guo Y-D, Strugnell S, Back DW, Jones G. Transfected human liver cytochrome P-450 hydroxylates vitamin D_3 analogs at different side-chain positions. Proc Natl Acad Sci. USA 1993; 90:8668–8672.
26. Setoguchi T, Salen,G, Tint GS, Mosbach EH. A biochemical abnormality in cerebrotendinous xanthomatosis. J Clin Invest 1974; 53:1393–1401.
27. Cali JJ, Hsieh, C-L, Francke V, Russell DW. Mutation in the bile acid biosynthetic enzyme sterol 27-hydroxylase underlie cerebrotendinous xanthomatosis. J Biol Chem 1991; 266:7779–7783.
28. Leitersdorf E, Reshef A, Meiner V, Levitzki R, Schwartz SP, Dann E, Berkman N, Cali JJ, Kapholz L, Berginer VM. Frame-shift and splice-junction mutation in the sterol-27-hydroxylase gene cause cerebrotendinous xanthomatosis in Jews of Moroccan origin. J Clin Invest 1993; 91:2488–2496.
29. Berginer VM, Shany S, Alkalay D, Berginer J, Dekel S, Salen G, Gazit D. Osteoporosis and increased bone fractures in cerebrotendinous xanthomatosis. Metab Clin Exp 1992; 42:69–74.
30. Skerede S, Börkhem I, Kvittigen EA, Buchmann MS, Lie SO, East C, Grundy S. Determination of 26-hydroxylation of C27-steroids in human skin fibroblasts and a deficiency of this activity in cerebrotendinous xanthomatosis. J Clin Invest 1986; 78:729–735.

31. Tucker G III, Gagnon RE, Haussler MR. Vitamin D_3-25-hydroxylase. Tissue occurrence and apparent lack of regulation. Arch Biochem Biophys 1973; 155:47–57.
32. Ponchon G, Kennan AL, DeLuca HF. Activation of vitamin D by the liver. J Clin Invest 1969; 48: 2032–2037.
33. Saarem K, Pedersen JI. Sex difference in the hydroxylation of cholecalciferol and of 5β-cholestane-3α,7α,12α-triol. Biochem J 1987; 247:73–78.
34. Axén E, Postlind H, Sjöberg H, Wikvall K. Liver mitochondrial cytochrome P450 CYP27 and recombinant-expressed human CYP27 catalyzes 1α-hydroxylation of 25-hydroxyvitamin D_3. Proc Natl Acad Sci USA 1994; 91:10,014–10,018.
35. Henry HL, Norman AW. Renal 25-hydroxyvitamin D_3-1α-hydroxylase. Involvement of cytochrome P450 and other properties. J Biol Chem 1974; 249:7529–7535.
36. Takeuchi A, Okano T, Kobayashi T. The existence of 25-hydroxyvitamin D_3-1α-hydroxylase in the liver of carp and bastard halibut. Life Sci 1991; 48:275–282.
37. Takeuchi A, Okano T, Sekino H, Kobayashi T. The enzymatic formation of 1α,25-dihydroxyvitamin D_3 from 25-hydroxyvitamin D_3 in the liver of fetal rats. Comp Biochem Physiol 1994; 109C:1–7.
38. Hollis BW. 25-Hydroxyvitamin D_3-1α-hydroxylase in porcine hepatic tissue. Proc Natl Acad Sci USA 1990; 87:6009–6013.
39. Axén E. Purification from pig kidney of a microsomal cytochrome P450 catalyzing 1α-hydroxylation of 25-hydroxyvitamin D_3. FEBS Lett 1995; 375:277–279.
40. Suda T, DeLuca HF, Schnoes HK, Ponchon G, Tanaka Y. Dihydroxycholecalciferol, a metabolite of vitamin D_3 preferentially active on bone. Biochemistry 1970; 9:2917–2922.
41. Lam HY, Schnoes HK, DeLuca HF, Chen TC. 24,25-Dihydroxyvitamin D_3: synthesis and biological activity. Biochemistry 1973; 12:4851–4855.
42. Tanaka Y, DeLuca HF, Ikekewa N, Monsaki M, Koizumi N. Determination of stereo-chemical configuration of the 24-hydroxyl group of 24,25-dihydroxyvitamin D_3 and its biological importance. Arch Biochem Biophys 1976; 170:620–626.
43. Boyle IT, Gray RW, DeLuca HF. Regulation by calcium of in vivo synthesis of 1,25-dihydroxycholecalciferol and 21,25-dihydroxycholecalciferol. Proc Natl Acad Sci USA 1971; 68:2131–2134.
44. Nakamura T, Suzuki K, Hirai T, Kurokawa T, Orimo H. Increased bone volume and reduced bone turnover in vitamin D-replete rabbits by the administration of 24,25-dihydroxyvitamin D_3. Calcif Tissue Int 1992; 50:221–227.
45. Shinki T, Jin CH, Nishimura, Nagai Y, Ohyama Y, Noshiro M, Okuda K-I, Suda T. Parathyroid hormone inhibits 25-hydroxyvitamin D_3-24-hydroxylase mRNA expression stimulated by 1α,25-dihydroxyvitamin D_3 in rats kidney but not in intestine. J Biol Chem 1992; 267:13,757–13,762.
46. Armbrecht HJ, Boltz MA. Expression of 25-hydroxyvitamin D 24-hydroxylase cytochrome P450 in kidney and intestine. FEBS Lett 1991; 292:17–20.
47. Chen K-S, Prahl JM, DeLuca HF. Isolation and expression of human 1,25-dihydroxyvitamin D_3 24-hydroxylase cDNA. Proc Natl Acad Sci USA 1993; 90:4543–4547.
48. Akiyoshi-Shibata M, Sakaki T, Ohyama Y, Noshiro M, Okuda K-I, Yabusaki, Y. Further oxidation of hydroxycholecalciferol by calcidiol 24-hydroxylase. Eur J Biochem 1994; 224:335–343.
49. Ohyama Y, Noshiro M, Eggertsen G, Gotoh O, Kato Y, Björkhem I, Okuda K-I. Structural characterization of the gene encoding rat 25-hydroxyvitamin D_3 24-hydroxylase. Biochemistry 1993; 32:76–82.
50. Iwata K, Yamamoto A, Satoh S, Ohyama Y, Tashiro Y, Setoguchi T. Quantitative immunoelectron microscopic analysis of the localization and induction of 25-hydroxyvitamin D_3 24-hydroxylase in rat kidney. J Histochem Cytochem 1995; 43:255–262.
51. Zierold C, Darwish H, DeLuca HF. Two vitamin D response elements function in the rat 1,25-dihydroxyvitamin D_3 24-hydroxylase promoter. J Biol Chem 1995; 270:1675–1678.
52. Chen ML, Heinrich G, Ohyama Y, Okuda K-I, Omdahl JL, Chen TC, Holick MF. Expression of 25-hydroxyvitamin D_3 24-hydroxylase mRNA in cultured human keratinocytes. Proc Soc Exp Biol Med 1994; 207:57–61.
53. Roy S, Martel J, Ma S, Tennenhouse HS. Increased renal 25-hydroxyvitamin D_3-24-hydroxylase messenger ribonucleic acid and immunoreactive protein in phosphate-deprived HYP mice. Endocrinology 1994; 134:1761–1767.
54. Nishimura A, Shinki T, Cheng HJ, Ohyama Y, Noshiro M, Okuda K-I, Suda T. Regulation of messenger ribonucleic acid expression of 1α,25-dihydroxyvitamin D_3-24-hydroxylase in rat osteoblast. Endocrinology 1994; 134:1794–1799.

55. Iida K, Taniguchi S, Kurokawa K. Distribution of 1,25-dihydroxyvitamin D_3 receptor and 25-hydroxyvitamin D_3-24-hydroxylase mRNA expression along rat nephron segments. Biochem Biophys Res Commun 1993; 194:659–664.
56. Iida K, Shinki T, Yamaguchi A, DeLuca HF, Kurokawa K, Suda T. A possible role of vitamin D receptors in regulating vitamin D activation in the kidney. Proc Natl Acad Sci USA 1995; 92: 6112–6116.
57. Omura T, Sato R. The carbon monoxide-binding pigment of liver microsomes. J Biol Chem 1964; 239:2370–2378.
58. Seldin DW, Giebisch G. The Kidney; Physiology and Pathology, 2nd ed. New York: Raven Press, 1992; 715.

6 The Vitamin D Binding Protein and Its Clinical Significance

John G. Haddad

1. INTRODUCTION

The plasma protein that transports vitamin D and its metabolites [vitamin D binding protein (DBP)] was first recognized as a postalbumin component of human sera during electrophoretic analyses *(1)*. Subsequently, the identity of this group-specific component (Gc-globulin) and plasma DBP was discovered *(2)* and confirmed *(3)*. The protein is synthesized in the liver and displays features homologous with albumin and α-fetoprotein *(4)*. Neither of the latter proteins binds vitamin D sterols with high affinity, however. DBP is 122 amino acids shorter than albumin and α-fetoprotein, and the genes of all three of these proteins map to human chromosome 4q11-22 *(4)*.

An inter-α globulin of 458 amino acids, DBP is a single chain polypeptide containing abundant Asp and Gln residues and a high number of Cys residues. As with albumin, there are few glycosyl residues. Three common phenotypes are observed by electrophoretic or isoelectric focusing techniques *(5)*, reflecting differences in exon 11 whereby residues 416 and 420 contain either Asp or Gln and Thr or Lys, respectively. To date, no functional differences among these common phenotypes are consistently recognized in the protein's D-sterol or G-actin binding.

2. STEROL TRANSPORT

2.1. Vitamin D Metabolites

All the major vitamin D sterols in plasma are bound avidly by DBP, but highest affinity (K_d 10^{-8} M) is shown toward 25-hydroxyvitamin D and the 24, 25, and 25,26 dihydroxy metabolites *(3,4,6)*. Vitamin D and 1,25-dihydroxyvitamin D are bound at somewhat lower affinity (K_d 10^{-7} M). All the sterol ligands are bound mole per mole by DBP. Secondary plasma carriers of vitamin D include albumin and lipoproteins, and albumin appears to be the major secondary carrier for 25(OH)D and 1,25(OH)$_2$D *(3,6)*.

2.2. DBP Occupancy by Sterols and Concentrations in Sera

Human serum concentrations of DBP are usually in the 4–8×10^{-6} M range (approx 250–500 µg/mL). When considered in relation to the usual plasma concentrations of

From: *Vitamin D: Physiology, Molecular Biology, and Clinical Applications*
Edited by: M. F. Holick © Humana Press Inc., Totowa, NJ

Table 1
Features of Human Vitamin D Binding Protein (DBP)

Feature	Measure
Isoelectric point	4.7
Molecular weight	58 kDa
Plasma concentration	4–8 µM
Plasma half-life	2.5–3 d
D-sterol binding sites	mol/mol
Affinity for sterol ligands (K_A)	
25(OH)D	$5 \times 10^8 \, M^{-1}$
1,25(OH)$_2$D	$4 \times 10^7 \, M^{-1}$
25(OH)D binding capacity	5 µM (2 mg/L)
Normal % of DBP occupied by vitamin D sterols	2
Actin binding sites	mol/mol
Affinity for G-actin (K_A)	$2 \times 10^9 \, M^{-1}$
G-actin binding capacity	5 µM (225 mg/L)

vitamin D sterols, there is normally a remarkable excess of DBP unoccupied by D-sterols (approx 95%). This has led observers to speculate that DBP constitutes a circulating reservoir for D-sterols, thereby contributing to the body's economic handling of these sterols (3). Some of the features of DBP are shown in Table 1. A variety of immunoassay techniques are available that can easily quantitate human serum DBP concentrations (6).

DBP concentrations in serum are remarkably stable in healthy persons (6). Lower concentrations can be seen in patients with severe (often end-stage) liver diseases (poor DBP production) and in patients with marked proteinuria (excessive losses) (6). Higher titers of serum DBP are seen during pregnancy and in subjects receiving oral estrogens (6,7). These increases are similar to those seen with other, hepatically synthesized plasma proteins and are thought to be due to an effect of high estrogen concentrations reaching the liver via the portal vein. Transcutaneous or skin patch delivery of estrogens does not usually result in high DBP titers. At approximately 1 month following pregnancy or cessation of oral estrogen use, serum DBP levels return to normal range.

The plasma half-life of DBP in humans is 2.5–3.0 d, and the daily production rate of DBP is approximately 10 mg/kg body weight (8). The plasma clearance of DBP is apparently not affected by D-sterol occupancy (9). Normal DBP levels in serum are seen in vitamin D-deficient, -sufficient, and -excess states (6). The protein appears to be catabolized extensively since no intermediate plasma forms are seen, and only small DBP peptides are found in urine (10).

2.3. "Free Sterol" Hypothesis

The free hormone hypothesis states that the biologic activity of a given hormone is effected by its unbound (free) rather than protein-bound concentration in the plasma (11). The vitamin D ligands are poorly soluble in aqueous media, and their binding to DBP promotes their solubility and transport. The proportion of total vitamin D sterol content in plasma that is unbound or free is dependent on the total DBP and total sterol concentrations, as well as their binding affinity. Estimates of free sterol content were made by

Table 2
D-Sterol Binding in Plasma

	25(OH)D	1,25(OH)$_2$D
Normal plasma concentration	$10^7 M^{-1}$ (40 ng/mL)	$10^{10} M^{-1}$ (40 pg/mL)
Fraction of total sterol on vitamin D binding protein (%)	88	85
Fraction on albumin (%)	12	15
Percent of total sterol in unbound (free) form	0.04	0.4
DBP binding competition by		
Saturated fatty acids	no	no
Cholesterol	no	no
Mono- and polyunsaturated fatty acids	yes	yes

applying the law of mass action *(6)*. Subsequently, a centrifugofiltration technique permitted direct analyses of free sterol content in given serum samples *(7,12–14)*. Some of these findings are shown in Table 2.

For practical purposes, serum DBP concentrations in the normal range do not require that a separate assessment be made of vitamin D sterol concentrations. However, low DBP concentrations permit a higher percentage of total sterol to be in a free or cell-accessible form *(14)*. Conversely, high DBP concentrations are associated with a lower percentage of total sterol in the free form *(7)*. Direct analyses of the free forms of 25(OH)D and 1,25(OH)$_2$D in human sera containing low DBP (liver disease) or high DBP (pregnancy or oral estrogen therapy) indicated that the physicochemical equilibrium between DBP and sterol ligands was altered as described above *(7,12–14)*. Selective removal of DBP from human serum also permitted an analysis of the contribution of the weaker sterol binding by albumin *(13)*. Also, high serum concentrations of the more avidly bound ligand, 25(OH)D, results in a higher percentage of the less avidly bound ligand, 1,25(OH)$_2$D in the free form, as described in vitamin D intoxication *(15)*.

2.4. Secondary Sterol Carriers

In addition to DBP, albumin and chylomicrons are recognized to be associated with vitamin D sterols *(3,6)*. Under conditions of oral vitamin D administration, the sterol is transported in the chyle in association with chylomicrons. On entering the subclavian vein blood, chylomicron remnants in association with the sterol are probably taken up by the liver via hepatic receptors *(16)*. Following the physiologic production of vitamin D$_3$ in human skin during ultraviolet light exposure, however, the ingress of vitamin D to blood from skin appears to be mostly on the DBP carrier *(17)*.

Measurements of plasma 25(OH)D following orally administered vitamin D revealed a faster increase of plasma 25(OH)D that was of shorter duration than that seen after parenterally administered (depot injections in oil vehicle) vitamin D *(18)*. The greater association of vitamin D with DBP after cutaneous production may provide for a longer retention of vitamin D in plasma and slower hepatic ingress of the sterol, thereby facilitating a more economic bioavailability of the parent vitamin for 25-hydroxylation *(3,19,20)*, rather than for esterification and removal in urine.

3. PLASMA ACTIN SCAVENGER SYSTEM

3.1. Actin Binding by DBP

DBP is recognized to associate avidly with the important cytoskeletal protein globular actin *(21,22)*. This curious association between a plasma protein and an intracellular protein was initially thought to be a nonspecific artifact created by disruption of tissues in the presence of plasma *(3)*. However, the binding affinity of these proteins is quite high, and DBP has been shown in vivo to be the dominant plasma sequestrant of monomeric or G-actin *(23)*. Vitamin D sterol occupancy of DBP does not alter the protein's avidity or capacity to bind G-actin mole per mole *(21)*.

3.2. Teamwork with Gelsolin

Another plasma protein, gelsolin, is recognized to bind both fibrous and globular actin moieties *(14)*. Gelsolin can sever actin filaments into oligomers. The actions of actin filament severing by gelsolin and monomeric actin sequestration by DBP coordinately enhance depolymerization of F-actin and prevent polymerization of G-actin. There is experimental evidence that this actin-scavenger system in plasma is saturable *(23)*. A wide variety of experimental and clinical conditions has been associated with cell disruption and actin entry into blood *(22)*. It appears that DBP and gelsolin constitute a plasma actin scavenger system. Overloading this system can lead to intravascular actin filament formation, thromboemboli, and vascular endothelial disruption *(23)*. In vitro analyses of the effects of these two proteins have confirmed their complementary activity *(24,25)*. The severing of actin polymers into oligomers increases the number of sites for the equilibrium between monomeric actin assembly and disassembly. The action of DBP to trap or sequester actin monomers shifts this equilibrium toward disassembly. Actin monomer sequestration can also be accomplished by the intracellular proteins DNAase-I and profilin *(21,26)*. Selective hepatic removal of plasma G-actin (Kuppfer cells) and F-actin (sinusoidal endothelial cells) has been demonstrated *(27)*.

3.3. Clinical Studies

The presence of actin–DBP and gelsolin–DBP complexes has been reported in sera from patients with conditions involving tissue injury and/or inflammation *(22)*. During severe hepatic necrosis and lung injury, actin association with plasma DBP and gelsolin has been seen *(28,29)*. Also, the actin-induced activation of platelet aggregation can be inhibited by plasma DBP and gelsolin *(24)*. At present, saturation of the actin-scavenger system during illness has not been demonstrated. DBP has been identified in lower respiratory tract secretions, presumably functioning as a cochemotaxin *(30)*.

4. OTHER FUNCTIONS

4.1. Cell-Associated DBP

DBP association with a variety of cell types, including lymphocytes, macrophages, trophoblasts and neurophils, has been reported *(4,31)*. Studies have indicated that the cell surface DBP was acquired from an extracellular source *(32)*, but the nature of the binding of DBP to the cells was not well understood. Recently, DBP binding to two binding sites (K_d 10^{-7} and 10^{-6} M) on normal and malignant B lymphocytes has been reported *(33)*.

A variety of biologic actions have been proposed for cell-associated DBP, including macrophage activation, modulation of chemotaxis, and surface inmunoglobulin disposition *(4,34–36)*. It now appears that modified (glycosidase-treated) DBP can bind to cells and enhance their sensitivity to C5a-activated chemotaxis. Earlier reports of a binding association between DBP and C5a have not clearly been confirmed *(34)*.

4.2. Osteoclastogenesis

In the op mutation of an osteopetrotic rat, a deficiency of B-lymphocyte β-galactosidase is recognized. These rats have deficient and dysmorphic osteoclast populations. In the ia (incisor absent) mutation, increased numbers of dysmorphic osteoclasts are found. Both osteopetrotic mutants have cellular and immune system deficits, including deficient superoxide production by leukocytes and defective macrophage oxidative metabolism *(37)*. Since a common lineage for macrophages and osteoclasts is recognized, Schneider and colleagues *(38)* treated both mutants with sialidase and β-galactosidase-treated DBP and found enhanced osteoclast function and enhanced osteoclast superoxide production. These and other *(37)* provocative studies point to common pathways among macrophage and osteoclast development, and a possible role for a modified DBP in these processes.

4.3. C5a Cochemotaxis

Considerable evidence points to the ability of DBP to enhance the sensitivity to and action of C5a and C5a-desArg on chemotaxis *(34,39)*. Initially, DBP was thought to bind either to C5a itself *(35)* or to a large chemotactic factor inactivator in serum *(40)*. Recent studies have indicated that an earlier cell association with DBP may suffice for the enhancement of C5a action on neutrophils *(34)*. Further studies have indicated that DBP may be processed by these cells and undergo internalization *(39)*. Such findings may complement those of Esteban et al. *(33)*, who found DBP uptake, endosomal location, and degradation of the protein. Yamamoto and colleagues *(37,41,42)*, have provided evidence that sialidase and β-galactosidase treatments of DBP will modify DBP into a macrophage activating factor *(37,41,42)*, effects possibly mediated by T- and B-lymphocytes, respectively.

5. SUMMARY

dBP is a multifunctional plasma protein of the albumin and α-fetoprotein family that possesses specific differences in structure and function. The protein provides a high-affinity, high-capacity binding reservoir for vitamin D sterols, transporting endogenously synthesized vitamin D_3, and regulating cellular access of vitamin D and its metabolites. At its carboxy terminus, DBP avidly binds G-actin and coordinates its actin monomer sequestration action with the actin filament-severing action of gelsolin to constitute the plasma scavenger system for actin. The DBP protein, possibly in modified form, can be found on cell surfaces, where its presence is linked to enhanced macrophage and neutrophil functions, C5a cochemotaxin activity, and possibly a key role in osteoclast development. It now seems clear that DBP, as with other plasma steroid binding proteins *(43)*, functions in several sophisticated ways in addition to its recognized regulation of the transport and cellular access of vitamin D sterols.

REFERENCES

1. Weitkamp LR, Rucknagel DL, Gershowitz H. Genetic linkage between structural loci for albumin and group specific component (Gc). Am J Hum Genet 1966; 18:559–571.
2. Daiger SP, Schanfield MS, Cavalli-Sforza LL. Human group-specific component (Gc) proteins bind vitamin D and 25-hydroxy vitamin D. Proc Natl Acad Sci USA 1975; 72:2076–2080.
3. Haddad JG. Traffic, binding and cellular access of vitamin D sterols. In: Bone and Mineral Research, vol. 5. Peck WA, ed. New York: Elsevier, 1987; 281–308.
4. Cooke NE, Haddad JG. Vitamin D binding protein (Gc-globulin). Endocr Rev 1989; 10:294–307.
5. Braun A, Bichlmaier R, Cleve H. Molecular analysis of the gene for the human vitamin D-binding protein (Gc): allelic differences of the common genetic Gc types. Hum Genet 1992; 89:401–406.
6. Haddad JG. Clinical aspects of measurements of plasma vitamin D sterols and the vitamin D binding protein. In: Disorders of Bone and Mineral Metabolism. Coe FL, Favus MJ, eds. New York: Raven, 1992; 195–216.
7. Bikle, DD, Gee E, Halloran B, Haddad JG. Free 1, 25 $(OH)_2D$ levels in serum from normal subjects, pregnant subjects and subjects with liver disease. J Clin Invest 1984; 74:1966–1971.
8. Kawakami M, Blum CB, Ranakrishman R, Dell RB, Goodman DS. Turnover of the plasma binding protein for vitamin D and its metabolites in normal human subjects. J Clin Endocrinol Metab 1981; 53:1110–1116.
9. Haddad JG, Fraser DR, Lawson DEM. Vitamin D binding protein: turnover and fate in the rabbit. J Clin Invest 1981; 67:1550–1560.
10. Harper KD, McLeod JF, Kowalski MA, Haddad JG. Vitamin D binding protein sequesters monomeric actin in the circulation of the rat. J Clin Invest 1987; 79:1365–1370.
11. Mendel CM. The free hormone hypothesis: a physiologically based mathematical model. Endocr Rev 1989; 10:232–274.
12. Bikle DD, Siiteri BK, Ryzen E, Haddad JG. Serum protein binding of $1,25-OH)_2D$: a re-evaluation by direct measurement of free metabolite levels. J Clin Endocr Metab 1985; 61:969–975.
13. Bikle DD, Gee E, Halloran BP, Kowalski MA, Ryzen E, Haddad JG. Assessment of the free fraction of 25-OHD in serum and its regulation by albumin and the vitamin D binding protein. J Clin Endocrinol Metab 1986; 63:954–959.
14. Bikle DD, Halloran BP, Ryzen E, Kowalski MA, Haddad JG. Free 25-OHD levels are normal in subjects with liver disease and reduced total 25-OHD levels. J Clin Invest 1986; 78:748–752.
15. Pettifor JM, Bikle DD, Cavaleros M, Zachen D, Kamdar MC, Ross FP. Serum levels of free $1,25(OH)_2D$ in vitamin D toxicity. Ann Intern Med 1995; 122:511–513.
16. Dueland S, Pedersen JI, Helgerud P, Drevon CA. Absorption, distribution and transport of vitamin D in the rat. Am J Physiol 1983; 245:E463–469.
17. Haddad JG, Matsuoka LY, Hollis BW, Hu YZ, Wortsman J. Plasma transport of endogenously synthesized cholecalciferol. J Clin Invest 1993; 91:2551–2555.
18. Whyte M, Haddad JG, Waters DD, Stamp TCB. Vitamin D bioavailability: serum 25-OHD levels in man after oral, subcutaneous, intramuscular and intravenous vitamin D administration. J Clin Endocrinol Metab 1979; 48:906–911.
19. Stanbury SW, Mawer EB. Vitamin D metabolism in man: contributions from clinical studies. In: Clinical Disorders of Bone and Mineral Metabolism. Frame B, Potts JT, eds. Amsterdam: Excerpta Medica, 1983; 72–76.
20. Haddad JG, Stamp TCB. Circulating 25-OHD in man. Am J Med 1974; 57:57–62.
21. McLeod J, Kowalski MA, Haddad JG. Interactions among serum vitamin D binding protein, monomeric actin, profilin and profilactin. J Biol Chem 1989; 264:1260–1267.
22. Lee WM, Galbraith RM. The extracellular actin-scavenger system and actin toxicity. N Engl J Med 1992; 326:1335–1341.
23. Haddad JG, Harper KD, Guoth M, Pietra GG, Sanger JW. Angiopathic consequences of saturating the plasma scavenger system for actin. Proc Natl Acad Sci USA 1990; 87:1381–1385.
24. Vasconcellos CA, Lind SE. Coordinated inhibitors of actin-induced platelet aggregation by plasma gelsolin and DBP. Blood 1993; 12:3648–3657.
25. Lees A, Haddad JG, Lin S. Brevin and DBP comparison of the effects of two serum proteins on actin assembly and disassembly. Biochemistry 1984; 23:3038–3047.
26. Korn ED. Actin polymerization and its regulation by proteins from nonmuscle cells. Physiol Rev 1982; 62:672–737.

27. Hermannsdoerfer AJ, Heeb GT, Fenstel PJ, Estes JE, Keenan CJ, Minnear FL, Selden L, Giunta C, Flor JR, Blumenstock FA. Vascular clearance and organ uptake of G- and F-actin in the rat. Am J Physiol 1993; 265:G1071–1081.
28. Young WO, Goldschmidt-Clermont PJ, Emerson DL, Lee WM, Jollow DJ, Galbraith RM. Correlation between extent of liver damage in fulminant hepatic necrosis and complexing of group-specific component (DBP). J Lab Clin Med 1987; 110:83–90.
29. Lind SE, Smith DB, Janmey PA, Stossel TP. Depression of gelsolin levels and detection of gelsolin-actin complexes in plasma of patients with acute lung injury. Am Rev Respir Dis 1988; 138:429–434.
30. Metcalf JP, Thompson AB, Gossman GL, Nelson KJ, Koyama S, Rennard SI, Robbins RA. Gc-globulin functions as a cochemotoxin in the lower respiratory tract. Am Rev Respir Dis 1991; 143: 844–849,
31. Cooke NE, Haddad JG. Vitamin D-binding protein (Gc-globulin): update 1995. Endocrinol Rev 1995; 4:125–128.
32. Guoth M. Murgia A, Smith R, Prystowsky M, Cooke N, Haddad J. Cell surface vitamin D binding protein (Gc-globulin) is acquired from plasma. Endocrinology 1990; 127:2313–2321.
33. Esteban C, Geuskens M, Ena JM, Mishal Z, Macho A, Torres JM, Uriel J. Receptor-mediated uptake and processing of vitamin D-binding protein in human 13-lymphoid cells. J Biol Chem 1992; 267: 10,177–10,183.
34. Kew RR, Mollison KW, Webster RO. Binding of Gc globulin (DBP) to C5a or C5a des Arg is not necessary for co-chemotactic activity. J Leukoc Biol 1995; 58:55–58.
35. Kew RR, Webster RO. Gc-globulin (DBP) enhances the neutrophil chemotactic activity of C5a and C5a des Arg. J Clin Invest 1988; 82:364–369.
36. Petrini M, Emerson DL, Galbraith RM. Linkage between surface immunoglobulin and cytoskeleton of B-lymphocytes may involve Gc-protein. Nature 1983; 306:73–75.
37. Yamamato N, Lindsay DD, Naraparaju R, Ireland RA, Popoff SN. A defect in the inflammation-primed macrophage-activation cascade in osteopetrotic rats. J Immunol 1994; 152:100–107.
38. Schneider GB, Benis KA, Flay NW, Ireland RA, Popoff SN. Effects of vitamin D binding protein-macrophage activating factor (DBP-MAF) infusion on bone resorption in two osteopetrotic mutations. Bone 1995; 6:657–662.
39. Kew RR, Fisher JA, Webster RO. Co-chemotactic effect of Gc-globulin (DBP) for C5a. transient conversion into an active co-chemotaxin by neutrophils. J Immunol 1995; 155:5369–5374.
40. Robbins RA, Hamel FG. Chemotactic factor inactivator interaction with Gc-globulin (DBP). a mechanism of modulating the chemotactic activity of C5a. J Immunol 1990; 14:2371–2376.
41. Naraparaju VR, Yamamoto N. Roles of β-galactosidase of B lymphocytes and sialidase of T lymphocytes in inflammation-primed activation of macrophages. Immunol Lett 1994; 43:143–148.
42. Yamamoto N, Homma S, Haddad JG, Kowalski MA. Vitamin D binding protein required for in vitro activation of macrophages after alkylglycerol treatment of mouse peritoneal cells. Immunology 1991; 74:420–424.
43. Hammond GL. Potential functions of plasma steroid-binding proteins. Trends Endocrinol Metab 1995; 6:298–304.

7
Molecular Biology of the Vitamin D Receptor

Paul N. MacDonald

1. INTRODUCTION

Vitamin D was discovered as a micronutrient that is essential for normal skeletal development and for maintaining bone integrity. Its importance in bone physiology is most apparent in the deficiency state, in which the lack of vitamin D produces rickets in children and osteomalacia in adults. However, vitamin D is more appropriately classified as a hormone, and it is the vitamin D endocrine system that regulates skeletal homeostasis. In response to hypocalcemia and elevated parathyroid hormone, the kidney synthesizes and releases 1,25-dihydroxyvitamin D_3 [$1,25(OH)_2D_3$], the bioactive, hormonal form of vitamin D. $1,25(OH)_2D_3$ acts on mineral-regulating target tissues such as intestine, bone, kidney, and parathyroid glands to maintain normal calcium and mineral homeostasis. Its predominant role is to enhance the intestinal absorption of dietary calcium and phosphorus. Thus, vitamin D preserves skeletal calcium by ensuring that adequate absorption of dietary calcium takes place. In addition to this calciotropic role, vitamin D functions in a plethora of cellular actions, perhaps the most fundamental of which is cellular differentiation *(1)*. In skeletal tissue, $1,25(OH)_2D_3$ increases osteoclast number *(2)* possibly by inducing the differentiation of preosteoclasts into mature bone-resorbing cells *(3)*. Vitamin D also acts directly on the osteoblast, in which one well-established effect is stimulating the synthesis of several bone matrix proteins including osteocalcin and osteopontin. Thus vitamin D is thought to preserve and maintain the integrity of the bony tissues via an integrated series of diverse effects.

The biologic effects of $1,25(OH)_2D_3$ are mediated through a soluble receptor protein termed the vitamin D receptor (VDR). VDR binds $1,25(OH)_2D_3$ with high affinity and high selectivity. In the target cell, the interaction of the $1,25(OH)_2D_3$ hormone with VDR initiates a complex cascade of molecular events culminating in alterations in the rate of transcription of specific genes or gene networks (Fig. 1). Central to this mechanism is the requisite interaction of VDR with retinoid X receptor (RXR) to form a heterodimeric complex that binds to specific DNA sequence elements [vitamin D response element (VDREs)] in vitamin D-responsive genes and ultimately influences the rate of RNA polymerase II-mediated transcription. Thus an emerging concept in this mechanism is that of the VDR–RXR heterodimer serving as the functional transcriptional enhancer in

From: *Vitamin D: Physiology, Molecular Biology, and Clinical Applications*
Edited by: M. F. Holick © Humana Press Inc., Totowa, NJ

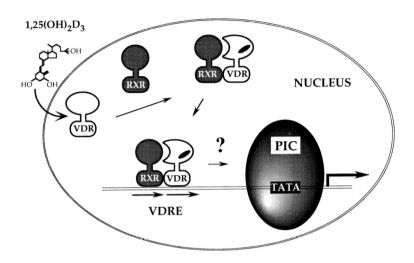

Fig. 1. Molecular mechanism of vitamin D action.

vitamin D-activated transcription. However, much less is known of this process following VDR–RXR interaction with the VDRE. Recent data suggest that protein–protein interactions between the VDR–RXR heterodimer and the transcription machinery are essential for the mechanism of vitamin D-mediated gene expression.

This chapter discusses the molecular biology of the VDR and focuses on various aspects of VDR function with an emphasis on the macromolecular interactions required for the ultimate transcriptional regulatory activity of the VDR. These macromolecular interactions include the association of VDR with the $1,25(OH)_2D_3$ ligand, the mechanisms required for specific, high-affinity interaction of VDR with DNA, the heterodimeric interaction of VDR with RXR, and finally the protein–protein contacts that may comprise the communications links between VDR and the transcription preinitiation complex. These contacts include direct or indirect interaction of VDR with components of the transcription complex including transcription factor IIB and putative coactivator/corepressor proteins. This last aspect is a burgeoning area of research in the nuclear receptor field, and its implications in the transcriptional mechanism of the VDR are profound.

2. CLONING AND CHARACTERIZATION OF VDR DNA SEQUENCES

2.1. Chromosomal Localization of the VDR Gene

The location of the VDR gene within the human genome was determined using Southern blot analysis of DNA from human–Chinese hamster cell hybrids that localized the VDR gene to human chromosome 12 *(4)*. This was later extended to 12q by means of somatic cell hybrid mapping *(5)* and further refined to the 12q13–14 region using *in situ* hybridization and linkage analysis *(6)*. Similar approaches identified the VDR sequence on chromosome 7 of the rat genome *(5)*. In both the human and rat genomes, the VDR gene is located in regions where distantly related DNA binding proteins also map, including transcription factor SP-1 *(5)* and the γ-isoform of the retinoic acid receptor

(RAR-γ) *(7,8)*. This clustering of related DNA binding proteins suggests the possibility that they were derived from some common ancestral gene.

Human chromosome 12q13–14 is also the location of the gene involved in pseudo-vitamin D-deficient rickets (PDDR) *(6)*. This autosomal recessive disorder is caused by impaired activity of the renal 1α-hydroxylase and results in insufficient levels of serum 1,25(OH)$_2$D$_3$. Although the physiologic significance of this link is not clear, it is intriguing that the two most crucial components of the vitamin D endocrine system, namely, VDR and the 1α-hydroxylase (or a component of the enzyme) map close to each other on the same region of human chromosome 12.

2.2. Organization of the VDR Gene

The gene encoding the hVDR has been isolated from a human liver genomic DNA library *(9)*. Although a detailed report on its structural organization and sequence has not been published,* several preliminary reports show that the hVDR gene is contained within approximately 45 kb of genomic DNA. It has a complex structure consisting of nine coding exon sequences interrupted by intronic sequences ranging in size from 0.2 to 13 kb *(9–11)*. The translation initiation codon is located in exon 2, which also encodes sequences for the first zinc finger of the DNA binding domain (*see below*). Exon 3 contains the sequence for the second zinc finger. The observation that the two motifs of the zinc-finger DNA binding domain are encoded by separate exons is a characteristic trait of the steroid receptor superfamily. Most of the C-terminal ligand binding domain is encoded by exons 8 and 9. Little is known of the promoter region of the VDR gene, presumably because of a rather large intron that separates the transcriptional start site from exon 2.

2.3. VDR cDNA Sequence

2.3.1. ISOLATION OF THE VDR cDNAs

The tools that were needed to isolate the VDR cDNA sequence were developed in the mid-1980s. Avian VDR (aVDR) was purified to near homogeneity, monoclonal antibodies directed against aVDR were generated *(12,13)*, and the antibodies were used to isolate the aVDR cDNA from a λgt 11 intestinal cDNA expression library *(14)*. The partial avian cDNA clone was then used to obtain the full-length human VDR cDNA *(15)*. At approximately the same time, the rat VDR cDNA was isolated independently by Burmester et al. *(16)* using similar strategies. More recently, the human and rat cDNAs were used as probes to isolate the bovine, murine, and *Xenopus laevis* VDR cDNAs (Fig. 2).

The full-length human VDR cDNA consists of 4605 bp containing a 115 bp noncoding leader sequence, a 1281-bp open reading frame, and 3209 bp of 3' noncoding sequence. The functional significance of this rather long 3' untranslated region is unknown, but it is a characteristic feature of the steroid receptor superfamily. The size of this cDNA agrees well with the predominant mRNA species observed in human extracts (approximately 4.6 kb).

The authenticity of the VDR cDNA clone was established when it was expressed in mammalian cells and the expressed receptor bound the 1,25(OH)$_2$D$_3$ ligand with the appropriate affinity and specificity *(14,15)*. Moreover, its expression in a cell line that does not contain endogenous VDR (CV-1 cells) was sufficient to confer vitamin D

---

*The structural organization of the human vitamin D receptor chromosomal gene was described recently by Miyamoto et al. Mol Endocrinol 1997; 11:1165–1179.

Fig. 2. Deduced amino acid sequence comparison of vitamin D receptors (VDRs) from several species.

responsiveness to the cells *(17)*. Finally, the amino acid sequence deduced from the rat cDNA was confirmed by immunopurifying native VDR and obtaining amino acid sequence information near the N-terminus and the C-terminus of the pure receptor *(18)*. Consequently, these studies showed that the isolated cDNAs encoded the VDR, and they demonstrated the essential role of VDR in mediating the transcriptional effects of $1,25(OH)_2D_3$.

2.3.2. Characteristics of the VDR cDNA and Sequence Comparisons

Sequence comparison of the VDR cDNAs showed striking similarity with the members of the superfamily of nuclear receptors for steroid and thyroid hormones (first reported in ref. *19*). An area of high sequence relatedness is found in the DNA binding

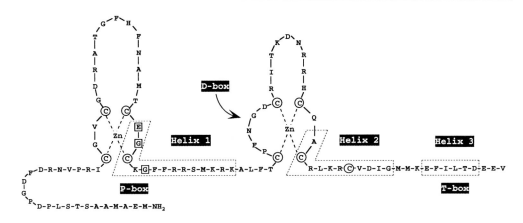

Fig. 3. The N-terminal, zinc-finger DNA binding motif of human vitamin D receptor.

domain (DBD) of these receptors. This 70-amino acid domain is rich in cysteine, lysine, and arginine residues and it is the region of the receptor that is responsible for high-affinity interaction with specific DNA sequence elements. An area of more limited homology is in the large C-terminal domain, which is responsible for high-affinity interaction with the various hormones, termed the ligand binding domain (LBD). Interestingly, distinct regions of high similarity are present in the LBDs of the various receptors despite the obvious structural nonrelatedness of their individual ligands. This suggests that, in addition to hormone interactions, this C-terminal domain functions in other, more related areas of receptor function (e.g., in protein–protein interactions; see below).

The deduced amino acid sequences of the VDR from several species are illustrated in Fig. 2. The human cDNA encodes 427 amino acids with a calculated molecular mass of approx 48 kDa, agreeing well with previous protein data. A comparison of the coding regions of the human, rat, mouse, bovine, and avian VDRs shows a high degree of sequence identity, particularly within the N-terminal portion (amino acids 16–116 of hVDR) that constitutes the DBD (97% identity) and the C-terminal domain (amino acids 226–427 of hVDR) that comprises most of the $1,25(OH)_2D_3$ binding domain (85% identity). The sequences tend to diverge in the hinge region of the receptors, which is the region located between the DBD and the LBD. The high degree of sequence conservation of VDR between species supports the fundamental roles that these domains serve in VDR function.

3. MOLECULAR ANALYSIS OF THE FUNCTIONAL DOMAINS OF THE VDR

3.1. The N-Terminal DNA Binding Domain

3.1.1. LOCALIZATION AND STRUCTURE

The location of the DBD of VDR was originally mapped to the N-terminal portion of the receptor using limited proteolysis of avian VDR preparations *(17)*. After the VDR cDNA was obtained, molecular approaches were used to do more refined mapping studies. McDonnell et al. *(17)* expressed C-terminal deletion mutants of hVDR in mammalian cells, monitored each mutant for DNA binding, and defined the location of the DBD to amino acids 1-113. Using similar techniques, Sone et al. *(21)* showed that the N-terminal 21 amino acids immediately preceding the first cysteine residue (Fig. 3) could be removed without affecting DNA binding or the transcriptional activation potential of the VDR. Thus, on the basis of these functional assays, the minimal domain of the VDR that mediates VDR–DNA interactions resides between amino acid residues 22 and 113 in the human sequence.

There are nine cysteine residues within the DBD that are conserved throughout the members of the superfamily of receptor proteins. The first eight of these cysteines (counting from the N-terminus) tetrahedrally coordinate two zinc atoms to form two zinc-finger DNA binding motifs (Fig. 3), a coordination scheme proposed originally for transcription factor IIIA of *Xenopus laevis (22,23)*. Mutagenesis studies of the VDR DBD provides strong support for this zinc coordination scheme. Mutation of the first eight of the nine cysteine residues to serines eliminated VDR binding to both nonspecific and specific DNA sequences and eliminated VDR-mediated transactivation *(21)*. A serine mutation at the ninth cysteine residue (C84S) had little effect on VDR function, suggesting that this residue is not functionally analogous to the first eight cysteines. These data showed the essential nature of the first eight cysteines in the overall organization and structural integrity of the DBD and led to the model illustrated in Fig. 3.

3.1.2. VDR–DNA Contacts

Much of what is known of the nature of VDR–DNA interactions is modeled after functional and structural data of other related nuclear receptors. Three important subdomains within the DBDs of nuclear receptors are required to recognize and bind specific nucleotide sequences of DNA (Fig. 3). The first region is an α-helical domain referred to as the proximal (P)-box, which confers target sequence selectivity for the glucocorticoid receptor (GR) and the estrogen receptor (ER) *(24–26)*. Changing three P-box amino acids (boxed residues in Fig. 3) in the ER to the corresponding amino acid residues in the GR is sufficient to change the specificity of the receptor *(25)*, that is, the resulting mutant ER no longer recognizes an estrogen-responsive element, but strongly activates transcription from a glucocorticoid-responsive construct in response to estrogen. A second region is known as the distal (D)-box, which is important for homodimerization of the GR subfamily of receptors *(26)*. A third α-helical region, referred to as the T-box, resides just C-terminal to the second zinc finger, and it mediates homodimer and monomer interactions with DNA *(27,28)*. Initial mutational analysis of these three regions of the VDR DBD provided insight into the similarities and differences in the mechanisms in which VDR and other nuclear receptors interact with their cognate response elements *(29)*. Mutations in the T-box of the VDR resulted in a dramatic reduction in VDR binding to DNA and in transactivation, indicating an important role for this α-helical domain in VDR function. However, altering the P-box or D-box residues of VDR to those of the GR did not confer GR target gene selectivity to the VDR *(29)*. Moreover, these mutations did not significantly affect VDR interaction with DNA or VDR-activated transcription. Thus, it is apparent that the specificity determinants for VDR are more complex than previously thought, perhaps owing to the heterodimeric nature in which VDR binds DNA compared with homodimeric interactions of ER or GR with their response elements (*see below*).

Three-dimensional structural analysis of the purified DBDs for the GR, ER, RARβ, and RXR-α has provided detailed insights into the mechanism of receptor–response element interaction *(27,30–33)*. The common structural feature of the DBDs for all these receptors is the folding of two α-helices in the carboxyl terminal portion of the each zinc finger into a single DNA binding domain. The first α-helix in the amino-terminal finger (denoted helix 1 in Fig. 3) lies across the major groove of DNA, making specific contacts with the DNA binding site; as mentioned above, it is this region that contains the crucial amino acids that determine response element specificity for some receptors. The second α-helix (denoted helix 2 in Fig. 3) folds across the first in a perpendicular arrangement. The DBD is rich in the positively charged amino acids, lysine and arginine, several of which form favorable electrostatic interactions with the negatively charged phosphate backbone of the DNA helix. The crystal structure of the TR–RXR heterodimer bound to DNA was recently solved, and structural aspects for VDR–RXR binding were predicted *(34)*. One key feature is that the T-box residues of VDR make direct contact with the D-box residues of RXR, providing additional support for the importance of the VDR T-box in binding to DNA.

3.2. The Multifunctional C-Terminal Domain

Although the large C-terminal domain of VDR (amino acids 116–427 of hVDR) clearly functions as the binding site for the $1,25(OH)_2D_3$ ligand, it also fulfills several other critical roles in VDR function. A prominent role for the LBD is that of a protein–protein interaction surface through which VDR contacts its heterodimeric partner RXR for high order binding to DNA. It is through this C-terminal domain that VDR also interacts with other proteins such as transcription factor (TF)IIB *(35,36)* and coactivators *(37,38)* that are important for the mechanism of VDR-mediated transcription. Finally, key serine residues in this domain serve as sites of phosphorylation that may be important in regulating the transcriptional activity of the VDR.

3.2.1. LIGAND BINDING DOMAIN

The LBD is responsible for high-affinity binding of $1,25(OH)_2D_3$, exhibiting equilibrium binding constants on the order of 10^{-10}–10^{-11} M *(39–42)*. Both the 1-hydroxyl and 25-hydroxyl moieties are crucial for efficient recognition by VDR *(43,44)*. For example, the monohydroxylated vitamin D_3 compounds, $1\alpha(OH)D_3$ and $25(OH)D_3$, bind with approx 100-fold less affinity than $1,25(OH)_2D_3$. The nonhydroxylated parent vitamin D_3 compound does not bind at all to the VDR. Modifications in both the side-chain structure and in the A-ring of the secosteroid also dramatically reduce the affinity of the LBD for the analog.

The location of the LBD in the C-terminal region of VDR was first demonstrated biochemically using limited digestion of the intact receptor with carboxypeptidase A *(45)*. More refined mapping, using a series of 5'- and 3'-deletion mutants of the hVDR expressed in COS-1 cells, defined the N-terminal boundary of the LBD between residues 114 and 166 and the C-terminal boundary between residues 373 and 425. Nakajima et al. *(46)* described a deletion mutant (Δ403–427) that bound $1,25(OH)_2D_3$ specifically but with reduced affinity, and further deletion to residue 382 eliminated hormone binding altogether. Therefore, between 200 and 300 amino acids in this C-terminal region are necessary for binding the $1,25(OH)_2D_3$ ligand, and the most C-terminal amino acids (i.e., between residues 403 and 425) may play an obligate role in high-affinity ligand interactions. A sequence comparison of various steroid receptors also showed that this most C-terminal region is unique to each individual receptor and therefore, this region was suggested to participate in hormone-specific interactions *(47)*. Support for this concept was realized in the crystal structures of unliganded RXR compared with liganded RAR or liganded TR *(48–50)*. It was suggested that this C-terminal tail is responsible for folding over and "closing off" the ligand binding pocket of some nuclear receptors *(49,50)*.

The specific amino acid residues that are directly involved in $1,25(OH)_2D_3$ interactions are not known. Kristjansson et al. *(51)* reported an R274L point mutation in a patient with hereditary vitamin D-resistant rickets that decreased the affinity of the mutant VDR for the $1,25(OH)_2D_3$ ligand by a factor of 1000, suggesting that this basic residue may participate in hormone interactions. More recently, Nakajima et al. *(52)* demonstrated the essential role of cysteine residues, in particular C288 in the human VDR, in high-affinity binding of the $1,25(OH)_2D_3$ ligand. Whether these particular residues actually contact the ligand or are required for maintaining the structural integrity of the hydrophobic binding pocket will obviously require the eventual crystallization of the unliganded and liganded VDR complexes and x-ray diffraction analysis. Moreover, the development of new affinity analogs to identify systematically those residues that comprise the $1,25(OH)_2D_3$ binding pocket will be useful *(53)*.

3.2.2. Heterodimerization with RXR

In addition to hormone binding, the LBD is required for several other aspects of receptor function, including a major role in mediating protein–protein interactions. One important protein–protein contact is the heterodimerization of VDR with receptor auxiliary factors (RAFs) such as RXR. As mentioned previously, VDR–RXR heterodimer formation is generally required for high-affinity interaction of the receptor with VDREs, and at least three putative regions in the LBD of VDR mediate protein–protein contacts with RXRs and RAFs *(43,51)*. A predominant, C-terminal heterodimerization domain resides between residues 382 and 403 in the hVDR sequence *(46)*. Mutagenesis of several specific residues in this domain (at Lys 382, Met 383, and Glu 385) completely disrupted VDR–RAF and VDR–RXR interaction in vitro and eliminated the transcriptional activation activity of the VDR. A second putative interaction domain was identified between residues 318 and 339 *(46)*. These regions correspond to heptads 9 and 4, respectively, which are motifs that were proposed to function in protein–protein interactions by Forman and Samuels *(47)*. A third putative heterodimerization surface exists in the amino-terminal segment of the LBD between amino acids 244 and 263 *(54)*. Selected point mutations within this region do not interfere with ligand binding, but they affect the ability of the VDR to heterodimerize with RAFs or RXRs and disable transcription from vitamin D-responsive constructs. One important outcome of these studies is that, in all the receptor mutants examined, heterodimerization of VDR with RXR was required for VDRE interaction and for $1,25(OH)_2D_3$-/VDR-mediated transcriptional activation.

By drawing analogies to other related receptors, the crystal structure of the RXR–RXR homodimeric complex provides some insight into what may be the crucial heterodimerization surface of the VDR–RXR complex. The RXRα crystal structure revealed an α-helix-rich LBD (65%) consisting of 11 α-helices organized into what has been termed a three-layer antiparallel sandwich *(48)*. The RXR dimer is symmetrically arranged, with the interaction surface being formed mainly by helix 10 and, to a lesser extent, by helix 9. Helix 10 of RXR corresponds to the C-terminal region of VDR identified by Nakajima et al. *(46)* as being crucial for heterodimer formation (heptad 9, amino acids 382–403). The other heptad repeats described by Forman and Samuels *(47)* do not contribute to the interaction surface, but instead are involved in intramolecular contacts that may be important in maintaining the overall structural integrity of the receptor. Thus, assuming a structural relatedness between VDR and RXR, the C-terminal region of hVDR that includes residues 382–403 (helix 10) may directly contact helix 10 of RXR to comprise

Chapter 7 / Molecular Biology of VDR

415 - T	P	L	V	L	E	V	F	G	N	E	I	S		hVDR
446 - P	P	L	F	L	E	V	F	E	D					hTRβ
396 - P	P	L	F	L	E	V	F	E	D	Q	E	V		hTRα
407 - P	P	L	I	Q	E	M	L	E	N	S	E	G	L	hRARα
400 - P	P	L	I	Q	E	M	L	E	N	S	E	G	H	hRARβ
409 - P	P	L	I	R	E	M	L	E	N	P	E	M	F	hRARγ
448 - D	T	F	L	M	E	M	L	E	A	P	H	Q	M	hRXRα
537 - Y	D	L	L	L	E	M	L	D	A	H	R	L	H	hER

Fig. 4. Amino acid sequence comparison of the AF-2 domain of vitamin D receptor and related nuclear receptors.

the major interaction surface, which would generate a structurally symmetrical VDR–RXR heterodimeric complex.

3.2.3. TRANSACTIVATION

Transcriptional activation domains are generally defined through mutations that selectively affect the transcriptional activity of the receptor without disrupting other receptor functions such as ligand binding, response element interaction, or nuclear localization. These domains are often transferable, that is, fusing the activation domain itself to a heterologous DNA binding domain such as Gal 4 (amino acids 1–147) imparts high-order transcription to the heterologous fusion protein. The ligand-activated nuclear receptors express at least two general activation domains. The AF-1 domain is located in the N-terminal region of the receptors, and in many cases it is a constitutive (i.e., hormone-independent) activation domain. The AF-2 domain acts in a ligand-dependent manner and is located at the extreme C-terminus of the receptors. This domain is highly conserved throughout the hormone receptor superfamily (Fig. 4) and its main structural feature is that of an amphipathic α-helix.

The N-terminus of the VDR is truncated compared with other nuclear receptors, and thus it is unlikely that an analogous constitutive AF-1 domain exists N-terminal to the DBD in VDR. Indeed, as previously mentioned, removing the N-terminal 22 amino acids preceding the DBD had no affect on VDR-activated transcription. Whether a constitutive activation domain resides elsewhere in the VDR sequence is unknown. However, based on sequence similarity (Fig. 4) and mutagenesis studies, the C-terminal, ligand-dependent AF-2 domain exists and is functional in the VDR. Removing 25 amino acids from the C-terminus of hVDR (Δ403–427), which contains the AF-2 domain, resulted in a complete loss of 1,25(OH)$_2$D$_3$-/VDR-activated transcription *(46)*. This loss of function was not due to altered binding of RXR, VDRE, or hormone, and the mutant receptor was appropriately targeted to the cell nucleus. Thus, the AF-2 domain of VDR, which corresponds to helix 12 in the RXR, RAR, and TR crystal structures *(48–50)*, serves an important role in VDR-mediated transcription. Based on the observation that the AF-2 activity of a particular receptor is interfered with (or squelched) by overexpressing another AF-2 domain of the same or different receptor *(55,56)*, the AF-2 domain is proposed to be a protein interaction site for transcriptional mediators or intermediary factors that are required for nuclear receptor-dependent transcription *(see below)*.

3.2.4. PHOSPHORYLATION

Phosphorylation is widely regarded as a key means of regulating cellular processes. Most of the steroid/thyroid hormone receptors are known to be phosphorylated, including the VDR. In fact, VDR present in mouse 3T6 cells is hyperphosphorylated in response to physiologic concentrations of $1,25(OH)_2D_3$ *(57)*. Ligand-dependent phosphorylation has also been demonstrated in ROS 17/2.8 cells *(55)* and in chick duodenal organ culture *(59)*, two relevant target systems for vitamin D action. In this last system, the effect is observed within 15 min following the addition of $1,25(OH)_2D_3$. Since this precedes most other cellular responses to $1,25(OH)_2D_3$, the authors suggested the possibility that phosphorylation of VDR might play an initiating event in the transcriptional processes mediated by the VDR.

The major phosphorylated residues of the VDR have been determined. Using domain-specific antibodies, Brown and Deluca *(60)* mapped a major phosphorylation site(s) to the N-terminal region of the LBD in porcine VDR. Studies in ROS 17/2.8 cells revealed that the main phosphorylated domain of hVDR resided between Met197 and Val234 in hVDR *(61)*. Within this domain is a cluster of serine residues, many of which resemble consensus sites for casein kinase II. Indeed, hVDR is an effective substrate for in vitro phosphorylation by purified casein kinase II *(62)*. Site-directed mutagenesis defined Ser 208 as the site phosphorylated by casein kinase II in vitro and as a major phosphorylated residue in vivo when VDR was transiently expressed in COS-7 cells *(62)*. Furthermore, coexpression of casein kinase II in this system augmented VDR phosphorylation of Ser 208, showing that this kinase could also phosphorylate Ser 208 in the cell *(63)*. Hilliard et al. *(64)* systematically identified this same Ser 208 residue as the main phosphorylated residue of VDR using phosphopeptide mapping studies. Interestingly, in this study, phosphorylation at Ser 208 was augmented eightfold when the cells were treated for 4 h with $1,25(OH)_2D_3$. Thus, Ser 208 is the major phosphorylated residue of VDR and probably represents the hormone-dependent phosphorylation site observed in earlier studies. A second alternate site of phosphorylation is Ser 51, which resides between the two zinc-finger motifs in the DBD of the hVDR *(65)*. Ser 51 is a consensus site for protein kinase C (PKC), and it is selectively phosphorylated by the PKC-β isoform in vitro and in vivo.

Although the global phosphorylated state of the cell clearly affects VDR-mediated transcriptional activity *(66,67)*, demonstrating that VDR phosphorylation *per se* is important for vitamin D-mediated transcription has been elusive. For example, mutations that disrupt phosphorylation at Ser 208 and Ser 51 do not affect VDR-activated transcription *(63,68)*. One caveat here is that mutations in one serine residue may actually promote phosphorylation of an adjacent serine residue to compensate *(64)*. Thus, more detailed studies are required to define the precise functional roles of Ser 208 and Ser 51 phosphorylation by casein kinase II and by protein kinase C in VDR function.

4. MOLECULAR MECHANISM OF TRANSCRIPTIONAL CONTROL BY VDR

4.1. VDR Interaction with Vitamin D-Responsive Elements

4.1.1. VDRES

Nuclear receptors modulate transcription by binding to specific DNA elements in the promoter regions of hormone-responsive genes. The specific VDR-interactive promoter sequences are termed vitamin D-responsive elements. VDREs from a variety of vitamin

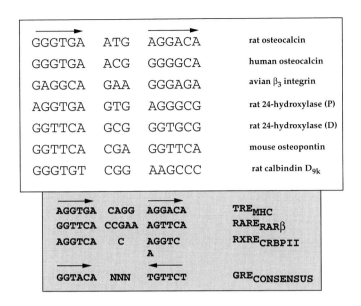

Fig. 5. Comparison of vitamin D response elements (VDREs) from several vitamin D-responsive genes and other hormone receptor response elements.

D-responsive genes have been identified (Fig. 5) on the basis of several functional criteria including (1) deletion or mutation of the element resulted in the loss of promoter responsiveness to $1,25(OH)_2D_3$, (2) the sequence alone conferred vitamin D responsiveness to an otherwise unresponsive, heterologous promoter, and (3) the element served as a high-affinity binding site for VDR in vitro.

On the basis of this limited number of natural response elements, the VDRE is generally described as an imperfect direct repeat of a core hexanucleotide sequence, G/A G G T G/C A, with a spacer region of three-nucleotides separating each half-element (also termed DR-3 for direct repeat with a three-nucleotide spacer). This direct repeat motif is analogous to DNA elements that mediate retinoic acid and thyroid hormone responsiveness (RAREs and TREs), and it contrasts with responsive elements that mediate glucocorticoid or estrogen (GREs and EREs) responsive genes, which are generally palindromic or inverted repeat sequences (Fig. 5, bottom).

It is apparent in Figure 5 that some degree of plasticity exists in the sequence of each half-element of the VDRE, suggesting that there is flexibility in the precise sequence that will mediate vitamin D responsiveness. However, there is little question that the nucleotide sequence of the element is critical for receptor-mediated transcription. Modest changes in the nucleotides of either half-element in the rat osteocalcin VDRE had disrupted VDR–VDRE interactions and compromised VDR-dependent transactivation *(69)*. An additional determinant of response specificity for this class of nuclear receptors is the length of the spacer region *(26)*. In general, VDR, thyroid hormone receptor (TR), and RAR all recognize similar direct repeat sequences, but with differing spacer regions of three, four, and five nucleotides, respectively. This phenomenon, termed the "3-4-5 rule" and later expanded to the "1–5 rule" *(19,70)* illustrates that half-site spacing plays a major role in determining selective hormonal response. With regard to the VDRE, it is also clear that the actual nucleotide sequence of the spacer region is also important since mutations in the spacer of the rat osteocalcin VDRE disrupt VDR binding and transactivation *(69)*.

The natural VDREs identified thus far provide only a snapshot of the DNA sequences that mediate the transcriptional effects of the VDR. The elements in Figure 5 are certainly a limited sample, and one should avoid generalizations. Variations on the DR-3 motif for VDREs have been identified in the elements that mediate vitamin D responsiveness in the calbindin D9k and calbindin D28k genes *(71,72)*. Moreover, several synthetic elements with large spacer regions and inverted arrangements can mediate vitamin D responsiveness under certain conditions *(73)*. It is likely that the affinity of VDR for these atypical elements may vary from that of the classic DR-3 motif, adding yet another level of regulatory complexity to the process of VDR-mediated gene expression. In the future, it will be important to examine systematically VDR interaction with the variety of VDREs identified thus far, to perhaps classify VDREs into weak versus strong elements.

4.1.2. VDR–RXR Heterodimers

The mechanism of VDR binding to VDREs is reflected in the direct repeat nature of the element. For example, the class I members of the nuclear receptor superfamily (e.g., glucocorticoid or progesterone receptor) bind to palindromic response elements as symmetrical homodimers (see ref. *74* for a review). By contrast, the class II receptors (of which VDR is a member) generally bind to direct repeat elements as asymmetrical heterodimers. For VDR, this was originally documented in experiments showing that purified VDR alone does not bind to a VDRE with high affinity *(75)*, rather, an unidentified coreceptor (designated NAF or RAF) is required for high-affinity binding of VDR to VDREs *(75)*. This study provided the first evidence that the VDR binds to the osteocalcin VDRE, not as a homodimer, but as a heteromeric complex in association with another nuclear factor.

Substantial evidence now indicates that RAF is RXR, one class of nuclear receptors for vitamin A that mediate the actions of 9-*cis* retinoic acid *(76,77)*. This is based on several observations: (1) RXRs substitute effectively for RAF in VDR/VDRE binding assays *(78–80)*, (2) highly purified RAF contains RXR immunoreactivity *(79,81)*, (3) RXRs augment vitamin D-mediated transcription *(73,79,80)*, (4) numerous VDR mutants that do not interact with RXR also fail to activate transcription in vivo *(46)*, and (5) in a yeast system that lacks proteins analogous to the mammalian nuclear receptors, both VDR and RXR are required to elicit activated transcription from a VDRE-driven reporter gene construct *(82)*. Taken together, these data support the concept of the VDR–RXR heterodimer as the functional enhancer in vitamin D-activated transcription.

The direct repeat motif is asymmetrical, and the VDR–RXR heterodimer binds to the VDRE with a defined polarity. VDR and RXR bind to the DR-3 element with RXR occupying the 5' half site and VDR occupying the 3' half site *(82,83)*. Modeling based on the crystal structure of TR-RXR suggests that the DBDs of a VDR–RXR heterodimer on a DR-3 are arranged in an asymmetrical, head-to-tail assembly with the VDR T-box forming a direct contact with the RXR D-box *(34)*. By contrast, the predominant interaction interface in the receptor C-termini would place the LBDs of VDR and RXR in a symmetrical arrangement in the heterodimeric complex (*see* Section 3.2.2.). To rationalize this discrepancy, the DBDs of the nuclear receptors are proposed to have an inherent flexibility, presumably a function of the hinge region of the receptors, that permits a high degree of rotational freedom in the DBD (up to 180°) that would accommodate such a binding scheme *(84)*.

4.1.3. A ROLE FOR THE 1,25(OH)$_2$D$_3$ LIGAND

Insight into a role for the 1,25(OH)$_2$D$_3$ ligand in the transactivation process has emerged from studies examining VDR, RXR, and VDRE interactions. The 1,25(OH)$_2$D$_3$ ligand dramatically enhances VDR–RXR heterodimerization, both the direct interaction of VDR with RXR in solution *(21,37)* and the interaction of the VDR–RXR heterodimer with the VDRE *(21,75,79)*. Surface plasmon resonance quantitated the binding constants for these interactions and showed a clear 1,25(OH)$_2$D$_3$-dependent decrease in VDR interaction with itself (i.e., VDR homodimers) and a concomitant increase in VDR heterodimerization with RXR *(85)*. These ligand-induced changes in VDR–RXR interactions are probably caused by altered conformations of the VDR in the absence and presence of 1,25(OH)$_2$D$_3$ *(86)*. Thus one putative role for the 1,25(OH)$_2$D$_3$ in VDR-mediated transcription may be to induce a distinct conformational change in VDR that disrupts weak homodimers of unliganded VDR and promotes liganded VDR heterodimerization with RXR. The interaction of VDR and RXR generates a heterodimeric complex that is highly competent to bind DR-3 like VDREs and subsequently affect the transcriptional process.

4.2. Communication Between VDR and the Transcriptional Machinery

An obvious gap in our understanding of VDR function is the sequence of events that follow VDR–RXR heterodimer binding to the VDRE and the penultimate influence of the heterodimeric complex on RNA polymerase II-directed transcription. Specifically, what is the communication process that links the VDR to transcriptional preinitiation complex?

Central to the process of activated transcription are the general transcription factors and the ordered assembly of the preinitiation complex (PIC) (reviewed in ref. *87*). PIC assembly begins with TATA binding protein (TBP, a subunit of TFIID) binding to the TATA element of class II promoters in a process that is facilitated by TFIIA. Then, in what may be the rate-limiting step, TFIIB enters the complex by direct interaction with TBP. RNA pol II, in association with TFIIF, binds to this early complex by contacting TFIIB. Thus, TFIIB serves as a bridging protein between TBP and RNA pol II. The further association with TFIIE and other general factors results in a complex capable of accurately initiating RNA synthesis. Transcription initiated by these minimal components represents basal level transcription, which can be stimulated (or repressed) by sequence-specific, *trans*-acting factors, such as the VDR. One current hypothesis for activated transcription suggests that the activator communicates with the PIC through extensive protein–protein contacts. This interaction may (1) facilitate one or more steps in the assembly of the transcriptional machinery, (2) cause a conformational change that would trigger initiation by RNA pol II, (3) enhance the promoter clearance phase, or (4) comprise a combination of the three (reviewed in ref. *88*). Which, if any, of these mechanisms apply to vitamin D-mediated transcription is presently unknown.

Transactivator interaction with the PIC may occur via adapter proteins such as the TBP-associated factors (TAFs) *(89,90)* or via direct interaction with other core transcription factors. In this regard, TFIIB may be a key target since it is known to interact directly with a variety of *trans*-acting factors. An illustrative example is VP16 (a herpes simplex virion protein). The VP16 activation domain interacts directly with TFIIB, and this

7. Ishikawa T, Umesono K, Mangelsdorf D, Aburatani H, Stanger B, Shibasaki Y, Imawari M, Evans R, Takaku F. A functional retinoic acid receptor encoded by the gene on human chromosome 12. Mol Endocrinol 1990; 4:837–844.
8. Mattei MG, Riviere M, Krust A, Ingvarsson S, Vennstrom B, Islam MQ, Levan G, Kautner P, Zelent A, Chambon P, Szirer J, Szirer C. Chromosomal assignment of retinoic acid receptor (RAR) genes in the human, mouse, and rat genomes. Genomics 1991; 10:1061–1069.
9. Pike JW, Kesterson RA, Scott RA, Kerner SA, McDonnell DP, O'Malley BW. Vitamin D_3 receptors: molecular structure of the protein and its chromosomal gene. In: Vitamin D: Molecular, Cellular and Clinical Endocrinology. Norman AW, Schaefer K, Grigoleit H-G, nd von Herrath D, eds. Berlin: Walter de Gruyter, 1988; 215–224.
10. Sone T, Marx SJ, Liberman UA, Pike J W. A unique point mutation in the human vitamin D receptor chromosomal gene confers hereditary resistance to 1,25-dihydroxyvitamin D_3. Mol Endocrinol 1990; 4:623–631.
11. Hughes MR, Malloy PJ, Kieback DG, Kesterson RA, Pike JW, Feldman D, O'Malley BW. Point mutations in the human vitamin D receptor gene associated with hypocalcemic rickets. Science 1988; 242:1702–1705.
12. Pike JW. Monoclonal antibodies to chick intestinal receptors for 1,25-dihydroxyvitamin D_3. J Biol Chem 1984; 259:1167–1173.
13. Pike JW, Marion SL, Donaldson CA, Haussler MR. Serum and monoclonal antibodies against the chick intestinal receptor for 1,25-dihydroxyvitamin D_3. J Biol Chem 1983; 258:1289–1296.
14. McDonnell DP, Mangelsdorf DJ, Pike JW, Haussler MR, O'Malley BW. Molecular cloning of complementary DNA encoding the avian receptor for vitamin D. Science 1987; 235:1214–1217.
15. Baker AR, McDonnell DP, Hughes M, Crisp TM, Mangelsdorf DJ, Haussler MR, Pike JW, Shine J, O'Malley BW. Cloning and expression of full-length cDNA encoding human vitamin D receptor. Proc Natl Acad ScI USA 1988; 85:3294–3298.
16. Burmester JK, Wiese RJ, Maeda N, DeLuca HF. Structure and regulation of the rat 1,25-dihydroxyvitamin D_3 receptor. Proc Natl Acad Sci USA 1988; 85:9499–9502.
17. McDonnell DP, Scott RA, Kerner SA, O'Malley BW, Pike JW. Functional domains of the human vitamin D_3 receptor regulate osteocalcin gene expression. Mol Endocrinol 1989; 3:635–644.
18. Brown TA, Prahl JM, DeLuca HF. Partial amino acid sequence of porcine 1,25-dihydroxyvitamin D_3 receptor isolated by immunoaffinity chromatography. Proc Natl Acad Sci USA 1988; 85: 2454–2458.
19. Leid M, Kastner P, Chambon P. Multiplicity generates diversity in the retinoic acid signalling pathways. Trends Biochem Sci 1992; 17:427–433.
20. Allegretto EA, Pike JW, Haussler MR. Immunochemical detection of unique proteolytic fragments of the chick 1,25-dihydroxyvitamin D_3 receptor. Distinct 20 kDa DNA-binding and 45 kDa hormone-binding species. J Biol Chem 1987; 262:1312–1319.
21. Sone T, Kerner S, Pike JW. Vitamin D receptor interaction with specific DNA: association as a 1,25-dihydroxyvitamin D_3-modulated heterodimer. J Biol Chem 1991; 266:23,296–23,305.
22. Brown RS, Sander C, Argos P. The primary structure of transcription factor IIIA has 12 consecutive repeats. FEBS Lett 1985; 186:271–274.
23. Miller J, McClachlin AD, Klug A. Repetitive zinc-binding domains in the protein transcription factor IIIA from *Xenopus* oocytes. EMBO J 1985; 4:1609–1614.
24. Danielson M, Hinck L, Ringold GM. Two amino acids within the knuckle of the first zinc finger specify response element activation by the glucocorticoid receptor. Cell 1989; 57:1131–1138.
25. Mader S, Kumar V, deVereneuil H, Chambon P. Three amino acids of the oestrogen receptor are essential to its ability to distinguish an oestrogen from a glucocorticoid-responsive element. Nature 1989; 338:271–274.
26. Umesono K, Evans RM. Determinants of target gene specificity for steroid/thyroid hormone receptors. Cell 1989; 57:1139–1146.
27. Lee MS, Kliewer SA, Provencal J, Wright PE, Evans RM. Structure of the retinoid X receptor α DNA binding domain: a helix required for homodimeric DNA binding. Science 1993; 260:1117–1121.
28. Wilson TE, Paulsen RE, Padgett KA, Milbrandt J. Participation of non-zinc finger residues in DNA binding by two nuclear orphan receptors. Science 1992; 256:107–110.
29. Hsieh J-C, Jurutka PW, Selznick SH, Reeder MC, Haussler CA, Whitfield GK, Haussler, MR. The T-box near the zinc fingers of the human vitamin D receptor is required for heterodimeric DNA binding and transactivation. Biochem Biophys Res Commun 1995; 215:1–7.

30. Härd T, Kellenbach E, Boelens R, Maler B, Dahlman K, Freedman LP, Carlstedt-Duke J, Yamamoto KR, Gustafsson J-Å, Kaptein R. Solution structure of the glucocorticoid receptor DNA-binding domain. Science 1990; 249:157–160.
31. Katahira M, Knegtel RM, Boelens R, Eib D, Schilthuis JG, van der Saag PT, Kaptein R. Homo- and heteronuclear NMR studies of the human retinoic acid receptor beta DNA binding domain: sequential assignments and identification of secondary structural elements. Biochemistry 1992; 31:6474–6480.
32. Luisi BF, Xu WX, Otwinowski Z, Freedman LP, Yamamoto KR, Sigler PB. Crystallographic analysis of the interaction of the glucocorticoid receptor with DNA. Nature 1991; 352:497–505.
33. Schwabe JWR, Neuhaus D, Rhodes D. Solution structure of the DNA-binding domain of the oestrogen receptor. Nature 1990; 348:458–461.
34. Rastinejad F, Perlmann T, Evans RM, Sigler PB. Structural determinants of nuclear receptor assembly on DNA direct repeats. Nature 1995; 375:203–211.
35. Blanco JCG, Wang I-M Tsai SY, Tsai M-J, O'Malley BW, Jurutka PW, Haussler MR, Ozato K. Transcription factor TFIIB and the vitamin D receptor cooperatively activate ligand-dependent transcription. Proc Natl Acad Sci USA 1995; 92:1535–1539.
36. MacDonald PN, Sherman DR, Dowd DR, Jefcoat SC, DeLisle RK. The vitamin D receptor interacts with general transcription factor IIB. J Biol Chem 1995; 270:4748–4752.
37. Le Douarin B, Zechel C, Garnier J-M, Lutz Y, Tora L, Pierrat B, Heery D, Gronemeyer H, Chambon P, Losson R. The N-terminal part of TIF1, a putative mediator of the ligand-dependent activation function (AF-2) of nuclear receptors, is fused to B-raf in the oncogenic protein T18. EMBO J 1995; 14:2020–2033.
38. vom Baur E, Zechel C, Heery D, Heine MJS, Garneir JM, Vivat V, LeDouarin B, Gronemeyer H, Chambon P, Losson R. Differential ligand-dependent interactions between the AF-2 activating domain of nuclear receptors and the putative transcriptional intermediary factors mSUG1 and TIF1. EMBO J 1996; 15:110–124.
39. Brumbaugh PF, Haussler MR. 1,25-Dihydroxycholecalciferol receptors in the chick intestine. II. Temperature dependent transfer of the hormone to chromatin via a specific cytosol receptor. J Biol Chem 1974; 249:1258–1262.
40. Brumbaugh PF, Haussler MR. Specific binding of 1α,25-dihydroxycholecalciferol to nuclear components of chick intestine. J Biol Chem 1975; 250:1588–1594.
41. Mellon W, DeLuca HF. An equilibrium and kinetic study of 1,25-dihydroxyvitamin D_3 binding to chicken intestinal cytosol employing high specific activity 1,25-dihydroxy[^{3}H-26,27]vitamin D_3. Arch Biochem Biophys 1979; 197:90–95.
42. Wecksler WR, Norman AW. A kinetic and equilibrium binding study of 1α,25-dihydroxyvitamin D_3 with its cytosol receptor from chick intestinal mucosa. J Biol Chem 1980; 255:3571–3574.
43. Procsal DA, Okamura WH, Norman AW. Structural requirements for the interaction of 1α,25-(OH)$_2$-vitamin D_3 with its chick intestinal receptor system. J Biol Chem 1975; 250:8382–8388.
44. Wecksler WR, Okamura WH, Norman AW. Studies on the mode of action of vitamin D: XIV. Quantitative assessment of the structural requirements for the interaction of 1α,25-dihydroxyvitamin D_3 with its chick intestinal mucosa receptor system. J Steroid Biochem 1978; 9:929–937.
45. Allegretto EA, Pike JW, Haussler MR. C-terminal proteolysis of the avian 1,25-dihydroxyvitamin D_3 receptor. Biochem Biophys Res Commun 1987; 147:479–485.
46. Nakajima S, Hsieh J-C, MacDonald PN, Galligan MA, Haussler CA, Whitfield GK, Haussler MR. The C-terminal region of the vitamin D receptor is essential to form a complex with a receptor auxiliary factor required for high affinity binding to the vitamin D-responsive element. Mol Endocrinol 1994; 8:159–172.
47. Forman BM, Samuels HH. Interactions among a subfamily of nuclear hormone receptors: the regulatory zipper model. Mol Endocrinol 1990; 4:1293–1301.
48. Bourguet W, Ruff M, Chambon P, Gronemeyer H, Moras D. Crystal structure of the ligand-binding domain of the human nuclear receptor RXR-α. Nature 1995; 375:377–382.
49. Renaud J-P, Rochel N, Ruff M, Vivat V, Chambon P, Gronemeyer H, Moras D. Crystal structure of the RAR-γ ligand-binding domain bound to all-trans retinoic acid. Nature 1995; 378:681–689.
50. Wagner RL, Apriletti JW, McGrath ME, West BL, Baxter JD, Fletterick RJ. A structural role for hormone in the thyroid hormone receptor. Nature 1995; 378:690–697.
51. Kristjansson K, Rut AR, Hewison M, O'Riordan JLH, Hughes MR. Two mutations in the hormone binding domain of the vitamin D receptor cause tissue resistance to 1,25 dihydroxyvitamin D_3. J Clin Invest 1993; 92:12–16.

52. Nakajima S, Hsieh J-C, Jurutka PW, Galligan MA, Haussler CA, Whitfield GK, Haussler MR. Examination of the potential functional role of conserved cysteine residues in the hormone binding domain of the human 1,25-dihydroxyvitamin D_3 receptor. J Biol Chem 1996; 271:5143-5149.
53. Ray R, MacDonald PN, Swamy N, Ray S, Haussler MR, Holick MF. Affinity labeling of the 1α,25-dihydroxyvitamin D3 receptor. J Biol Chem 1996; 271:2012-2017.
54. Whitfield GK, Hsieh J-C, Nakajima S, MacDonald PN, Thompson PD, Jurutka PW, Haussler CA, Haussler MR. A highly conserved region in the hormone-binding domain of the human vitamin D receptor contains residues vital for heterodimerization with retinoid X receptor and for transcriptional activation. Mol Endocrinol 1995; 9:1166-1179.
55. Meyer M-E, Gronemeyer H, Turcotte B, Bocquel M-T, Tasset D, Chambon P. Steroid hormone receptors compete for factors that mediate their enhancer function. Cell 1989; 57:433-442.
56. Shemshedini L, Ji J, Brou C, Chambon P, and Gronemeyer H. In vitro activity of the transcription activation functions of the progesterone receptor. J Biol Chem 1992; 267:1834-1839.
57. Pike JW, Sleator NM. Hormone-dependent phosphorylation of the 1,25-dihydroxyvitamin D_3 receptor in mouse fibroblasts. Biochem Biophys Res Commun 1985; 131:378-385.
58. Haussler MR, Terpening CM, Jurutka PW, Meyer J, Schulman BA, Haussler CA, Whitfield GK, Komm B S. Vitamin D hormone receptors: structure, regulation and molecular function. In: Progress in Endocrinology. Imura H, Shizume K, Yoshida S, eds. Amsterdam: Elsevier Science Publishers, 1988; 763-770.
59. Brown TA, DeLuca HF. Phosphorylation of the 1,25-dihydroxyvitamin D_3 receptor: A primary event in 1,25-dihydroxyvitamin D_3 action. J Biol Chem 1990; 265:10,025-10,029.
60. Brown TA, DeLuca HF. Sites of phosphorylation and photoaffinity labeling of the 1,25-dihydroxyvitamin D3 receptor. Arch Biochem Biophys 1991; 286:466-472.
61. Jones BB, Jurutka PW, Haussler CA, Haussler MR, Whitfield GK. Vitamin D receptor phosphorylation in transfected ROS 17/2.8 cells is localized to the N-terminal region of the hormone-binding domain. Mol Endocrinol 1991; 5:1137-1146.
62. Jurutka PW, Hsieh J-C, MacDonald PN, Terpening CM, Haussler CA, Haussler MR, Whitfield GK. Phosphorylation of serine 208 in the human vitamin D receptor. The predominant amino acid phosphorylated by casein kinase II, *in vitro*, and identification as a significant phosphorylation site in intact cells. J Biol Chem 1993; 268:6791-6799.
63. Jurutka PW, Hsieh J-C, Nakajima S, Haussler CA, Whitfield GK, Haussler MR. Human vitamin D receptor phosphorylation by casein kinase II at Ser-208 potentiates transcriptional activation. Proc Natl Acad Sci USA 1996; 93:3519-3524.
64. Hilliard GM, Cook RG, Weigel NL, Pike JW. 1,25-Dihydroxyvitamin D_3 modulates phosphorylation of Ser 205 in the human vitamin D receptor: site-directed mutagenesis of this residue promotes alternative phosphorylation. Biochemistry 1994; 33:4300-4311.
65. Hsieh J-C, Jurutka PW, Galligan MA, Terpening CM, Haussler CA, Samuels DS, Shimizu Y, Shimizu N, Haussler MR. Human vitamin D receptor is selectively phosphorylated by protein kinase C on serine 51, a residue crucial to its trans-activation function. Proc Natl Acad Sci USA 1991; 88: 9315-9319.
66. Desai RK, van Wijnen AJ, Stein JL, Stein GS, Lian JB. Control of 1,25-dihydroxyvitamin D3 receptor-mediated enhancement of osteocalcin gene transcription: effects of perturbing phosphorylation pathways by okadaic acid and staurosporine. Endocrinology 1995; 136:5685-5693.
67. Matkovits T, Christakos S. Ligand occupancy is not required for vitamin D receptor and retinoid receptor-mediated transcriptional activation. Mol Endocrinol 1995; 9:232-242.
68. Hsieh J-C, Jurutka PW, Nakajima S, Galligan MA, Haussler CA, Shimizu Y, Shimizu N, Whitfield GK, Haussler MR. Phosphorylation of the human vitamin D receptor by protein kinase C. Biochemical and functional evaluation of the serine 51 recognition site. J Biol Chem 1993; 268: 15,118-15,126.
69. Demay MB, Kiernan MS, DeLuca HF, Kronenberg HM. Characterization of 1,25-dihydroxyvitamin D_3 receptor interactions with target sequences in the rat osteocalcin gene. Mol Endocrinol 1992; 6: 557-562.
70. Mangelsdorf DJ, Umesono K, Kliewer SA, Borgmeyer U, Ong ES, Evans RM. A direct repeat in the cellular retinol-binding protein type II gene confers differential regulation by RXR and RAR. Cell 1991; 66:555-561.
71. Darwish HM, DeLuca HF. Identification of a 1,25-dihydroxyvitamin D_3-response element in the 5'-flanking region of the rat calbindin D-9k gene. Proc Natl Acad Sci USA 1992; 89:603-607.

72. Gill, RK, Christakos S. Identification of sequence elements in mouse calbindin-D_{28k} gene that confer 1,25-dihydroxyvitamin D_3- and butyrate-inducible responses. Proc Natl Acad Sci USA 1993; 90: 2984–2988.
73. Carlberg C, Bendik I, Wyss A, Meier E, Sturzenbecker LJ, Grippo JF, Hunziker W. Two nuclear signalling pathways for vitamin D. Nature 1993; 361:657–660.
74. Glass CK. Differential recognition of target genes by nuclear receptor monomers, dimers, and heterodimers. Endocr Rev 1994; 15:391–407.
75. Liao J, Ozono K, Sone T, McDonnell DP, Pike JW. Vitamin D receptor interaction with specific DNA requires a nuclear protein and 1,25-dihydroxyvitamin D_3. Proc Natl Acad Sci USA 1990; 87: 9751–9755.
76. Levin AA, Sturzenbecker J, Kazmer S, Bosakowski T, Huselton C, Allenby G, Speck J, Kratzeisen CL, Rosenberger M, Lovey A, Grippo JF. 9-Cis retinoic acid stereoisomer binds and activates the nuclear receptor RXRα. Nature 1992; 355:359–361.
77. Mangelsdorf DJ, Ong ES, Dyck JA, Evans RM. Nuclear receptor that identifies a novel retinoic acid response pathway. Nature 1990; 345:224–229.
78. Kliewer SA, Umesono K, Mangelsdorf DJ, Evans RM. Retinoid X receptor interacts with nuclear receptors in retinoic acid, thyroid hormone, and vitamin D_3 signalling. Nature 1992; 355:446–449.
79. MacDonald PN, Dowd DR, Nakajima S, Galligan MA, Reeder MC, Haussler CA, Ozato K, Haussler MR. Retinoid X receptors stimulate and 9-cis retinoic acid inhibits 1,25-dihydroxyvitamin D_3-activated expression of the rat osteocalcin gene. Mol Cell Biol 1993; 13:5907–5917.
80. Yu VC, Delsert C, Andersen B, Holloway JM, Devary OV, Naar AM, Kim SY, Boutin J-M, Glass CK, Rosenfeld MG. RXRβ: a coregulator that enhances binding of retinoic acid, thyroid hormone, and vitamin D receptors to their cognate response elements. Cell 1991; 67:1251–1266.
81. Munder M, Herzberg IM, Zierold C, Moss VE, Hanson K, Clagett-Dame M, DeLuca HF. Identification of the porcine intestinal accessory factor that enables DNA sequence recognition by vitamin D receptor. Proc Natl Acad Sci USA 1995; 92:2795–2799.
82. Jin CH, Pike JW. Human vitamin D receptor-dependent transactivation in *Saccharomyces cerevisiae* requires retinoid X receptor. Mol Endocrinol 1996; 10:196–205.
83. Schrader M, Nayeri S, Kahlen JP, Muller KM, Carlberg C. Natural vitamin D_3 response elements formed by inverted palindromes: polarity-directed ligand sensitivity of vitamin D_3 receptor-retinoid X receptor heterodimer-mediated transactivation. Mol Cell Biol 1995; 15:1154–1161.
84. Mangelsdorf DJ, Evans RM. The RXR heterodimers and orphan receptors. Cell 1995; 83:841–850.
85. Cheskis B, Freedman LP. Modulation of nuclear receptor interactions by ligands: kinetic analysis using surface plasmon resonance. Biochemistry 1996; 35:3309–3318.
86. Peleg S, Sastry M, Collins ED, Bishop JE, Norman AW. Distinct conformational changes induced by the 20-epi analogues of 1α,25-dihydroxyvitamin D_3 are associated with enhanced activation of the vitamin D receptor. J Biol Chem 1995; 270:10,551–10,558.
87. Zawel L, Reinberg D. Advances in RNA polymerase II transcription. Curr Opin Cell Biol 1992; 4: 488–495.
88. Sheldon M, Reinberg D. Tuning-up transcription. Curr Biol 1995; 5:43–46.
89. Gill G, Pascal E, Tseng ZH, Tjian R. A Glutamine-rich hydrophobic patch in transcription factor Sp1 contacts the $dTAF_{ii}110$ component of the *Drosophila* tfiid complex and mediates transcriptional activation. Proc Natl Acad Sci USA 1994; 91:192–196.
90. Goodrich JA, Hoey T, Thut CJ, Admon A, Tjian R. Drosophila $TAF_{II}40$ interacts with both a VP16 activation domain and the basal transcription factor TFIIB. Cell 1993; 75:519–530.
91. Choy B, Green MR. Eukaryotic activators function during multiple steps of preinitiation complex assembly. Nature 1993; 366:531–536.
92. Lin Y-S, Green MR. Mechanism of action of an acidic transcriptional activator in vitro. Cell 1991; 64:971–981.
93. Roberts SGE, Choy B, Walker SS, Lin Y-S, Green MR. A role for activator-mediated TFIIB recruitment in diverse aspects of transcriptional regulation. Curr Biol 1995; 5:508–516.
94. Roberts SGE, Green MR. Activator-induced conformational change in general transcription factor TFIIB. Nature 1994; 371:717–720.
95. Cavailles V, Dauvois S, L'Horset F, Lopez G, Hoare S, Kushner PJ, Parker MG. Nuclear factor RIP140 modulates transcriptional activation by the estrogen receptor. EMBO J 1995; 14:3741–3751.
96. Chen JD, Evans RM. A transcriptional co-repressor that interacts with nuclear hormone receptors. Nature 1995; 377:454–457.

97. Horlein AJ, Naar AM, Heinzel T, Torchia J, Gloss B, Kurokawa R, Ryan A, Kamei Y, Soderstrom M, Glass CK, Rosenfeld MG. Ligand-independent repression by the thyroid hormone receptor mediated by a nuclear receptor co-repressor. Nature 1995; 377:397–404.
98. Kamei Y, Xu L, Heinzel T, Torchia J, Kurokawa R, Gloss B, Lin S-C, Heyman RA, Rose DW, Glass CK, Rosenfeld MG. A CBP integrator complex mediates transcriptional activation and AP-1 inhibition by nuclear receptors. Cell 1996; 85:403–414.
99. Lee JW, Ryan F, Swaffield JC, Johnston SA, Moore DD. Interaction of thyroid-hormone receptor with a conserved transcriptional mediator. Nature 1995; 374:91–94.
100. Onate SA, Tsai SY, Tsai M-J, O'Malley BW. Sequence and characterization of a coactivator for the steroid hormone receptor superfamily. Science 1995; 270:1354–1357.
101. Saijo T, Ito M, Takeda E, Mahbubul Huq AHM, Naito E, Yokota I, Sone T, Pike JW, Kuroda Y. A unique mutation in the vitamin D receptor gene in three Japanese patients with vitamin D-resistant rickets type II: utility of single-strand conformation polymorphism analysis for heterozygous carrier detection. Am J Hum Genet 1991; 49:668–673.

8 Vitamin D Receptor Translocation

Julia Barsony

1. INTRODUCTION

In eukaryotic cells, a constant exchange of ions, lipids, RNA, and proteins is carried out between the cytoplasm and nucleus. The nuclear envelope separates the nucleus from the cytoplasm, and the nuclear pore complex (NPC) is the site of exchange between the two compartments. Small molecules may pass through the NPC by diffusion, but macromolecules that are larger than 20 kDa cannot enter the nucleus by passive diffusion. Transport of large proteins, such as the steroid receptors, into the nucleus is a multistep, highly regulated, time- and energy-dependent process [recently reviewed by Jans and Hubner *(1)*]. Protein import involves transport from the cytoplasm to the NPC, translocation through the NPC, and intranuclear targeting (Fig. 1).

Proteins are directed to the nucleus by their nuclear localization sequences (NLSs). NLSs are short peptide sequences that function via interactions with the recently discovered NLS receptor named importinα/karyopherinα/Srp1 *(2)*. There is no general consensus sequence for NLS, but all known NLSs are very hydrophilic. A number of proteins possess two or more NLSs that are required in concert to achieve exclusive nuclear localization. Multiple copies of an NLS appear to be more efficient than a single copy. One form of multiple NLS is the bipartite NLS, which consists of two series of basic residues separated by a 10- to 12-amino acid spacer. Multiple NLSs have been defined for the glucocorticoid receptor (GR) *(3–5)*. A bipartite NLS is within the hinge region between residues 497 and 524 (NL1), and another NLS is in the hormone binding domain between residues 540 and 795 (NL2). Other steroid receptors possess similar NLSs *(6–9)*. For the VDR, a bipartite NLS has been found in the hinge region between residues 76 and 102 *(10)*. Another NLS between the two DNA-binding zinc fingers (residues 49–55) has been recently documented *(10a)*.

Nuclear proteins are synthesized in the cytoplasm and often do not go immediately to the nucleus. They can remain in the cytoplasm due to interactions with "anchor" proteins or with proteins that mask NLS, or because of phosphorylations that mask NLS *(11)*. The fully processed nuclear proteins can be released rapidly from the docking complex upon hormone binding or activation of the cells. This mechanism provides a form of transcriptional regulation by controlling the number of transcription factors in the nucleus.

Steroid receptors are nuclear proteins that can also remain in the cytoplasm before hormone binding, bound to a hetero-oligomeric protein complex. Within this complex, the GR has been shown to associate with the 90-kDa heat shock protein (hsp90), hsp56, hsp70, 56-kDa immunophilin, and p23 *(12–17)*. Hsp90 has been shown to mask the bipartite NLS of GR, androgen (AR), progesterone (PR), mineralocorticoid (MC), and estrogen

From: *Vitamin D: Physiology, Molecular Biology, and Clinical Applications*
Edited by: M. F. Holick © Humana Press Inc., Totowa, NJ

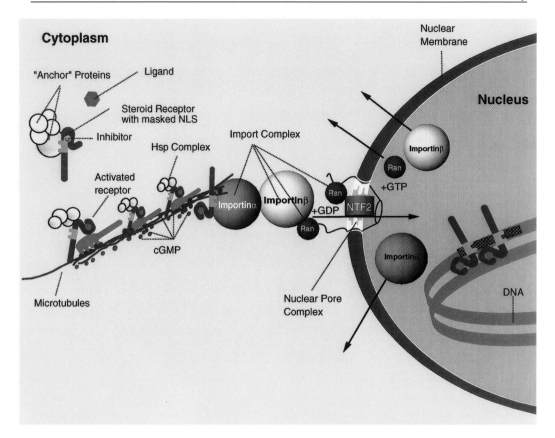

Fig. 1. Schematic model for translocation of ligand-dependent nuclear proteins through the nuclear pore. Ligand binding induces a conformational change, which results in the release of the protein from the "anchor" complex. This "anchor" complex may contain heat shock proteins (hsps) and calreticulin. After phosphorylation, proteins move along microtubules, escorted by molecular chaperones such as hsp. After dissociation of hsp near the nuclear pore, the nuclear localization sequence (NLS) is unmasked, and the protein binds to the NLS receptor importinα. Importinα can bind to pore-associated or cytoplasmic importinβ. The importin heterodimer binds to Ran-guanosine diphosphate (GDP), which binds to nucleoporins at the tip of the cytoplasmic fibers that extend from the nuclear pore complex. Ran also binds to nuclear transport factor 2 (NTF2), and this import complex translocates through the NPC. The import complex disassembles at the nuclear side of the NPC, after which the transport proteins are exported back to the cytoplasm, and the NLS-containing protein is targeted to specific nuclear subcompartments.

receptors [recently reviewed by Pratt and Scherrer (18)]. Calreticulin (19,20) or calmodulin (21,22) binding to steroid receptors can also serve cytoplasmic "docking" functions.

After release of the nuclear protein from its cytoplasmic docking site, a rapid nuclear accumulation occurs, usually within 5 to 60 min. In frog oocytes, synthetic peptides with NLS have been translocated into the nucleus with a $T_{1/2}$ of 5 min (23). The GR translocates into the nucleus within 5–15 min (24–26). This movement includes an active transport through the nuclear pore, and it also includes traveling time through the cytoplasm to reach the nuclear pore. This movement is generally believed to be driven by diffusion, but the expected rate of transport by diffusion is much slower, with a $T_{1/2}$ of 1–3 h. Active transport along cytoskeletal components would better explain this rapid movement of steroid receptors than diffusion.

Many current studies suggest that the translocation of GR involves an interaction with microtubules. Immunocytologic studies have shown the colocalization of GR and other components of the hetero-olygomeric complex with microtubules in cultured cells *(12,27–30)*. An association of activated GR with tubulin is further supported by in vitro experiments, showing that GR converts from soluble to particulate form under conditions that favor microtubule polymerization *(31–33)*. Experiments with overexpressed receptors, however, showed a microtubule-independent and hormone-independent nuclear localization of GR and PR *(34,35)*. The transport process of receptors from the docking site to the nuclear pore is probably more complex. It may also involve other protein–protein interactions, such as receptor interactions with hsps and immunophilins *(30,36,37)*.

The import process through the NPC is mediated by transport factors. The 56-kDa protein, importinα is primarily responsible for NLS recognition. After steroid receptors bind to importinα, importinα binds to a 97-kDa protein, importinβ. This import complex docks at fibers that extend from the cytoplasmic surface of the NPC (Fig. 1). Translocation of the importin–protein complex through the nuclear pore involves the small guanosine triphosphatase (GTPase) Ran, which has to be in the guanosine diphosphate (GDP)-bound form *(38–40)*. Ran-GDP binds to another cytoplasmic transport factor, nuclear transport factor 2 (NTF2 or p10) *(41,42)*. This import complex travels through the nuclear pore by sequential binding to proteins of the NPC. When the NLS protein-receptor complex reaches the nuclear side of the NPC, Ran binds to GTP and dissociates from importin. The importins dissociate from the receptor, and the receptor interacts with putative intranuclear targeting proteins.

2. SUBCELLULAR DISTRIBUTION

The vitamin D receptor (VDR) belongs to the v-erb-A superfamily of ligand-activated transcription factors. The hormonal form of vitamin D, like other steroid hormones, acts through its receptor to regulate the transcription of target genes and thus modulates a variety of cell functions. These hormones also exert rapid, so-called nongenomic actions that take place outside of the nucleus. The mechanisms of these extranuclear actions and their physiological importance are not well understood.

Although it is evident that transcriptional regulation is mediated through nuclear VDR, it is still controversial whether the unliganded VDRs reside in the cytoplasm or in the nucleus.

The use of radioligands led to the classical model for steroid receptor activation that placed the receptors in the cytoplasm *(43)*. Parallel studies suggested that steroid hormone binding initiated receptor translocation into the nucleus *(44)*. Later, antibodies were raised against steroid receptors, and immunocytology suggested that even unactivated steroid receptors reside exclusively in the nucleus. The evidence for receptor translocation was widely criticized as an artifact *(45)*. Since then, as the sensitivity and specificity of immunohistochemical techniques have developed, evidence has accumulated showing that GR, MC, and AR reside both in the cytoplasm and nucleus and that hormone exposure leads to translocation of these receptors *(46,47)*.

Initially, several studies with immunocytology on aldehyde-fixed cells indicated that VDRs reside exclusively in the nucleus *(48)*. Subsequently, technical developments, such as microwave energy fixation of tissues *(49)*, high-efficiency immunodetection systems, confocal microscopy, and image processing, resulted in sufficient temporal and spatial resolution and sensitivity to detect VDR in the cytoplasm and to see the rapid effects of hormone binding on VDR distribution.

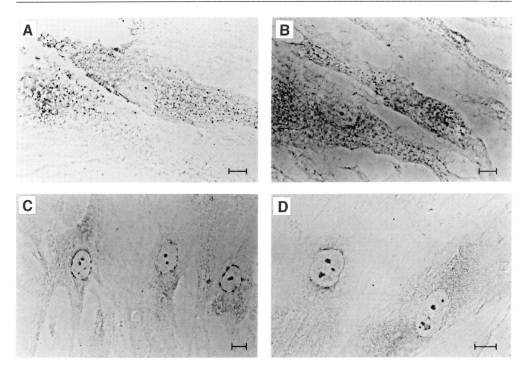

Fig. 2. Calcitriol (10 n*M*) induced rapid reorganization of vitamin D receptor (VDR) in dermal fibroblasts. Cells were cultured without serum or phenol red for 48 h. VDR immunoreactivity was detected on microwave-fixed cells with 9A7 rat monoclonal antibody (1:800 dilution) and with a phycoerythrin-labeled second antibody. (**A**) No hormone added. (**B**) 30 s with hormone (clumping of VDRs and alignment of VDR clumps along fibrils, directed toward the nucleus). (**C**) 1 min with hormone (perinuclear and intranuclear accumulation of clumped VDRs). (**D**) 3 min with hormone (VDR accumulation within foci in the nucleus). Pictures were taken from a Zeiss Photomicroscope III equipped for epifluorescence analysis, with a built-in camera. Scale bars = 10 μm.

Using microwave fixation, we found VDR in the cytoplasm of cultured fibroblasts *(50)*. Compartmentalization of VDR in the absence of calcitriol was regulated by serum or estrogen. VDRs were mainly cytoplasmic in cells cultured without serum or phenol red, but they were predominantly intranuclear when serum or an estrogen was added to the culture medium for at least 18 h. Microwave fixation and immunocytology also revealed a rapid, hormone-dependent translocation of cytoplasmic VDR. After addition of calcitriol, we detected clumping of VDR within 15–45 s, the alignment of these clumps along cytoplasmic fibers within 30–45 s, a perinuclear accumulation of the clumps within 45–90 s, and an intranuclear accumulation of the clumps within 1–3 min (Fig. 2). These sequential pattern changes show a calcitriol dose dependency and a calcitriol analog specificity characteristic of the VDR. Translocation was complete within 10 min when fibroblasts were in exponential growth, but much slower after the cells became confluenced (unpublished results). The finding that wheat germ agglutinin, which blocks protein transport through the nuclear pores, also blocked the calcitriol-dependent translocation of VDR supported the reality of a translocation process.

Detailed analysis ruled out the possibility that much of the cytoplasmic VDR signal might be nonspecific. First, when anti-VDR antibody was omitted or replaced with a nonimmune rat ascites fluid, there was negligible signal. Second, two additional anti-

VDR antibodies directed against different epitopes on VDR gave a similar signal distribution. Third, the amount of VDR immunoreactivity in different cell lines correlated with the number of receptors predicted by biochemical studies, including the virtual absence of VDR signal in CV1 cells. The physiologic significance and specificity of rapid, hormone-dependent VDR reorganization are reinforced by experimental results in dermal fibroblasts from patients with hereditary resistance to calcitriol (mutant cells). These cells harbor homozygous mutations in the VDR gene, causing various defects in VDR functions. These functional defects fall into three categories: hormone binding defects, DNA binding defects, or translocation defects. In a cell line with hormone binding defect (P8), VDR mRNA level was undetectable *(51)*, and VDR immunoreactivity was also absent. In another cell line with hormone binding defect (P10), VDR mRNA level was normal, but a point mutation causing early termination of translation of VDR mRNA deleted the hormone binding region of VDR. In this P10 cell line, VDR immunoreactivity was present, but hormone addition did not induce any change in the pattern of VDR distribution. Defects in the DNA binding region (P3 and P7) affected the intranuclear distribution of VDR immunoreactivity: focal intranuclear accumulation was not present after hormone exposure. In cells with defects in VDR translocation (P1a and P2a), hormone addition caused clumping of VDR, but the clumps accumulated in the perinuclear region, without any intranuclear accumulation. These studies suggested that hormone-induced VDR pattern changes truly represent VDR translocation into the nucleus, as opposed to an artifact, since both an inhibitor of nuclear import and mutations in the VDR gene prevented the pattern changes.

Immunocytology on microwave fixed cells thus became a reliable tool to study VDR translocation. Meanwhile, improved immunocytology and microscopy techniques revealed cytoplasmic VDR in aldehyde-fixed cells as well *(52–58)*. The hormone- induced VDR translocation was also detectable in osteoblasts of parietal bone *(52)*, in which a perinuclear accumulation was detected 1 min after hormone exposure and an intranuclear accumulation 10 min after hormone exposure. These data show that the cytoplasmic localization and translocation of VDR are reproducible, and significant.

3. DISTRIBUTION AND TRANSLOCATION IN LIVING CELLS

Recently we developed a pharmacologically relevant fluorescent ligand for VDR, 3β-BODIPY-calcitriol (BP-calcitriol) *(59)*. To obtain BP-calcitriol, the fluorescent dye BODIPY is attached to the A-ring of calcitriol by an ester link at the 3β-position. This reagent is stable, is resistant to endogenous esterases, freely enters living cells, and retains most of the biologic activity of the parent hormone and its affinity to the receptor. The high quantum efficiency of the BODIPY dye provides sufficient brightness at physiologic hormone concentrations to differentiate signals from noise. The best feature of BP-calcitriol is that its fluorescence emission increases when bound to VDR, but does not increase when bound to other proteins. This allows the use of low hormone doses for binding studies and visualization of VDR. Binding studies with BP-calcitriol in cytosolic extracts and in whole cells showed comparable results to binding studies with radio-labeled calcitriol. The specificity of BP-calcitriol binding to VDR has been established by several criteria. First, the relative potencies of calcitriol analogs to compete with BP-calcitriol are similar to their potencies to compete with the unlabeled calcitriol. Second, the hormone-free BODIPY dye is not retained by cultured cells. Third, the affinity of

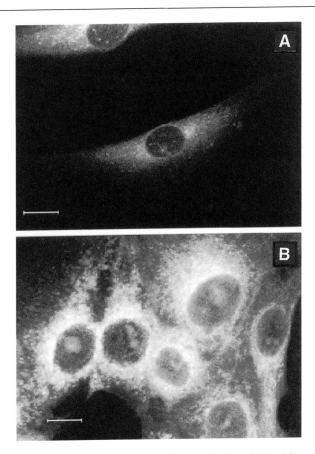

Fig. 3. Subcellular distribution of BP-calcitriol binding sites in dermal fibroblasts (**A**) and in LLC-PK1 kidney epthelial cells (**B**). VDR was detected in the cytoplasm, in the nuclear membrane, and in discrete foci within the nucleus in both cell lines. In the cytoplasm, VDR was predominantly in the perinuclear region and along fibrillar structures. Cells were treated with 10 nM labeled hormone for 30 min and then washed with buffer. Images were taken from a Zeiss Axioplan microscope equipped with a Bio-Rad MRC-600 confocal laser scanning unit. Scale bars = 10 µm.

BP-calcitriol to VDR is influenced by the position of the BODIPY labels on the steroid molecule. Fourth, cells that do not express VDR (CV1, mutant fibroblasts) do not have displacable BP-calcitriol binding *(59,60)*.

BP-calcitriol has provided an important new tool for studying the translocation and subcellular distribution of VDR in its physiologic setting of living cells rather than fixed cells. With BP-calcitriol, it became possible to measure hormone binding even in single cells and within cell compartments. Hormone uptake studies with BP-calcitriol demonstrated a detectable cytoplasm to nuclear translocation of VDR within 5 min, and a maximal nuclear accumulation within 1 h. Further increases in nuclear VDR reflected hormone-induced receptor upregulation. Studies with BP-calcitriol also confirmed our previous observation with immunocytology on microwave-fixed cells, showing that VDR resides not only in the nucleus but also in the cytoplasm (Fig. 3). Cytoplasmic VDRs were along fibrillar structures. The nuclear VDR signal was not only diffuse but also localized within discrete accumulation foci. We found specific disruptions in BP-calcitriol binding patterns in mutant fibroblasts, similar to the pattern changes found with immunocytology *(60)*.

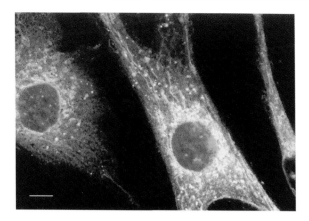

Fig. 4. Colocalization of BP-calcitriol-bound VDR with rhodamine-brefeldin A (rBFA)-labeled endoplasmic reticulum (ER) and Golgi membranes in living cells. The red signal is from rBFA and shows ER and Golgi membranes. The green signal is from BP-calcitriol and shows VDR. The yellow signal shows colocalization of the two signals. Fibroblasts were incubated with 0.37 µg/mL rBFA and 10 nM BP-calcitriol for 30 min at 37°C. The picture was taken from living cells using a Nikon Optiphot microscope equipped with a Bio-Rad MRC-1024 confocal laser scanning unit with appropriate excitation lines and emission filters for the two dyes. Merging of digitized images was done by Laser-Sharp software from Bio-Rad. Scale bars = 10 µm.

4. DOCKING SITES IN THE ENDOPLASMIC RETICULUM

Studies with BP-calcitriol also revealed that the cytoplasmic docking site for VDR is in the endoplasmic reticulum (ER). BP-calcitriol binding sites in the ER appeared as fibrillar structures, forming three-way junctions and free-ended tubules. Colocalization studies in living and fixed cells further supported the presence of VDRs in the ER. Colocalization studies in living cells recently became possible with the development of rhodamine-brefeldin A (rBFA) *(61)*. Brefeldin A binds to ER membranes, inhibits export from ER to Golgi membranes, and causes spreading and tubulation of ER *(62)*. In living cells, rBFA gives a red colored signal in swollen fibril-like structures; this signal colocalizes with specific BP-calcitriol binding sites (Fig. 4).

In fixed cells, immunocytology shows colocalization of VDR immunoreactivity with calreticulin in the ER (Fig. 5). Calreticulin is a major calcium-sequestering protein of the ER lumen. Recently, calreticulin has been shown to bind to the DNA binding region of GR, retinoid X receptor (RXR), and AR and inhibit their ability to activate transcription *(63–65)*. Calreticulin also inhibits vitamin D signal transduction by interacting with a protein motif in the DNA binding domain of the VDR *(66,67)*. Interestingly, after calcitriol exposure, a portion of VDR remained in the ER and nuclear envelope with calreticulin 30 min after hormone addition. This points to a significant difference in the time-course of VDR translocation between cell types and culture conditions (unpublished observation). Calreticulin and VDR did not colocalize inside the nucleus (Fig. 5).

A docking site for VDRs in the ER and a hormone-mediated binding to calreticulin in the ER are strongly supported by a series of elegant in vitro experiments on rat osteosarcoma cell membranes *(68)*. These experiments show that tritiated calcitriol-VDR complex binds to ER membranes and that this binding is competitively inhibited by either a peptide corresponding to the calreticulin binding portion of VDR, or an anticalreticulin antibody. Hormone binding to VDR did not abolish the binding of VDR

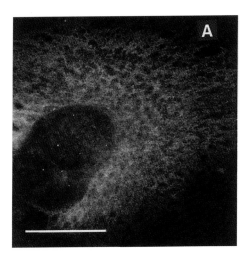 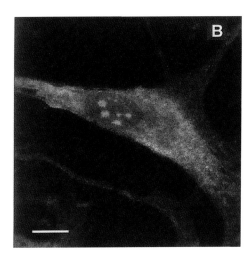

Fig. 5. VDR colocalizes with calreticulin within the endoplasmic reticulum (ER). Areas where the VDR (green) and calreticulin (red) colocalize appear as a yellow signal. (**A**) Without hormone, VDR and calreticulin colocalize in a honeycomb pattern characteristic for ER. There are ER structures without VDR in the peripheral region (red signal) as well. (**B**) After calcitriol exposure, VDR accumulates within nuclear foci, but VDR does not colocalize with calreticulin inside the nucleus (note green spots in the nucleus). Dermal fibroblasts were cultured without serum and phenol red for 24 h, then incubated with a buffer containing 10 µL/mL ethanol (A) or 10 nM calcitriol (B) for 30 min, and then fixed with 2.1s microwave irradiation. Immunocytology was done as described *(60)*. Briefly, an anti-VDR primary, and a lissamine–rhodamine-labeled anti-rat secondary antibody incubation was followed by anti-calreticulin primary and CY5-labeled anti-rabbit secondary antibody incubations. Pictures were taken from a Nikon Optiphot microscope equipped with a Bio-Rad MRC-1024 confocal laser scanning unit with appropriate excitation lines and emission filters. Merging of digitized image files was done by Laser-Sharp software from Bio-Rad. Scale bars = 10 µm.

to ER membranes *(68)*. Taken together, these results suggest that VDRs reside in the ER and bind to calreticulin in the ER during activation.

5. INTERACTION WITH MICROTUBULES

After hormone binding, VDR is released from the ER. During this early activation, a series of events is initiated to set VDR into motion toward the nucleus. Our studies with immunocytology on microwave-fixed cells showed a temporary association of reorganizing VDR with cytoplasmic fibers (Fig. 2B) *(50)*. Similar fibrillar structures appeared in living fibroblasts 15s after BP-calcitriol addition (Fig. 6) *(60)*. Immunostaining indicated that the VDR-associated fibers react with antitubulin antisera and have the characteristic microtubule pattern *(60,69)*. We showed first in fixed cells, and then in living cells, that tubulin-disruptive drugs (nocodazole, colchicine, vinblastine) prevented the alignment of VDR *(60,69)*. Similar findings have been reported in MCT3T fibroblasts, in which 1 min of hormone exposure induced a colocalization of VDR with microtubules. The fibrillar pattern of VDR in that system was also sensitive to colchicine *(52)*. What is more important, in normal human monocytes, disruption of microtubule assembly inhibited genomic functions of VDR *(70)*. Calcitriol-induced increases of 24-hydroxylase mRNA level, increases in production of 24-hydroxylated vitamin D metabolites, and downregulation of calcitriol synthesis were all sensitive to microtubule depolymeriza-

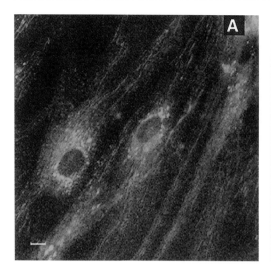 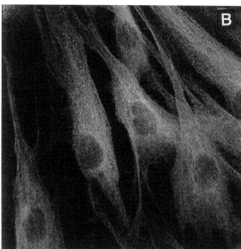

Fig. 6. Colocalization of VDR with microtubules in living (A) and in fixed (B) cells. (A) Dermal fibroblasts were incubated with 10 n*M* BP-calcitriol for 15 s, and a picture was taken immediately from living cells using a Zeiss Axiovert 10 microscope, 40 ↔ 1.3 NA objective and a Photometrics cooled CCD camera. (B) Cells were incubated with 10 n*M* BP-calcitriol for 30 min and then fixed with Bouin's solution and permeabilized with 0.1 mg/mL digitonin. Fixed cells were incubated with a mouse antitubulin primary antibody followed by a CY5-conjugated anti-mouse secondary antibody as described *(60)*. The picture was taken from a Nikon Optiphot microscope equipped with a Bio-Rad MRC-1024 confocal laser scanning unit, 63 ↔ 1.3 NA objective, and appropriate excitation filters for CY5 or BODIPY. Merging of digitized images was done by Laser-Sharp software from Bio-Rad, showing tubulin immunoreactivity (red) and BP-calcitriol (green) partially colocalized (yellow). Scale bars = 10 µm.

tion. These results suggest that the interaction between microtubules and VDR could have physiologic importance. Nevertheless, colocalization of VDR with microtubule-like structures could reflect association with ER or microfilaments as well. It is a concern that microtubule-depolymerizing drugs affect various other cell functions. These problems will be addressed by ongoing in vitro experiments in our laboratory, showing that VDR copolymerizes with microtubule-associated protein (MAP)-containing purified tubulin.

It is well known that motor MAPs such as dynein transport proteins toward the cell center *(71)*. Immunoabsorption experiments showed that the hsp56 immunophilin component of the untransformed GR complex is associated with both intermediate and heavy chains of dynein *(12)*. Although hsp56 could be one of the proteins that play a role in the interaction between VDR and microtubules, further studies are necessary to determine whether this or other proteins have a chaperone function for VDR.

Molybdate is a well-known inhibitor of steroid receptor activation. It stabilizes the association of untransformed receptors with the hsp90 heterocomplex and prevents receptor binding to DNA. Although the VDR does not appear to be permanently associated with hsp90 *(72)*, it does reside in a large protein complex. Hormone-free VDR sediments in a 5.5 S complex, and it turns into a 3.2 S form after activation *(73)*. Molybdate stabilizes VDR as well, and preserves VDR in a conformation that prevents its binding to DNA *(74)*. The components of the VDR-associated protein complex are not known, and a temporary association with hsp90 cannot be excluded *(75)*. We demonstrated that pretreatment of cultured dermal fibroblasts with molybdate causes a permanent association of VDR with microtubules and prevents VDR translocation into the nucleus (69). In agreement with our results on molybdate inhibition of VDR transport, it has recently been shown that molybdate treatment traps both GR and progesterone receptors in the cytoplasm of hormone-treated cells (16). These findings suggested to us that dissociation of hsp90 from steroid receptors is not required for interaction with microtubules (69). Current results support this hypothesis, showing that hsp90 facilitates the interaction of GR with the nuclear transport machinery (36). This supports the hypothesis that a transient dissociation-association of hsp90 contributes to steroid receptor movement along microtubules.

6. A ROLE OF RAPID cGMP ACCUMULATION

We found that one early action of calcitriol is to increase the intracellular cyclic guanosine monophosphate (cGMP) concentration in normal human skin fibroblasts. Calcitriol increased cGMP by two- to threefold, in a dose-dependent manner, with a detectable response to 0.1nM of hormone and a maximal response to 100 nM hormone. The elevation was detectable as early as 1 min and reached a maximum after 6–8 min. Other steroid hormones caused similar effects. Mutant cells with defective VDR functions did not exhibit calcitriol-induced cGMP accumulation, although they responded normally to dihydrotestosterone (76). This finding suggested that this rapid action of calcitriol is mediated through activation of VDR, not through a VDR-independent activation of a plasma membrane component. Immunocytology on microwave-fixed cells revealed rapid and agonist-specific changes in the subcellular accumulation patterns of cGMP (49). Using this method, we showed that calcitriol induces accumulation of cGMP in cytoplasmic aggregates within 15 s (77). Double immunostaining for cGMP and for VDR showed that the initial cGMP accumulation after calcitriol addition colocalized with clumped cytoplasmic VDR. At 30 s after calcitriol treatment, both VDRs and cGMP aligned along microtubules (77).

We also found that molybdate is a highly potent inducer of cGMP accumulation in dermal fibroblasts and in many other cell lines (69). Molybdate not only stabilized VDR association with microtubules but also stabilized cGMP accumulation along the same fibrillar structures (69). The addition of a high dose of 8-Br-cGMP to fibroblasts mimicked the effect of molybdate, preventing the intranuclear accumulation of VDR. Similar findings have been reported regarding estrogen receptor stabilization by cGMP and molybdate (78,79). Based on these results, it appears likely that cGMP plays a role in the interaction of VDR with microtubules.

The molecular mechanisms of interaction between VDR and microtubules through cGMP are not yet known. An involvement for cGMP is suggested by experiments showing that cGMP increases tubulin polymerization and promotes microtubule function in leukocytes (80). cGMP has also been observed to stimulate dynein-mediated particle movement (79) and to phosphorylate proteins in the dynein complex extracted from *Paramecium* (81,82). These data suggest that the interaction of VDR with tubulin could be mediated by cGMP-dependent phosphorylation of MAPs. Whereas cGMP accumulation is likely to play a role in microtubule-mediated movement of VDR, hormone-bound VDR itself does not have guanylate cyclase activity. Presumably, activation of guanylate cyclase after calcitriol exposure is indirect.

Accumulating data suggest that the molecular events initiating VDR translocation originate in the ER and involve activation of protein kinase C (PKC), release of calcium,

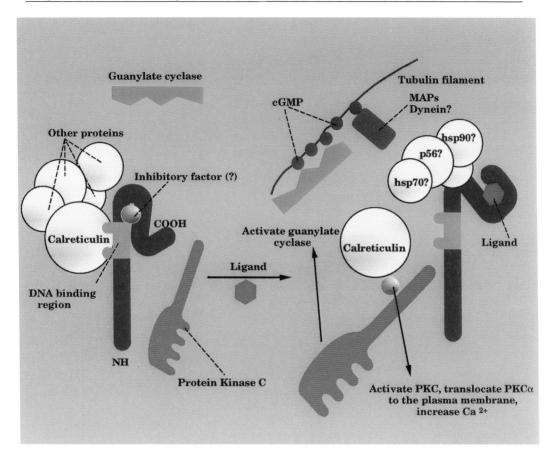

Fig. 7. A schematic model for the early steps of VDR activation. Unliganded receptor resides in a docking complex bound to calreticulin and other proteins. This receptor complex may also contain a low molecular weight inhibitory factor (93). Hormone binding changes the conformation of VDR and results in dissociation of the inhibitory factor from the receptor. The free factor then activates protein kinase C, which activates guanylate cyclase and increases intracellular calcium. After its release from the endoplasmic reticulum, VDR transiently associates with heat shock proteins (hsps). Hsp and cGMP facilitate vitamin D receptor (VDR) interactions with microtubules to deliver VDR to the vicinity of the nuclear pore complex.

and activation of guanylate cyclase (Fig. 7). This sequence of events was first indicated by experiments on rat colonocytes *(83)*. These experiments showed that calcitriol induces activation of particulate guanylate cyclase via a PKC-dependent mechanism. Several other articles showed that at physiologic concentrations, calcitriol first induces a rapid (15 s) release of inositol triphosphate and diacylglycerol, then activates PKC, then causes PKCα to translocate from the cytoplasm to the plasma membrane, and finally induces an increase in intracellular free calcium *(83–88)*. It has been demonstrated that PKC *(89)*, guanylate cyclase *(90)*, and cGMP-dependent protein kinase *(91)* are present in the ER. It has also been shown that the activation of PKC and guanylate cyclase in the ER facilitates protein export from the ER. Thus, these rapid actions of calcitriol could be important for VDR release from the ER *(92)*. Frequently, the activation of PKC is interpreted as an action independent of receptor activation. The low hormone dose exerting this effect, the analog specificity, and the requirement for functional VDR in the cell

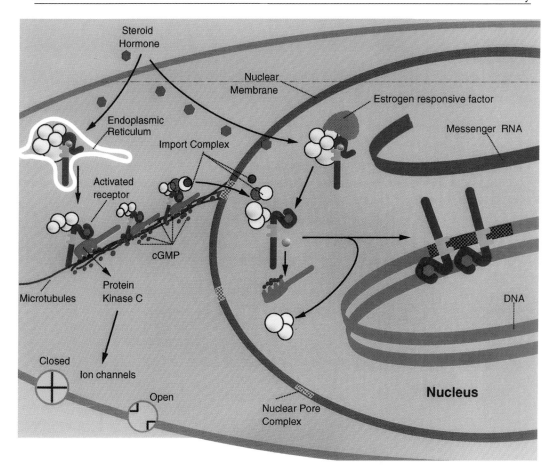

Fig. 8. A schematic model for vitamin D receptor (VDR) activation and translocation into the nucleus. VDR can reside both in the nucleus and in the endoplasmic reticulum within a heteroolygomeric complex. Nuclear retention of VDR is dependent on an estrogen-responsive factor. Hormone binding changes the conformation of VDR and initiates motion of VDR. This motion takes cytoplasmic VDR to the nuclear pore complex (NPC). This process involves activation of protein kinase C (PKC) and then activation of guanylate cyclase by PKC. Accumulation of cGMP facilitates VDR interaction with microtubules, and a temporary association with heat shock proteins maintains a piggyback movement along microtubules. Near the nuclear pore, VDR binds to the importin complex, which interacts with cytosolic import factors [Ran, nuclear transport factor 2 (NTF2)] to pass through the NPC. Inside the nucleus, VDR dissociates from the import complex and reaches the target genes through an intranuclear transport process.

allow for an alternative explanation: mediation through the VDR. A possible link between steroid receptors and PKC could be the steroid receptor-bound low molecular weight inhibitory factor that was shown to inhibit the translocation and activation of GR, AR, and estrogen receptors *(93,94)*. This modulator is an ether aminophosphoglyceride that appears to be the "endogenous molybdate factor." When this factor is released upon hormone binding, it can stimulate PKC activity *(95–97)*. Thus, both morphological and functional results suggest that after VDR is released from the ER, it could translocate into the nucleus through a hormone-dependent activation process that involves stimulation of PKC and guanylate cyclase. However, information about protein interactions that determine directional movement of VDR along microtubules is still missing.

7. SUMMARY: MECHANISMS

Figure 8 summarizes our current working model for the mechanism of VDR translocation. Although the information on VDR translocation is still fragmented, it is clear that at least a portion of VDR resides in the cytoplasm. The most likely docking site for VDR is in the ER, and calreticulin may be part of the docking complex. Upon hormone exposure, molecular events lead to the association of VDR with microtubules and with the nuclear import machinery. Figure 7 shows the initial activation steps: release of a low molecular weight inhibitor, stimulation of PKC activity by the inhibitor, activation of guanylate cyclase by PKC, and phosphorylation of MAPs. Once VDR reaches the vicinity of the nuclear pore, it can associate with importins and translocate through the nuclear pore by a Ran-mediated, energy-dependent process (Fig. 1). This model does not exclude the possibility that VDR can also reside in the nucleus in its unactivated form. This "nuclear" VDR would associate with importins without hormone binding. Further studies will clarify the exact mechanism for the association of the different VDR forms with the nuclear import complex.

REFERENCES

1. Jans DA, Hubner S. Regulation of protein transport to the nucleus: central role of phosphorylation. Physiol Rev 1996; 76:651–685.
2. Adam SA, Gerace L. Cytosolic proteins that specifically bind nuclear localization signals are receptors for nuclear import. Cell 1991; 66:837–847.
3. Picard D, Yamamoto KR. Two signals mediate hormone-dependent nuclear localization of the glucocorticoid receptor. EMBO J 1987; 6:3333–3340.
4. Picard D, Salser SJ, Yamamoto KR. A movable and regulable inactivation function within the steroid binding domain of the glucocorticoid receptor. Cell 1988; 54:1073–1080.
5. Cadepond F, Gasc JM, Delahaye F, Jibard N, Schweizer-Groyer G, Segard-Maurel I, et al. Hormonal regulation of the nuclear localization signals of the human glucocorticosteroid receptor. Exp Cell Res 1992; 201:99–108.
6. Guiochon-Mantel A, Delabre K, Lescop P, Milgrom E. The Ernst Schering Poster Award. Intracellular traffic of steroid hormone receptors. J Steroid Biochem Mol Biol 1996; 56:3–9.
7. Zhou ZX, Sar M, Simental JA, Lane MV, Wilson EM. A ligand-dependent bipartite nuclear targeting signal in the human androgen receptor. Requirement for the DNA-binding domain and modulation by NH2-terminal and carboxyl-terminal sequences. J Biol Chem 1994; 269:13,115–13,123.
8. LaCasse EC, Lochnan HA, Walker P, Lefebvre YA. Identification of binding proteins for nuclear localization signals of the glucocorticoid and thyroid hormone receptors [published erratum appears in Endocrinology 1993; 133:2760]. Endocrinology 1993; 132:1017–1025.
9. Ylikomi T, Bocquel MT, Berry M, Gronemeyer H, Chambon P. Cooperation of proto-signals for nuclear accumulation of estrogen and progesterone receptors. EMBO J 1992; 11:3681–3694.
10. Vancurova I, Jochova-Rupes J, Lou W, Paine PL. Distinct phosphorylation sites differentially influence facilitated transport of an NLS-protein and its subsequent intranuclear binding. Biochem Biophys Res Commun 1995; 217:419–427.
10a. Hsieh JC, Simizu Y, Minosima S, Simizu N, Haussler CA, Jurutka PW et al. Novel nuclear localization signal between the two DNA-binding zinc fingers in the human vitamin D receptor. J Cell Biochem 1998; 70:94–109.
11. Kuiper GG, Brinkmann AO. Steroid hormone receptor phosphorylation: is there a physiological role? Mol Cell Endocrinol 1994; 100:103–107.
12. Czar MJ, Owens-Grillo JK, Yem AW, Leach KL, Deibel MR, Welsh MJ, et al. The hsp56 immunophilin component of untransformed steroid receptor complexes is localized both to microtubules in the cytoplasm and to the same nonrandom regions within the nucleus as the steroid receptor. Mol Endocrinol 1994; 8:1731–1741.
13. Hutchison KA, Stancato LF, Owens-Grillo JK, Johnson JL, Krishna P, Toft DO, et al. The 23-kDa acidic protein in reticulocyte lysate is the weakly bound component of the hsp foldosome that is required for assembly of the glucocorticoid receptor into a functional heterocomplex with hsp90. J Biol Chem 1995; 270:18,841–18,847.

14. Hutchison KA, Dittmar KD, Pratt WB. All of the factors required for assembly of the glucocorticoid receptor into a functional heterocomplex with heat shock protein 90 are preassociated in a self-sufficient protein folding structure, a "foldosome". J Biol Chem 1994; 269:27,894–27,899.
15. Tai PK, Chang H, Albers MW, Schreiber SL, Toft DO, Faber LE. P59 (FK506 binding protein 59) interaction with heat shock proteins is highly conserved and may involve proteins other than steroid receptors. Biochemistry 1993; 32:8842–8847.
16. Yang J, DeFranco DB. Assessment of glucocorticoid receptor-heat shock protein 90 interactions in vivo during nucleocytoplasmic trafficking. Mol Endocrinol 1996; 10:3–13.
17. Smith DF. Dynamics of heat shock protein 90-progesterone receptor binding and the disactivation loop model for steroid receptor complexes. Mol Endocrinol 1993; 7:1418–1429.
18. Pratt B, Scherrer L. Heat shock proteins and the cytoplasmic-nuclear trafficking of steroid receptors. In: Steroid Hormone Receptors: Basic and Clinical Aspects. Moudgil VK, ed. Boston: Birkhauser, 1994; 215–246.
19. Shago M, Flock G, Leung Hagesteijn CY, Woodside M, Grinstein S, Giguere V, et al. Modulation of the retinoic acid and retinoid X receptor signaling pathways in P19 embryonal carcinoma cells by calreticulin. Exp Cell Res 1997; 230:50–60.
20. Michalak M, Burns K, Andrin C, Mesaeli N, Jass GH, Busaan JL, et al. Endoplasmic reticulum form of calreticulin modulates glucocorticoid-sensitive gene expression. J Biol Chem 1996; 271: 29,436–29,445.
21. Ning YM, Sanchez ER. In vivo evidence for the generation of a glucocorticoid receptor-heat shock protein-90 complex incapable of binding hormone by the calmodulin antagonist phenoxybenzamine. Mol Endocrinol 1996; 10:14–23.
22. Ning YM, Sanchez ER. Evidence for a functional interaction between calmodulin and the glucocorticoid receptor. Biochem Biophys Res Commun 1995; 208:48–54.
23. Goldfarb DS, Gariepy J, Schoolnik G, Kornberg RD. Synthetic peptides as nuclear localization signals. Nature 1986; 322:641–644.
24. Htun H, Barsony J, Renyi I, Gould D, Hager G. Visualization of glucocorticoid receptor translocation and intranuclear organization in living cells with a green fluorescent protein chimera. Proc Natl Acad Sci USA 1996; 93:4845–4850.
25. Richards SA, Lounsbury KM, Carey KL, Macara IG. A nuclear export signal is essential for the cytosolic localization of the Ran binding protein, RanBP1. J Cell Biol 1996; 134:1157–1168.
26. Carey KL, Richards SA, Lounsbury KM, Macara IG. Evidence using a green fluorescent protein-glucocorticoid receptor chimera that the Ran/TC4 GTPase mediates an essential function independent of nuclear protein import. J Cell Biol 1996; 133:985–996.
27. Akner G, Sundqvist KG, Denis M, Wikstrom AC, Gustafsson JA. Immunocytochemical localization of glucocorticoid receptor in human gingival fibroblasts and evidence for a colocalization of glucocorticoid receptor with cytoplasmic microtubules. Eur J Cell Biol 1990; 53:390–401.
28. Akner G, Mossberg K, Wikstrom A, Sundquist K, Gustafsson J. Evidence for colocalization of glucocorticoid receptor with cytoplasmic microtubules in human gingival fibroblasts, using two different monoclonal anti-GR antibodies, confocal laser scanning microscopy, and image analysis. J Steriod Biochem Mol Biol 1991; 39:419–432.
29. Perrot-Applanat M, Cibert C, Geraud G, Renoir JM, Baulieu EE. The 59 kDa FK506-binding protein, a 90 kDa heat shock protein binding immunophilin (FKBP59-HBI), is associated with the nucleus, the cytoskeleton and mitotic apparatus. J Cell Sci 1995; 108:2037–2051.
30. Sanchez ER, Redmond T, Scherrer LC, Bresnick EH, Welsh MJ, Pratt WB. Evidence that the 90-kilodalton heat shock protein is associated with tubulin-containing complexes in L cell cytosol and in intact PtK cells. Mol Endocrinol 1988; 2:756–760.
31. Scherrer LC, Pratt WB. Energy-dependent conversion of transformed cytosolic glucocorticoid receptors from soluble to particulate-bound form. Biochemistry 1992; 31:10,879–10,886.
32. Scherrer LC, Pratt WB. Association of the transformed glucocorticoid receptor with a cytoskeletal protein complex. J Steroid Biochem Mol Biol 1992; 41:719–721.
33. Akner G, Wikstrom AC, Stromstedt PE, Stockman O, Gustafsson JA, Wallin M. Glucocorticoid receptor inhibits microtubule assembly in vitro. Mol Cell Endocrinol 1995; 110:49–54.
34. Perrot-Applanat M, Lescop P, Milgrom E. The cytoskeleton and the cellular traffic of the progesterone receptor. J Cell Biol 1992; 119:337–348.
35. Szapary D, Barber T, Dwyer N, Blanchette-Mackie E, Simons Jr. S. Microtubules are not required for glucocorticoid receptor mediated gene induction. J Steroid Biochem Mol Biol 1994; 51: 143–148.

36. Owens-Grillo JK, Czar MJ, Hutchison KA, Hoffmann K, Perdew GH, Pratt WB. A model of protein targeting mediated by immunophilins and other proteins that bind to hsp90 via tetratricopeptide repeat domains. J Biol Chem 1996; 271:13,468–13,475.
37. Selkirk JK, Merrick BA, Stackhouse BL, He C. Multiple p53 protein isoforms and formation of oligomeric complexes with heat shock proteins Hsp70 and Hsp90 in the human mammary tumor, T47D, cell line. Appl Theor Electrophor 1994; 4:11–18.
38. Chi NC, Adam EJ, Visser GD, Adam SA. RanBP1 stabilizes the interaction of Ran with p97 nuclear protein import. J Cell Biol 1996; 135:559–569.
39. Weis K, Dingwall C, Lamond AI. Characterization of the nuclear protein import mechanism using Ran mutants with altered nucleotide binding specificities. EMBO J 1996; 15:7120–7128.
40. Paschal BM, Delphin C, Gerace L. Nucleotide-specific interaction of Ran/TC4 with nuclear transport factors NTF2 and p97. Proc Natl Acad Sci USA 1996; 93:7679–7683.
41. Clarkson WD, Kent HM, Stewart M. Separate binding sites on nuclear transport factor 2 (NTF2) for GDP-Ran and the phenylalanine-rich repeat regions of nucleoporins p62 and Nsp1p. J Mol Biol 1996; 263:517–524.
42. Corbett AH, Silver PA. The NTF2 gene encodes an essential, highly conserved protein that functions in nuclear transport in vivo. J Biol Chem 1996; 271:18,477–18,484.
43. Jensen EV, Szuzuki T, Kawashima T, Stumpf W, Jungblut P. A two step mechanism for the interaction of estradiol with rat uterus. Proc Natl Acad Sci USA 1968; 59:632–638.
44. Williams D, Gorski J. Kinetic and equilibrium analysis of estradiol in uterus: a model of binding-site distribution in uterine cells. Proc Natl Acad Sci USA 1972; 69:3464–3468.
45. Gorski J, Welshons W, Sakai D, Hansen J, Kassis J, Shull J, et al. Evolution of a model of estrogen action. Recent Prog Horm Res 1986; 42:297–329.
46. Jensen EV. Steroid hormones, receptors, and antagonists. Ann NY Acad Sci 1996; 784:1–17.
47. Pratt WB, Sanchez ER, Bresnick EH, Meshinchi S, Scherrer LC, Dalman FC, et al. Interaction of the glucocorticoid receptor with the Mr 90,000 heat shock protein: an evolving model of ligand-mediated receptor transformation and translocation. Cancer Res 1989; 49:2222s–2229s.
48. Walters MR. Newly identified actions of the vitamin D endocrine system. Endocr Rev 1992; 13:719–764.
49. Barsony J, Marx SJ. Immunocytology on microwave-fixed cells reveals rapid and agonist-specific changes in subcellular accumulation patterns for cAMP or cGMP [published erratum appears in Proc Natl Acad Sci USA 1990; 87:3633]. Proc Natl Acad Sci USA 1990; 87:1188–1192.
50. Barsony J, Pike JW, DeLuca HF, Marx SJ. Immunocytology with microwave-fixed fibroblasts shows 1alpha,25-dihydroxyvitamin D3-dependent rapid and estrogen-dependent slow reorganization of vitamin D receptors. J Cell Biol 1990; 111:2385–2395.
51. Ritchie HH, Hughes MR, Thompson ET, Malloy PJ, Hochberg Z, Feldman D, et al. An ochre mutation in the vitamin D receptor gene causes hereditary 1,25-dihydroxyvitamin D3-resistant rickets in three families. Proc Natl Acad Sci USA 1989; 86:9783–9787.
52. Amizuka N, Ozawa H. Intracellular localization and translocation of 1alpha, 25-dihydroxyvitamin D_3 receptor in osteoblasts. Arch Histol Cytol 1992; 55:77–88.
53. Jakob F, Gieseler F, Tresch A, Hammer S, Seufert J, Schneider D. Kinetics of nuclear translocation and turnover of the vitamin D receptor in human HL60 leukemia cells and peripheral blood lymphocytes—coincident rise of DNA-relaxing activity in nuclear extracts. J Steroid Biochem Mol Biol 1992; 42:11–16.
54. Liu L, Ng M, Iacopino AM, Dunn ST, Hughes MR, Bourdeau JE. Vitamin D receptor gene expression in mammalian kidney. J Am Soc Nephrol 1994; 5:1251–1258.
55. Johnson JA, Grande JP, Roche PC, Campbell RJ, Kumar R. Immunolocalization of calcitriol receptor, plasma membrane calcium pump and calbindin-D28k in the cornea and ciliary body of the rat eye. Ophthalmic Res 1995; 27:42–47.
56. Berdal A, Papagerakis P, Hotton D, Bailleul-Forestier I, Davideau JL. Ameloblasts and odontoblasts, target-cells for 1,25-dihydroxyvitamin D3: a review. Int J Dev Biol 1995; 39:257–262.
57. Reichrath J, Collins ED, Epple S, Kerber A, Norman AW, Bahmer FA. Immunohistochemical detection of 1,25-dihydroxyvitamin D_3 receptors (VDR) in human skin. A comparison of five antibodies. Pathol Res Pract 1996; 192:281–289.
58. Johnson JA, Grande JP, Roche PC, Kumar R. Immunohistochemical detection and distribution of the 1,25-dihydroxyvitamin D_3 receptor in rat reproductive tissues. Histochem Cell Biol 1996; 105:7–15.
59. Barsony J, Renyi I, McKoy W, Kang HC, Haugland RP, Smith CL. Development of a biologically active fluorescent-labeled calcitriol and its use to study hormone binding to the vitamin D receptor. Anal Biochem 1995; 229:68–79.

60. Barsony J, Renyi I, McKoy W. Subcellular distribution of normal and mutant vitamin D receptors in living cells. J Biol Chem 1997; 272:5774–5782.
61. Deng Y, Bennink JR, Kang HC, Haugland RP, Yewdell JW. Fluorescent conjugates of brefeldin A selectively stain the endoplasmic reticulum and Golgi complex of living cells. J Histochem Cytochem 1995; 43:907–915.
62. Feiguin F, Ferreira A, Kosik KS, Caceres A. Kinesin-mediated organelle translocation revealed by specific cellular manipulations. J Cell Biol 1994; 127:1021–1039.
63. Dedhar S, Rennie PS, Shago M, Hagesteijn CY, Yang H, Filmus J, et al. Inhibition of nuclear hormone receptor activity by calreticulin. Nature 1994; 367:480–483.
64. Burns K, Duggan B, Atkinson EA, Famulski KS, Nemer M, Bleackley RC, et al. Modulation of gene expression by calreticulin binding to the glucocorticoid receptor. Nature 1994; 367:476–480.
65. Winrow CJ, Miyata KS, Marcus SL, Burns K, Michalak M, Capone JP, et al. Calreticulin modulates the in vitro DNA binding but not the in vivo transcriptional activation by peroxisome proliferator-activated receptor/retinoid X receptor heterodimers. Mol Cell Endocrinol 1995; 111:175–179.
66. Wheeler DG, Horsford J, Michalak M, White JH, Hendy GN. Calreticulin inhibits vitamin D_3 signal transduction. Nucleic Acids Res 1995; 23:3268–3274.
67. St-Arnaud R, Prud'homme J, Leung-Hagesteijn C, Dedhar S. Constitutive expression of calreticulin in osteoblasts inhibits mineralization. J Cell Biol 1995; 131:1351–1359.
68. Kim YS, Macdonald PN, Dedhar S, Hruska KA. Association of 1alpha,25-dihydroxyvitamin D3-occupied vitamin D receptors with cellular membrane acceptance sites. Endocrinology 1996; 137: 3649–3658.
69. Barsony J, McKoy W. Molybdate increases intracellular 3',5'-guanosine cyclic monophosphate and stabilizes vitamin D receptor association with tubulin-containing filaments. J Biol Chem 1992; 267:24,457–24,465.
70. Kamimura S, Gallieni M, Zhong M, Beron W, Slatopolsky E, Dusso A. Microtubules mediate cellular 25-hydroxyvitamin D_3 trafficking and the genomic response to 1,25-dihydroxyvitamin D_3 in normal human monocytes. J Biol Chem 1995; 270:22,160–22,166.
71. Wang C, Asai DJ, Robinson KR. Retrograde but not anterograde bead movement in intact axons requires dynein. J Neurobiol 1995; 27:216–226.
72. Whitfield GK, Hsieh JC, Nakajima S, Macdonald PN, Thompson PD, Jurutka PW, et al. A highly conserved region in the hormone-binding domain of the human vitamin D receptor contains residues vital for heterodimerization with retinoid X receptor and for transcriptional activation. Mol Endocrinol 1995; 9:1166–1179.
73. Hirst M, Feldman D. Salt-induced activation of 1,25-dihydroxyvitamin D_3 receptors to a DNA binding form. J Biol Chem 1987; 262:7072–7075.
74. Nakada M, Simpson RU, DeLuca HF. Molybdate and the 1,25-dihydroxyvitamin D_3 receptor from chick intestine. Arch Biochem Biophys 1985; 238:517–521.
75. Holley SJ, Yamamoto KR. A role for Hsp90 in retinoid receptor signal transduction. Mol Biol Cell 1995; 6:1833–1842.
76. Barsony J, Marx SJ. Receptor-mediated rapid action of 1alpha,25-dihydroxycholecalciferol: increase of intracellular cGMP in human skin fibroblasts. Proc Natl Acad Sci USA 1988; 85:1223-1226.
77. Barsony J, Marx SJ. Rapid accumulation of cyclic GMP near activated vitamin D receptors. Proc Natl Acad Sci USA 1991; 88:1436–1440.
78. Fleming H, Blumenthal R, Gurpide E. Rapid changes in specific estrogen binding elicited by cGMP or cAMP in cytosol from human endometrial cells. Proc Natl Acad Sci USA 1983; 80:2486 2490.
79. Gurpide E, Blumenthal R, Fleming H. Regulation of estrogen receptor levels in endometrial cancer cells. Prog Clin Biol Res 1984; 142:145-165.
80. Strukov AI, Paukov VS, Kaufman OI. [Leukocyte cytoskeleton under normal and pathological conditions]. Arkh Patol 1983; 45:81–87.
81. Walczak CE, Nelson DL. In vitro phosphorylation of ciliary dyneins by protein kinases from Paramecium. J Cell Sci 1993; 106:1369–1376.
82. Travis SM, Nelson DL. Regulation of axonemal Mg^{2+}-ATPase from *Paramecium* cilia: effects of Ca^{2+} and cyclic nucleotides. Biochim Biophys Acta 1988; 966:84–93.
83. Khare S, Wilson DM, Tien XY, Dudeja PK, Wali RK, Sitrin MD, et al. 1,25-Dihydroxycholecalciferol rapidly activates rat colonic particulate guanylate cyclase via a protein kinase C-dependent mechanism. Endocrinology 1993; 133:2213–2219.

84. Tien XY, Brasitus TA, Qasawa BM, Norman AW, Sitrin MD. Effect of 1,25(OH)$_2$D$_3$ and its analogues on membrane phosphoinositide turnover and [Ca^{2+}]$_i$ in Caco-2 cells. Am J Physiol 1993; 265: G143–G148.
85. Slater SJ, Kelly MB, Taddeo FJ, Larkin JD, Yeager MD, McLane JA, et al. Direct activation of protein kinase C by 1alpha,25-dihydroxyvitamin D3. J Biol Chem 1995; 270:6639–6643.
86. Morelli S, de Boland AR, Boland RL. Generation of inositol phosphates, diacylglycerol and calcium fluxes in myoblasts treated with 1,25-dihydroxyvitamin D$_3$. Biochem J 1993; 289:675–679.
87. Bissonnette M, Wali RK, Hartmann SC, Niedziela SM, Roy HK, Tien XY, et al. 1,25-Dihydroxyvitamin D$_3$ and 12-O-tetradecanoyl phorbol 13-acetate cause differential activation of Ca$^{(2+)}$-dependent and Ca$^{(2+)}$-independent isoforms of protein kinase C in rat colonocytes. J Clin Invest 1995; 95: 2215–2221.
88. Berry DM, Antochi R, Bhatia M, Meckling-Gill KA. 1,25-Dihydroxyvitamin D$_3$ stimulates expression and translocation of protein kinase Calpha and Cdelta via a nongenomic mechanism and rapidly induces phosphorylation of a 33-kDa protein in acute promyelocytic NB4 cells. J Biol Chem 1996; 271:16,090–16,096.
89. Goodnight JA, Mischak H, Kolch W, Mushinski JF. Immunocytochemical localization of eight protein kinase C isozymes overexpressed in NIH 3T3 fibroblasts. Isoform-specific association with microfilaments, Golgi, endoplasmic reticulum, and nuclear and cell membranes. J Biol Chem 1995; 270:9991–10,001.
90. Davidoff M, Dimitrov N. Electron microscopical localization of guanylate cyclase activity in the neocortex of the guinea pig. Acta Histochem 1989; 85:109–116.
91. Pryzwansky KB, Kidao S, Wyatt TA, Reed W, Lincoln TM. Localization of cyclic GMP-dependent protein kinase in human mononuclear phagocytes. J Leukoc Biol 1995; 57:670–678.
92. Fabbri M, Bannykh S, Balch WE. Export of protein from the endoplasmic reticulum is regulated by a diacylglycerol/phorbol ester binding protein. J Biol Chem 1994; 269:26,848–26,857.
93. Bodine PV, Litwack G. Purification of the glucocorticoid receptor-mineralocorticoid receptor modulator-2 from rabbit liver. Receptor 1995; 5:133–143.
94. Celiker MY, Haas A, Saunders D, Litwack G. Specific regulation of male rat liver cytosolic estrogen receptor by the modulator of the glucocorticoid receptor. Biochem Biophys Res Commun 1993; 195: 151–157.
95. Hsu TC, Bodine PV, Litwack G. Endogenous modulators of glucocorticoid receptor function also regulate purified protein kinase C. J Biol Chem 1991; 266:17,573–17,579.
96. Bodine PV, Litwack G. The glucocorticoid receptor and its endogenous regulators. Receptor 1990; 1:83–119.
97. Miller-Diener A, Schmidt TJ, Litwack G. Protein kinase activity associated with the purified rat hepatic glucocorticoid receptor. Proc Natl Acad Sci USA 1985; 82:4003–4007.

9
Molecular Recognition and Structure–Activity Relations in Vitamin D-Binding Protein and Vitamin D Receptor

Rahul Ray

1. INTRODUCTION

1α,25-dihydroxyvitamin D_3 [1α,25(OH)$_2$$D_3$], a dihydroxylated metabolite of vitamin D_3, is a secosteroid with diverse biological actions *(1,2)*. Vitamin D_3 is synthesized in skin by the interaction of ultraviolet (UV) light from the sun with 7-dehydrocholesterol, a constituent of the epidermis. The product of this photolytic reaction is predominantly a ring-opened compound called previtamin D_3, which then isomerizes slowly by the body temperature to vitamin D_3. After the cutaneous synthesis, vitamin D_3 diffuses into the blood stream and is sequentially oxidized by specific P450-containing hydroxylases in the liver to form 25-hydroxyvitamin D_3 [25(OH)D_3], and in the kidney to form 1α,25(OH)$_2$ D_3, the active form of vitamin D hormone. After the biosynthesis, 1α,25(OH)$_2$ D_3 translocates into numerous target organs like intestine, bone, skin, pituitary, kidney, ovary, etc., where its various biologic functions manifest (*see below*) (Fig. 1). In addition to vitamin D_3 (synthesized in the skin), another chemical form of vitamin D_3 (called vitamin D_2, which is obtained primarily from diet and vitamin D supplements) and all of its metabolites exist in nature. Vitamin D_2 is metabolized to 25-hydroxyvitamin D_2 and 1α,25-dihydroxyvitamin D_2, similar to vitamin D_3 *(1)*.

It is well-known that vitamin D_3 and all of its metabolites, as well as many synthetic analogs, bind strongly to serum vitamin D-binding protein (DBP), a member of globulin gene family *(3,4)*. This suggests that DBP may act a carrier for vitamin D_3 and its metabolites in the blood. However, the physiologic significance of this function is yet to be defined.

Classically 1α,25(OH)$_2$ D_3 is a calcium and phosphorus homeostatic hormone. Recently several nonclassical functions of 1α,25(OH)$_2$ D_3 have been identified, e.g., influencing the secretion of peptide hormones such as prolactin, parathyroid hormone (PTH), and insulin, and modifying T-lymphocytes *(2)*. Additionally, 1α,25(OH)$_2$$D_3$ has been shown to have profound effects on growth and maturation of a broad range of normal and malignant cells *(5–11)*. It is well established that most biologic functions of 1α,25(OH)$_2$ D_3 are mediated by its strong and specific binding by a protein called vitamin D receptor (VDR), present in the nucleus of the target cell. Binding of 1α,25(OH)$_2$ D_3 by VDR

From: *Vitamin D: Physiology, Molecular Biology, and Clinical Applications*
Edited by: M. F. Holick © Humana Press Inc., Totowa, NJ

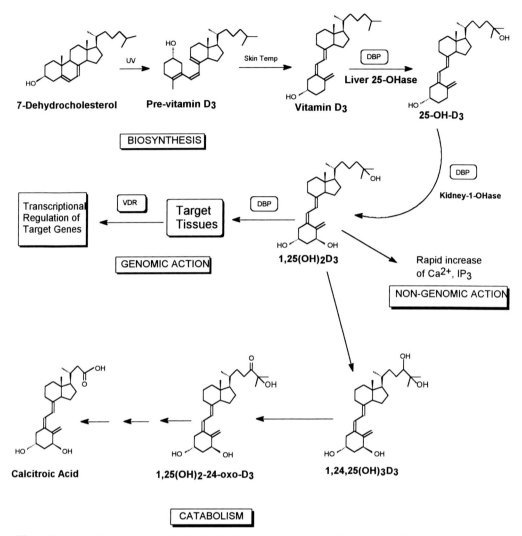

Fig. 1. Pathways for the biosynthesis, biologic actions, and catabolism of $1\alpha,25(OH)_2D_3$, the vitamin D hormone.

triggers a chain of genomic events that results in the transcriptional control of vitamin D-regulated genes *(12–14)*. After the completion of its biologic functions, $1,25(OH)_2D_3$ is catabolized by repeated oxidations, converted to calcitroic acid, and excreted (Fig. 1).

In addition to genomic actions, $1\alpha,25(OH)_2D_3$ is known to elicit rapid stimulation of Ca^{2+} in the intestine, changes in the metabolism in the liver, and uptake of Ca^{2+} in VDR-negative rat osteosarcoma cells *(15–17)*. These effects, which are not inhibited by cycloheximide and actinomycin D, but are inhibited by calcium channel blockers and inhibitors of protein kinase C, are thought to be initiated by the interaction of $1\alpha,25(OH)_2D_3$ with a membrane-bound vitamin D receptor distinct from the genomically active nuclear VDR.

2. BINDING PROTEINS FOR VITAMIN D_3 AND ITS METABOLITES

The processes of metabolism, transportation, biologic actions, and catabolism are mediated through different proteins. For example, binding and transportation of different

Chapter 9 / Molecular Recognition and Structure-Activity Relations

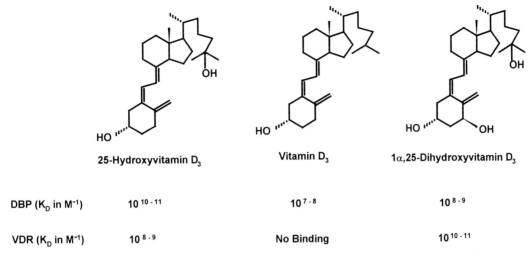

Fig. 2. Binding constants of vitamin D_3, 25(OH)D_3, and 1α,25(OH)$_2D_3$ toward DBP and VDR.

metabolites, including 1α,25(OH)$_2$ D_3, are mediated by DBP. The genomic and non-genomic actions occur through nuclear VDR (nVDR) and membrane VDR (mVDR), respectively. On the other hand, enzymatic conversions of vitamin D_3 to various metabolites, as well as catabolic inactivation of 1α,25(OH)$_2$ D_3 are mediated by several P450-hydroxylases in the liver, kidney, and other organs. Thus vitamin D_3 and its metabolites serve as ligands/substrates for different classes of binding proteins/receptors and enzymes, and specific binding triggers diverse actions like trafficking, as in the case of DBP, catalysis, as in the case of hydroxylases, transcriptional regulation, as in the case of VDR, and rapid calcemic response, as in the case of membrane-bound VDR. The question is, what makes these receptors and enzymes specifically recognize and bind vitamin D_3 and its metabolites in a large pool of diverse biomolecules?

3. MOLECULAR RECOGNITION AND BINDING OF VITAMIN D_3 AND ITS METABOLITES BY DBP AND VDR

The process of molecular recognition in a biologic system generally involves specific interaction between a large macromolecule such as a protein (receptor, enzyme) and a relatively small molecule (ligand, substrate). This interaction often leads to highly specific binding of the ligand (for a receptor) or the substrate (for an enzyme), usually at a specific site/area/domain of the protein called a binding site (for a receptor or binding protein) or catalytic active site (for an enzyme). The effect of this specific binding process ultimately manifests in the observed biologic function of the ligand or the substrate.

It is well recognized that the forces involved in specific recognition and subsequent binding events include hydrophobic, van der Waals, and hydrogen-bonding interactions; the efficiency of such binding is usually reflected in the dissociation constant (K_d) of the ligand–receptor complex. In the vitamin D endocrine system, for example, serum DBP has the highest binding affinity for 25(OH)D_3 ($K_d = 10^{-10-11} M$) *(3)*. The affinity, however, drops significantly for vitamin D_3 and 1α,25(OH)$_2$ D_3 with one less and one more hydroxyl group, respectively, than 25(OH)D_3 (Fig. 2).

In the case of VDR, binding affinity is approx 100-fold more for 1α,25(OH)$_2$ D_3 than for 25(OH)D_3 and virtually nil for vitamin D_3 (Fig. 2). Binding of vitamin D metabolites by VDR

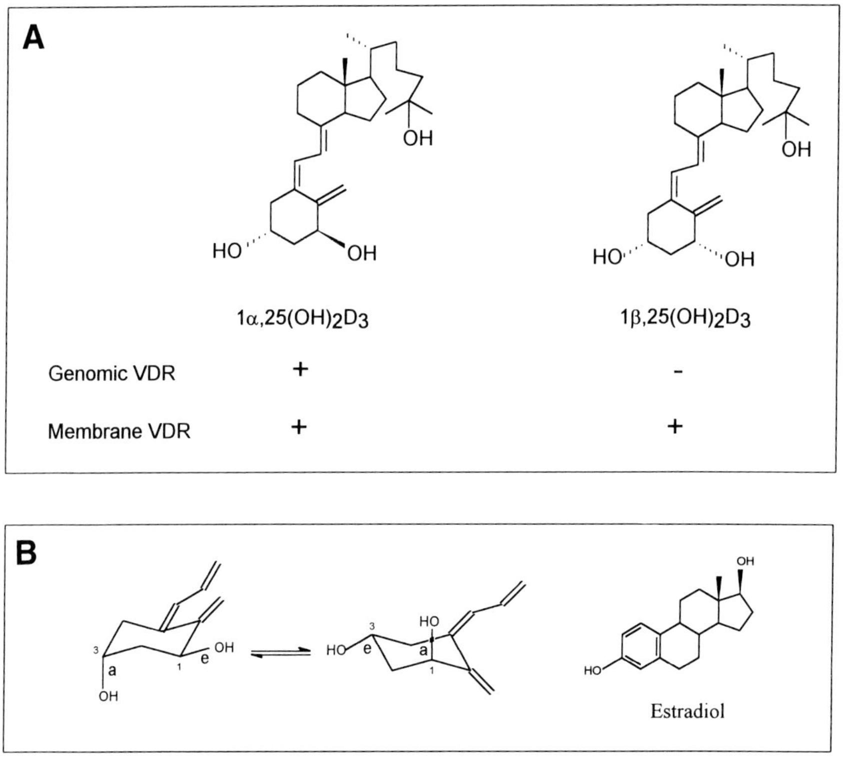

Fig. 3. (A) Binding characteristics of 1α,25-dihydroxyvitamin D_3 [1α,25(OH)$_2$D$_3$], the natural hormone and 1β,25-dihydroxyvitamin D_3, the synthetic 1-epimer of 1α,25-dihydroxyvitamin D_3 for genomic and membrane VDRs. **(B)** Conformational flexibility of the A-ring of 1α,25-dihydroxyvitamin D_3 (two extreme structures are shown) compared with that of estradiol.

is also stereospecific in nature, so that 1β,25-dihydroxyvitamin D_3 [1β,25(OH)$_2$ D_3] the 1-epimer of the natural metabolite 1α,25(OH)$_2$ D_3, containing the OH group in the opposite orientation, shows no binding for VDR (Fig. 3A). Thus the hydroxyl group at 1-position, with correct orientation (1α-OH), is an important recognition marker for VDR binding and subsequent genomic events leading to biologic activities. The nongenomic membrane VDR, however, binds 1α,25(OH)$_2$ D_3 and 1β,1,25(OH)$_2$ D_3 with almost equal efficiency *(18,19)*.

The above picture is, however, somewhat clouded by the conformational flexibility of the A-ring of 1α,25(OH)$_2$ D_3, so that the 1-OH group can assume a "pseudoaxial" or a "pseudoequatorial" position, or anything in between (Fig. 3B; two extreme conformations are shown). This raises the possibility that different conformations of 1α,25(OH)$_2$ D_3 may be responsible for its binding to VDRs from a myriad of target tissues. This scenario is in sharp contrast to that of estrogen and several other steroid hormones, in which the steroidal A-rings are conformationally rigid (Fig. 3B).

4. STRUCTURAL REQUIREMENTS FOR THE BINDING BETWEEN VITAMIN D_3 METABOLITES AND THEIR RECEPTORS

During the past two decades a well-orchestrated effort by synthetic organic chemists, biochemists, and biologists has resulted in the synthesis and functional characterization of a plethora of synthetic analogs of vitamin D_3 and its metabolites *(20)*. This interdis-

Chapter 9 / Molecular Recognition and Structure–Activity Relations

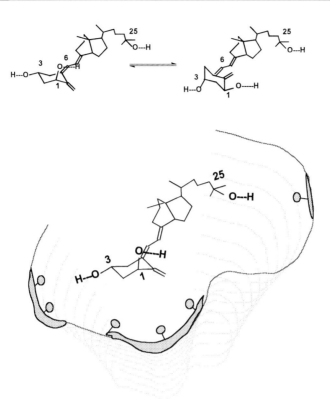

Fig. 4. Schematic representing probable processes involving docking and binding of vitamin D_3, 25-hydroxyvitamin D_3, and $1\alpha,25$-dihydroxyvitamin D_3 into the vitamin D sterol binding pockets of DBP and VDR. The shaded areas represent probable areas of contact. The lollipops are amino acid residues responsible for hydrogen bonding and hydrophobic interactions.

ciplinary effort has identified some important structural aspects of vitamin D_3 and its metabolites in terms of DBP and nVDR binding, and subsequent biologic functions including Ca^{2+}-transport, cell proliferation/differentiation, etc. However, a clear and consistent picture guiding structure-function correlations in the vitamin D endocrine system is still lacking. Broader treatment of this subject is beyond the scope of this article, and any interested reader is directed to a recently published review *(20)*.

Structure of the ligand, however, is only half the picture in the molecular recognition and subsequent binding process. The other half is provided by the three-dimensional structure of the receptor or enzyme, particularly that of the ligand-binding pocket or catalytically active site. Thus, a proper understanding of the molecular recognition and subsequent binding process is crucial for structure–activity studies of the macromolecule and its ligand.

An imaginary picture of the binding between $25(OH)D_3$ and/or $1\alpha,25(OH)_2 D_3$ with DBP and/or VDR is depicted in Fig. 4. According to this picture, the steroidal side chain containing the OH group (25-OH group), a known recognition marker, possibly aids in the docking and anchoring of the steroid molecule inside the binding pocket by a combination of hydrophobic interactions and hydrogen bonding *(21)*. The presence of an amino acid residue with a polarizable group in the binding pocket is an important feature in this picture. Support for this hypothesis comes from the fact that DBP binds vitamin D_3 (without the 25-OH group) with the lowest binding affinity, and VDR does not display any specific binding affinity for vitamin D_3. Furthermore, DBP and VDR do not bind cholesterol, which contains a similar steroidal side chain without the 25-OH group.

The 3-OH group, which is present in vitamin D_3 and all its metabolites, may also be important in initial anchoring of the steroid inside the binding pocket. However, the 3-OH group is also present in cholesterol, a nonspecific binder of DBP and VDR, and hence its importance is much less compared with the 25-OH group, as has been shown by earlier structure–function studies *(20)*. Presumably once the initial docking and anchoring steps are complete, the rest of the ligand snaps into the binding pocket by a combination of hydrophobic interactions and hydrogen bonding.

According to Fig. 4, the presence of a suitably positioned amino acid residue, which can form an H-bond with the 1α-OH group, is vitally important in the interaction between VDR and $1\alpha,25(OH)_2 D_3$. The 1-OH group in the opposite orientation [as in $1\beta,25(OH)_2 D_3$] presumably stays further away from the H-bonding amino acid, and as a result binding efficiency is drastically reduced. In the case of binding between DBP and vitamin D_3 and its metabolites, such H-bonding interaction is presumably less important. Hence the presence of the 1α-OH group [in $1\alpha,25(OH)_2 D_3$] actually reduces binding [compared with $25(OH)D_3$] possibly because of steric crowding and electrostatic repulsion between the polar 1-OH group and hydrophobic amino acids in close proximity.

A drawback of the above model is that it highlights the H-bonding interactions, but ignores the role of hydrophobic interaction between the hydrocarbon skeleton of the secosteroid with the amino acid backbone of the ligand-binding pocket of the protein. We recently observed that modification of the C-6 position [of $1\alpha,25(OH)_2 D_3$] completely obliterated VDR binding of the analog, but the DBP binding of the corresponding $25(OH)D_3$ analog was not significantly impaired *(22)*. These results indicated that the hydrophobic interaction previously mentioned is important for VDR binding, but less so for DBP binding.

The above model also does not account for the possibility that nonconnected parts of the receptor may fold over and touch the bound ligand. Recently Wurtz et al. *(23)* have proposed a three-dimensional model of the $1\alpha,25(OH)_2 D_3$ binding domain of VDR, in which a C-terminal domain folded over the ligand like a trap-door. This model was based on the ligand-binding domain structure for RARγ *(24)*. Although a strong homology between the two domains is expected, confirmation will require actual structural determination of this domain, possibly by X-ray crystallography and/or nuclear magnetic resonance (NMR) analysis.

Although the picture depicted in Fig. 4 is oversimplified, two features emerge as crucial in the recognition and binding processes, namely, the presence of certain amino acid residues in the binding pocket (contact points) and a steric requirement for this pocket. For at least two decades, identification of the essential amino acid residues at the catalytically active site of enzymes, as well as determination of the topography of this domain, has been in the forefront of enzyme research. More recently similar studies have been carried out on receptor proteins to shed light on the structure–function aspects of these macromolecules and their respective ligands. The methods used include modification of specific amino acid residues at the binding pocket chemically, or by site-directed mutagenesis, affinity/photoaffinity labeling, X-ray crystallography and NMR studies.

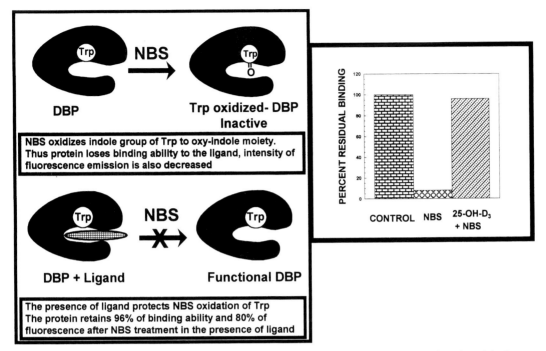

Fig. 5. Covalent modification of the single tryptophan residue in human serum vitamin D binding protein (DBP) by N-bromosuccinimide (NBS) in the absence or presence of an excess of 25(OH)D$_3$ and the effects of such treatment in ^{3}H-25(OH)D$_3$ binding activity. Control, untreated DBP, incubated with[^{3}H]25(OH)D$_3$; NBS, DBP, treated with NBS, followed by incubation with [^{3}H]25(OH)D$_3$; 25(OH)D$_3$ + NBS, DBP, treated with 25(OH)D$_3$ (excess) and NBS, followed by the removal of 25(OH)D$_3$ and NBS and incubation with [^{3}H]25(OH)D$_3$ (protection assay).

5. MUTATIONS OF SPECIFIC AMINO ACIDS IN DBP AND VDR: EFFECTS ON LIGAND-BINDING

Chemical modification of specific amino acids in a protein under controlled conditions was the maiden attempt by protein chemists and enzymologists to identify amino acids that are important for the biologic functions of the protein in question. Chemical treatment of the protein is usually followed by functional test/tests to determine whether chemical mutation of a particular amino acid is reflected in the function/functions of the protein.

In 1979 Wecksler et al. *(25)* treated rat intestinal cytosol, containing VDR, with iodoacetamide and N-ethylmaleimide, well-known modifying agents of Cys, and observed that 1α,25(OH)$_2$D$_3$ binding was adversely affected, suggesting that there are essential Cys residues at the 1α,25(OH)$_2$D$_3$ binding domain of VDR. Coty *(26)* also made similar observations using Cys-modifying organomercurials. Similarly, it was observed that the DNA binding domain of VDR contains an essential Cys residue(s) that is modified by p-chloromercuribenzenesulfonate *(27)*.

Recently we reported that treatment of human serum DBP with N-bromosuccinimide (NBS), a specific modifier of Trp, drastically reduced its ability to bind 25(OH)D$_3$, which was largely restored in the presence of a large excess of 25(OH)D$_3$ *(28)*. As shown in the schematic in Fig. 5, chemical modification of Trp 145, the only Trp in human DBP, changed the conformation of the 25(OH)D$_3$ binding pocket significantly, resulting in a sharp reduction of its intrinsic fluorescence emission and a drastic decrease in 25(OH)D_3 binding. However, preincubation of DBP with an excess of 25(OH)D_3 protected the Trp residue that was seen and chemically mutated by NBS. As a result, 25(OH)D_3 binding, after removal of NBS and 25(OH)D_3, was similar to that of untreated DBP (Fig. 5). These studies strongly emphasized a vital role of Trp 145, the only Trp in human DBP, in vitamin D sterol binding. Similar results were obtained when DBP was treated with diethylpyrocarbonate, a specific modifier of His, in the absence or presence of an excess of 25(OH)D_3. These results were coupled with UV spectrophotometric studies to suggest that one His residue out of a total of six plays an important role in 25(OH)D_3-binding *(28)*.

In recent years chemical modification of amino acids (in proteins) has largely been replaced by protein engineering techniques by which any amino acid can be either changed or deleted by site-directed mutagenesis. This powerful technique has also been used to obtain simultaneously mutated proteins as well as truncated proteins for structure–function analysis. The primary requirements for this technique are availability of the genomic DNA for the protein under consideration and a suitable system to express the protein in functional form.

Recently our laboratory has developed a bacterial expression system to obtain human DBP (the Gc2 isomorph) in functional form *(29)*. Using the same expression system, we changed the aforementioned Trp 145 to Ser and observed that the 25(OH)D_3 binding ability of the mutant protein (Ser 145) was reduced by approximately 80% of the wild variety (unpublished results). This result agreed well with our earlier observation that chemical mutagenesis of this Trp with NBS adversely effected 25(OH)D_3 binding of DBP *(28)*.

Since Trp 145 is located in the N-terminal area of the human serum DBP, these results suggested that vitamin D sterol binding of DBP may be restricted to the N-terminal region of the protein. To obtain support for this observation, we expressed deletion mutants 1–221 and 277–458 of human DBP and observed that, whereas the N-terminal mutant (1–221) displayed normal 25(OH)D_3 binding activity, the C-terminal mutant (277–458) had none *(30)*. These results, in combination, strongly suggested that vitamin D sterol binding (by DBP) takes place exclusively through the N-terminal domain of the protein.

In the case of human VDR, Nakajima et al. *(31)* observed that mutation of Cys 288 to Gly caused significant loss of 1α,25(OH)$_2$ D_3 binding in cotransfected COS-7 cells. It was shown earlier by the same group that mutations of Ser 51 and Ser 208 led to significant loss of protein kinase C-β and casein kinase II activities, respectively *(32,33)*. Recently it has been shown that mutation of His 305 caused serious loss of 1α,25(OH)$_2$ D_3 binding *(34)*.

6. STRATEGIES FOR THE COVALENT MODIFICATION OF THE LIGAND-BINDING DOMAIN OF DBP AND VDR: AFFINITY AND PHOTOAFFINITY LABELING

Affinity and photoaffinity labeling techniques are classical methods that have provided valuable information about the organelle distribution, polymorphic structure, ligand interaction, and binding domain structure of a wide array of enzymes as well as steroid and peptide hormone receptors *(35)*. In this technique a specific hormone receptor binding event is captured at a strategic time when an analog of the ligand (affinity/photoaffinity label) is in intimate contact with a specific hormone binding site by hydrophobic, noncovalent interaction (Fig. 6). In the case of affinity labeling, a chemical bond is formed

Chapter 9 / Molecular Recognition and Structure-Activity Relations

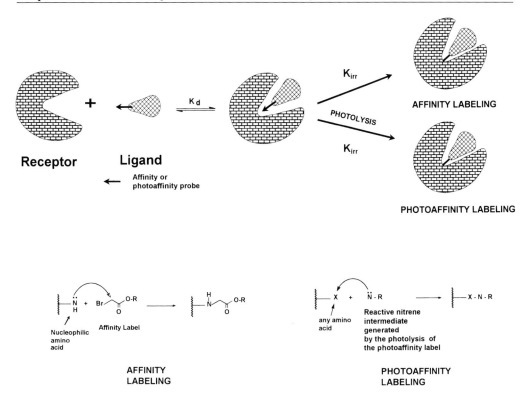

Fig. 6. (**Top**) Schematic describing the affinity and photoaffinity labeling processes; (**Bottom**) chemical reactions involved in the crosslinking of the affinity and photoaffinity labeling reagents to the ligand binding pocket of the receptor.

between a nucleophilic amino acid residue at the binding site with a reactive group (electrophile) in the affinity label (Fig. 6, top). In the case of photoaffinity labeling, light activation of the photoaffinity label produces a reactive species that rapidly forms a covalent bond with an amino acid residue in close proximity, often in the ligand-binding pocket (Fig. 6, bottom). Thus, the binding site becomes labeled by this magic bullet. A combination of site-specific cleavages of the labeled protein, isolation of the peptide bearing the covalently attached label (often radioactive), and sequencing of the fragment often leads to the amino acid sequence of the binding domain. In addition, an amino acid residue(s) that is modified by the affinity/photoaffinity labels (contact points) can often be identified by the gap in the amino acid sequence.

Although these techniques are conceptually very attractive, their use has been restricted because of stringent requirements in chemical synthesis as well as the biologic activity of the analogs. To begin with, synthesis of analogs, and their radiolabeled counterparts, should be simple enough, so that several analogs can be synthesized relatively easily and screened for their binding/functional activities. Furthermore, these analogs should possess reasonably good binding affinity for the protein under consideration to avoid nonspecific labeling. After this screening, the desired analogs should be radiochemically tagged for their detection after covalent labeling of the protein under consideration.

For almost a decade, several laboratories, including ours, have devoted considerable effort toward synthesizing and developing affinity and photoaffinity labeling analogs of $25(OH)D_3$ (for probing the vitamin D sterol binding site of DBP) and $1\alpha,25(OH)_2D_3$ (for

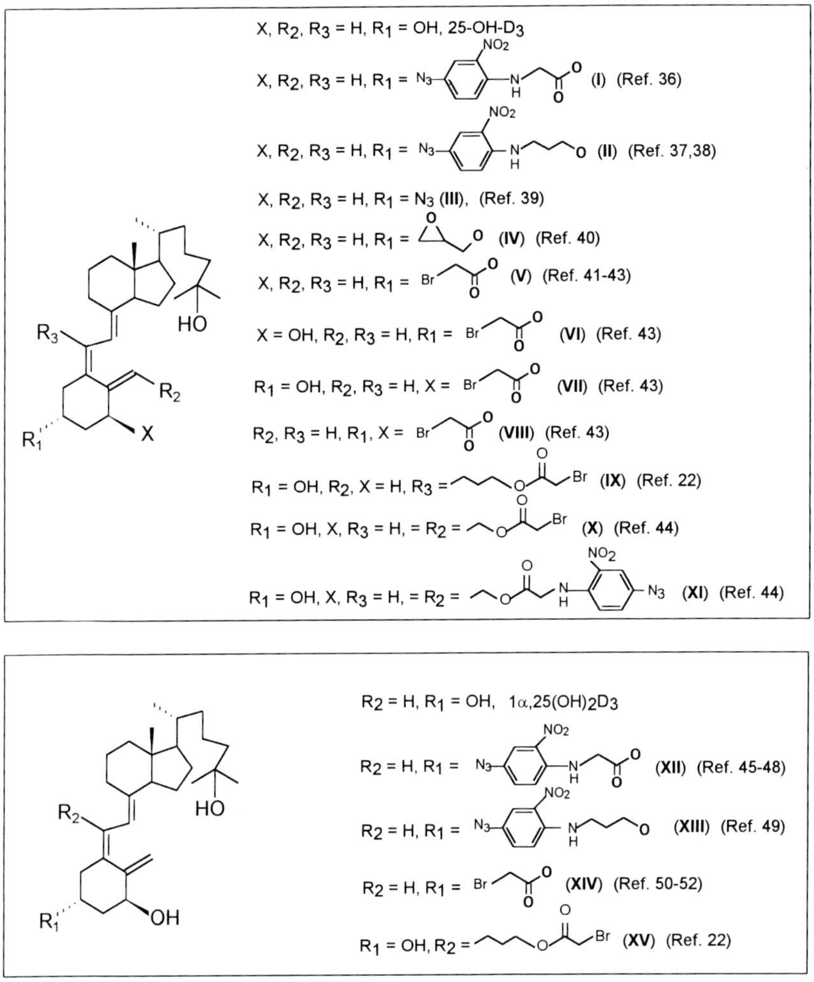

Fig. 7. Structures of the various affinity and photoaffinity labeling analogs of 25(OH)D$_3$ and 1α,25(OH)$_2$D$_3$.

probing the hormone binding sites of genomic and membrane VDRs). Structures of these reagents and corresponding references are shown in Fig. 7.

During the past few years several affinity and photoaffinity analogs have been synthesized to probe the ligand binding pocket of DBP (Fig. 7, top). Competitive binding analysis of these analogs with DBP showed that the protein bound all the analogs (I–XI) in a specific manner. However, binding efficiencies varied considerably among the analogs, possibly reflecting the steric requirement and electrostatic characteristic of each molecule. Furthermore, radiolabeled counterparts of the analogs IV–VI covalently labeled DBP on incubation with rat or human sera (affinity labeling), and analogs I–III did the same on incubation followed by photolysis (photoaffinity labeling). In each case, the extent of labeling was significantly reduced when incubation and incubation/photolysis were carried out in the presence of an excess of 25(OH)D$_3$, indicating that, in each case, the 25(OH)D$_3$ binding site was covalently modified. Interestingly, analogs VII–XI did not label DBP, in the presence or absence of 25(OH)D$_3$. These results demonstrated that in the case of analogs VII–X, the bromoacetate affinity probe is not in close enough

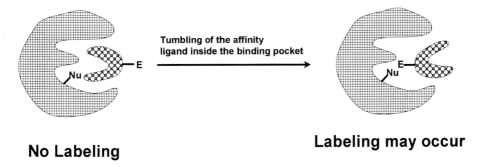

Fig. 8. Schematic depicting the "tumbling" of the ligand (affinity label) inside the binding pocket, and the probable outcome of such an event.

proximity with a nucleophilic amino in the ligand binding pocket to form a covalent bond (Fig. 6, bottom). On the other hand, failure of analog XI to label DBP by photoaffinity indicated that the probe probably sits outside the binding pocket and away from any part of the protein. These results further showed that the ligands have a very tight fit inside the binding pocket of DBP; once a ligand is placed in this pocket, it can not wiggle, bend, or rotate, as shown in the schematic in Fig. 8.

The vitamin D sterol binding site (of human serum DBP) was further probed by proteolytic cleavage of purified human DBP, covalently modified with tritium-labeled analog II (Fig. 7), using cyanogen bromide. Such treatment produced a 10.8-kDa peptide fragment containing almost all the radioactivity (associated with labeled DBP). Judged from the amino acid sequence of full-length DBP, this fragment represented a 1–108 fragment of DBP, indicating that binding of analog A, and by the same token 25(OH)D$_3$, is restricted to the N- terminus of the protein *(38)*. Recently Haddad et al. *(41)* used an affinity analog, ^{14}C-25-hydroxyvitamin D$_3$-3-bromoacetate (Fig. 7, ^{14}C-V), to label human DBP covalently, and to identify a labeled decapeptide in the N-terminal region of the protein after proteolytic cleavage of the labeled protein. Thus, results of the affinity and photoaffinity labeling studies agreed remarkably well with chemical and genetic mutation studies in which Trp 145 was mutated chemically or by site-directed mutagenesis (discussed earlier). This holds true even though affinity and photoaffinity labeling studies involved chemically modified natural ligand [25(OH)D$_3$].

7. COVALENT MODIFICATION OF THE LIGAND BINDING DOMAIN OF VDR BY AFFINITY AND PHOTOAFFINITY LABELING

In contrast to DBP, molecular probing studies of VDR by affinity/photoaffinity labeling have been seriously restricted due to difficulty in the synthesis of 1α,25(OH)$_2$ D$_3$ derivatives. Furthermore, extremely low abundance of natural VDR (from chick and pig intestinal mucosa, and calf thymus) required synthesis of radiochemically labeled 1α,25(OH)$_2$ D$_3$ derivatives with very high specific activity, which imposed further restraints on these studies. Synthesis and biochemical characterization of the first photoaffinity labeling agent for VDR (analog XII, Fig. 7) were reported in 1985 *(45–48)*. Although the corresponding ^{3}H-labeled analog (^{3}H-XII) specifically labeled VDR from chick intestinal cytosol, the efficiency of labeling was disappointingly low, despite a reasonably good VDR binding efficiency of analog VI *(47,48)*. This insufficiency (of labeling) could possibly be caused by a combination of low VDR abundance and a relative inability of the reactive intermediate, generated by the photolysis of the photo-affinity reagent (analog XII), to crosslink to amino acid residues in VDR (possibly because of the intramolecular rearrangement of the intermediate *(35)*.

Brown and DeLuca *(50)* attempted to exploit the intrinsic UV instability of the triene structure in $1\alpha,25(OH)_2 D_3$ to label pig intestinal VDR by photoaffinity with ^3H-$1\alpha,25(OH)_2 D_3$ of very high specific activity. Although this offered the best candidate for binding site labeling, because a natural and unmodified ligand was used as the covalently modifying agent, labeling efficiency was very poor.

Recently our laboratory has developed $1\alpha,25$-dihydroxyvitamin D_3-3-bromoacetate (analog XIV, Fig. 7) as a highly efficient affinity labeling agent for VDR *(51–53)*. We observed that ^{14}C-labeled analog XIV covalently modified the $1\alpha,25(OH)_2 D_3$ binding site in native VDRs from calf thymus and rat osteosarcoma cells as well as recombinant human VDR from the baculovirus expression system *(52)*. Furthermore, we have recently located the label in the C-terminal region of human VDR by a combination of affinity labeling and chemical cleavage of the labeled protein *(53)*. The availability of radiolabeled $1\alpha,25$-dihydroxyvitamin D_3-3-bromoacetate and a substantial quantity of recombinant VDR has potentially opened the door for mapping the hormone binding site of VDR by a combination of affinity labeling and site-directed mutagenesis, as described earlier for DBP. Such studies will be highly complementary to the future determination of the three-dimensional structure of the ligand binding domain of VDR by x-ray crystallography (*see above*).

8. COVALENT LABELING OF A MEMBRANE VITAMIN D RECEPTOR

Although the existence of a membrane-bound nongenomic VDR, which is distinct from classical genomic VDR, was predicted more than a decade ago, proper characterization of this factor has been an arduous task *(54)*. As a result, nongenomic VDR has remained an elusive molecule. Recently Baran et al. *(55)* have reported covalent labeling of a 36-kDa protein with ^{14}C-$1\alpha,25$-dihydroxyvitamin D_3-3-bromoacetate from ROS 17/2.8 cells, which lack genomic VDR. Similar labeling of a protein in the basal–lateral membrane with ^{14}C-$1\alpha,25$-dihydroxyvitamin D_3-3-bromoacetate has also been reported recently *(56)*. It must be emphasized that these are the first ever reports of identification of membrane-bound VDR, made possible by the availability of the affinity labeling analog of $1\alpha,25(OH)_2 D_3$ in the radiolabeled form. It is expected that these studies will ultimately lead to proper characterization (sequencing, cloning, etc.) of the nongenomic VDR, which in turn will delineate the physiologic significance of the nongenomic actions of $1\alpha,25(OH)_2 D_3$.

9. CRYSTALLOGRAPHIC APPROACH TO THE THREE-DIMENSIONAL STRUCTURE OF DBP

In recent years protein crystallography has taken a giant leap forward in determining three-dimensional structures of macromolecules, particularly because of the availability of recombinant proteins in large enough quantities for crystallization/x-ray studies. The crystallographic method has enabled structural biologists to probe deep into the catalytically active site/ligand binding sites of enzymes/receptors, identifying the amino acids important for catalysis, and determining the interactions involved in the molecular recognition and structure–function aspects of these proteins and their respective substrates/ligands.

The first application of x-ray crystallography in the vitamin D endocrine system was reported in 1985 by Koszelak et al. *(57)*, who crystallized human serum DBP in the presence of 25(OH)D$_3$. Later Vogelaar et al. *(58)* and Verboven et al. *(59)* described crystallization of DBP with 3.4- and 2.3-A resolutions, respectively. However, these crystals were not deemed to be of high enough quality for determining the three-dimensional structure of the protein. It was also observed that the crystallization process was adversely effected by the presence of various isoforms of DBP in pooled serum *(60)*. These observations delineated serious restrictions in obtaining highly diffractable crystals from a natural source to be used for determining the three-dimensional structure of the protein. Recently our laboratory has reported bacterial expression of human DBP in a functional form *(29)*. This recombinant variety, representing a single isomorph (Gc2), does not contain any carbohydrate and may be an ideal candidate for future crystallographic studies of this protein and determination of its three-dimensional structure.

The crystal structures of the ligand binding domains of retinoic acid receptor γ and thyroid hormone receptor have been determined recently, with far-reaching implications for hormone binding by VDR *(60,61)*. However, it should be noted that the ligand binding domain of VDR contains a flexible loop, which sets it apart from other members of the steroid receptor superfamily. Thus, determination of the three-dimensional structure of the hormone binding domain remains a coveted prize.

10. CONCLUSIONS

During the past decade considerable research has been carried out in an attempt to understand various functions of vitamin D and its metabolites and correlate them with their structures as well as structures of the respective receptor molecules. However, the return has so far been fairly modest, which emphasizes the inherent difficulty in these studies. Future research in DBP should include proper characterization of the binding processes between DBP and its various ligands, evaluation of interdependence among these ligands, and ultimately their physiologic significance. In the case of genomic VDR, a better understanding of the structure–function aspects of the ligand and the receptor should be pivotal in developing future generations of antiproliferative and anticancer drugs. Characterization of the nongenomic VDR should elucidate the role of $1\alpha,25(OH)_2$ D$_3$ in cell signaling pathways. Finally, various hydroxylases, which are also targets of vitamin D action, can and should be subjected to structure–function studies as described for other members of this endocrine system. This should provide a better understanding of the metabolic and catabolic regulations of the vitamin D hormone.

ACKNOWLEDGMENTS

This work was supported in part by grants DK 44337 and DK 47418 from the National Institute of Diabetes, Digestive and Kidney Diseases of the National Institutes of Health.

REFERENCES

1. Holick MF. Vitamin D: biosynthesis, metabolism and mode of action. In: Endocrinology, vol 2. DeGroot LJ, Besser GM, et al, eds. Philadelphia: WB Saunders, 1994; 902–926.
2. Bikle DD. A bright future for the sunshine hormone. Sci Am Sci Med 1995; March/April:58–67.
3. Cooke NE, Haddad JG. Vitamin D binding protein (Gc-globulin). Endocrinol Rev 1989; 10:294–307.
4. Ray R. Molecular recognition in vitamin D-binding protein. Proc Soc Exp Biol Med 1996; 212:305–312.

5. Binderup L. Immunological properties of vitamin D analogues and metabolites. Biochem Pharmacol 1992; 43:1885–1892.
6. Colston KW, MacKay AG, Janus SY, Binderup L, Chander S, Coombes RC. EB 1089: a new vitamin D analogue that inhibits the growth of breast cancer cells in vivo and in vitro. Biochem Pharmacol 1992; 44:2273–2280.
7. Smith E, Pincus SH, Donovan L, Holick MF. A novel approach for the evaluation and treatment of psoraisis: oral or topical use of 1,25-dihydroxyvitamin D_3 can be a safe and effective therapy for psoraisis. J Acad Dermatol 1988; 19:516–528.
8. Christakos S. Vitamin D and breast cancer (review). Adv Exp Med Biol 1994; 364:115–118.
9. Feldman D, Skowronski RJ, and Peehl DM. Vitamin D and prostate cancer (review). Adv Exp Med Biol 1995; 375:53–63.
10. Habahang M, Buras RR, Davoodi F, Schumaker LM, Nauta RJ, Uskokovic MR, Brenner RV, Evans SR. Growth inhibition of HT-29 human colon cancer cells by analogues of 1,25-dihydroxyvitamin D_3. Cancer Res 1994; 54: 4057–4064.
11. Jung SJ, Lee YY, Pakkala S, deVos S, Elstner E, Norman AW, Green J, Uskokovic MR, Koeffler HP. 1,25$(OH)_2$-16-ene-vitamin D_3 is a potent antileukemic agent with low potential to cause hypercalcemia. Leuk Res 1994; 18:453–463.
12. Haussler MR, Whitfield GK, Haussler CA, Hsieh J-C, Thompson PD, Selznick SH, Dominguez CE, Jurutka PW. The nuclear vitamin D receptor: biological and molecular regulatory properties revealed. J Bone Miner Res 1998; 13:325–349.
13. Haussler MR, Haussler CA, Jurutka PW, Thompson PD, Hsieh J-C, Remus LS, Selznick SH, Whitfield GK. The vitamin D hormone and its nuclear receptor: molecular actions and disease states. J Endocrinol 1997; 154:5S7–S73.
14. Darwish H, DeLuca HF. Vitamin D-regulated gene expression. Crit Rev Eukaryot Gene Expr 1993; 3:89–116.
15. Baran DT, Sorensen AM, Shalhoub V, Owen T, Oberdorf, A, Stein, G, Lian J. 1α,25-Dihydroxyvitamin D_3 rapidly increases cytosolic calcium in clonal rat osteosarcoma cells lacking the vitamin D receptor. J Bone Miner Res 1991; 6:1269–1275.
16. Farach-Carson MC, Sergeev IN, Norman AW. Nongenomic actions of 1α,25-dihydroxyvitamin D_3 in rat osteosarcoma cells: structure-function studies using ligand analogs. Endocrinology 1991; 129: 1876–1884.
17. Norman AW, Nemere I, Zhou LX, Bishop JE, Lowe KE, Maiyar AC, Collins ED, Taoka T, Sergeev I, Farach-Carson MC. 1α,25-$(OH)_2$-Vitamin D_3, a steroid hormone that produces biologic effects via both genomic and nongenomic pathways. J Steroid Mol Biol 1992; 41:231–240.
18. Baran DT, Ray R, Sorensen AM, Honeyman T, Holick MF. Binding characteristics of a membrane receptor that recognizes 1α,25-dihydroxyvitamin D_3 and its epimer, 1β,25-dihydroxyvitamin D_3. J Cell Biochem 1994; 56:510–517.
19. Norman AW, Bouillon R, Farach-Carson MC, Bishop JE, Zhou LX, Nemere I, Zhao J, Muralidharan KR, Okamura WH. Demonstration that 1β,25-dihydroxyvitamin D_3 is an antgonist of the nongenomic but not genomic biological responses and biological profile of the three A-ring diastereoisomers of 1α,25-dihydroxyvitamin D_3. J Biol Chem 1993; 268:2022–2030.
20. Bouillon R, Okamura WH, Norman AW. Structure-function relationships in the vitamin D endocrine system. Endocr Rev 1995; 16: 200–257.
21. Norman AW, Bishop JE, Collins E-GS, Satchell DP, Dormanen MC, Zanello SB, Farcah-Carson MC, Bouillon R, Okamura WH. Differing shapes of 1α,25-dihydroxyvitamin D_3 function as ligands for the D-binding protein, nuclear receptor: a staus report. J Steroid Biochem Mol Biol 1996; 56: 13–22.
22. Addo JK, Swamy N, Ray R. Novel C-6 functionalized analogs of 25-hydroxyvitamin D_3 and 1α,25-dihydroxyvitamin D_3: synthesis and binding analysis with vitamin D-binding protein and vitamin D receptor. 1998, submitted.
23. Wurtz J-M, Guillot B, Moras D. Model of the ligand binding domain of the vitamin D nuclear receptor based on the crystal structure of holo RARγ. In: Vitamin D Chemistry, Biology and Clinical Applications of the Steroid Hormone. Norman AW, Bouillon R, Thomasset M, eds. Riverside, CA: University of California, 1997; 165–172.
24. Wurtz JM, Bourguet W, Renaud JP, Vivat V, Chambon P., Moras D, Gronenmeyer H. A canonical structure for the ligand-binding domain of nuclear receptors. Nature Struct Biol 1996; 3:87–94.
25. Wecksler WR, Ross FP, Norman AW. Characterization of the 1α,25-dihydroxyvitamin D_3 receptor from rat intestinal cytosol. J Biol Chem 1979; 254:9488–9491.

26. Coty WA. Reversible dissociation of steroid hormone receptor complexes by mercurial reagents. J Biol Chem 1980; 255:8035-8037.
27. Pike JW. Evidence for a reactive sulhydryl in the DNA binding domain of the 1α,25-dihydroxyvitamin D_3 receptor. Biochem Biophys Res Commun 1981; 100:1713–1719.
28. Swamy N, Brisson M, Ray R. Trp-145 is essential for the binding of 25-hydroxyvitamin D_3 to human serum vitamin D-binding protein. J Biol Chem 1995; 270:2636–2639.
29. Swamy N, Ghosh S, Ray R. Bacterial expression of human vitamin D-binding protein (Gc2) in functional form. Protein Expr Purif 1997; 10:115–122.
30. Swamy N, Paz N, Ray R. Expression of 25-OH-D_3-binding domain (LBD) of human vitamin D-binding protein: C-terminal domain does not bind 25-OH-vitamin D_3. In:. Vitamin D Chemistry, Biology and Clinical Applications of the Steroid Hormone. Norman AW, Bouillon R, Thomasset M, eds. Riverside, CA: University of California, 1997; 114–115.
31. Nakajima S, Hsieh J-C, Jurutka PW, Galligan MA, Haussler CA Whitfield GK, Haussler MR. Examination of the potential functional role of conserved cysteine residues in the hormone binding domain of the human 1,25-dihydroxyvitamin D_3 receptor. J Biol Chem 1996; 271:5143–5149.
32. Hsieh J-C, Jurutka PW, Nakjima S, Galligan MA, Haussler CA, Shimizu Y, Shimizu N, Whitfield GK, Haussler MR. Phosphorylation of the human vitamin D receptor by protein kinase C: biochemical and functional evaluation of the serine 51 recognition site. J Biol Chem 1993; 268:1511–1512.
33. Jurutka PW, Hsieh J-C, MacDonald PN, Terpening CM, Haussler, CA, Haussler MR, Whitfield GK. Phosphorylation of serine 208 in the human vitamin D receptor: the predominant amino acid phosphorylated by casein kinase II, *in vitro*, and identification as a significant phosphorylation site in intact cells. J Biol Chem 1993; 268:6791–6799.
34. Malloy PJ, Eccleshall TR, Gross C, Van Maldergem L, Bouillon R, Feldman D. Hereditary vitamin D resistant rickets caused by a novel mutation in the vitamin D receptor that results in decreased affinity for hormone and cellular responsiveness. J Clin Invest 1997; 99:297–304.
35. Sweet F, Murdock GL. Affinity labeling of hormone-specific proteins. Endocr Rev 1987; 8:154–184.
36. Ray R, Holick SA, Hanafin N, Holick MF. Photoaffinity labeling of the rat plasma vitamin D binding protein with [26,27-^{3}H]-25-hydroxyvitamin D_3-3-[N-(4-amido-2-nitrophenyl)glycinate]. Biochemistry 1986; 25:4729–4733.
37. Ray R, Bouillon R, Van Baelen HG, Holick MF. Photoaffinity labeling of rat plasma vitamin D binding protein with a second generation photoaffinity analog of 25-hydroxyvitamin D_3. Biochemistry 1991; 36:4809–4813.
38. Ray R, Bouillon R, Van Baelen HG, Holick MF. Photoaffinity labeling of human serum vitamin D binding protein, and chemical cleavages of the labeled protein: identification of a 11.5 KDa peptide, containing the putative 25-hydroxyvitamin D_3-binding site. Biochemistry 1991; 30:7638–7642.
39. Link R, Kutner A, Schnoes HK, DeLuca HF. Photoaffinity labeling of serum vitamin D binding protein by 3-deoxy-3-azido-25-hydroxyvitamin D_3. Biochemistry 1987; 26:3957–3964.
40. Swamy N, Ray R. 25-Hydroxy[26,27-methyl-^{3}H]vitamin D_3-3β-(1,2-epoxypropyl)ether: an affinity labeling reagent for human vitamin D-binding protein. Arch Biochem Biophys 1995; 319:504–507.
41. Haddad JG, Hu YZ, Kowalski MA, Laramore C, Ray K, Robzyk P, Cooke NE. Identification of the sterol- and actin-binding domains of plasma vitamin D binding protein (Gc-globulin). Biochemistry 1992; 31:7174–7181.
42. Swamy N, Ray R. Affinity labeling of rat serum vitamin D binding protein. Arch Biochem Biophys 1996; 333:139–144.
43. Swamy N, Dutta A, Ray R. Roles of structure and orientation of ligands and ligand-mimicks inside the ligand-binding pocket of vitamin D-binding protein. Biochemistry 1997; 36:7432–7436.
44. Addo JK, Ray R. Synthesis and binding analysis of 5E-[19-(2-bromoacetoxy)methyl]25-hydroxyvitamin D_3 and 5E-25-hydroxyvitamin D_3-19-methyl[(4-azido-2-nitro)phenyl]glycinate: novel C_{19}-modified affinity and photoaffinity analogs of 25-hydroxyvitamin D_3. Steroids 1998, in press.
45. Ray R, Holick SA, Holick MF. Synthesis of a photoaffinity-labelled analogue of 1,25-dihydroxyvitamin D_3. J Chem Soc Chem Commun 1985; 702,703.
46. Ray R, Rose SR, Holick SA, Holick MF. Evaluation of a photolabile derivative of 1,25-dihydroxyvitamin D_3 as a photoaffinity probe for 1,25-dihydroxyvitamin D_3 receptor. Biochem Biophys Res Commun 1985; 132:198–203.
47. Ray R, Holick MF. The synthesis of a radiolabeled photoaffinity analog of 1,25-dihydroxyvitamin D_3. Steroids 1988; 51:623–630.
48. Ray R, Ray S, Holick MF. Photoaffinity labeling of chick intestinal 1α,25-dihydroxyvitamin D_3 receptor. Steroids 1993; 58:462–465.

49. Roy A, Ray R. Aminopropylation of vitamin D hormone (1α,25-dihydroxyvitamin D_3), its biological precursors, and other steroidal alcohols: an anchoring moiety for affinity studies of sterol. Steroids 1995; 60:530–533.
50. Brown TA, DeLuca HF. Photoaffinity labeling of the 1,25-dihydroxyvitamin D_3 receptor. Biochim Biophys Act 1991; 1073:324–328.
51. Ray R, Ray S, Holick MF. 1α,25-Dihydroxyvitamin D_3-3-deoxy-3β-bromoacetate, an affinity labeling analog of 1α,25-dihydroxyvitamin D_3. Bioorg Chem 1994; 22: 276–283.
52. Ray R, Swamy N, MacDonald PN, Ray S, Haussler MR, and Holick MF. Affinity labeling of 1α,25-dihydroxyvitamin D_3 receptor. J Biol Chem 1996; 271:2012–2017.
53. Swamy N, Kounine M, Ray R. Identification of the subdomain in the nuclear receptor for the hormonal form of vitamin D_3, 1α,25-dihydroxyvitamin D_3, vitamin D receptor, that is covalently modified by an affinity labeling reagent. Arch Biochem Biophys 1997; 348:91–95.
54. Norman AW, Roth J, Orci L. The vitamin D endocrine system: steroid metabolism, hormone receptors and biological response (calcium binding proteins). Endocr Rev 1982; 3:331–366.
55. Baran D, Merriman H, Ray R, Sorensen A, Quail J. Characteristics of an osteoblast protein that recognizes 1α,25-dihydroxyvitamin D_3. J Bone Miner Res 1995; 10(Suppl 1):S569.
56. Nemere I, Ray R, Jia Z. Further characterization of the putative basal-lateral membrane receptor for 1,25$(OH)_2D_3$. J Bone Miner Res 1996; 11(Suppl 1):M522.
57. Koszelak S, McPherson A, Bouillon R, Van Baelen H. Crystallization and preliminary x-Ray analysis of the vitamin D-binding protein from human serum. J Steroid Biochem 1985; 23:1077,1078.
58. Vogelaar NJ, Lindberg U, Schutt CE. Crystallization and preliminary X-Ray analysis of Gc, the vitamin D-binding protein in serum. J Mol Biol 1991; 220:545–547.
59. Verboven CC, De Bondt HL, De Ranter C, Bouillon R, Van Baelen H. Crystallization and x-Ray investigation of vitamin D-binding protein from human serum. Identification of the crystal content. J Steroid Biochem Mol Biol 1995; 54:11–14.
60. Wagner RL, Apriletti JW, McGrath ME, West BL, Baxter JD, Fletterick RJ. A structural role for hormone in the thyroid hormone receptor. Nature 1995; 378:690–697.
61. Renaud J-P, Rochel N, Ruff M, Vivat V, Chambon P, Gronenmeyer H, Moras D. Crystal structure of the RAR-γ ligand-binding doman bound to all-trans retinoic acid. Nature 1995; 378:681–689.

10 Mechanism of Action of 1,25-Dihydroxyvitamin D_3 on Intestinal Calcium Absorption and Renal Calcium Transport

Mihali Raval-Pandya, Angela R. Porta, and Sylvia Christakos

1. INTRODUCTION

In mammals the plasma calcium concentration under normal conditions is maintained at 2.5 mM or 10 mg/dL. Ionized calcium represents approx 45% of the total plasma calcium. An equal portion of the plasma calcium is bound to proteins, and approx 10% is complexed with small anions. Vitamin D is a principle factor that maintains the plasma calcium level within the normal range, and the intestine, kidney, and bone are the three target organs of vitamin D action primarily responsible for maintaining calcium homeostasis *(1)*. This chapter focuses on how vitamin D [and specifically the active form of vitamin D, 1,25-dihydroxyvitamin D_3 (1,25(OH)$_2$$D_3$)] acts at times of increased calcium demand to increase the efficiency of calcium absorption from the intestine and to enhance the tubular reabsorption of calcium from the kidney.

2. INTESTINAL CALCIUM ABSORPTION

2.1. Overview

The generally accepted view of intestinal calcium absorption is that it is a phenomenon comprised of two different modes of calcium transport, the saturable process, which functions at calcium concentrations between 1 and 10 mM and is mainly transcellular, and a nonsaturating component, which occurs only at high concentrations of intraluminal calcium (10–50 mM) and is paracellular. The transcellular process is comprised of three events that result in net calcium absorption: (1) the entry of calcium from the lumen of the intestine across the intestinal brush border membrane, (2) the transcellular movement of calcium through the cytosol of the intestinal cell or enterocyte, and (3) the energy-requiring extrusion of calcium against a concentration gradient at the basolateral membrane of the enterocyte into the lamina propria and eventually into the plasma *(2–5)*. The paracellular path is believed to be due to passive movement of calcium across tight junctions and intracellular spaces and is directly related to the concentration of calcium in tion gradient. Recently, the intestinal plasma membrane calcium pump (PMCA) and PMCA mRNA and transcription have been shown to be stimulated by $1,25(OH)_2D_3$, suggesting for the first time that the intestinal calcium absorptive process may involve a direct effect of $1,25(OH)_2D_3$ on calcium pump expression *(19–22)*. It has been suggested that calbindin, which increases calcium diffusion toward the basolateral membrane in the second phase of the intestinal calcium absorptive process, can indirectly stimulate calcium extrusion by increasing the local calcium adjacent to the pump *(15)*. Diets deficient in calcium or phosphorus result in increased synthesis of $1,25(OH)_2D_3$, increased synthesis of calbindin and the PMCA, and therefore increased intestinal calcium absorption *(20,21,23)*. Although it has been estimated that the sodium/calcium exchanger (which is also present at the basolateral membrane of the enterocyte) may have a role in extrusion of approx 20% of calcium from the duodenum *(24)*, it should be noted that this cotransporter is not $1,25(OH)_2D_3$ inducible *(24)*.

2.3. Other Models of Intestinal Calcium Transport

Other models, which have been suggested but need to be further defined, are the vesicular transport model and the very rapid stimulation of calcium absorption by $1,25(OH)_2D_3$ (within minutes as opposed to hours required for transcellular calcium transport) termed *transcaltachia (25–27)*. In the vesicular calcium transport model, calcium inside the lumen is believed to be taken up in the endocytic vesicles, which fuse with lysosomes and extrude calcium via exocytosis *(25,26)*. Although calbindin is predominantly localized in the cytoplasm, recent evidence suggests that some vitamin D-dependent calbindin can be localized inside of small vesicles and lysosome structures of the intestinal cell, thus implicating a role for calbindin in the vesicular transport model *(27–29)*.

Transcaltachia, a nongenomic mechanism involving the very rapid stimulation of intestinal calcium transport by $1,25(OH)_2D_3$, has been observed in normal chicks *(27)*, and a specific binding protein for $1,25(OH)_2D_3$ in the basolateral membrane of chick intestinal epithelium has been implicated in this process *(30)*. Further studies are needed to determine the significance of the vesicular transport model as well as transcaltachia in relation to overall intestinal calcium absorption.

2.4. $1,25(OH)_2D_3$ and Intestinal Phosphorus Absorption

It should be noted that although the major biologic function of vitamin D is to maintain calcium homeostasis, $1,25(OH)_2D_3$ can also enhance the intestinal absorption of dietary phosphorus, principally from the jejunum and the ileum. It has been suggested that $1,25(OH)_2D_3$ acts by affecting sodium-dependent phosphorus influx into the brush border membrane *(31)*. However, since significant phosphorus absorption from the jejunum has been reported when $1,25(OH)_2D_3$ levels are markedly decreased, it has been suggested that the concentration of dietary phosphorus [and not $1,25(OH)_2D_3$] is the major determinant of net phosphorus absorption *(32)*.

3. RENAL CALCIUM TRANSPORT

3.1. Overview

In addition to the intestine, the kidney is also a major target tissue involved in the regulation by $1,25(OH)_2D_3$ of calcium homeostasis. Almost 98% of the calcium filtered

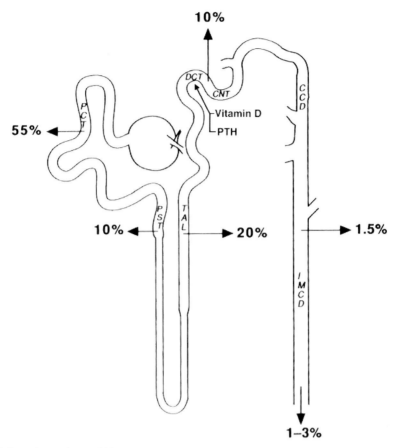

Fig. 2. Calcium absorption at different nephron sites and hormonal site of action. The percentages refer to the percentage of filtered calcium absorbed at different nephron segments. PCT, proximal convoluted tubule; PST, proximal straight tubule; TAL, thick ascending limb of Henle's loop; DCT, distal convoluted tubule; CNT, connecting tubule; CCD, cortical collecting duct; IMCH, inner medullary collecting duct; PTH, parathyroid hormone; vitamin D, 1,25-dihydroxyvitamin D_3. (Reproduced with permission from ref. *35*.)

by the glomerulus is reabsorbed along the nephron. Only plasma calcium that is not bound to proteins is filtered in the glomerulus. The different regions of the nephron involved in the reabsorption process and the hormonal sites of action are schematically represented in Fig. 2. The proximal tubule absorbs about 65% of the filtered calcium and the loop of Henle absorbs another 20% *(33–36; see* refs. *35* and *36* for reviews). Calcium absorption in the proximal tubule is passive and proceeds via a paracellular pathway, whereas both passive and active transport pathways contribute to calcium aborption from Henle's loop [the ascending medullary and cortical thick limbs are the parts of Henle's loop where calcium absorption occurs, *(35,36)*]. The remaining 10% of the filtered calcium is absorbed by the distal convoluted tubule and the connecting tubules. In the distal nephron, calcium absorption involves active transport and proceeds via a transcellular pathway *(35,36)*. It is the distal nephron that is believed to be the key site for calcium regulation. It is here that the calcium is dually regulated by $1,25(OH)_2D_3$ and parathyroid hormone (PTH) *(35,36)*. In this part of the chapter, the role of vitamin D in the kidney is discussed, including the factors involved in this process such as the vitamin D-dependent calcium binding proteins calbindin-D_{28k} and calbindin-D_{9k} and the PMCA. Besides the role of 1,25$(OH)_2D_3$ in the tubular reabsorption of calcium, another important effect in the kidney, which is also discussed, is the effect of 1,25$(OH)_2D_3$ on the production of renal vitamin D hydroxylases.

3.2. Effect of 1,25$(OH)_2D_3$ on Renal Calcium Transport

3.2.1. CURRENT UNDERSTANDING OF THE ROLE OF 1,25$(OH)_2D_3$

Although there has been some controversy concerning the role of 1,25$(OH)_2D_3$ in renal calcium transport, data obtained using discrete nephron segments as well as studies using renal tubule cells have indicated that vitamin D metabolites enhance calcium transport in the distal nephron *(37–39)*. In addition, vitamin D deficiency has been reported to decrease the distal tubular reabsorption of calcium *(40,41)* and to decrease the stimulatory effect of PTH on calcium reabsorption *(40)*. Using mouse distal convoluted tubule cells, 1,25$(OH)_2D_3$ has been shown to accelerate PTH-dependent calcium uptake significantly *(38)*. In the presence of 1,25$(OH)_2D_3$ the time required for PTH to induce membrane hyperpolarization (required for stimulation of calcium entry into the distal tubule cells) as well as to increase intracellular calcium and ^{45}Ca uptake is significantly reduced *(38)*. More recent evidence has demonstrated that 1,25$(OH)_2D_3$ increases PTH receptor mRNA and binding activity, suggesting that this effect may be involved in the acceleration by 1,25$(OH)_2D_3$ of PTH-dependent calcium entry *(42)*. Transcellular calcium transport in distal tubular cells involves calcium entry through the apical (luminal) plasma membrane, diffusion through the cytosol, and active extrusion across the opposing basolateral membrane by two calcium transporters mediating efflux, the PMCA and the Na^+/Ca^+ exchanger *(35,36)*. Since transcellular calcium transport in the distal tubule cell is a multiple-step process, similar to transcellular calcium transport in the enterocyte, 1,25$(OH)_2D_3$ may result in enhanced calcium transport by affecting various steps in this process, as outlined below.

3.2.2. ROLE OF VITAMIN D-DEPENDENT CALCIUM BINDING PROTEINS IN RENAL CALCIUM TRANSPORT

Consistent with the calcium transport studies localizing the site of action of 1,25$(OH)_2D_3$ to the distal nephron *(37,38)*, autoradiographic data have demonstrated that the nuclear uptake of [^3H]1,25$(OH)_2D_3$ is localized predominantly in the distal nephron *(43,44)*. In addition, the exclusive localization of the vitamin D-dependent calcium binding proteins calbindin-D_{28k} and calbindin-D_{9k} is in the distal nephron [the distal convoluted tubule, the connecting tubule, and the cortical collecting duct 12, 13 *(45–48)*]. Recent kinetic studies have suggested that the two proteins affect renal calcium absorption by different mechanisms. Calbindin-D_{28k} was found to stimulate calcium transport from the apical (luminal) membrane, which is mediated by calcium channels *(49)*, and calbindin-D_{9k} was found to enhance the ATP-dependent calcium transport of the basolateral membrane *(50)*. It has also been suggested that calbindin-D_{28k} may facilitate the diffusion of calcium through the cytosol of the distal tubule cell *(51)*. These findings provide evidence for a role for the calbindins in these calcium transport processes and suggest mechanisms whereby 1,25$(OH)_2D_3$ may act, via the induction of the calbindins, to enhance calcium transport in the distal tubule.

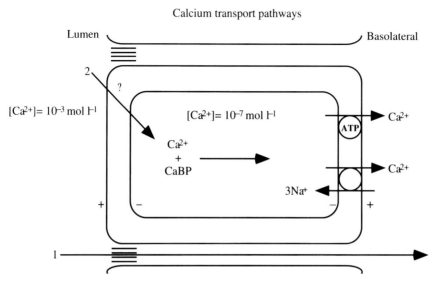

Fig. 3. Model of transcellular Ca^{2+} transport in the distal convoluted tubule. CaBP, calcium binding protein (calbindin); 1, paracellular pathway; 2, transcellular pathway, which is the mode of calcium absorption in the distal nephron. (Reproduced with permission from ref. *70*.)

3.3.3. ROLE OF THE PLASMA MEMBRANE CALCIUM PUMP

In addition to the calbindins, the PMCP has also been localized exclusively to the distal tubule and to the collecting duct *(51–53)*. However, the interrelationship between the renal calcium pump and $1,25(OH)_2D_3$ is not clear at this time. Consistent with the finding that calbindin-D_{9k} enhances ATP-dependent calcium transport at the renal basolateral membrane, a calbindin-D_{9k} binding domain has been identified in the PMCP *(54)*. Thus $1,25(OH)_2D_3$ may not directly regulate the activity of the renal calcium pump but may do so indirectly by increasing calbindin-D_{9k} *(55)*. A model of distal tubule renal calcium transport (involving the vitamin D-dependent calcium binding proteins) similar but not identical to the model of intestinal calcium transport is proposed (Fig. 3). Presumably calbindin-D_{28k} increases the influx of calcium at the apical (luminal) membrane, acts as a diffusional carrier of calcium to the basolateral membrane where calbindin-D_{9k} binds calcium, and stimulates the basolateral extrusion of calcium via the PMCP. It should be noted that besides the PMCP, the sodium calcium exchanger is also localized in the distal tubule and is thought to contribute to calcium extrusion *(56,57)*. The exact role of the Na^+/Ca^{2+} exchanger in distal tubular calcium efflux is a topic of current investigation.

3.3. Other Effects of $1,25(OH)_2D_3$ in the Kidney

Besides the role of $1,25(OH)_2D_3$ in enhancing the tubular reabsorption of calcium, another important effect of $1,25(OH)_2D_3$ in the kidney is its ability to regulate its own production *(58,59)*. 1,25-Dihydroxyvitamin D_3 decreases its own production by inhibiting the renal $25(OH)D_3$ 1α hydroxylase enzyme, which hydroxylates $25(OH)D_3$ at the α position of carbon 1 of the A-ring, resulting in the formation of $1,25(OH)_2D_3$. 1,25-Dihydroxyvitamin D_3 also stimulates the renal 24-hydroxylase enzyme, which hydroxylates the 24 position of both $25(OH)D_3$ and $1,25(OH)_2D_3$. The 24-hydroxylation of 1,25(OH)$_2$D$_3$ is thought to be the first step in the catabolism of 1,25(OH)$_2$D$_3$ *(60)*. The predominant localization of these two enzymes is in the proximal convoluted tubule *(61)*. Although the precise mechanism by which 1,25(OH)$_2$D$_3$ reciprocally regulates these two enzymes is not known, 1,25(OH)$_2$D$_3$-mediated genomic responses appear to be involved. Recently both the 1α hydroxylase and 24-hydroxylase genes have been cloned *(62,63)*, and two vitamin D response elements have been identified in the promoter of both the human and rat 24-hydroxylase genes that allow VDR binding and subsequent transcriptional activation *(64–67)*.

Besides enhancement of calcium transport in the distal nephron, modulation of the 25(OH)$_2$D$_3$ hydroxylases, and a few reports concerning the inhibition of phosphate transport in the proximal tubule *(68)*, little is known concerning other possible effects of 1,25(OH)$_2$D$_3$ in the kidney. Studies using immunolocalization and reverse transcriptase polymerase chain reaction have recently shown that the vitamin D receptor is localized not only in the distal nephron, as previously suggested *(43,44)*, but also in the glomeruli, proximal tubules, and collecting duct *(69)*. These findings suggest multiple genomically mediated actions of 1,25(OH)$_2$D$_3$ within the kidney. Further research is needed to provide new insight concerning additional renal effects of 1,25(OH)$_2$D$_3$.

REFERENCES

1. Hurwitz S. Homeostatic control of plasma calcium concentration. Crit Rev Biochem Mol Biol 1996; 31:41–100.
2. Johnson JA, Kumar R. Renal and intestinal calcium transport: roles of vitamin D and vitamin D dependent calcium binding proteins. Semin Nephrol 1994; 14:119–128.
3. Pansu D, Bellaton C, Bronner F. Effect of calcium intake on saturable and non-saturable components of duodenal calcium transport. Am J Physiol 1981; 240:G32–37.
4. Bronner F, Pansu D, Stein WD. An analysis of intestinal calcium transport across the rat intestine. Am J Physiol 1986; 250:G561–569.
5. Wasserman RH, Fullmer CS. Vitamin D and intestinal calcium transport: facts, speculations and hypotheses. J Nutr 1995; 125:S1971–1979.
6. Karbach U. Paracellular calcium transport across the small intestine. J Nutr 1992; 122:672–677.
7. Weinstein RS, Underwood JL, Hutson MS, DeLuca HF. Bone histomorphometry in vitamin D deficient rats infused with calcium and phosphorous. Am J Physiol 1984; 146:E499–E505.
8. Chandra S, Fullmer CS, Smith CA, Wasserman RH, Morrison GH. Ion microscopic imaging of calcium transport in the intestinal tissue of vitamin-D deficient and vitamin-D replete chicks: a [44]Ca stable isotope study. Proc Natl Acad Sci USA 1990; 87:5715–5719.
9. Bikle DD, Munson S, Christakos SC, Kumar R, Buckendahl P. Calmodulin binding to the intestinal brush-border membrane: comparison to other calcium binding proteins. Biochim Biophys Acta 1989; 1010:122–127.
10. Kaune R, Munson S, Bikle DD. Regulation of calmodulin binding to the ATP extractable 110 kDa protein (myosin I) from chicken duodenal brush border by 1,25(OH)$_2$D$_3$. Biochim Biophys Acta 1994; 1190:329–336.
11. Mooseker MS, Wolenski JS, Coleman TR, Hayden SM, Cheney RD, Espreafico, E, Heintzelman MB, Peterson MD. Structural and functional dissection of a membrane-bound mechanoenzyme: brush border myosin I. Curr Topics Membr 1991; 33:31–55.
12. Christakos S, Gabrielides C, Rhoten WB. Vitamin D-dependent calcium-binding proteins: chemistry, distribution, functional considerations and molecular biology. Endocr Rev 1989; 10:3–26.
13. Christakos S. Vitamin D-dependent calcium binding proteins: chemistry, distribution, functional considerations and molecular biology. Update 1995. Endocr Rev Monogr 1995; 4:108–110.
14. Pansu D, Bellaton C, Roche C, Bronner F. Theophylline inhibits active Ca transport in rat intestine by inhibiting Ca binding by CaBP. Prog Clin Biol Res 1988; 252:115–120.
15. Kretsinger RH, Mann JE, Simmonds JG. Model of facilitated diffusion of calcium by the intestinal calcium binding protein. In: Vitamin D, Chemical, Biochemical and Clinical Endocrinology of

Calcium Metabolism. Norman AW, Schaefer K, von Herrath D, Grigoleit HG, eds. Berlin: de Gruyter, 1982; 233–248.
16. Feher JJ, Fullmer CS, Wasserman RH. The role of facilitated diffusion of calcium by calbindin in intestinal calcium absorption. Am J Physiol 1992; 262:C517–526.
17. Feher JJ. Facilitated calcium diffusion by intestinal calcium binding protein. Am J Physiol 1983; 244: 303–307.
18. Glenney JR, Glenney P. Comparison of Ca^{++} regulated events in the intestinal brush border. J Cell Biol 1985; 100:754–763.
19. Zelinski JM, Sykes DE, Weiser MM. The effect of vitamin D on rat intestinal plasma membrane Ca-pump mRNA. Biochem Biophys Res Commun 1991; 179:749–755.
20. Wasserman RH, Smith CA, Brindak ME, DeTalamoni N, Fullmer CS, Penniston JT, Kumar R. Vitamin D and mineral deficiencies increase the plasma membrane calcium pump of chicken intestine. Gastroenterology 1992; 102:886–894.
21. Cai Q, Chandler JS, Wasserman RH, Kumar R, Penniston JT. Vitamin D and adaptation to dietary calcium and phosphate deficiency increase intestinal plasma membrane calcium pump gene expression. Proc Natl Acad Sci USA 1993; 90:1345–1249.
22. Pannabecker TL, Chandler JS, Wasserman RH. Vitamin D dependent transcriptional regulation of the intestinal plasma membrane calcium pump. Biochem Biophys Res Commun 1995; 213:499–505.
23. Henry HL, Norman AW. Vitamin D: metabolism and biological actions. Annu Rev Nutr 1984; 4: 493–520.
24. Ghijsen WEJM, DeJong MD, VanOs CH. Kinetic properties of Na^{2+}/Ca^{2+} exchange in basolateral plasma membrane of rat small intestine. Biochim Biophys Acta 1983; 730:85–94.
25. Nemere I, Norman AW. 1,25 Dihydroxyvitamin D_3-mediated vesicular calcium transport in intestine: dose-response studies. Mol Cell Endocrinol 1989; 67:47–53.
26. Nemere I. Vesicular calcium transport in chick intestine. J Nutr 1992; 122:657–661.
27. Nemere I, Norman AW. Transcaltachia, vesicular calcium transport and microtubule associated calbindin-D_{28k}: emerging views of 1,25dihydroxyvitamin D_3 mediated intestinal calcium absorption. Miner Electrolyte Metab 1990; 16:109–114.
28. Nemere I, Leathers V, Norman AW. 1,25 Dihydroxyvitamin D_3 mediated calcium transport across the intestine biochemical identification of lysosomes containing calcium and calcium binding protein (calbindin-D 28k). J Biol Chem 1986; 261:16,106–16,114.
29. Nemere I, Leathers VL, Thompson BS, Luben BA, Norman AW. Distribution of calbindin-D_{28k} in chick intestine in response to calcium transport. Endocrinology 1991; 129:2972–2984.
30. Nemere I. Dormanen MC, Hammond MW, Okamura WH, Norman AW. Identification of a specific binding protein for 1 α-25-dihydroxyvitamin D_3 in basolateral membranes in chick intestinal epithelium and relationship to transcaltachia. J Biol Chem 1994; 269:23,750–23,756.
31. Favus M J. Intestinal absorption of calcium, magnesium and phosphorus In:. Disorders of Bone and Mineral Metabolism. Coe FL, Favus MJ, eds. New York, Raven Press, 1992; 51–81.
32. Lemann J. Intestinal absorption of calcium, magnesium and phosphorus In:. Primer on Metabolic Bone Diseases and Disorders of Mineral Metabolism, 2nd ed. Favus MJ, ed. New York, Raven Press, 1993; 46–50.
33. LeGrimellec C. Micropuncture study along the proximal convoluted tubule. Electrolyte reabsorption in first convolutions. Pfluegers Arch 1975; 354:133–150.
34. Lassiter WE, Gottschalk CW, Mylle, M. Micropuncture study of renal tubular reabsorption of calcium in normal rodents. Am J Physiol 1963; 204:771–775.
35. Friedman PA, Gesek FA. Calcium transport in renal epithelial cells. Am J Physiol 1993; 264:F181–F198.
36. Friedman PA, Gesek FA. Cellular calcium transport in renal epithelia: measurement, mechanisms and regulation. Physiol Rev 1995; 75:429–471.
37. Winaver, J, Sylk DB, Robertson JS, Chen TC, Puschett JB. Micropuncture study of the acute renal effects of 25hydroxyvitamin D_3 in the dog. Miner Electrolyte Metab 1980; 4:178–188.
38. Friedman PA, Gesek FA. Vitamin D_3 accelerates PTH-dependent calcium transport in distal convoluted tubule cells. Am J Physiol 1993; 265:F300–F308.
39. Yamamoto M, Kawanobe Y, Takahashi H, Shimazawa E, Kimura S, Ogata E. Vitamin D-deficiency and renal calcium transport in the rat. J Clin Invest 1984; 74:507–513.
40. Bouhtiauy I, Lajeunesse D, Brunette MG. Effect of vitamin D depletion on calcium transport by the luminal and basolateral membranes of the proximal and distal nephrons. Endocrinology 1993; 132: 115–120.

41. Bindels RJM, Hartog A, Timmermans J, Van Os CH. Active Ca^{++} transport in primary cultures of rabbit kidney CCD: stimulation by 1,25-dihydroxyvitamin D_3 and PTH. Am J Physiol 1991; 261: F799–F807.
42. Sneddon WB, Gesek FA, Friedman PA. $1,25(OH)_2$ vitamin D_3 up-regulates the expression of the parathyroid hormone receptor in distal convoluted tubule cells. J Am Soc Nephrol 1993; 4:729.
43. Stumpf, WE, Sar M, Reid FA, Tanaka Y, DeLuca HF. Target cells for 1,25-dihydroxyvitamin D_3 in intestinal tract, stomach, kidney, skin, pituitary and parathyroid. Science 1979; 206:1188–1190.
44. Stumpf WE, Sar M, Narbaitz R, Reid FA, DeLuca HF, Tanaka Y. Cellular and subcellular localization of $1,25(OH)_2D_3$ in rat kidney: comparison with localization of parathyroid hormone and estradiol. Proc Natl Acad Sci USA 1980; 77:1149–1153.
45. Taylor AN, McIntosh JE, Bordeau JE. Immunocytochemical localization of vitamin D-dependent calcium binding protein in renal tubules of rabbit, rat, and chick. Kidney Int 1982; 21:765–773.
46. Rhoten WB, Christakos S. Immunocytochemical localization of vitamin D-dependent calcium binding protein in mammalian nephron. Endocrinology 1981; 109:981–983.
47. Roth J, Brown D, Norman AW, Orci L. Localization of the vitamin D-dependent calcium binding protein in mammalian kidney. Am J Physiol 1982; 243:F243–F252.
48. Rhoten WB, Bruns ME, Christakos S. Presence and localization of two vitamin D-dependent calcium binding proteins in kidneys of higher vertebrates. Endocrinology 1985; 117:674–683.
49. Bouhtiauy I, Lajeunesse D, Christakos S, Brunette MG. Two vitamin D-dependent calcium binding proteins increase calcium reabsorption by different mechanisms. II. Effect of $CaBP_{28k}$. Kidney Int 1994; 45:461–468.
50. Bouhtiauy I, Lajeunesse D, Christakos S, Brunette MG. Two vitamin D-dependent calcium binding proteins increase calcium reabsorption by different mechanisms. II. Effect of $CaBP_{9k}$. Kidney Int 1994; 45:469–474.
51. Koster HPG, Hartog, A, VanOs CN, Bindels RJM. Calbindin-D_{28k} facilitates cytosolic calcium diffusion without interfering with calcium signaling. Cell Calcium 1995; 18:187–196.
52. Borke, JL, Caride A, Verma AK, Penniston JT, Kumar R. Plasma membrane calcium pump and 28-kDa calcium binding protein in cells of rat kidney distal tubules. Am J Physiol 1989; 257:F842–F849.
53. Borke JL, Minami J, Verma AK, Penniston JT, Kumar R. Monoclonal antibodies to human erythrocyte membrane Ca^{2+}-Mg^{2+} adenosine triphosphatase pump recognize an epitope in the basolateral membrane of human kidney distal tubule cells. J Clin Invest 1987; 80:1225–1231.
54. James P, Vorherr, T, Thulin E, Forsen S, Carafoli E. Identification and primary structure of a calbindin$_{9k}$ binding domain in the plasma membrane Ca^{2+} pump. FEBS Lett 1991; 278:155–159.
55. Walters JR, Howard A, Charpin MV, Gniecko KC, Brodin P, Thulin E, Forsen S. Stimulation of intestinal bastolateral membrane calcium-pump activity by recombinant synthetic calbindin-D_{9k} and specific mutants. Biochem Biophys Res Commun 1990; 170:603–608.
56. Yu ASL, Hebert SC, Lee S, Brenner BM, Lyttan J. Identification and localization of renal Na^+-Ca^+ exchanger by polymerase chain reaction. Am J Physiol 1992; 263:F680–F685.
57. Bourdeau JE, Taylor AN, Iacopino AM. Immunocytochemical localization of sodium calcium exchanger in canine nephron. J Am Soc Nephrol 1993; 4:105–110.
58. Henry HL. Vitamin D hydroxylases. J Cell Biochem 1992; 49:4–9.
59. Kumar R. Metabolism of 1,25-dihydroxyvitamin D_3. Physiol Rev 1984; 64:478–504.
60. Shinki T, Jin CH, Nishimura A, Nagai Y, Ohyama Y, Noshiro M, Okuda K, Suda T. Parathyroid hormone inhibits 25-hydroxyvitamin D_3-24-hydroxylase mRNA expression stimulated by 1α25-dihydroxyvitamin D_3 in rat kidney but not in intestine. J Biol Chem 1992; 267:13,757–13,762.
61. Kawashima H, Torikai S, Kurokawa K. Localization of 25-hydroxyvitamin D_3 1αhydroxylase and 24-hydroxylase along rat nephron. Proc Natl Acad Sci USA 1981; 78:1199–1203.
62. St.-Arnaud R, Moir M, Messerlain S, Glorieux FH. Molecular cloning and characterization of a cDNA for vitamin D 1αhydroxylase. J Bone Miner Res 1996; 11(Suppl 1): S124.
63. Ohyama Y, Noshiro M, Okuda K. Cloning and expression of cDNA encoding 25-hydroxyvitamin D_3 24-hydroxylase. FEBS Lett 1991; 278:195–198.
64. Zierold C, Darwish HM, DeLuca HF. Identification of a vitamin D response element in the rat calcidiol (25-hydroxyvitamin D_3) 24-hydroxylase gene. Proc Natl Acad Sci USA 1994; 91:900–902.
65. Ohyama Y, Ozono K, Uchida M, Shinki T, Kato S, Suda T, Yamamoto U, Noshiro M, Kato Y. Identification of a vitamin D-responsive element in the 5' flanking region of the rat 25-hydroxyvitamin D_3 24-hydroxylase gene. J Biol Chem 1994; 269:10,545–10,550.

66. Hahn CN, Kerry DM, Omdahl JL, May BK. Identification of a vitamin D responsive element in the promoter of the rat cytochrome $P450_{24}$ gene. Nucleic Acids Res 1994; 22:2410–2416.
67. Chen K-S, DeLuca HF. Cloning of human $1\alpha 25$-hydroxyvitamin D_3 24-hydroxylase gene promoter and identification of two vitamin D responsive elements. Biochim Biophys Acta 1995; 1263:1–9.
68. Bonjour JP, Preston C, Fleisch H. Effect of 1,25-dihydroxyvitamin D_3 on renal handling of P_i in thyroparathyroidectomized rats. J Clin Invest 1977; 60:1419–1428.
69. Liu L, Khastgir A, McCauley J, Dunn ST, Morrissey JH, Christakos S, Hughes MR, Bourdeau JE. RT-PCR microlocalization of mRNAs for calbindin-D_{28k} and vitamin D receptor in murine nephron. Am J Physiol 1996; 270:F677–F681.
70. Bindels RJM. Calcium handling by the mammalian kidney. J Exp Biol 1993; 184:89–104.

11 Biologic and Molecular Effects of Vitamin D on Bone

Jane B. Lian, Ada Staal, André van Wijnen, Janet L. Stein, and Gary S. Stein

1. INTRODUCTION

The skeleton provides rigid mechanical support to the body, protects vital organs, and serves as a reservoir of ions, especially for calcium and phosphate required for serum homeostasis. The integrity of the skeleton is maintained by continuous remodeling of bone tissues throughout life in response to a broad spectrum of physiologic signals. As described in Chapter 9, the active hormone 1,25-dihydroxyvitamin D_3 (vitamin D) plays a key role in the maintenance of calcium and phosphate blood levels. In response to reduced serum calcium levels, calcium transport is stimulated across the gut and from the renal tubular lumen into the bloodstream. At the same time, calcium is mobilized from bone. Vitamin D actively promotes the release of bone mineral into the circulation by direct effects on the several cellular populations that reside in bone. The hormone influences differentiation and activity of cells of the osteoblast lineage, which form the mineralized bone matrix and cells of the osteoclast lineage, which resorb the mineralized bone (Fig. 1). Vitamin D exerts its effects on these cells by modulating the transcription of a broad spectrum of genes related to these bone cell phenotypes (1). How vitamin D mediates resorption of the bone matrix and subsequent bone formation through complex interactions between different populations of bone cells and at the level of regulation of gene expression is the primary subject of this chapter. The molecular mechanisms contributing to vitamin D-dependent transcription of the bone-specific osteocalcin gene have provided new insights for understanding steroid hormone responses in relation to a broad spectrum of physiologic conditions and phenotypic properties of a cell.

2. BONE MATRIX ORGANIZATION AND THE INFLUENCE OF VITAMIN D

Bone is the connective tissue characterized by an extensive extracellular matrix that is mineralized. Crystals of hydroxyapatite [$Ca_{10}(PO4)_6(OH)_2$] deposit within the bone, a process that is in part dependent on adequate intake, absorption, and retention of dietary calcium and phosphate mediated by vitamin D. Deficiencies in vitamin D lead to decreased mineral deposition in the skeleton, resulting in rachitic bone in children or

From: *Vitamin D: Physiology, Molecular Biology, and Clinical Applications*
Edited by: M. F. Holick © Humana Press Inc., Totowa, NJ

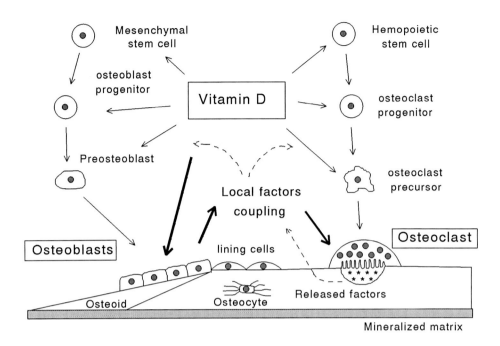

Fig. 1. Schematic representation of the role of vitamin D in development of the bone cell types and regulation of bone remodeling. Lineage progression during development of osteoblasts and osteoclasts is influenced by vitamin D at various maturational stages (thin arrows). Osteoblasts are derived from pluripotent mesenchymal cells and lie together with the lining cells. Osteoblasts form new bone on previously resorbed bone by synthesizing bone matrix (osteoid), which is subsequently mineralized. Finally, when osteoblasts are trapped in the mineralized matrix and change their functional and morphologic characteristics, they are called osteocytes. Osteoclasts are derived from progenitors of hemopoietic origin and mononuclear precursors, which fuse to mature multinucleated osteoclasts. During resorption, an acidic and proteolytic microenvironment is created (*) by the osteoclast to release bone mineral and a number of bone matrix-bound factors (broken arrows). Some of these factors are important for the coupling of bone resorption to formation by recruiting osteogenic precursors and inhibiting osteoclast differentiation (broken arrows). Vitamin D supports tissue remodeling initially by influencing retraction of the osteoblast lining cells from the bone surface, and then by modulating the synthesis of several osteoblast products, which directly affect osteoclast activity (bold arrows) or become a component of the newly reformed bone matrix.

osteomalacia in adults (*see* Chapters 18 and 19). Histologic sections of bone from vitamin D-deficient patients or animals show increased osteoid (unmineralized matrix) width. The ordered deposition of hydroxyapatite in bone is initiated through modifications of cell membranes forming matrix vesicles and propagated by the binding of ions to organic matrix components *(2)*. Hydroxyapatite crystals become incorporated within the fibrillar structure of collagen *(3)*. Collagen type I is the major component of the organic bone matrix, accounting for as much as 90% of the total protein in adult bone. In addition, numerous noncollagenous proteins residing in the bone matrix may function in either promoting or inhibiting mineralization *(4)*. Many of these are specialized calcium and phosphate binding proteins, such as the bone sialoprotein, osteocalcin, which is a γ-carboxyglutamate (Gla) protein, or osteopontin, a protein containing *O*-phosphoserine. These proteins are synthesized selectively or predominantly in mineralized tissues. Also important to the mineralization process are elevated levels of the plasma membrane-bound enzyme alkaline phosphatase *(5)*. Alkaline phosphatase activity and synthesis of several structural proteins of the bone matrix (e.g., collagen, osteopontin, and osteocalcin) are known to be regulated by vitamin D *(6,7)*. Physiologic levels of the hormone contribute to normal bone formation, and pharmacologic doses of vitamin D that elevate serum calcium in vivo stimulate synthesis and accumulation of these proteins, but may also lead to abnormal bone formation *(8–11)*.

3. CELLS OF BONE AND THE ROLE OF VITAMIN D IN COUPLING BONE RESORPTION TO FORMATION

The cellular composition of bone is quite heterogeneous, with two distinct cell lineages giving rise to the bone-forming osteoblast and the bone-resorbing osteoclast. These differentiated cells are derived from progenitors that progress through discrete stages of maturation characterized by the expression of phenotypic markers. Vitamin D is known to influence lineage progression at various stages of cellular differentiation.

Osteoblasts are derived from pluripotent mesenchymal cells arising from the connective tissue mesenchyme or the bone marrow stroma, which is distinct from the hemopoietic compartment. Pluripotent stem cells are capable of producing several tissue-specific cells, including, for example, reticular, fibroblastic, osteogenic, chondrogenic, and adipose cells, depending on environmental stimuli. In vitro studies of cultured adherent marrow cells have demonstrated that glucocorticoids promote differentiation of stromal cells through the osteoblast lineage, whereas vitamin D can promote adipocyte formation. Progenitor cells or immature osteoblasts are found near the outer bone surface, either in the periosteum or along the endosteum. When bone formation is triggered, some of the preosteoblasts can divide, migrate into the interior of the bone, and differentiate. Osteoblasts lie together with the lining cells, which are inactive osteoblasts on the endosteal surfaces of bone, and these cells are responsive to elevated parathyroid hormone (PTH) and vitamin D levels when bone resorption is induced (Fig. 1). Finally, when osteoblasts are trapped in the mineralized matrix and change their functional and morphologic characteristics, they are called osteocytes. Osteocytes have numerous cell processes that reach out through lacunae in bone tissue forming a network of cells, which facilitates their mechanosensor function in bone *(12)*. The growth and differentiation of these osteoblast populations are maintained and is stringently regulated through the activity of cytokines, growth factors, and hormones *(13–15)*. The antiproliferative effects of vitamin D lead to an induction of osteoblast phenotypic properties. In this context vitamin D influences osteoblast differentiation. Vitamin D also supports tissue remodeling by modulating the synthesis of several osteoblast products, which directly affect osteoclast activity or are stored in the bone matrix (Fig. 1).

Osteoclasts are of hemopoietic origin, and the progenitors can be recruited from marrow, spleen, and blood *(16,17)*. Immature hemopoietic cells, circulating monocytes, and some tissue macrophages are capable of differentiating into osteoclasts. The recruitment of progenitors and the fusion of mononuclear precursors to mature multinucleated osteoclasts, as well as activation of osteoclast activity on the bone surface, are processes mediated by vitamin D (Fig. 1). Evidence has been provided from in vitro culture models and in vivo animal models. Characteristic features of the actively resorbing osteoclast include the clear zone and ruffled border *(18)*. The clear zone serves to attach osteoclasts to the bone surface and separates the bone resorption area from the unresorbed bone. Interestingly, the matrix protein osteopontin and the integrin αvβ3 are also involved in the attachment of the osteoclast to bone, and synthesis of osteopontin and the β3-subunit are increased in response to vitamin D *(19–22)*. The ruffled border, a structure of deeply infolded plasma membrane, is surrounded by the clear zone. This creates an acidic microenvironment suitable for resorbing mineral as a consequence of osteoclast-secreted protons and proteases. During bone resorption, besides release of the mineral, chemotactic and mitogenic factors stored in the bone matrix become active [for example, transforming growth factor-β (TGF-β)] *(23)* and are important for the coupling of bone resorption to formation. These released factors can recruit osteogenic precursors and inhibit osteoclast differentiation (Fig. 1), thereby providing a negative feedback mechanism for bone resorption *(13,16,17)*.

Vitamin D contributes to the coupling of osteoclast and osteoblast activities at two stages of the remodeling process, thereby mediating completion of the bone remodeling sequence. Initially, vitamin D targets osteoblasts and lining cells to retract from the bone surface and secrete vitamin D-responsive factors, which induce osteoclast activity. At the same time, vitamin D directly promotes osteoclast formation from mononuclear cells. Following the resorption phase, vitamin D can stimulate synthesis of cytokines for preosteoblast recruitment and growth and expression of osteoblast proteins, which form the bone matrix. In this manner, vitamin D contributes to the completion of the bone remodeling sequence.

4. VITAMIN D REGULATION OF GENE EXPRESSION DURING BONE FORMATION AND IN OSTEOBLASTS IN VITRO

Cultured normal diploid osteoblasts provide a model for identifying selective effects of vitamin D on the expression of cell growth and tissue-specific genes that are dependent on the differentiated state of the bone cell *(15)*. Osteoblasts in vitro reinitiate proliferation and undergo a developmental sequence of growth and differentiation, leading to the formation of bone-like tissue. Thus, preosteoblast-like cells, active osteoblasts, and osteocyte-like cells in a mineralized matrix, analogous to in vivo bone cell populations, can be studied. Developmental periods are characterized by a stage-specific expression of cell growth and bone phenotype-related genes. The sequential expression of these cell growth- and tissue-specific genes has been mapped during progressive development of the bone cell phenotype by the combined application of Northern blot analysis, *in situ* hybridization, nuclear run-on transcription, and histochemistry (Fig. 2A) *(24,25)*.

The temporal sequence of gene expression defines three principal developmental periods of osteoblast differentiation. Initially, proliferation supports an increase in osteoblast number and synthesis of collagen type I bone extracellular matrix. Osteoblasts can multilayer within a collagenous matrix (Fig. 3). Following the initial proliferation period, expression of genes associated with maturation and competency of the extracellular bone matrix for mineralization is upregulated (e.g., alkaline phosphatase). During osteoblast maturation, the cellular representation of family members of transcription factors, important to osteoblast differentiation, changes. Progression from the proliferation to the differentiation stages involves a downregulation of c-fos and c-jun, but an increase in fra2 *(26)* in the matrix maturation period. The Msx-2 homeodomain protein expression declines *(27)*, whereas the homeodomain Dlx-5 protein appears postprolif-

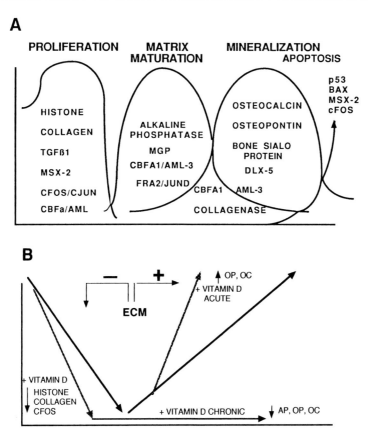

Fig. 2. Development of the osteoblast phenotype. (**A**) The temporal expression of genes (phenotypic markers and transcription factors) characterizing three stages of differentiation, proliferation, matrix maturation, and mineralization stages are indicated. Some apoptosis is associated with cells in the mineralizing nodule. Maximal levels of genes related to stimulation of growth and matrix production during the proliferation period are indicated. A cessation of proliferation leads to induction of phenotypic markers that characterize the subsequent postproliferative differentiation stages. The second stage is characterized by maximal levels of alkaline phosphatase, considered an early marker of the differentiated osteoblast together with expression of CBFA1/AML-3. The genes represented in the mineralization increase in expression in relation to calcium deposition, and then decline when apoptosis markers (Bax, p53, Msx-2, c-fos) become induced. (**B**) Schematic illustration of the reciprocal relationship between growth and differentiation and signaling pathways. Formation of the extracellular matrix (ECM) contributes to an important transition point, the downregulation of cell growth necessary for expression of differentiation markers, represented by the heavy arrow. Stippled arrows illustrate acute effects of vitamin D, which contributes to inhibition of growth in the proliferation period and consequently can block differentiation under chronic vitamin D conditions. However, in the postproliferative period, vitamin D can increase the expression of differentiation markers.

ertively *(28)*. AML-3/CBFA1 [an acute myelogenous leukemia factor, also named core binding factor], a member of the *runt* homology domain family, is a critical determinant of bone formation and is increased during osteoblast differentiation *(29–31)*.

The third developmental period involves gene expression related to the ordered deposition of hydroxyapatite. Genes encoding hydroxyapatite binding proteins, e.g., bone sialoprotein, osteocalcin, and osteopontin, are maximally expressed during the third developmental period. Osteocalcin, which is expressed only postproliferatively in the

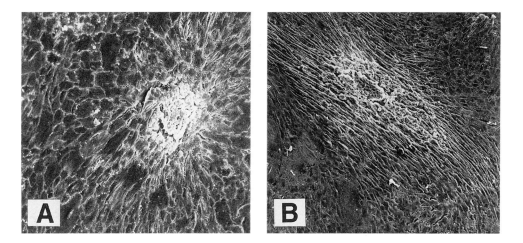

Fig. 3. Scanning electron micrograph of bone tissue-like formation in vitro and the effects of vitamin D. (**A**) Calvaria osteoblasts from 21-d fetal rats cultured for 21 d produce nodules (multilayered cells within a mineralized matrix). (**B**) Cultures exposed to $10^{-8}\,M$ 1,25(OH)$_2$D$_3$ from d 14 to 21 exhibited elongated osteoblasts associated with the nodule, whereas osteoblasts in monolayers (surrounding the nodules) were unaltered morphologically.

mature osteoblast, is upregulated by vitamin D only when basal transcription is ongoing *(15)*. Osteopontin, by contrast, can be regulated by vitamin D both in the growth period and in the mineralization stage. We have shown by *in situ* hybridization that the nodule-associated cells are the mature osteoblasts competent to express osteocalcin *(32)*. These cells respond morphologically to vitamin D and, together with involvement of osteocalcin and osteopontin in bone resorption *(33)*, may be of significance during the bone remodeling sequence. Later, during the mineralization period, collagenase gene expression is increased, reflecting continued organization of the collagenous matrix. Finally, genes expressed earlier (e.g., alkaline phosphatase, osteocalcin, osteopontin) become downregulated in the mature osteocytes of the mineralizing nodule. Apoptosis (programmed cell death) occurs within the mineralized nodules, and compensatory proliferative activity and collagen gene expression are evident. Clearly apparent is the reciprocal relationship between proliferation and the expression of phenotypic markers of the differentiated osteoblast (Fig. 2B).

Expression of both cell growth and bone cell phenotype-related genes are developmentally responsive to vitamin D *(14,15,34,35)*. The effects of vitamin D on osteoblast phenotype development are related to the maturational stage of the osteoblast *(14,15)* and the duration of hormone exposure. Notably, in the proliferative preosteoblast stage, vitamin D inhibits growth and downregulates some genes (such as collagen), but not others (e.g., osteopontin) within 24 h. By contrast, in mature osteoblasts in a mineralized matrix, vitamin D stimulates proliferation and collagen expression *(15,36)*. In human osteoblast cultures that derive from mature adult bone specimens, vitamin D always stimulates collagen synthesis *(37)*. If osteoblasts are continuously treated with vitamin D from the proliferation stage, vitamin D blocks osteoblast differentiation. This occurs because vitamin D inhibits cell growth, collagen type I synthesis, and alkaline phosphatase activity in this rat model. Consequently bone nodule formation, mineralization, and osteocalcin expression are impaired. However, in the postproliferative stage of nodule formation when cultures are treated with vitamin D, osteocalcin is now upregulated four- to fivefold, reflecting increasing maturation of the osteoblast (Fig. 2B). Other studies show that vitamin D added to mature osteoblast cultures can promote mineral deposition *(38)*. The molecular mechanisms that contribute to these complex and diverse effects of vitamin D on gene expression are just beginning to be understood from studies defining vitamin D regulation of osteocalcin gene transcription *(39)*.

5. TRANSCRIPTIONAL CONTROL OF OSTEOCALCIN GENE EXPRESSION

Characterization of the prototypical osteocalcin gene promoter has provided many new insights for understanding gene-specific vitamin D-mediated transcriptional control. Modularly organized regulatory elements in the promoter of the bone-specific osteocalcin gene render the gene competent to support expression during skeletal development and during bone remodeling, as well as for skeletal involvement in calcium homeostasis. However, physiologic responsiveness of gene transcription requires that the promoter provide more than a series of regulatory elements. Control of gene transcription must consider multiple levels of regulation. Structure and organization of gene regulatory sequences must be related to the extent to which the gene is transcribed. Control of gene transcription may require modulation of DNA binding protein levels and protein–protein interactions at promoter elements related to tissue-specific development or hormonal requirements. The overlap of binding domains within gene regulatory sequences reflects options for utilization of multiple factors. Mutual exclusive occupancy of a binding domain by functionally different transcription factors allows several regulatory pathways to increase the potential of responsiveness. Crosstalk between regulatory elements and the basal transcription machinery reflects another level of transcription control, which is supported by modification in chromatin structure and the nuclear matrix. Steroid hormones and other physiologic regulations may interact directly with their receptors and cognate elements or function indirectly by affecting the synthesis and/or modification (e.g., phosphorylation) of other transcription factors for DNA binding domains. The vitamin D response element (VDRE) in the rat and human osteocalcin gene promoters is a particularly noteworthy example of these multiple levels of transcriptional control.

6. OSTEOCALCIN GENE PROMOTER REGULATORY SEQUENCES

A series of regulatory sequences within the osteocalcin promoter contributes to tissue-specific developmental and steroid hormone-mediated transcriptional control of osteocalcin gene expression (Fig. 4). These include sequences that contribute to basal expression: (1) a TATA motif, which is a sequence in the promoter that binds a multisubunit complex containing a DP1/NF-Y/CBF-related CCAAT factor complex; (2) the osteocalcin box (OC box I), a 24 nucleotide element with a homeodomain motif as the central core; and (3) a second conserved sequence (OC box II), originally designated site C in the rat promoter, which bound a bone-specific nuclear matrix protein (NMP2) *(40,41)* and OSE2 in mouse *(42)*. These complexes were subsequently identified as *runt* homology domain factors related to the AML/CBF and the homologous mouse PEBP2α (polyoma enhancer binding protein) proteins. In the osteocalcin promoter, two additional AML recognition motifs occur in the distal promoter flanking the VDRE *(40,43)*, sites A and B (Figs. 4 and 5). These motifs may contribute to the three-dimensional organization of the promoter because of the association of AML transcription factors with the

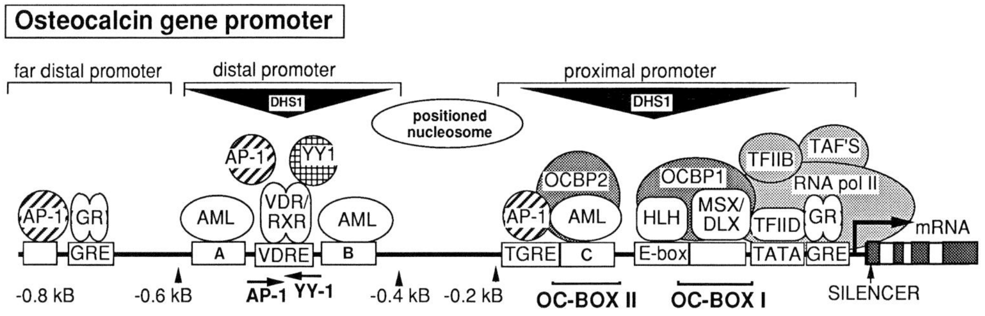

Osteocalcin gene promoter

Fig. 4. Organization of osteocalcin promoter regulatory sequences. Promoter regulatory domains and cognate transcription factors that reside in −800 nt of the 5' flanking sequence are shown. The OC boxes I and II are the primary tissue-specific transcriptional elements that bind homeodomain proteins (MSX and DLX) or *runt* homology domain transcription factors (AML-3), respectively, and osteoblast-specific complexes, OCBP1 and OCBP2. Fos-jun-related proteins interact as heterodimers with AP-1 sites, which overlap with or are in proximity to TGFβ (TGRE), glucocorticoid (GRE), and vitamin D_3 response elements (VDRE). HLH proteins bind to the E box motif contiguous with the OC box I. There are three AML/CBFA1 recognition sites (A, B, and C), which bind the osteoblast-specific AML-3/CBFA1 transcription factor. The occupancy of OC promoter regulatory elements results in recruitment of RNA pol II, TFIIB, TFIID, and associated factors (TAFs) to the site of transcription initiation. DNase I-hypersensitive sites (DHS), which are enhanced by vitamin D when the gene is transcribed, are indicated.

nuclear matrix (*see* Section 8.). Activation of osteocalcin expression in nonosseous cells can occur by forced expression of AML factors *(29,41)*. AML-3/CBFA1 is preferentially expressed in differentiating osteoblasts *(29)* and is essential for osteoblast differentiation *(29,31)* and bone formation *(30)*. Both OC boxes I and II are required for bone tissue-specific osteocalcin expression *(44)*. Mutations in OC box I, which alleviate homeodomain protein binding, lead to osteocalcin promoter activity in nonosseous cells *(45)*. Interestingly, vitamin D regulates expression of the Msx-2 homeodomain protein *(27,46)* and the bone-restricted AML-3/CBFA1 transcription factor in mouse *(47)*. Thus, vitamin D contributes indirectly to tissue-specific expression of the osteocalcin gene.

Other regulatory sequences in the osteocalcin gene promoter include a series of glucocorticoid response elements, AP-1 sites *(26)*, and sequences mediating TGF-β *(48)* and fibroblast growth factor-2 (FGF-2) responsiveness *(49)*. The AP-1 sites, which bind homo- or heterodimers of the fos and jun family of proteins, contribute to both enhancer and suppressor activity. c-fos/c-jun complexes decrease osteocalcin transcription, whereas fra2/jun D increases expression *(26)*.

7. PROTEIN–DNA AND PROTEIN–PROTEIN INTERACTIONS AT THE VDRE

VDREs consist of two direct repeats of palindromic hexameric elements [e.g., PuG(G/T)TCA] separated by a three (e.g., rat osteocalcin), four [e.g., mPit-1 *(50)*], or six [e.g., in the c-fos *(51)*] nucleotide spacing. However, this rule may not apply to some of the recently identified VDREs, for example, in the c-fos *(51)* and collagen genes *(52)*. In most promoters, a single VDRE contributes to vitamin D regulation, but in others multiple pairs

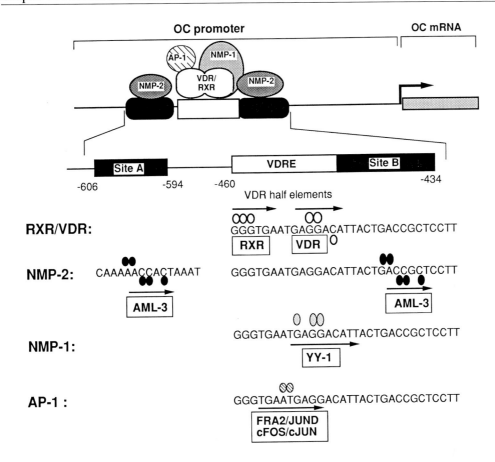

Fig. 5. Regulatory elements within the vitamin D response element domain. (**Top**) The YY-1 and AP-1 factors that interact with the VDRE and the two nuclear matrix protein (NMP-2) interacting sites, which flank the vitamin D response element. (**Bottom**) The specific nucleotides involved in these protein–DNA interactions. The polarity of the RXR–VDR heterodimeric complex is indicated; the bone-specific nuclear matrix complex NMP-2 is an AML-3-related transactivation factor; NMP-1, which is the YY-1 transcription factor, interacts with nucleotides that overlap the proximal VDR binding steroid half-element. Ovals indicate the nucleotide involved in the protein–DNA interactions established by methylation interference analyses. The AP-1 recognition motif can bind either c-fos/c-jun or fra2/jun D heterodimers. Mutations of the two nucleotides in the AP-1 sequence indicated by the ovals do not interfere with VDR–RXR heterodimeric binding to the half-steroid elements and have established that interactions with AP-1 factors are necessary for complete vitamin D enhancement of OC transcription.

of half-steroid element sequences may be involved, e.g., 24-hydroxylase *(53,54)* and β_3 integrin promoters *(55,56)*. For enhancement of osteocalcin or osteopontin transactivation, the VDRE must bind a heterodimeric vitamin D receptor (VDR)–retinoic X receptor (RXR) complex *(19,57–61)*. Polarity of binding VDR–RXR complexes to the VDRE occurs with VDR association to the proximal VDRE half-element and RXR binding to the distal half-element *(62–65)* (Fig. 5).

Multiple factors appear to contribute to the formation of the VDR complex in vivo. Upon binding of its ligand, 1,25-dihydroxyvitamin D_3, the conformation of the VDR is changed, and the VDR is phosphorylated *(66)*. Ligand binding appears to be involved in stabilization of the receptor and is required to detect receptor complex–DNA interactions by in vivo footprinting analyses *(67)*. Phosphorylation of the VDR is required for its binding to the VDRE and for its transactivation function *(68)*. The interaction of the VDR with accessory proteins or coactivators provides an additional level of regulation *(51)*. VDR interactions with RXR thyroid receptor and other nuclear factors are known *(69–72)*. 9-*cis* retinoic acid inhibits VDR-mediated activation by decreasing the availability of its heterodimeric partner *(73,74)*. Physiologic factors can affect the formation of heterodimers by regulating receptor levels of either receptor in the heterodimeric complex. For example, tumor necrosis factor-α, glucocorticoids, and TGF-β have been reported to affect VDR levels. Finally, VDR gene polymorphisms may affect VDR function *(75–77)*. However, to date, this issue is controversial. VDR gene polymorphisms have only been studied in relation to osteoporosis.

Variations in the nucleotide sequence of the VDREs in vitamin D-regulated genes reflect options for different responsiveness to vitamin D stimulation (Fig. 5). The rat osteocalcin VDRE domain contains functionally active AP-1 and YY-1 regulatory sequences. YY-1, originally identified as a nuclear matrix binding protein (NMP-1) *(43)*, binds to nucleotides overlapping the proximal steroid half-site (Fig. 5). Recently, we have shown that YY-1 suppresses vitamin D enhancement of transcription in the rat osteocalcin gene *(78)*. Overexpression of YY-1 in osteoblasts inhibits vitamin D-stimulated transcription of osteocalcin. By contrast, the AP-1 sites contiguous to the human OC VDRE *(62)* and overlapping the distal steroid half-element of the rat OC VDRE *(79)* contribute to enhancement of vitamin stimulation by a mechanism unknown to date. Indeed, other transcription factors have been reported to contribute to vitamin D enhancement of transcription using artificial constructs *(71)*. These findings point to the significance of crosstalk through protein–protein interactions involving VDR–RXR receptor complexes. The necessity for such interactions can be understood in the context of in vivo expression of osteocalcin during bone development and during osteoblast differentiation.

The VDR–RXR heterodimeric complex functions as an enhancer of osteocalcin gene transcription only when basal transcription is ongoing in postproliferative osteoblasts (Fig. 6). To explain the absence of vitamin D inducibility when the osteocalcin gene is inactive, that is, not expressing in proliferating osteoblasts or nonosseous cells, a model of phenotype suppression was proposed. Originally, this model postulated that high AP-1 activity in proliferating cells interferes with basal and VDR-mediated transactivation of the osteocalcin gene *(80,81)*. However, it is now understood that other transcription factors can suppress osteocalcin transcription, including homeodomain proteins (Msx-2, Dlx-5) and the multifunctional transcription factor YY-1, which directly interferes with VDR–RXR binding to the VDRE as well as with transcription factor IIB (TFIIB)–VDR interactions *(82,83)*.

TFIIB is a TATA binding factor that interacts with the VDR. This interaction supports a mechanism for regulation of vitamin D enhancer activity by mediating crosstalk between the VDR and basal transcriptional machinery. YY-1–TFIIB protein–protein interactions have also been demonstrated *(78,84)*. Thus, a plausible mechanism of the inhibitory activity of YY-1 on vitamin D enhancer activity is that YY-1–VDRE interactions compete with the VDRE–VDR–TFIIB interaction (Fig. 7). The model indicates that displacement of YY-1 by the VDR after vitamin D stimulation leads to VDRE–VDR–TFIIB interactions and an increase in osteocalcin gene transcription (Fig. 7). Thus, the overlapping and contiguous organization of regulatory elements, as illustrated by the AP-1–YY-1–VDRE recognition sequences, provides a basis for combined activities that

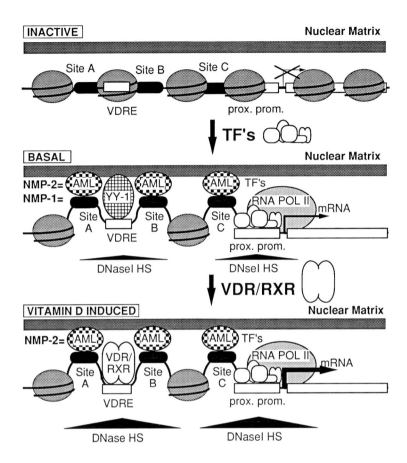

Fig. 6. Schematic representation of the osteocalcin gene promoter organization and occupancy of regulatory elements by cognate transcription factors paralleling and supporting functional relationships to inactive, basal, or vitamin D-induced states. Inactive, suppression of transcription in proliferating osteoblasts; basal, activation of expression in differentiated cells contributed in part by association of the gene with nuclear matrix-bound bone-specific AML-3 transcription factor; vitamin D-induced, enhancement of transcription by vitamin D; here, ligand (vitamin D) and DNA binding of the VDR–RXR receptor complex displaces the suppressor YY-1 factor. In each panel, the placement of nucleosomes is indicated. Remodeling of chromatin structure and nucleosome organization to support suppression, basal, and vitamin D-induced transcription of the osteocalcin gene is shown. The representation and magnitude of DNase I-hypersensitive sites are designated by solid triangles, and gene–nuclear matrix interactions that facilitate transactivation are shown.

support responsiveness to physiologic mediators. Notably, in the mouse osteopontin promoter, the AP-1 site does not overlap the VDRE, and the YY-1 sequence is not present *(78)*. These characteristics of the VDRE sequence may account for expression of osteopontin and vitamin D regulation in proliferating osteoblasts.

The subtle sequence variations between the osteocalcin VDRE (GGGTGAnnnAGGACA) and the osteopontin VDRE (GG<u>TTC</u>Annn<u>GGTT</u>CA) also appear to alter the conformation of the transcription factor complex *(85)*. This conformational change may be required for alternative protein–protein interactions during crosstalk with other binding elements within the promoter region. In summary, unique sequence features of the rat

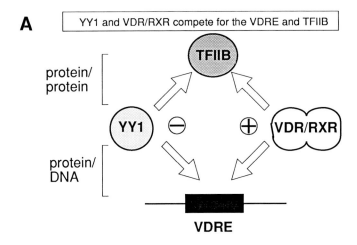

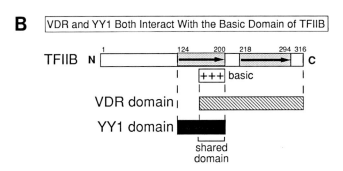

Fig. 7. Regulation of vitamin D enhancement of osteocalcin gene transcription by YY-1. (**A**) The model illustrates the mechanism by which YY-1 exerts its inhibitory effect by interfering with the binding of the VDR–RXR heterodimer to VDRE, as well as by interfering with the interaction between DNA-bound VDR–RXR and TFIIB. (**B**) The TFIIB domains of YY-1 required for interaction with VDR (crosshatched bars) are indicated. YY-1 binds to the basic region of TFIIB and competes with the VDR–RXR interaction for TFIIB.

osteocalcin VDRE contribute to the inability of vitamin D to induce transcription of the osteocalcin gene in proliferating osteoblasts in the absence of ongoing basal expression.

Differences in vitamin D regulation of osteocalcin transcription among the species (human, rat, and mouse) are accounted for by nucleotide variations of the steroid half-elements and flanking sequences. Notably, whereas rat or human promoter–reporter constructs expressed in transgenic mice (86,87) or expressed in mouse MC3T3-E1 cells (88,89) respond to vitamin D with a three- to sixfold increase in osteocalcin transcription, the endogenous mouse osteocalcin gene is downregulated by vitamin D. These experiments indicate that differences in regulation of the mouse vs rat/human osteocalcin genes are not caused by cell type (species) differences in the availability of transcription factors. The mouse OC VDRE, GGGCAAnnnAGGACA, is identical to the rat OC VDRE in its proximal half-site, but differs in the fourth and fifth nucleotide of the distal half-site from both the rat and human OC VDRE. The mouse VDRE domain is competent to bind the

VDR–RXR complex [reported by Lian et al. *(90)*, but not in the studies of Ducy's group *(47)*] and functions on a heterologous promoter *(90)*. However, decreased activity of the mouse VDRE-tk promoter was observed in response to vitamin D *(90)* and not enhancer activity as occurs when using the rat VDRE sequence in a tk reporter construct *(88)*. This indicates that dinucleotide differences in the VDRE composition may lead to different responsiveness of the gene. Notably, the VDRE of the PTH gene functions in mediating repressor activity *(60)* GGG<u>CA</u>AnnnAGGACA. Thus, it can be appreciated that a multiplicity of regulatory factors impinge on vitamin D receptor-mediated transcriptional control of gene expression. Direct protein–DNA interaction modifications at the VDRE (dictated by subtle sequence variations among VDREs for selective occupancy of nonreceptor factors) or modifications in protein–protein interactions (dictated by either cellular protein levels or chromatin conformation mediating such interactions) contribute to precise transcriptional control in support of physiologic responsiveness.

8. NUCLEAR STRUCTURE INFLUENCES VITAMIN D REGULATION OF GENE TRANSCRIPTION

A fundamental question is the mechanism by which activities of the modularly organized transcription regulatory elements of the osteocalcin gene promoter are functionally integrated. Nuclear architecture provides a basis for support of stringently regulated modulation of cell growth- and tissue-specific transcription, which is necessary for the onset and progression of cellular differentiation.

Chromatin structure (reflected by DNase 1 hypersensitive sites and nucleosome organization revealed by micrococcal nuclease digestions of the DNA) regulates transcription by (1) modulating accessibility of transcription factors to regulatory sequences, and (2) mediating interaction between modular components of transcriptional control by reducing the distance between promoter elements (Fig. 8). Figures 6 and 8 show positions of nuclease-hypersensitive sites that flank both the basal domain and the VDRE element, reflecting increased availability of sites for protein–DNA interactions *(91)*. We observed a parallel relationship between the intensity of the hypersensitive sites and the extent to which the osteocalcin gene is transcribed. Nucleosomes in the osteocalcin gene block access of transcription factors to promoter sequences to in nonosseous and nonexpressing osteoblasts *(92)*. With gene activation, a remodeling of chromatin occurs, represented by changes in nucleosome number and specific placement. Further changes in the chromatin of the osteocalcin promoter occur with developmental upregulation of the gene in osteoblasts. These changes in chromatin organization accommodate activities of both basal and steroid hormone-responsive regulatory elements, the OC box and VDRE.

The nuclear matrix is another level of nuclear architecture that contributes to transcriptional control. The nuclear matrix is defined as the anastomosing network of filaments remaining after salt extraction of soluble factors. This component of nuclear structure contributes to transcriptional control of genes by mediating changes in gene conformation and supporting the concentration and targeting of transcription factors to gene-regulating elements, both of which are present at very low concentrations relative to the vast amount of protein and DNA in the cell. Sequence-specific DNA binding proteins have been shown to be selectively associated with the nuclear matrix, and a striking difference exists in the nuclear matrix composition between proliferating and differentiated cells, reflecting the modifications in gene expression *(93)*. Thus, the association

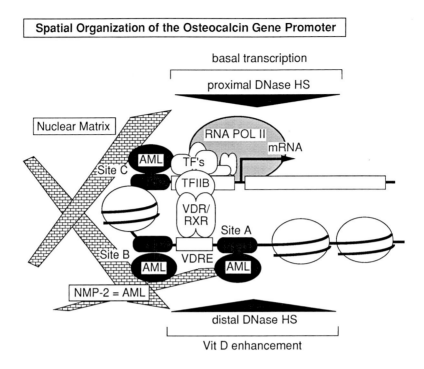

Fig. 8. Three-dimensional spatial organization of the rat osteocalcin gene promoter. A schematic model is presented for the spatial organization of the rat osteocalcin gene promoter based on evidence for nucleosome placement and the interaction of DNA binding sequences with the nuclear matrix. These components of chromatin structure and nuclear architecture restrict mobility of the promoter and impose physical constraints that reduce distances between proximal and distal promoter elements. Such postulated organization of the osteocalcin gene promoter can facilitate cooperative interactions for crosstalk between elements that mediate transcription factor binding and consequently determine the extent to which the gene is transcribed. The positioned nucleosome contributes to the three-dimensional organization of the promoter to promote protein–protein interactions (e.g., VDR–RXR and TFIIB) between the proximal and distal regulatory domains.

of the AML/CBFA and YY-1 transcription factors with the nuclear matrix is of functional significance in contributing to tissue specific and steroid hormone-mediated osteocalcin gene expression. All AML/CBFA factors are targeted to the nuclear matrix by a conserved 31-amino acid nuclear matrix targeting signal (NMTS) in the C-terminus (94). We postulate that the AML/CBFA proximal and the two distal binding sites, which flank the VDRE, together with the positioned nucleosome, contribute to the three-dimensional organization of the promoter (Fig. 8). This dynamic model facilitates not only the binding of tissue-specific transactivation factors but protein–protein interactions between the VDR and the basal transcriptional machinery.

Taken together, these components of nuclear architecture facilitate biologic requirements for physiologically responsive modifications in gene expression within the context of (1) homeostatic control involving short-term responsiveness; (2) developmental control, which is progressive and stage specific; and (3) differentiation-related control, which is associated with long-term phenotypic commitments to gene expression for support of structural and functional properties of cells and tissues.

In closing, studies to date have shown that vitamin D regulation of gene transcription interfaces with numerous physiologic modulators of bone formation and resorption. The molecular mechanisms that allow for such complexities can now be addressed at several levels: from the standpoints of nucleotide recognition motifs, cellular levels of the DNA binding proteins, subnuclear location of transcriptional activation complexes, and structural parameters of the gene.

REFERENCES

1. Hannah SS, Norman AW. 1-Alpha,25(OH)$_2$ vitamin D$_3$-regulated expression of the eukaryotic genome. Nutr Rev 1994; 52:376–382.
2. Anderson HC. Molecular biology of matrix vesicles. Clin Orthop 1995; 314:266–280.
3. Boskey AL. Mineral-matrix interactions in bone and cartilage. Clin Orthop 1992; 281:244–274.
4. Gehron Robey P. Bone matrix proteoglycans and glycoproteins. In: Principles of Bone Biology. Bilezikian JP, Raisz LG, Rodan GA, eds. San Diego: Academic Press, 1996; 155–165.
5. Hui M, Li SQ, Holmyard D, Cheng P. Stable transfection of nonosteogenic cell lines with tissue nonspecific alkaline phosphatase enhances mineral deposition both in the presence and absence of beta-glycerophosphate: possible role for alkaline phosphatase in pathological mineralization. Calcif Tissue Int 1997; 60:467–472.
6. Lian JB, Stein GS. Vitamin D regulation of osteoblast growth and differentiation. In: Nutrition and Gene Expression. Berdanier CD, Hargrove JL, eds. Boca Raton, FL: CRC Press, 1993; 391–429.
7. Boyan BD, Schwartz Z, Swain LD. In vitro studies on the regulation of endochondral ossification by vitamin D. Crit Rev Oral Biol Med 1992; 3:15–30.
8. Wittenberg JB, Stein JL. Hemoglobin in the symbiont-harboring gill of the marine gastropod *Alviniconcha hessleri*. Biol Bull 1995; 188:5-7.
9. Marie PJ, Hott M, Garba MT. Contrasting effects of 1,25-dihydroxyvitamin D$_3$ on bone matrix and mineral appositional rates in the mouse. Metabolism 1985; 34:777–783.
10. Hock JM, Gunness-Hey M, Poser J, Olson H, Bell NH, Raisz LG. Stimulation of undermineralized matrix formation by 1,25 dihydroxyvitamin D$_3$ in long bones of rats. Calcif Tissue Int 1986; 38:79–86.
11. Wronski TJ, Halloran BP, Bikle DD, Globus RK, Morey-Holton ER. Chronic administration of 1,25-dihydroxyvitamin D$_3$: increased bone but impaired mineralization. Endocrinology 1986; 119: 2580–2585.
12. Burger EH, Klein-Nulend J, van der Plas A, Nijweide PJ. Function of osteocytes in bone—their role in mechanotransduction. J Nutr 1995; 125:2020S–2023S.
13. Mundy GR, Boyce B, Hughes D, Wright K, Bonewald L, Dallas S, Harris S, Ghosh-Choudhury N, Chen D, Dunstan C, et al. The effects of cytokines and growth factors on osteoblastic cells. Bone 1995; 17:71S–715S.
14. Gerstenfeld LC, Zurakowski D, Schaffer JL, Nichols DP, Toma CD, Broess M, Bruder SP, Caplan AI. Variable hormone responsiveness of osteoblast populations isolated at different stages of embryogenesis and its relationship to the osteogenic lineage. Endocrinology 1996; 137:3957–3968.
15. Owen TA, Aronow MS, Barone LM, Bettencourt B, Stein GS, Lian JB. Pleiotropic effects of vitamin D on osteoblast gene expression are related to the proliferative and differentiated state of the bone cell phenotype: dependency upon basal levels of gene expression, duration of exposure, and bone matrix competency in normal rat osteoblast cultures. Endocrinology 1991; 128:1496–1504.
16. Suda T, Takahashi N, Etsuko A. Role of vitamin D in bone resorption. J Cell Biochem 1992; 49:53–58.
17. Roodman GD. Advances in bone biology: the osteoclast. Endocr Rev 1996; 17:308–332.
18. Baron R, Neff L, Louvard D, Courtoy PJ. Cell-mediated extracellular acidification and bone resorption: evidence for a low pH in resorbing lacunae and localization of a 100-kD lysosomal membrane protein at the osteoclast ruffled border. J Cell Biol 1985; 101:2210–2222.
19. Noda M, Vogel R, L., Craig AM, Prahl J, DeLuca HF, Denhardt D. Identification of a DNA sequence responsible for binding of the 1,25-dihydroxyvitamin D$_3$ receptor and 1,25-dihydroxyvitamin D$_3$ enhancement of mouse secreted phosphoprotein 1 (Spp-1 or osteopontin) gene expression. Proc Natl Acad Sci USA 1990; 87:9995–9999.
20. Cao X, Ross FP, Zhang L, MacDonald PN, Chappel J, Teitelbaum SL. Cloning of the promoter for the avian integrin beta 3 subunit gene and its regulation by 1,25-dihydroxyvitamin D$_3$. J Biol Chem 1993; 268:27,371–27,380.

21. Reinholt FP, Hultenby K, Oldberg A, Heinegard D. Osteopontin—a possible anchor of osteoclasts to bone. Proc Natl Acad Sci USA 1990; 87:4473–4475.
22. Grano M, Zigrino P, Colucci S, Zambonin G, Trusolino L, Serra M, Baldini N, Teti A, Marchisio PC, Zallone AZ. Adhesion properties and integrin expression of cultured human osteoclast-like cells. Exp Cell Res 1994; 212:209–218.
23. Oursler MJ. Osteoclast synthesis and secretion and activation of latent transforming growth factor beta. J Bone Miner Res 1994; 9:443–252.
24. Lian JB, Stein GS. Osteoblast biology. In: Osteoporosis. Marcus R, Feldman D, Kelsey J, eds. San Diego: Academic Press, 1996; 23–60.
25. Aubin JE, Liu F. The osteoblast lineage. In: Principles of Bone Biology. Bilezikian JP, Raisz LG, Rodan GA, eds. San Diego: Academic Press, 1996; 51–68.
26. McCabe LR, Kockx M, Lian J, Stein J, Stein G. Selective expression of fos- and jun-related genes during osteoblast proliferation and differentiation. Exp Cell Res 1995; 218:255–262.
27. Hoffmann HM, Catron KM, van Wijnen AJ, McCabe LR, Lian JB, Stein GS, Stein JL. Transcriptional control of the tissue-specific, developmentally regulated osteocalcin gene requires a binding motif for the Msx family of homeodomain proteins. Proc Natl Acad Sci USA 1994; 91:12,887–12,891.
28. Ryoo H-M, Hoffmann HM, Beumer TL, et al. Stage specific expression of Dlx-5 during osteoblast differentiation: involvement in regulation of osteocalcin gene expression. Mol Endocrinol 1997; 11: 1681–1694.
29. Komori T, Yagi H, Nomura S, Yamaguchi A, Sasaki K, Deguchi K, Shimizu Y, Bronson RT, Gao Y-H, Inada M, Sato M, Okamoto R, Kitamura Y, Yoshiki S, Kishimoto T. Targeted disruption of *Cbfa1* results in a complete lack of bone formation owing to maturational arrest of osteoblasts. Cell 1997; 89:755–764.
30. Banerjee C, McCabe LR, Choi J-Y, Hiebert SW, Stein JL, Stein GS, Lian JB. Runt homology domain proteins in osteoblast differentiation: AML-3/CBFA1 is a major component of a bone specific complex. J Cell Biochem 1997; 66:1–8.
31. Ducy P, Zhang R, Geoffroy V, Ridall AL, Karsenty G. Osf2/Cbfa1: a transcriptional activator of osteoblast differentiation. Cell 1997; 89:747–754.
32. Pockwinse SM, Stein JL, Lian JB, Stein GS. Developmental stage-specific cellular responses to vitamin D and glucocorticoids during differentiation of the osteoblast phenotype: interrelationship of morphology and gene expression by in situ hybridization. Exp Cell Res 1995; 216:244–260.
33. Glowacki J, Rey C, Glimcher MJ, Cox KA, Lian J. A role for osteocalcin in osteoclast differentiation. J Cell Biochem 1991; 45:292–302.
34. Ishida H, Bellows CG, Aubin JE, Heersche JN. Characterization of the 1,25-$(OH)_2D_3$-induced inhibition of bone nodule formation in long-term cultures of fetal rat calvaria cells. Endocrinology 1993; 132:61–66.
35. Lian JB, Stein GS. The developmental stages of osteoblast growth and differentiation exhibit selective responses of genes to growth factors (TGF beta 1) and hormones (vitamin D and glucocorticoids). J Oral Implantol 1993; 19:95–105.
36. Candeliere GA, Prud'homme J, St-Arnaud R. Differential stimulation of fos and jun family members by calcitrol in osteoblastic cells. Mol Endocrinol 1991; 5:1780–1788.
37. Wong M-M, Rao LG, Ly H, Hamilton L, Tong J, Sturtridge W, McBroom R, Aubin JE, Murray TM. Long-term effects of physiologic concentrations of dexamethasone on human bone-derived cells. J Bone Miner Res 1990; 5:803–813.
38. Matsumoto T, Igarashi C, Takeuchi Y, Harada S, Kikuchi T, Yamato H, Ogata E. Stimulation by 1,25-dihydroxyvitamin D_3 of in vitro mineralization induced by osteoblast-like MC3T3-E1 cells. Bone 1991; 12:27–32.
39. Stein GS, Lian JB. Molecular mechanisms mediating proliferation-differentiation interrelationships during progressive development of the osteoblast phenotype: update. 1995. Endocr Rev 1995; 4: 290–297.
40. Merriman HL, van Wijnen AJ, Hiebert S, Bidwell JP, Fey E, Lian J, Stein J, Stein GS. The tissue-specific nuclear matrix protein, NMP-2, is a member of the AML/CBF/PEBP2/runt domain transcription factor family: interactions with the osteocalcin gene promoter. Biochemistry 1995; 34:13,125–13,132.
41. Banerjee C, Hiebert SW, Stein JL, Lian JB, Stein GS. An AML-1 consensus sequence binds an osteoblast-specific complex and transcriptionally activates the osteocalcin gene. Proc Natl Acad Sci USA 1996; 93:4968–4973.

42. Ducy P, Karsenty G. Two distinct osteoblast-specific cis-acting elements control expression of a mouse osteocalcin gene. Mol Cell Biol 1995; 15:1858–1869.
43. Bidwell JP, van Wijnen AJ, Fey EG, Dworetzky S, Penman S, Stein JL, Lian JB, Stein GS. Osteocalcin gene promoter-binding factors are tissue-specific nuclear matrix components. Proc Natl Acad Sci USA 1993; 90:3162–3166.
44. Lian JB, Stein GS, Stein JL, van Wijnen AJ, McCabe L, Banerjee C, Hoffmann H. The osteocalcin gene promoter provides a molecular blueprint for regulatory mechanisms controlling bone tissue formation: role of transcription factors involved in development. Conn Tissue Res 1996; 35:15–21.
45. Hoffmann HM, Beumer TL, Rahman S, McCabe LR, Banerjee C, Aslam F, Tiro JA, van Wijnen AJ, Stein JL, Stein GS, Lian JB. Bone tissue-specific transcription of the osteocalcin gene: role of an activator osteoblast-specific complex and suppressor hox proteins that bind the OC box. J Cell Biochem 1996; 61:310–324.
46. Hodgkinson JE, Davidson CL, Beresford J, Sharpe PT. Expression of a human homeobox-containing gene is regulated by 1,25(OH)$_2$D$_3$ in bone cells. Biochim Biophys Acta 1993; 1174:11–116.
47. Zhang R, Ducy P, Karsenty G. 1,25-dihydroxyvitamin D$_3$ inhibits osteocalcin expression in mouse through an indirect mechanism. J Biol Chem 1997; 272:110–116.
48. Banerjee C, Stein JL, van Wijnen AJ, Frenkel B, Lian JB, Stein GS. Transforming growth factor-beta 1 responsiveness of the rat osteocalcin gene is mediated by an activator protein-1 binding site. Endocrinology 1996; 137:1991–2000.
49. Newberry EP, Boudreaux JM, Towler DA. The rat osteocalcin fibroblast growth factor (FGF)-responsive element: an okadaic acid-sensitive, FGF-selective transcriptional response motif. Mol Endocrinol 1996; 10:1029–1040.
50. Rhodes SJ, Chen R, DiMattia GE, Scully KM, Kalla KA, Lin SC, Yu VC, Rosenfeld MG. A tissue-specific enhancer confers Pit-1-dependent morphogen inducibility and autoregulation on the pit-1 gene. Genes Dev 1993; 7:913–932.
51. Candeliere GA, Jurutka PW, Haussler MR, St-Arnaud R. A composite element binding the vitamin D receptor, retinoid X receptor alpha, and a member of the CTF/NF-1 family of transcription factors mediates the vitamin D responsiveness of the c-fos promoter. Mol Cell Biol 1996; 16:584–592.
52. Pavlin D, Bedalov A, Kronenberg MS, Kream BE, Rowe DW, Smith CL, Pike JW, Lichtler AC. Analysis of regulatory regions in the COL1A1 gene responsible for 1,25-dihydroxyvitamin D$_3$-mediated transcriptional repression in osteoblastic cells. J Cell Biochem 1994; 56:490–501.
53. Ohyama Y, Ozono K, Uchida M, Shinki T, Kato S, Suda T, Yamamoto O, Noshiro M, Kato Y. Identification of a vitamin D-responsive element in the 5'-flanking region of the rat 25-hydroxyvitamin D$_3$ 24-hydroxylase gene. J Biol Chem 1994; 269:10,545–10,550.
54. Hahn CN, Kerry DM, Omdahl JL, May BK. Identification of a vitamin D responsive element in the promoter of the rat cytochrome P450(24) gene. Nucleic Acids Res 1994; 22:2410–2416.
55. Medhora MM, Teitelbaum S, Chappel J, Alvarez J, Mimura H, Ross FP, Hruska K. 1 Alpha,25-dihydroxyvitamin D$_3$ up-regulates expression of the osteoclast integrin alpha v beta 3. J Biol Chem 1993; 268:1456–1461.
56. Mimura H, Cao X, Ross FP, Chiba M, Teitelbaum SL. 1,25-Dihydroxyvitamin D$_3$ transcriptionally activates the beta 3-integrin subunit gene in avian osteoclast precursors. Endocrinology 134: 1994; 1061–1066.
57. Yu VC, Delsert C, Andersen B, Holloway JM, Devary OV, Naar AM, Kim SY, Boutin JM, Glass CK, Rosenfeld MG. RXR beta: a coregulator that enhances binding of retinoic acid, thyroid hormone, and vitamin D receptors to their cognate response elements. Cell 1991; 67:1251–1266.
58. Kliewer SA, Umesono K, Mangelsdorf DJ, Evans RM. Retinoid X receptor interacts with nuclear receptors in retinoic acid, thyroid hormone and vitamin D$_3$ signalling. Nature 1992; 355:446–449.
59. Kerner SA, Scott RA, Pike JW. Sequence elements in the human osteocalcin gene confer basal activation and inducible response to hormonal vitamin D$_3$. Proc Natl Acad Sci USA 1989; 86:4455–4459.
60. Demay MB, Kiernan MS, DeLuca HF, Kronenberg HM. Sequences in the human parathyroid hormone gene that bind the 1, 25-dihydroxyvitamin D$_3$ receptor and mediate transcriptional repression in response to 1,25-dihydroxyvitamin D$_3$. Proc Natl Acad Sci USA 1992; 89:8097–8101.
61. Markose ER, Stein JL, Stein GS, Lian JB. Vitamin D-mediated modifications in protein-DNA interactions at two promoter elements of the osteocalcin gene. Proc Natl Acad Sci USA 1990; 87:1701–1705.

62. Ozono K, Liao J, Kerner SA, Scott RA, Pike JW. The vitamin D-responsive element in the human osteocalcin gene. Association with a nuclear proto-oncogene enhancer. J Biol Chem 1990; 265: 21,881–21,888.
63. Mader S, Chen JY, Chen Z, White J, Chambon P, Gronemeyer H. The patterns of binding of RAR, RXR and TR homo- and heterodimers to direct repeats are dictated by the binding specificites of the DNA binding domains. EMBO J 1993; 12:5029–5041.
64. Perlmann T, Rangarajan PN, Umesono K, Evans RM. Determinants for selective RAR and TR recognition of direct repeat HREs. Genes Dev 1993; 7:1411–1422.
65. Kurokawa R, Yu VC, Naar A, Kyakumoto S, Han Z, Silverman S, Rosenfeld MG, Glass CK. Differential orientations of the DNA-binding domain and carboxy-terminal dimerization interface regulate binding site selection by nuclear receptor heterodimers. Genes Dev 1993; 7:1423–1435.
66. Arbour NC, Prahl JM, DeLuca HF. Stabilization of the vitamin D receptor in rat osteosarcoma cells through the action of 1,25-dihydroxyvitamin D3. Mol Endocrinol 1993; 7:1307–1312.
67. Breen EC, van Wijnen AJ, Lian JB, Stein GS, Stein JL. In vivo occupancy of the vitamin D responsive element in the osteocalcin gene supports vitamin D-dependent transcriptional upregulation in intact cells. Proc Natl Acad Sci USA 1994; 91:12,902–12,906.
68. Hsieh JC, Jurutka PW, Galligan MA, Terpening CM, Haussler CA, Samuels DS, Shimizu Y, Shimizu N, Haussler MR. Human vitamin D receptor is selectively phosphorylated by protein kinase C on serine 51, a residue crucial to its trans-activation function. Proc Natl Acad Sci USA 1991; 88: 9315–9319.
69. Schräder M, Muller KM, Nayeri S, Kahlen JP, Carlberg C. Vitamin D_3-thyroid hormone receptor heterodimer polarity directs ligand sensitivity of transactivation. Nature 1994; 370:382–386.
70. Liao J, Ozono K, Sone T, McDonnell DP, Pike JW. Vitamin D receptor interaction with specific DNA requires a nuclear protein and 1,25-dihydroxyvitamin D_3. Proc Natl Acad Sci USA 1990; 87: 9751–9755.
71. Liu M, Freedman LP. Transcriptional synergism between the vitamin D_3 receptor and other nonreceptor transcription factors. Mol Endocrinol 1994; 8:1593–1604.
72. St-Arnaud R, Prud'homme J, Leung-Hagesteijn C, Dedhar S. Constitutive expression of calreticulin in osteoblasts inhibits mineralization. J Cell Biol 1995; 131:1351–1359.
73. MacDonald PN, Dowd DR, Nakajima S, Galligan MA, Reeder MC, Haussler CA, Ozato K, Haussler MR. Retinoid X receptors stimulate and 9-*cis* retinoic acid inhibits 1,25-dihydroxyvitamin D_3-activated expression of the rat osteocalcin gene. Mol Cell Biol 1993; 13:5907–5917.
74. Zhang XK, Lehmann J, Hoffmann B, Dawson MI, Cameron J, Graupner G, Hermann T, Tran P, Pfahl M. Homodimer formation of retinoid X receptor induced by 9-cis retinoic acid. Nature 1992; 358: 587–591.
75. Eisman JA. Vitamin D receptor gene alleles and osteoporosis: an affirmative view [editorial]. J Bone Miner Res 1995; 10:1289–1293.
76. Morrison NA, Qi JC, Tokita A, Kelly PJ, Crofts L, Nguyen TV, Sambrook PN, Eisman JA. Prediction of bone density from vitamin D receptor alleles [see comments]. Nature 1994; 367:284–287.
77. Uitterlinden AG, Pols HA, Burger H, Huang Q, Van Daele PL, Van Duijn CM, Hofman A, Birkenhager JC, Van Leeuwen JP. A large-scale population-based study of the association of vitamin D receptor gene polymorphisms with bone mineral density. J Bone Miner Res 1996; 11:1241–1248.
78. Guo B, Aslam F, van Wijnen AJ, Roberts SGE, Frenkel B, Green M, DeLuca H, Lian JB, Stein GS, Stein JL. YY1 regulates VDR/RXR mediated transactivation of the vitamin D responsive osteocalcin gene. Proc Natl Acad Sci USA 1997; 94:121–126.
79. Aslam F, McCabe LR, Frenkel B, van Wijnen AJ, Stein GS, Lian JB, and Stein JL. Convergence of AP-1 and vitamin D receptor (VDR) signalling pathways at the rat osteocalcin VDR element: requirement of the internal AP-1 site for vitamin D-mediated transactivation. Submitted.
80. Lian JB, Stein GS, Bortell R, Owen TA. Phenotype suppression: a postulated molecular mechanism for mediating the relationship of proliferation and differentiation by Fos/Jun interactions at AP-1 sites in steroid responsive promoter elements of tissue-specific genes. J Cell Biochem 1991; 45:9–14.
81. Owen TA, Bortell R, Yocum SA, Smock SL, Zhang M, Abate C, Shalhoub V, Aronin N, Wright KL, van Wijnen AJ, Stein JL, Curran T, Lian JB, Stein GS. Coordinate occupancy of AP-1 sites in the vitamin D responsive and CCAAT box elements by Fos-Jun in the osteocalcin gene: model for phenotype suppression of transcription. Proc Natl Acad Sci USA 1990; 87:9990–9994.
82. Blanco JCG, Wang I-M, Tsai SY, Tsai M-J, O'Malley BW, Jurutka PW, Haussler MR, Ozato K. Transcription factor TFIIB and the vitamin D receptor cooperatively activate ligand-dependent transcription. Proc Natl Acad Sci USA 1995; 92:1535–1539.

83. MacDonald PN, Sherman DR, Dowd DR, Jefcoat SCJ, DeLisle K. The vitamin D receptor interacts with general transcription factor IIB. J Biol Chem 1995; 270:4748–4752.
84. Usheva A, Shenk T. TATA-binding protein-independent initiation: YY1, TFIIB, and RNA polymerase II direct basal transcription on supercoiled template DNA. Cell 1994; 76:1115–1121.
85. Staal A, van Wijnen AJ, Birkenhäger JC, Pols HAP, Prahl J, DeLuca H, Gaub M-P, Lian JB, Stein GS, van Leeuwen JPTM, Stein JL. Distinct conformations of VDR/RXRα heterodimers are specified by dinucleotide differences in the vitamin D responsive elements of the osteocalcin and osteopontin genes. Mol Endocrinol 1996; 10:1444–1456.
86. Kesterson RA, Stanley L, DeMayo F, Finegold M, Pike JW. The human osteocalcin promoter directs bone-specific vitamin D-regulatable gene expression in transgenic mice. Mol Endocrinol 1993; 7:462–467.
87. Baker AR, Hollingshead PG, Pitts-Meek S, Hansen S, Taylor R, Stewart TA. Osteoblast-specific expression of growth hormone stimulates bone growth in transgenic mice. Mol Cell Biol 1992; 12:5541–5547.
88. Towler DA, Bennett CD, Rodan GA. Activity of the rat osteocalcin basal promoter in osteoblastic cells is dependent upon homeodomain and CP1 binding motifs. Mol Endocrinol 1994; 8:614–624.
89. Uchida M, Ozono K, Pike JW. Activation of the human osteocalcin gene by 24R,25-dihydroxyvitamin D_3 occurs through the vitamin D receptor and the vitamin D-responsive element. J Bone Miner Res 1994; 9:1981–1987.
90. Lian JB, Shalhoub V, Aslam F, Frenkel B, Green J, Hamrah M, Stein GS, Stein JL. Species-specific glucocorticoid and 1,25-dihydroxyvitamin D responsiveness in mouse MC3T3-E1 osteoblasts: dexamethasone inhibits osteoblast differentiation and vitamin D downregulates osteocalcin gene expression. Endocrinology 1997; 138:2117–2127.
91. Montecino M, Pockwinse S, Lian J, Stein G, Stein J. DNase I hypersensitive sites in promoter elements associated with basal and vitamin D dependent transcription of the bone-specific osteocalcin gene. Biochemistry 1994; 33:348–353.
92. Montecino M, Lian J, Stein G, Stein J. Changes in chromatin structure support constitutive and develomentally regulated transcription of the bone-specific osteocalcin gene in osteoblastic cells. Biochemistry 1996; 35:5093–5102.
93. Dworetzky SI, Fey EG, Penman S, Lian JB, Stein JL, Stein GS. Progressive changes in the protein composition of the nuclear matrix during rat osteoblast differentiation. Proc Natl Acad Sci USA 1990; 87:4605–4609.
94. Zeng C, van Wijnen AJ, Stein JL, Meyers S, Sun W, Shopland L, Lawrence JB, Penman S, Lian JB, Stein GS, Hiebert SW. Identification of a nuclear matrix targeting signal in the leukemia and bone-related AML/CBFα transcription factors. Proc Natl Acad Sci USA 1997; 94:6746–6751.

12 Nongenomic Rapid Effects of Vitamin D

Daniel T. Baran

1. INTRODUCTION

1α,25-Dihydroxyvitamin D_3 [1α,25(OH)$_2$$D_3$] is the most potent vitamin D metabolite. The mechanism(s) by which the hormone alters cell growth and differentiation is largely unknown. Specific nuclear vitamin D receptors (nVDRs) for this secosteroid have been found in numerous organs and cell lines *(1)*. The binding of 1α,25(OH)$_2$$D_3$ to the nVDR and subsequent binding of the hormone–receptor complex to selected DNA sequences were thought to be the mechanisms explaining all the hormone's actions.

The nVDR for 1α,25(OH)$_2$$D_3$ is a member of a nuclear transacting receptor superfamily and shares amino acid sequence homology with the thyroid hormone receptor, other steroid hormone receptors, and the viral oncogene *erb* A product *(2)*. The nVDR is mainly cytoplasmic in the absence of 1α,25(OH)$_2$$D_3$, but after addition of the hormone, there is a rapid reorganization *(3)*, with intranuclear accumulation after 1–3 min *(4)*. This intranuclear accumulation appears to be dependent on interaction with microtubules *(5)*. The 1α,25(OH)$_2$$D_3$ occupied receptor undergoes a covalent modification within the nucleus by phosphorylation at different sites catalyzed by several kinases *(6–8)*. This suggests that binding of the 1α,25(OH)$_2$$D_3$–nVDR complex to chromatin is not the only factor regulating transcription *(9)*. Our observation that the rapid actions of 1α,25(OH)$_2$$D_3$ are able to modulate certain of the hormone's genomic effects has led to the hypothesis that the rapid actions play a role in 1α,25(OH)$_2$$D_3$-induced alterations in transcription *(10)*.

Recently, a variety of cell types including osteoblasts *(11)*, osteoblast-like cells *(12)*, intestine *(13)*, kidney *(14)*, parathyroid cells *(15)*, hematopoietic cells *(16)*, muscle *(17)*, chondrocytes *(18)*, fibroblasts *(19)*, hepatocytes *(20)*, keratinocytes *(21)*, and insulinoma *(22)* have been shown to respond rapidly (seconds to minutes) to 1α,25(OH)$_2$$D_3$, with increases in intracellular calcium, pH, and cyclic nucleotides, and alterations in phospholipid metabolism and protein kinase C activity (*see* refs. *23* and *24* for review). Indeed, present studies suggest that other steroid hormones including estrogen *(25)*, testosterone *(26)*, aldosterone *(27)*, and progesterone *(28)* exert some of their effects by a nongenomic action (*see* ref. *29* for review). Like 1α,25(OH)$_2$$D_3$, testosterone *(26)*, estradiol *(30)*, and progesterone *(30)* have been shown to induce transmembrane signaling pathways in rat osteoblasts. The ability of 1α,25(OH)$_2$$D_3$ and a number of other steroid hormones to exert rapid effects on a variety of tissues suggests that these actions are functionally significant and are not limited to one steroid. This chapter focuses on two of the target tissues for

From: *Vitamin D: Physiology, Molecular Biology, and Clinical Applications*
Edited by: M. F. Holick © Humana Press Inc., Totowa, NJ

vitamin D, osteoblasts and intestine, the mechanisms mediating the rapid actions in these tissues, and their significance.

2. NONGENOMIC EFFECTS OF $1\alpha,25(OH)_2D_3$ ON OSTEOBLASTS

Early studies demonstrated that $1\alpha,25(OH)_2D_3$ stimulated ^{45}Ca accumulation in rat osteoblast-like cells *(31)*. Treatment of the cells for times as short as 15 min with $1\alpha,25(OH)_2D_3$, 2 p*M*, increased ^{45}Ca accumulation by 10–20%. The first-generation fluorescent Ca^{2+} indicator Quin 2AM was used to assess vitamin D-induced changes in cellular calcium in mouse osteoblasts *(11,32)*. $1\alpha,25(OH)_2D_3$, 10–100 p*M*, increased intracellular calcium by 50%, an effect that was dependent on the presence of extracellular calcium. Calcium channel blockers also inhibited the vitamin D effects on intracellular calcium, suggesting that membrane voltage-gated calcium channels were involved in the process. These voltage-gated channels, modulated within 1 min by $1\alpha,25(OH)_2D_3$, were subsequently confirmed by electrophysiologic studies, which showed that $1\alpha,25(OH)_2D_3$ has the ability to regulate voltage-sensitive calcium channels through a direct receptor-mediated action *(33,34)*. The fact that synthetic analogs of $1\alpha,25(OH)_2D_3$ caused effects on membrane calcium channels and showed pharmacologic specificity different from the binding to the nVDR suggested that the hormone's effect on calcium influx was mediated by a distinct signaling system *(34)*.

The rapid effects of $1\alpha,25(OH)_2D_3$ are not limited to interaction with the plasma membrane. The hormone rapidly increased nuclear calcium levels in both intact cells and isolated nuclei, suggesting that rapid increases in nuclear calcium may also play a role in the regulation of osteoblastic activity *(35)*. Using digital microscopy, changes in cytosolic and nuclear region fluorescence were observed in single osteoblast-like cells within 3 min of exposure to $1\alpha,25(OH)_2D_3$. The direct effects of $1\alpha,25(OH)_2D_3$ on the nuclear envelope were confirmed by observations that the hormone increased calcium levels in isolated nuclei *(35)*. At present, it is not known whether the receptor that mediates the rapid effects of $1\alpha,25(OH)_2D_3$ on the nuclear envelope is identical to the plasma membrane receptor (pmVDR).

Numerous hormones have been shown to increase intracellular calcium rapidly by a process involving phospholipid metabolism. $1\alpha,25(OH)_2D_3$ was also shown to increase phospholipase C activity rapidly, resulting in the generation of inositol triphosphate (IP_3), a recognized mediator of hormone-induced cellular calcium increments *(12,36)*. $1\alpha,25(OH)_2D_3$ treatment of isolated nuclei also increases IP_3 levels in the nuclear envelope *(37)*. The phospholipid environment is thought to be important in the regulation of nuclear function *(38-40)*. The $1\alpha,25(OH)_2D_3$-induced changes in nuclear phospholipid metabolism and the resultant changes in nuclear calcium movement are likely to play a role in the hormone's physiologic effects.

Although the rapidity of these vitamin D actions suggested direct interaction with a membrane signaling system, it was unclear whether the nVDR was necessary for the hormone to exert its rapid effects. Utilizing a rat osteoblast-like cell line that lacked the nVDR (ROS 24/1), we *(41)* and Civitelli et al. *(12)* have demonstrated that $1\alpha,25(OH)_2D_3$ was able to increase cellular calcium. Using Northern blots, these cells were shown to lack the mRNA for the nVDR *(41)*. These Northern blots hybridized with the full-length cDNA probe showed absence of smaller size bands, indicating that the pmVDR is not an alternatively spliced product of the nVDR gene. In addition, the functional absence of nVDR in these cells was demonstrated by the lack of nVDR in the nuclear extracts that was capable of binding to the vitamin D response element (VDRE) of the osteocalcin (OC) gene *(41)*. Moreover, the 1β-epimer of 1α,25(OH)$_2$D$_3$, which does not interact with the nVDR *(42)* and does not bind to the VDRE of the OC gene *(41)*, inhibited the rapid effects of the hormone in osteoblast cells that both possess and lack the nVDR. This suggested that the epimer bound to but did not activate the signaling system mediating the rapid actions of 1α,25(OH)$_2$D$_3$ and that this signaling system was independent of the nVDR *(41)*. This has allowed the use of the 1β-epimer as a specific biologic antagonist of the rapid actions of 1α,25(OH)$_2$D$_3$. The presence of a distinct 1α,25(OH)$_2$D$_3$ signaling system was confirmed using vitamin D$_3$ analogs *(43)*. The results indicated that there are distinct nuclear and plasma membrane-associated receptors for 1α,25(OH)$_2$D$_3$ that are involved in genomic and rapid actions in osteoblasts. A structural hierarchy of vitamin D$_3$ analogs was demonstrated with regard to their efficacy as transducers of the genomic and rapid pathways *(44,45)*.

2.1. Membrane Binding

1α,25(OH)$_2$D$_3$ has very rapid effects on membrane physiologic processes, suggesting that the hormone may interact with a membrane signaling system. The presence of membrane receptors for other steroid hormones has been reported *(29)*. Steroid binding to plasma membranes has been noted for aldosterone *(46)*, corticosterone *(47)*, dexamethasone *(48)*, estradiol *(49,50)*, progesterone *(51)*, and pregnenolone sulfate *(52)*. Utilizing a plasma membrane preparation derived from ROS 24/1 cells, the rat osteoblast-like cell line lacking the nVDR, 1α,25(OH)$_2$D$_3$ and its 1β-epimer have been shown to displace [^{3}H]1α,25(OH)$_2$D$_3$ binding from the membranes, whereas 25-hydroxyvitamin D$_3$ does not *(53)*. This demonstrated that the plasma membrane binding of 1α,25(OH)$_2$D$_3$ was not due to the presence of the nVDR and that binding was specific.

The apparent K_D for 1α,25(OH)$_2$D$_3$ in membrane preparations of ROS 24/1 cells was 810 nM. The calculated apparent K_D for the 1β-epimer was 480 nM *(53)*. These apparent dissociation constants are considerably higher than the binding of hormone to the nVDR, 0.1 nM. The K_D is also higher than the concentration of 1α,25(OH)$_2$D$_3$ needed to elicit rapid actions in biologic systems. The hormone has been shown to alter cellular calcium in the ROS 24/1 cells at concentrations between 0.2 and 10 nM. These differences between the K_D of 1α,25(OH)$_2$D$_3$ binding to membranes and the concentrations that elicit rapid effects may reflect the low specific activity of the vitamin D ligands or the difficulties in dealing with lipid-soluble ligands. Because vitamin D is lipid-soluble and there is lipid in the membrane preparation, the free monomeric concentration of the vitamin D analogs may be much lower than that assumed for K_D calculation. This would lead to a much lower calculated affinity. The observations that the 1β-epimer, but not 25-hydroxyvitamin D$_3$, can displace [^{3}H]1α,25(OH)$_2$D$_3$ binding to membranes, coupled with the previous findings that the 1β-epimer is a specific inhibitor of the rapid actions of 1α,25(OH)$_2$D$_3$ *(12,39)*, suggest that this binding activity may be involved in the receptor signaling system that mediates the rapid nongenomic effects of vitamin D.

To characterize this binding activity further, we have used 1α,25(OH)$_2$D$_3$-bromoacetate, which has been reported to bind covalently to vitamin D receptors *(54)*. Both 1α,25(OH)$_2$D$_3$ and the bromoacetate analog significantly increased DNA synthesis, OC mRNA levels, and cellular Ca^{2+} in ROS 17/2.8 cells *(55)*. Because both the genomic and rapid actions of the analog are identical to those of 1α,25(OH)$_2$D$_3$, analog-treated plasma membranes have been used to isolate proteins that bind [^{14}C]1α,25(OH)$_2$D$_3$-bromoacetate.

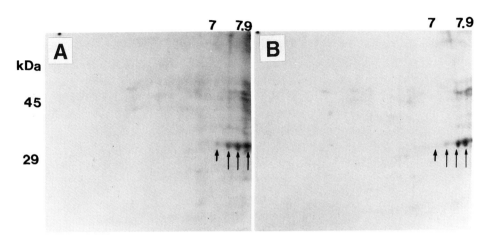

Fig. 1. Preparative two-dimensional gel electrophoresis of ROS 24/1 plasma membrane exposed to [^{14}C]1α,25(OH)$_2$D$_3$-bromoacetate for 90 min in the absence (**A**) or presence (**B**) of 1β,25(OH)$_2$D$_3$. The binding of the [^{14}C] analog to a protein of MW 34–36 kDa, p*I* 7.0–7.1 (A, small arrow), is abolished by 1β,25(OH)$_2$D$_3$ (B, small arrow). 1β,25 (OH)$_2$D$_3$ also decreases binding to other proteins in this area of MW 34–36 kDa, p*I* 7.1–7.9, which may represent isoforms or posttranslational modifications (large arrows).

The covalent binding of the bromoester allows separation of membrane proteins by sodium dodecyl sulfate polyacrylamide gel electrophoresis (SDS-PAGE) without displacement of the hormone. 1β,25(OH)$_2$D$_3$ has been shown to displace the binding of [^{14}C]1α,25(OH)$_2$D$_3$-bromoacetate from a protein of MW 34–36 kDa, p*I* ≈6.9–7.3 *(56)* (Fig. 1). Using liquid-phase isoelectric focusing, membrane proteins in the pI range 6.8–7.3 have been isolated. Binding of [^{14}C]1α,25(OH)$_2$D$_3$-bromoacetate to a protein of MW 34–36 kDa (Fig. 2, lane 1) is abolished by 1β,25(OH)$_2$D$_3$ (Fig. 2, lane 2). Binding to several other proteins is diminished, but not abolished, by 1β,25(OH)$_2$D$_3$ (Fig. 2, lane 2). The band to which binding is abolished by 1β,25(OH)$_2$D$_3$ aligns with a doublet stained with amido black (Fig. 2, lanes 3, 4). These studies suggest that 1α,25(OH)$_2$D$_3$, like other steroids, binds to a pmVDR in osteoblasts that is distinct from the nVDR. One report has been unable to document specific binding of [^{3}H]1α,25(OH)$_2$D$_3$ to the plasma membranes of ROS 17/2.8 cells, but the methodology was not provided *(57)*.

2.2. Physiologic Role of the Nongenomic Effects

To begin to define the physiologic role of the nongenomic actions of 1α,25(OH)$_2$D$_3$, we have employed the 1β-epimer, which specifically inhibits the rapid effects of the hormone but does not alter binding of the 1α,25(OH)$_2$D$_3$–VDR complex to the VDRE of the OC gene *(58)*. The epimer inhibited 1α,25(OH)$_2$D$_3$-induced increases in OC gene transcription and steady-state levels *(58)*. The biochemical changes resulting from the rapid nongenomic actions of 1α,25(OH)$_2$D$_3$ may modify subtle structural and/or functional properties of the nuclear vitamin–receptor DNA complex, or they may affect other protein–DNA interactions that support OC gene transcription. 1β,25(OH)$_2$D$_3$ does not alter binding of the *fos–jun* heterodimer to the AP-1 site(s) of the OC gene, so it is unlikely that modifications at the AP-1 site are responsible for the inhibition of transcription.

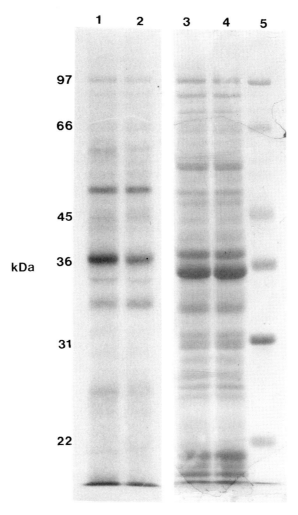

Fig. 2. Gel electrophoresis is of ROS 24/1 membrane fractions between p*I* 6.9 and 7.3 exposed to [^{14}C]1α,25(OH)$_2$D$_3$-bromoacetate for 90 min in the absence (lane A) or presence (lane B) of 1β,25(OH)$_2$D$_3$. The binding of the [^{14}C] analog to a protein of MW 34–36 kDa (lane 1) is abolished by 1β,25(OH)$_2$D$_3$ (lane 2). Binding to several other proteins is diminished, but not abolished, by 1β,25(OH)$_2$D$_3$. The band to which binding is abolished by 1β,25(OH)$_2$D$_3$ aligns with a doublet stained with amido black (lanes 3, 4).

Although one of the rapid actions of 1α,25(OH)$_2$D$_3$ is to increase cellular calcium, we do not believe the increases in cellular calcium *per se* are the sole signal for modulation of gene expression *(59)*. Substitution of extracellular sodium with choline prevents the 1α,25(OH)$_2$D$_3$-induced increases in intracellular pH and also OC and OPN mRNA levels. Thus, blocking the 1α,25(OH)$_2$D$_3$ rapid effect on cytosolic pH blocks the effect of the hormone on mRNA steady-state levels. These effects on gene expression occur even though cellular calcium rapidly increases in the sodium-free medium. Although the genomic effects of 1α,25(OH)$_2$D$_3$ appear to be modulated by the rapid actions, it does not appear that the rapid increases in cellular calcium are the sole regulators of these processes. This finding has recently been confirmed, namely, that 1α,25(OH)$_2$D$_3$-induced increases in cellular Ca^{2+} *per se* do not appear to regulate gene expression in osteoblasts *(60)*.

3. NONGENOMIC EFFECTS OF $1\alpha,25(OH)_2D_3$ ON THE INTESTINE

Early studies demonstrated that $1\alpha,25(OH)_2D_3$ increased calcium transport in vascularly perfused duodena of normal chicks within 2–14 min *(13,61)*. This rapid stimulatory effect of the hormone on calcium transport across the intestine was termed *transcaltachia*. The influx of calcium through voltage-gated calcium channels *(62)*, along with activation of protein kinase C and protein kinase A, appears to trigger transcaltachia *(63)*. Synthetic analogs of $1\alpha,25(OH)_2D_3$ caused effects on membrane calcium channels and Ca^{2+} transport with pharmacologic specificity different from the binding to the nVDR, suggesting that the hormone's effect on transcaltachia was mediated by a receptor distinct from the nVDR *(64–66)*. The rapid effects of $1\alpha,25(OH)_2D_3$ on vascularly perfused duodena are not limited to enhancement of calcium transport. Within 4–8 min, $1\alpha,25(OH)_2D_3$ increases phosphate transport in perfused duodena *(67)*. Moreover, removal of the phosphate from the lumen abolished the effect of $1\alpha,25(OH)_2D_3$ on rapid calcium transport. This suggests that the transcaltachic effect of $1\alpha,25(OH)_2D_3$ is dependent on the ionic content of the lumen.

$1\alpha,25(OH)_2D_3$ has subsequently been shown to increase calcium uptake rapidly in isolated rat enterocytes *(68–70)* and in the human colon cancer-derived cell line CaCo-2 *(71–73)*. The effects of $1\alpha,25(OH)_2D_3$ on colonocyte calcium transport appear to be mediated by a process involving phospholipid metabolism. Normal enterocytes respond to $1\alpha,25(OH)_2D_3$ with an increase in phosphatidylinositol breakdown within 5–15 s *(70,74)*. Similarly, $1\alpha,25(OH)_2D_3$ rapidly increases IP_3 formation in CaCo-2 cells *(71,73)*, an effect that appears to precede the increase in cellular Ca^{2+}.

At present, it is unclear whether these rapid effects of $1\alpha,25(OH)_2D_3$ on the intestine are mediated by the nVDR. Analogs of $1\alpha,25(OH)_2D_3$ that do not bind to the nVDR rapidly stimulate calcium absorption *(64–66)*, suggesting that a distinct signaling system is involved. Moreover, as in the osteoblast *(41)*, $1\beta,25(OH)_2D_3$, which does not interact with the nVDR *(42)*, is an antagonist of $1\alpha,25(OH)_2D_3$-stimulated transcaltachia *(46,75)*. Thus, the data suggest that in the intestine, as in the osteoblast, there are distinct nuclear and plasma membrane-associated receptors for $1\alpha,25(OH)_2D_3$ that are involved in genomic and rapid actions.

3.1. Membrane Binding

Early studies suggested that a functional receptor was required for $1\alpha,25(OH)_2D_3$ to exert rapid effects on the intestine *(74)*. The hormone had no effect on phosphatidylinositol metabolism in enterocytes isolated from neonatal rats, a time at which specific binding of $1\alpha,25(OH)_2D_3$ to the intestine is absent. Although this study suggested that a receptor molecule was necessary for the $1\alpha,25(OH)_2D_3$-mediated effects on membrane phospholipids in the intestine, it did not differentiate between a membrane receptor and the classical nVDR.

Recently, specific and saturable binding for $[^3H]1\alpha,25(OH)_2D_3$ with a K_D of 0.7 n*M* has been shown in detergent-solubilized basolateral membrane preparations of chick intestinal epithelial cells from which fat had been removed *(76)*. Fractions containing the highest specific $[^3H]1\alpha,25(OH)_2D_3$ binding (4500-fold enrichment) were resolved in SDS-PAGE showing one to three proteins with a MW >60 kDa. Functional correlations between the observed membrane binding and transcaltachia were noted in three experimental situations: (1) vitamin D deficiency, which suppresses transcaltachia, resulted in reduced specific binding; (2) $1\alpha,25(OH)_2D_3$ treatment resulted in downregulation of

Speculated Mechanisms of the Rapid Action of 1α,25 Dihydroxyvitamin D₃

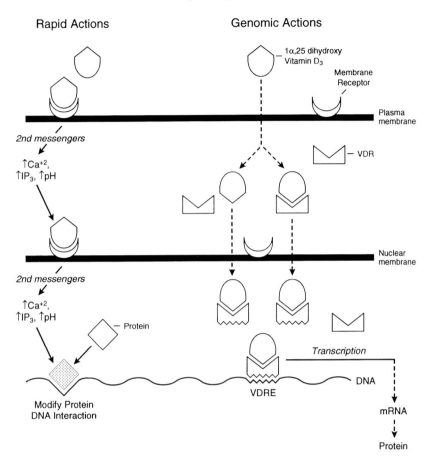

Fig. 3. The rapid actions of $1\alpha,25(OH)_2D_3$, which include increases in cellular Ca^{2+}, IP_3 levels, and pH, appear to be the result of interaction with a "receptor" located on the plasma and nuclear membranes. The precise mechanism by which the rapid actions modulate the hormone-induced changes in gene transcription in osteoblasts is unclear. Inhibition of the rapid actions of $1\alpha,25(OH)_2D_3$ prevents vitamin D-induced gene transcription without altering interaction of the hormone–VDR complex with the VDRE. Thus, the second messengers activated by binding to the membrane receptor may modify protein–DNA interactions that are necessary to support transcription, or may alter the interaction of DNA regulatory elements and the nuclear matrix. For clarity of presentation, the rapid effect response region is shown to be upstream from the VDRE. The rapid response region may be at the VDRE.

specific binding; and (3) the relative potencies of analogs to initiate transcaltachia paralleled their ability to reduce specific binding *(66)*.

3.2. Physiologic Role of the Nongenomic Effects

The physiologic role of the vitamin D-induced changes in transcaltachia and membrane phospholipids in the intestine is unknown. It has been suggested that the rapid, nongenomic responses allow the short-term modulation from minute to minute of intestinal absorptive and secretory functions, whereas the genomic nucleus pathway appears to regulate the long-term adaptation of the enterocyte to the needs of the body *(74)*.

4. SUMMARY

The vitamin D hormone, $1\alpha,25(OH)_2D_3$, has been shown to act on cellular processes through genomic and rapid pathways. The latter effects, which appear to be mediated by interaction with a pmVDR, include rapid changes in cellular calcium and pH, as well as alterations in phospholipase C activity. The requirement for the classic nuclear receptors in mediating the rapid effects is unclear. Indeed, a recent report suggests that some of the rapid effects may be the result of interaction of $1\alpha,25(OH)_2D_3$-occupied VDRs with membrane sites *(77)*. The binding of nVDR–ligand complexes to extranuclear proteins, as well as the binding of ligand to a pmVDR, may both contribute to the generation of the rapid effects. Other steroid hormones including estrogen, testosterone, progesterone, and aldosterone have been demonstrated to act via rapid, presumably nongenomic, mechanisms. The rapid effects of $1\alpha,25(OH)_2D_3$ modulate the hormone's effect on gene transcription and steady-state level in osteoblasts. The rapid pathway may serve a regulatory function by, in part, controlling the hormone's effects on genomic pathways (Fig. 3, see p. 201) and by allowing rapid adaptation at the cellular level *(78)*.

REFERENCES

1. Minghetti PP, Norman AW. $1\alpha,25$-$(OH)_2D_3$ Vitamin D_3 receptors: Gene regulation and genetic circuitry. FASEB J 1988; 2:3043–3053.
2. Evans RM. The steroid and thyroid hormone receptor superfamily. Science 1988; 240:889–895.
3. Haussler MR, Mangeldorf DJ, Komm BS, Terpening CM, Yamaoka K, Allegretto EA, Baker AR, Shine J, McDonnel DP, Hughes M, Weigel NL, O'Malley BW, Pike JW. Molecular biology of the vitamin D hormone. Rec Prog Horm Res 1988; 44:263–305.
4. Barsony J, Pike JW, DeLuca HF, Marx SJ. Immunocytology with microwave fixed fibroblasts show $1\alpha,25$-dihydroxyvitamin D_3-dependent rapid and estrogen-dependent slow reorganization of vitamin D receptors. J Cell Biol 1990; 111:28,355–28,395.
5. Barsony J, McKoy W. Molybdate increases intracellular 3', 5'-guanosine cyclic monophosphate and stabilizes vitamin D receptor association with tubulin-containing filaments. J Biol Chem 1992; 267: 24,457–465.
6. Hsieh JC, Jurutka PW, Nakajima S, Galligan MA, Haussler CA, Shimizu Y, Shimizu N, Whitfield GK, Haussler MR. Phosphorylation of the human vitamin D receptor by protein kinase C: biochemical and functional evaluation of the serine 51 recognition site. J Biol Chem 1993; 268:15,118–15,126.
7. Jurutka PW, Hsieh JC, MacDonald PW, Terpening CM, Haussler CA, Haussler MR, Whitfield GK. Phosphorylation of serine 208 in the human vitamin D receptor: the predominant amino acid phosphorylated by casein kinase II *in vitro*, and identification as a significant phosphorylation site. J Biol Chem 1993; 268:6791–6799.
8. Jurutka PW, Hsieh JC, Haussler MR. Phosphorylation of the human 1,25-dihydroxyvitamin D_3 receptor by cAMP dependent protein kinase *in vitro* and in transfected COS-7 cells. Biochem Biophys Res Commun 1993; 191:1089–1096.
9. Haussler MR, Jurutka PW, Hsieh JC, Thompson PD, Selznick SH, Haussler CA, Whitfield GK. Receptor mediated genomic actions of 1,25-$(OH)_2D_3$: modulation by phosphorylation. In: Proceedings of the Ninth Workshop on Vitamin D, Orlando, FL, May 28–June 2, 1994, pp. 209–216.
10. Baran DT, Sorensen AM. Rapid actions of $1\alpha,25$-dihydroxyvitamin D: physiologic role. Proc Soc Exp Biol Med 1994; 207:175–179.
11. Lieberherr M. Effects of vitamin D_3 metabolites on cytosolic free calcium in confluent mouse osteoblasts. J. Biol Chem 1987; 262:13,168–13,173.
12. Civitelli R, Kim YS, Gunsten SL, Fugimori A, Huskey M, Avioli LV, Hruska KA. Nongenomic activation of the calcium message system by vitamin D metabolites in osteoblast-like cells. Endocrinology 1990; 127:2253–2262.
13. Nemere I, Yoshimoto Y, Norman AW. Calcium transport in perfused duodena from normal chicks: enhancement within fourteen minutes of exposure to 1,25-dihydroxyvitamin D_3. Endocrinology 1984; 115:1476–1483.

14. Suzuki M, Kurihara S, Kawaguchi Y, Sakai O. Vitamin D_3 metabolites increase $[Ca^{2+}]_i$ in rabbit renal proximal straight tubule cells. Am J Physiol 1991; 260:F757–F763.
15. Boudreau A, Atmani F, Grosse B, Lieberherr M. Rapid effects of 1,25-dihydroxyvitamin D_3 and extracellular Ca^{2+} on phospholipid metabolism in dispersed porcine parathyroid cells. Endocrinology 1990; 127:2738–2743.
16. Desai SS, Appel MC, Baran DT. Differential effects of 1,25- dihydroxyvitamin D_3 on cytosolic calcium in two human cell lines (HL-60 and U-937). J Bone Miner Res 1986; 1:497–501.
17. Selles J, Boland R. Evidence on the participation of the 3', 5'-cyclic AMP pathway in the nongenomic action of 1,25-dihydroxyvitamin D_3 in cardiac muscle. Mol Cell Endocrinol 1991; 82:229–235.
18. Schwartz Z, Schlader DL, Swain LD, Boyan BD. Direct effects of 1,25-dihydroxyvitamin D_3 and 24,25-dihydroxyvitamin D_3 on growth zone and resting zone chondrocyte membrane alkaline phosphatase and phospholipase A_2 specific activities. Endocrinology 1988; 123:2878–2884.
19. Barsony J, Marx SJ. Receptor mediated rapid action of $1\alpha,25$-dihydroxycholecalciferol: increase of intracellular cGMP in human skin fibroblasts. Proc Natl Acad Sci USA 1988; 85:1223–1226.
20. Baran DT, Milne ML. 1,25-Dihydroxyvitamin D_3 increases hepatocyte cytosolic calcium levels: a potential regulator of vitamin D-25-hydroxylase. J Clin Invest 1986; 77:1622–1626.
21. Smith EL, Holick MF. The skin: the site of vitamin D_3 synthesis and a target tissue for its metabolite 1,25-dihydroxyvitamin D_3. Steroids 1987; 49:103–131.
22. Segrev IN, Rhoten WB. Video imaging of intracellular calcium in insulinoma cells: effects of 1,25-$(OH)_2D_3$. In: Proceedings of the Ninth Workshop on Vitamin D, Orlando, FL, May 28–June 2, 1994, pp. 355–356.
23. Walters MR. Newly identified actions of the vitamin D endocrine system. Endocr Rev 1992; 13: 719–764.
24. Duval D, Durant S, Homo-DeLarch F. Nongenomic effects of steroids: interactions of steroid molecules with membrane structures and functions. Biochim Biophys Acta 1983; 737:409–442.
25. Aronica SM, Kraus WL, Katzenellenbogen BS. Estrogen action via the cAMP signalling pathway: stimulation of adenylate cyclase and cAMP-regulated gene transcription. Proc Natl Acad Sci USA 1994; 91:8517–8521.
26. Lieberherr M, Grosse B. Androgens increase intracellular calcium concentration and inositol 1,4,5-triphosphate and diacylglycerol formation via a pertussis-sensitive G protein. J Biol Chem 1994; 269:7217–7223.
27. Wehling M, Ulsenheimer A, Schneider M, Neylon C, Christ M. Rapid effects of aldosterone on free intracellular calcium in vascular smooth muscle and endothelial cells: cellular localization of calcium elevation by single cell imaging. Biochem Biophys Res Commun 1994; 204:475–441.
28. Blackmore PF, Neulen J, Lattanzio F, Beebe SJ. Cell surface binding sites for progesterone mediate calcium uptake in human sperm. J Biol Chem 1991; 266:18,655–18,659.
29. Wehling M. Nongenomic actions of steroid hormones. Trends Endocr Metab 1994; 5:347–353.
30. Lieberherr M, Grosse B, Tassin M-T, Kachkache M, Bourdeau A. Transmembrane signal pathways induced by calcitriol, estradiol testosterone, and progesterone in osteoblasts. In: Proceedings of the Ninth Workshop on Vitamin D, Orlando, FL, May 28–June 2, 1994, pp. 315–323.
31. Kim YS, Birge SJ, Avioli LV, Miller R. Early manifestations of vitamin D effects in rat osteogenic sarcoma cells. Calcif Tissue Int 1987; 41:223–227.
32. Oshima J, Watanabe M, Hirosumi J, Orimo H. $1\alpha,25$-$(OH)_2D_3$ increases cytosolic Ca^{++} concentration of osteoblastic cells, clone MC3T3-E1. Biochem Biophys Res Commun 1987; 145:956–960.
33. Caffrey JM, Farach-Carson MC. Vitamin D_3 metabolites modulate dihydropyridine-sensitive calcium currents in clonal rat osteosarcoma cells. J Biol Chem 1989; 264:20,265–20,274.
34. Yukihiro S, Posner GH, Guggino SE. Vitamin D_3 analogs stimulate calcium currents in rat osteosarcoma cells. J Biol Chem 1994; 269:23,889–23,893.
35. Sorensen AM, Bowman D, Baran DT. $1\alpha,25$-Dihydroxyvitamin D_3 rapidly increases nuclear calcium levels in rat osteosarcoma cell. J Cell Biochem 1993; 52:237–242.
36. Grosse B, Bourdeau A, Lieberherr M. Oscillations in inositol 1,4,5,-triphosphate and diacylglycerol induced by vitamin D_3 metabolites in confluent mouse osteoblasts. J Bone Miner Res 1993; 8: 1059–1069.
37. Sorensen AM, Baran DT. $1\alpha,25$-dihydroxyvitamin D_3 rapidly alters phospholipid metabolism in the nuclear envelope of osteoblasts. J Cell Biochem 1995; 58:15–21.
38. Capitani S, Bertagnolo Y, Mazzoni M, Santi P, Previati M, Antonucci A, Manzoli FA. Lipid phosphorylation in isolated rat liver nuclei: synthesis of polyphosphoinositides at subnuclear level. FEBS Lett 1989; 254:194–198.

39. Payrastre B, Nievers M, Boonstra J, Breton M, Verkleij AJ, Van Bergen en Henegouwen PMP. A differential location of phosphoinositide kinases, diacylglycerol kinase, and phospholipase C in the nuclear matrix. J Biol Chem 1992; 267:5078–5084.
40. York JD, Majerus PW. Nuclear phosphatidylinositols decrease during S-phase of the cell cycle in HeLa cells. J Biol Chem 1994; 269:7847–7850.
41. Baran DT, Sorensen AM, Shalhoub V, Owen T, Oberdorf A, Stein G, Lian J. 1α,25-dihydroxyvitamin D_3 rapidly increases cytosolic calcium in clonal rat osteosarcoma cells lacking the vitamin D receptor. J Bone Miner Res 1991; 6:1269–1275.
42. Holick SA, Holick MF, MacLaughlin JA. Chemical synthesis of [1β-^{3}H] 1α,25-dihydroxyvitamin D_3 and [1α-^{3}H] 1β,25-dihydroxyvitamin D_3: biological activity of 1β,25-dihydroxyvitamin D_3. Biochem Biophys Res Commun 1980; 97:1031–1037.
43. Farach-Carson MC, Sergeev I, Norman AW. Nongenomic actions of 1,25-dihydroxyvitamin D_3 in rat osteosarcoma cells: structure-function studies using ligand analogs. Endocrinology 1991; 129: 1876–1884.
44. Norman AW, Okamura WH, Farach-Carson MC, Allewaert K, Branisteanu D, Nemers I, Muralidharan KR, Bouillon R. Structure function studies of 1,25-dihydroxyvitamin D_3 and the vitamin D endocrine system. J Biol Chem 1993; 268:13,811–13,819.
45. Norman AW, Bouillon R, Farach-Carson MC, Bishop JE, Zhou L- X, Nemere I, Zhao J, Muralidharan KR, Okamura WH Demonstration that 1β,25-dihydroxyvitamin D_3 is an antagonist of the nongenomic but not genomic biological responses and biological profile of the three A-ring diastereomers of 1α,25-dihydroxyvitamin D_3. J Biol Chem 1993; 268:20,022–20,030.
46. Christ M, Sippel K, Eisen C, Wehling, M. Nonclassical receptors for aldosterone in plasma membranes from pig kidney. J Mol Cell Endocrinol 1994; 99:R31–34.
47. Orchinik M, Murray TF, Moore FL. A corticosteroid receptor in neuronal membranes. Science 1991; 252:1848–1851.
48. Quelle FW, Smith RV, Hrycyna CA, Kaliban TD, Crooks JA, O'Brien JM. ^{3}H-dexamethasone binding to plasma membrane enriched fractions from liver of nonadrenalectomized rats. Endocrinology 1988; 123:1642–1651.
49. Bression D, Michard M. LeDafniet M, Pagesy P, Peillon F. Evidence for a specific estradiol binding site on rat pituitary membranes. Endocrinology 1986; 119:1048–1051.
50. Pappas TC, Gametchu B, Watso CS. Membrane estrogen receptors identified by multiple antibody labeling and impeded-ligand binding. FASEB J 1995; 9:404–410.
51. Ke FC, Ramirez VD. Binding of progesterone to nerve cell membranes of rat brain using progesterone conjugated to ^{125}I-bovine serum albumin as a ligand. J Neurochem 1990; 54:467–472.
52. Majewska MD, Demirogoren S, London ED. Binding pregnenolone sulfate to rat brain membranes suggests multiple sites of steroid action at the $GABA_A$ receptor. Eur J Pharmacol 1990; 189:307–315.
53. Baran DT, Ray R, Sorensen AM, Honeyman T, Holick MF, Baran DT. Binding characteristics of a membrane receptor that recognizes 1α,25-dihydroxyvitamin D_3 and its epimer, 1β,25-dihydroxyvitamin D_3. J Cell Biochem 1994; 56:510–517.
54. Ray R, Ray S, Holick MF. 1α,25-Dihydroxyvitamin D_3-3β-bromo acetate, an affinity labeling analog of 1α,25-dihydroxyvitamin D_3 receptor. Bioorganic Chem 1994; 22:276–283.
55. Van Auken M, Buckley D, Ray R, Holick MF, Baran DT. The effects of the vitamin D_3 analog 1α,25-dihydroxyvitamin D_3-3β-bromo acetate on rat osteosarcoma cells: comparison with 1α,25-dihydroxyvitamin D_3. J Cell Biochem 1996; 63:302–310.
56. Baran D, Merriman H, Ray R, Sorensen A, Quail J. Characteristics of an osteoblast membrane protein that recognizes 1α,25-$(OH)_2D_3$. J Bone Miner Res 1994; 10:S292.
57. Kim Y, Dedhar S, Hruska K. Binding of the occupied vitamin D receptor to extranuclear sites: a potential mechanism of nongenomic actions of 1α,25-$(OH)_2D_3$. In: Proceedings of the Ninth Workshop on Vitamin D, Orlando, FL, May 28–June 2, 1994, pp. 341–344.
58. Baran DT, Sorensen AM, Shalhoub V, Owen T, Stein GS, Lian JB. The rapid nongenomic actions of 1α,25-dihydroxyvitamin D_3 modulate the hormone-induced increments in osteocalcin gene transcription in osteoblast-like cells. J Cell Biochem 1992; 50:124–129.
59. Jenis LG, Lian JB, Stein GS, Baran DT. 1α,25-Dihydroxyvitamin D_3-induced changes in intracellular pH in osteoblast-like cells modulate gene expression. J Cell Biochem 1993; 53:234–239.
60. Khoury R, Ridall AL, Norman AW, Farach-Carson MC. Target gene activation by 1,25-dihydroxyvitamin D_3 in osteosarcoma cells is independent of calcium influx. Endocrinology 1994; 135:2446–2453.

61. Nemere I, Norman AW. Parathyroid hormone stimulates calcium transport in perfused duodena from normal chicks: comparison with the rapid (transcaltachic) effect of 1,25-dihydroxyvitamin D_3. Endocrinology 1986; 119:1406–1408.
62. deBoland AR, Nemere I, Norman AW. Ca^{2+} channel agonist bay k 8644 mimics $1,25(OH)_2$-vitamin D_3 rapid enhancement of Ca^{2+} transport in chick perfused duodena. Biochem Biophys Res Commun 1990; 166:217–222.
63. deBoland AR, Norman AW. Influx of extracellular calcium mediates 1,25-dihydroxyvitamin D_3 dependent transcaltachia (the rapid stimulation of duodenal Ca^{2+} transport). Endocrinology 1990; 127:2475–2480.
64. Zhou L-X, Nemere I, Norman AW. 1,25-Dihydroxyvitamin D_3 analog structure-function assessment of the rapid stimulation of intestinal calcium absorption (transcaltachia). J Bone Miner Res 1992; 7:457–463.
65. Dormanen MC, Bishop JE, Hammond MW, Okamura WH, Nemere I, Norman AW. Non-nuclear effects of the steroid hormone $1\alpha,25$-$(OH)_2$-vitamin D_3: analogs are able to functionally differentiate between nuclear and membrane receptors. Biochem Biophys Res Commun 1994; 201:394–401.
66. Norman AW, Bishop JE, Collins ED, Seo E-G, Satchell DP, Dormanen MC, Zanello SB, Farach-Carson MC, Bouillon R, Okamura WH. Differing shapes of $1\alpha,25$-dihydroxyvitamin D_3 functions as ligands for the D-binding protein, nuclear receptor, and membrane receptor: a status report. J Steroid Biochem Mol Biol 1996; 56:13–22.
67. Nemere I. Apparent non-nuclear regulation of intestinal phosphate transport: effects of 1,25-dihydroxyvitamin D_3, 24-25-dihydroxyvitamin D_3 and 25-hydroxyvitamin D_3. Endocrinology 1996; 137:2254–2261.
68. Nemere I, Szego CM. Early actions of parathyroid hormone and 1,25-dihydroxycholecalciferol isolated epithelial cells from rat intestine. Endocrinology 1981; 108:1450–1462.
69. Lucas PA, Roullet C, Duchambon P, Lacour B, Drueke T. Rapid stimulation of calcium uptake by isolated rat enterocytes by $1,25(OH)_2D_3$. Pflugers Arch 1989; 413:407–413.
70. Wali RK, Baum CL, Sitrin MD, Brasitus TA. $1,25(OH)_2$ vitamin D_3 stimulates membrane phosphoinositide turnover, activates protein kinase C, and increases cytosolic calcium in rat colonic epithelium. J Clin Invest 1990; 85:1296–1303.
71. Wali RK, Baum CL, Bolt MJG, Brasitus TA, Sitrin MD. 1,25-Dihydroxyvitamin D_3 inhibits Na+-H+ exchange by stimulating membrane phosphoinositide turnover and increasing cytosolic calcium in CaCo-2 cells. Endocrinology 1992; 131:1125–1133.
72. Tien X-Y, Katnik C, Qasawa BM, Sitrin MD, Nelson DJ, Brasitus TA. Characterization of the 1,25-dihydroxycholecalciferol-stimulated calcium influx pathway in CaCo-2 cells. J Membr Biol 1993; 136:159–168.
73. Tien X-Y, Brasitus TA, Qasawa BM, Norman AW, Sitrin MD. Effect of $1,25(OH)_2D_3$ and its analogues on membrane phosphoinositide turnover and $[Ca^{2+}]_i$ in CaCo-2 cells. Am J Physiol 1993; 265:G143–G148.
74. Lieberherr M, Grosse B, Duchambon P, Drueke T. A functional cell surface type receptor is required for the early action of 1,25-dihydroxyvitamin D_3 on the phosphoinositide metabolism in rat enterocytes. J Biol Chem 1989; 264:20,403–20,406.
75. Norman AW, Bouillon R, Farach-Carson MC, Bishop JE, Zhou L-X, Nemere I, Zhao J, Muralidharan KR, Okamura WH. Demonstration that $1\beta,25$-dihydroxyvitamin D_3 is an antagonist of the nongenomic but not genomic biological responses and biological profile of the three A-ring diastereomers of $1\alpha,25$-dihydroxyvitamin D_3. J Biol Chem 1993; 268:20,022–20,030.
76. Nemere I, Dormanen MC, Hammond MW, Okamura MW, Norman AW. Identification of a specific binding protein for $1\alpha,25$-dihydroxyvitamin D_3 in basal-lateral membranes of chick intestinal epithelium and relationship to transcaltachia. J Biol Chem 1994; 269:23,750–23,756.
77. Kim YS, MacDonald PN, Dedhar S, Hruska KA. Association of $1\alpha,25$-dihydroxyvitamin D_3-occupied vitamin D receptors with cellular membrane acceptance sites. Endocrinology 1996; 137:3649–3658.
78. Baran DT. Nongenomic actions of the steroid hormone $1\alpha,25$-dihydroxyvitamin D_3. J Cell Biochem 1994; 56:303–306.

13 Noncalcemic Actions of 1,25-Dihydroxyvitamin D_3 and Clinical Implications

Michael F. Holick

1. INTRODUCTION

Vitamin D is synonymous with bone health and calcium metabolism. It is the biologically active vitamin D metabolite, 1,25-dihydrovitamin D [$1,25(OH)_2D$; D without a subscript represents either D_2 or D_3] that is responsible for maintaining serum calcium levels in the normal range by increasing the efficiency of intestinal calcium absorption. When dietary calcium absorption is inadequate to satisfy the body's requirement, then $1,25(OH)_2D$ mobilizes stem cells and induces them to become mature osteoclasts, which in turn mobilize calcium stores from bone to maintain serum calcium in a physiologically acceptable range *(1–3)*.

2. HISTORICAL PERSPECTIVE

When $1,25(OH)_2D$ was discovered, it was assumed that specific vitamin D receptors were present in calcium-regulating organs including the intestine, bone, and kidney. In 1979, Stumpf et al. *(4)* reported on the localization of radiolabeled $1,25(OH)_2D$, [^3H]$1,25(OH)_2D_3$, in vitamin D-deficient rat tissues and found that the [^3H]$1,25(OH)_2D_3$ was localized in the nuclei of cells in the small intestine, kidney, and bone. Remarkably, they also found, by audioradiographic analysis of frozen sections of tissues, that [^3H]$1,25(OH)_2D_3$ was also present in cells in the gonads, thymus, pituitary gland, pancreas, stomach, breast, teeth, placenta, and skin *(4–6)*. This observation was the impetus for the identification of vitamin D receptors (VDRs) in all these tissues, as well as in several tumor cell lines of leukemia, breast cancer, melanoma, squamous cell carcinoma, colon cancer, and prostate cancer *(7–11)*. VDR activity was also detected in cells related to immunity including circulating monocytes, activated T- and B-lymphocytes, and macrophages *(12–16)* (Table 1).

3. BIOLOGIC FUNCTION OF $1,25(OH)_2D$ IN CANCER CELLS

In 1981, Abe et al. *(7)* provided the first insight regarding the possible noncalcemic function of $1,25(OH)_2D$. They reported that mouse myeloid leukemic cells (M-1) possessed VDR activity; when they were incubated with $1,25(OH)_2D_3$, the cells' proliferative

From: *Vitamin D: Physiology, Molecular Biology, and Clinical Applications*
Edited by: M. F. Holick © Humana Press Inc., Totowa, NJ

Table 1
Vitamin D Receptor Activity

Calcemic tissues
   Small intestine
   Bone
   Kidney
Noncalcemic tissues
   Pituitary
   Prostate
   Gonads
   Thymus
   Parathyroids
   Pancreas
   Breast
   Stomach
   Placenta
   Epidermis
   Melanocytes
   Hair follicles
   Dermis
   Monocytes
   Lymphocytes
   Myocytes
   Cardiac muscle

Reproduced with permission from ref. 58.

activity was markedly reduced, and the cells differentiated into functional macrophages (Fig. 1). Similar observations were made in human promyelocytic leukemic cells (HL 60) [17]. This prompted an experiment to determine whether $1,25(OH)_2D_3$ or its 25-deoxy analog 1α-hydroxyvitamin D_3 [$1(OH)D_3$] could prolong survival in mice with an M-1 myeloid leukemia. Honma et al. [18] observed that mice with an M-1 leukemia, when treated with $1,25(OH)_2D_3$ or $1(OH)D_3$, had markedly increased survival compared with the control group (Fig. 2). These observations were of great interest especially since Eisman et al. [19] reported that 80% of 54 different breast cancers had VDR activity and that when breast cancer cells in culture were exposed to $1,25(OH)_2D_3$ there was a marked decrease in their proliferative activity. Similar observations were made in other cultured cancer cell lines, such as melanoma, squamous cell carcinoma, and colon cancer [9–11].

4. BIOLOGIC FUNCTIONS OF $1,25(OH)_2D_3$ IN THE SKIN

Stumpf et al. [6] reported that $1,25(OH)_2D_3$ specifically localized in the nuclei of the germative cells of the epidermis (basal cells) as well as in the outer stratum spinosum and granulosum of the epidermis. In addition, the keratinocytes in the outer root sheath of the hair follicle concentrated [^{3}H]$1,25(OH)_2D_3$ [20]. Subsequently, VDR was identified in all these cells and in human skin fibroblasts [21]. When cultured human keratinocytes and fibroblasts were incubated with $1,25(OH)_2D_3$, the hormone inhibited their proliferative

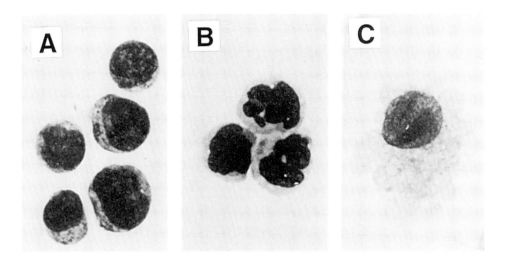

Fig. 1. Morphologic changes in HL-60 cells, treated with vehicle (**A**), $1.2 \times 10^{-8} M$ of $1\alpha,25(OH)_2D_3$ (**B**), or $1.0 \times 10^{-9} M$ or TPA (**C**) for 3 d. Wright-Giemsa stain. (Reproduced with permission from ref. 59.)

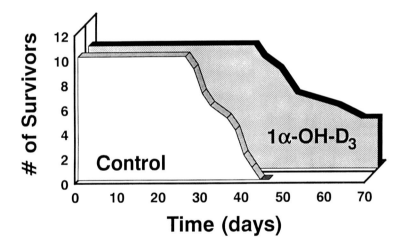

Fig. 2. Mice injected with M-1 cell leukemia had a prolongation in their survival after receiving the $1,25(OH)_2D_3$ analog $1\alpha(OH)D_3$ (40).

activity in a dose-dependent manner (Fig. 3). In addition, $1,25(OH)_2D_3$ induced terminal differentiation of cultured human keratinocytes (21–23). The antiproliferative effect of $1,25(OH)_2D_3$ was shown to be mediated through its VDR when it was observed that fibroblasts from patients who had either a defective or deficient VDR (vitamin D-dependent rickets type II) had a marked decrease or absent response to the antiproliferative activity of $1,25(OH)_2D_3$ (21).

It remains unclear whether normal melanocytes have VDR. None was detected in cultured human melanocytes using the standard binding assay, and they did not respond to the antiproliferative activity of $1,25(OH)_2D$ (24). An immunohistochemical analysis

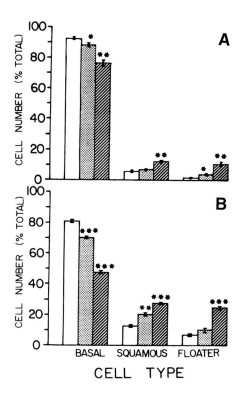

Fig. 3. Effect of $1\alpha,25(OH)_2D_3$ on the morphologic differentiation of cultured human keratinocytes. The proportion of different keratinocyte cell types after 1 wk (**A**) or 2 wk (**B**) of incubation with vehicle alone (open bar); $1\alpha,25(OH)_2D_3$ at $10^{-10}M$ (dotted bar); or $1\alpha,25(OH)_2D_3$ at $10^{-8}M$ (striped bar). Each bar represents the mean of triplicate determinations $\pm$ SEM. Student's t-test was used to assess level of significance (*, $p<0.05$; **, $p<0.01$; ***, $p<0.001$). (Reproduced with permission from ref. 22.)

of human skin suggested that there was VDR immunoreactivity in melanocytes *(25)*. It is unknown whether this is an artifact or whether the VDR antibody is recognizing some other protein than VDR. However, cultured melanoma cells that have VDR activity respond to the antiproliferative activity of $1,25(OH)_2D$ *(9)*.

5. BIOLOGIC FUNCTION OF $1,25(OH)_2D_3$ ON PROSTATE AND TESTES

The ovaries and testes have VDR activity. In cultured Sertoli cells, $1,25(OH)_2D_3$ enhanced rapid uptake of $^{45}Ca^{2+}$ *(26)*. Primary cultures of prostate cells derived from normal, benign prostatic hyperplasia, and prostate cancer tissues have VDR activity and, when exposed to $1,25(OH)_2D_3$ in culture, their proliferative activity is reduced *(27)*.

6. EFFECT OF $1,25(OH)_2D_3$ ON SKELETAL AND MYOCARDIAL SMOOTH MUSCLE

It is well recognized that children with rickets and adults with osteomalacia often display severe muscle weakness in association with their metabolic bone disease. This muscle weakness is not accounted for by hypocalcemia or hypophosphatemia. Myoblasts from chick embryo skeletal muscle have VDR activity *(28)*. When myoblast cells (G-8 and H9c2) were incubated with $1,25(OH)_2D_3$, the hormone inhibited cell proliferation in a dose-dependent fashion. $1,25(OH)_2D_3$ also induced terminal differentiation and, when the cells became fused microtubules, the VDR activity decreased *(28)*.

Rodent heart tissue contains VDR activity *(29)*. Cardiac muscle cells from rat heart responded to $1,25(OH)_2D$ by increasing calcium uptake in a time and dose-dependent manner *(29)*. It has been suggested that $1,25(OH)_2D_3$ may have an important role in controlling vascular tone and the ionotropy of the ventricular musculature *(30)*.

7. REGULATION OF PARATHYROID HORMONE SECRETION

The parathyroid glands are a target organ for $1,25(OH)_2D_3$. The chief cells in the parathyroid gland have VDR activity. $1,25(OH)_2D_3$ suppresses preproparathyroid hormone mRNA production in cultured bovine parathyroid cells *(31)* (*see* Chapter 14). It has been suggested that in patients with chronic renal failure who develop tertiary hyperparathyroidism [i.e., autonomous production of parathyroid hormone (PTH) that is unresponsive to serum calcium], nests of chief cells deficient in VDR are responsible for the unregulated overproduction of PTH *(32)*.

8. EFFECTS OF $1,25(OH)_2D_3$ ON THE IMMUNE SYSTEM

When resting T-lymphocytes were evaluated for VDR activity, none was observed. However, when they were activated with phytohemagglutinin, VDR activity was induced *(13,16)*. Once VDR was present, the cells became responsive to $1,25(OH)_2D$. In vitro $1,25(OH)_2D_3$ decreased interleukin-2 (IL-2) production *(14)*. VDR activity was also absent in resting B-lymphocytes but was induced once they were activated. $1,25(OH)_2D_3$ also inhibited DNA synthesis and immunoglobulin production in stimulated B-lymphocytes in vitro *(15,33)*.

The function of $1,25(OH)_2D_3$ in regulating the immune system is not well understood *(34)*. It is recognized that vitamin D-deficient patients have recurrent infections, mainly of the respiratory tract *(35)*. Furthermore, vitamin D-deficient patients have both depressed inflammatory and phagocytic responses that are corrected by making patients vitamin D sufficient *(35,36)*. Patients with absent or defective vitamin D receptors (vitamin D-dependent rickets type II; *see* Chapter 20) have a subtle defect in their monocyte responsiveness to stimulation with concanavalin A when compared with normal patients *(37)*. When patients who were vitamin D-deficient or who could not make $1,25(OH)_2D$ (renal failure) received $1,25(OH)_2D_3$ or $1(OH)D_3$, macrophage and lymphocyte functions were restored *(38–41)*.

Monocytes have VDR activity and, when exposed to $1,25(OH)_2D_3$, become mature functioning macrophages with Fc and C3 receptors along with enhanced lysosomal activity (Fig. 4), augmentation of IL-1 production, and an increase in OK II binding *(42)*.

$1,25(OH)_2D_3$ has a variety of in vitro and in vivo effects on the immune system *(43)*. However, the in vitro observations do not necessarily predict in vivo outcomes. For example, in vitro $1,25(OH)_2D_3$ inhibited IL-2 production, yet in vivo, it stimulated IL-2 production *(34,40)*. Thus, the multitude of effects of $1,25(OH)_2D_3$ on T- and B-lymphocytes, monocytes, and macrophages in combination may manifest in numerous ways. Of great

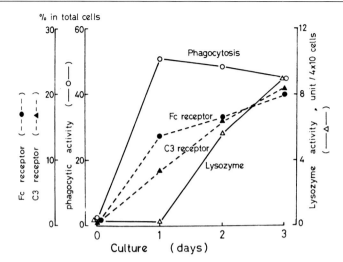

Fig. 4. Time-course of change in differentiation-associated properties of MI cells induced by $1\alpha,25(OH)_2D_3$. Cells were incubated with $1.2 \times 10^{-8} M$ of $1\alpha,25(OH)_2D_3$ for 3 d. At indicated times cells were harvested and their lysozyme activity was determined. In addition, the percentages of phagocytic cells and of cells with Fc and C3 rosettes within the treated culture were determined. (Reproduced with permission from ref. *59*.)

interest is that $1,25(OH)_2D_3$ reduces the development of autoimmune thyroiditis and encephalomyelitis *(44,45)*. $1,25(OH)_2D$ prolongs survival of transplanted skin allographs in mice *(46)* and reduces by >90% the incidence of autoimmune diabetes in NOD mice *(47)* (Fig. 5). Of great interest is the observation that $1,25(OH)_2D_3$ can prevent the onset of an autoimmune neuropathic disease in mice that is similar to multiple sclerosis *(48)*.

9. POTENTIAL CLINICAL USES FOR NONCALCEMIC ACTIVITIES OF $1,25(OH)_2D_3$

9.1. New Approach to Treating Cancer

The numerous observations that $1,25(OH)_2D_3$ was a potent inhibitor of tumor cell growth provided the rationale for using this hormone to treat patients with preleukemia. When 18 patients with myelodysplasia (preleukemia) received 2 µg of $1,25(OH)_2D_3$ for 12 weeks, it was found that most of them had a significant increase in their granulocyte, monocyte, and platelet counts. Coincident with this increase was a rise in the serum calcium up to 17 mg% (upper limit of normal 10.5 mg%). However, after 12 weeks of treatment, there was no significant difference in the blood count of granulocytes, monocytes, and platelets compared with baseline, and most patients progressed to acute myelocytic leukemia *(49)*. It has been speculated that clones of leukemic cells having either a defective or absent VDR were more blastic in their activity and thereby were selected out when the patients were treated with $1,25(OH)_2D_3$. Three patients with myelofibrosis who received $1,25(OH)_2D_3$ (0.5 µg daily) had some improvement in their blood count indices after therapy *(50)*. However, no follow-up report has been available to determine the long-term benefit. The availability of a broad spectrum of vitamin D analogs that possess potent antiproliferative activity and minimal calcemic activity *(51)* offers hope for the use of active vitamin D analogs to fulfill their promise to treat cancer.

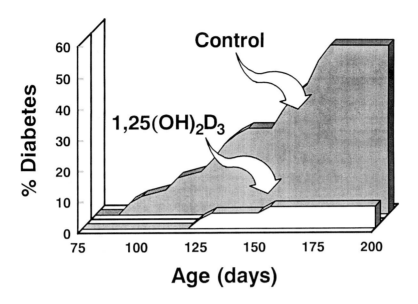

Fig. 5. The effect of 1,25(OH)$_2$D$_3$ on reducing the incidence of diabetes mellitus type I in NOD mice *(40)*.

9.2. Treatment of Proliferative Disorders of Skin and Hair

The potent antiproliferative and prodifferentiation activities of 1,25(OH)$_2$D$_3$ and its analogs have great use for nonmalignant hyperproliferative disorders such as the skin disease psoriasis (*see* Chapter 21). 1,25(OH)$_2$D$_3$ and its analogs, calcipotriene and 1,24-dihydrovitamin D$_3$, have been proved to be effective topical therapeutic agents for psoriasis *(52–54)*. Oral 1,25(OH)$_2$D$_3$ has also been effective for treating psoriasis as well as one of its complications, psoriatic arthritis *(55,56)*. Patients with cancer who receive chemotherapy have a major side effect that is psychologically devastating, i.e., hair loss. One study has suggested that 1,25(OH)$_2$D$_3$, when given to mice before chemotherapy, can prevent chemotherapy-induced alopecia *(57)*.

10. CONCLUSIONS

During the past two decades, there has been an enormous amount of research on the noncalcemic activities of 1,25(OH)$_2$D$_3$ and its analogs. Although the therapeutic efficacy of 1,25(OH)$_2$D$_3$ and its analogs for treating leukemia and other cancers has not fulfilled its promise, the use of 1,25(OH)$_2$D$_3$ and its active analogs has been extremely effective as a new therapeutic approach for treating the very difficult skin disease psoriasis. What is remarkable about the use of 1,25(OH)$_2$D$_3$ and its analogs in treating psoriasis is that one might expect that since 1,25(OH)$_2$D$_3$ possesses such potent antiproliferative activity its chronic topical application to the skin could potentially cause skin senescence similar to the topical application of steroids. However, it has been observed that 1,25(OH)$_2$D$_3$ remarkably transforms a hyperproliferative keratinocyte into a normal proliferative keratinocyte but does not cause it to become hypoproliferative in vivo *(58)*. Therefore, there is no evidence that 1,25(OH)$_2$D$_3$, when applied topically, has any adverse effect on skin health.

It is likely in the future that analogs of 1,25(OH)$_2$D$_3$ will be developed to interact specifically with the VDR or the recently observed membrane VDR (*see* Chapter 5).

Selective analogs that act on these different receptors could be engineered to have selected desired effects, either inhibiting proliferation or inducing terminal differentiation in a wide variety of disorders including cancer. Of great promise is the subtle effects $1,25(OH)_2D$ have on the immune system. They have great potential for treating autoimmune disorders, such as juvenile onset diabetes (type I) and multiple sclerosis.

ACKNOWLEDGMENTS

This work was supported in part by NIH grants RO1 AR36963 and MO1RR 00533.

REFERENCES

1. Holick MF. Vitamin D: photobiology, metabolism, mechanism of action, and clinical application. In: Primer on the Metabolic Bone Diseases and Disorders of Mineral Metabolism, 3rd ed. Favus MJ, ed. Philadelphia: Lippincott-Raven, 1996; 74–81.
2. Holick MF. Vitamin D: photobiology, metabolism, and clinical applications. In: Endocrinology, 3rd ed. DeGroot L, et al., eds. Philadelphia: WB Saunders, 1995; 990–1013.
3. Holick MF, Krane S, Potts JR Jr. Calcium, phosphorus, and and bone metabolism: calcium-regulating hormones. In: Harrison's Principles of Internal Medicine, 13th ed. Isselbacher KJ, Braunwald E, Wilson JD, et al., eds. New York: McGraw-Hill, 1994; 2137–2151.
4. Stumpf WE, Sar M, Reid FA, et al. Target cells for 1,25-dihydroxyvitamin D_3 in intestinal tract, stomach, kidney, skin, pituitary, and parathyroid. Science 1979; 206:1188–1190.
5. Kim YS, Stumpf WE, Clark SA, Sar M, DeLuca HF. Target cells for 1,25-dihydroxyvitamin D_3 in developing rat incisor teeth. J Dent Res 1983; 62:58,59.
6. Stumpf WE, Sar M, Narbaitz R, Huang S, DeLuca HF. Autoradiographic localization of 1,25, dihydroxyvitamin D_3 in rat placenta and yolk sac. Horm Res 1983; 18:215–220.
7. Abe E, Miyaura C, Sakagami H, et al. Differentiation of rat myc leukemia cells induced by 1,25-dihydroxyvitamin D_3. Proc Natl Acad Sci USA 1981; 78:4990–4994.
8. Eisman JA, Suva LJ, Sher E, Pearce PJ, Funder JW, Martin TJ. Frequency of 1,25-dihydroxyvitamin D_3 receptor in human breast cancer. Cancer Res 1981; 41:5121–5124.
9. Colston K, Colston MJ, Feldman D. 1,25-Dihydroxyvitamin D_3 and malignant melanoma: the presence of receptors and inhibition of cell growth in culture. Endocrinology 1981; 108:1083–1086.
10. Bikle DD, Pillai S, Gee E. Squamous carcinoma cell lines produce 1,25-dihydroxyvitamin D, but fail to respond to its prodifferentiating effect. J Invest Dermatol 1991; 97:435–441.
11. Halline AG, Davidson NO, Skarosi SF, Sitrin MD, Tietze C, Alpers DH, Brasitus TA. Effects of 1,25-dihydroxyvitamin D_3 on proliferation and differentiation of Caco-2 cells. Endocrinology 1994; 134: 1710–1717.
12. Manolagas SC, Provvedine DM, Murray EJ, Tsoukas CD, Deftos LJ. The antiproliferative effect of calcitriol on human peripheral blood mononuclear cells. J Clin Endocrinol Metab 1986; 63:394–400.
13. Provvedine DM, Tsoukaas CD, Deftos LJ, Manolagas SC. 1,25-dihydroxyvitamin D_3 receptors in human leukocytes. Science 1983; 221:1181.
14. Tsoukas CD, Provvedine DM, Manolagas SC. 1,25-Dihydroxyvitamin D_3, a novel immuno-regulatory hormone. Science 1984; 221:1438–1440.
15. Provvedine DM, Tsoukas CD, Deftos LJ, Manolagas SC. $1\alpha,25$-Dihydroxyvitamin D_3-binding macromolecules in human B lymphocytes: effects on immunoglobulin production. J Immunol 1986; 136: 2734–2739.
16. Bhalla AK, Clemens T, Amento E, Holick MF, Krane SM. Specific high-affinity receptors for 1,25-dihydroxyvitamin D_3 in human peripheral blood mononuclear cells: presence in monocytes and induction in T lymphocytes following activation. J Clin Endocrinol Metab 1983; 57:1308–1310.
17. Kuribayashi T, Tanaka H, Abe E, Suda T. Functional defect of variant clones of a human myeloid leukemia cell line (HL-60) resistant to $1\alpha,25$-dihydoxyvitamin D_3. Endocrinology 1983; 113: 1992–1998.
18. Honma Y, Hozumi M, Abe E, Konno K, Fukushima M, Hata S, Nishii Y, DeLuca HF. $1\alpha,25$-Dihydroxyvitamin D_3 and 1α-hydroxyvitamin D_3 prolong survival time of mice inoculated with myeloid leukemia cells. Proc Natl Acad Sci USA 1982; 80:201–204.

19. Eisman JA. 1,25-Dihydroxyvitamin D_3 receptor and role of 1,25-dihydroxyvitamin D_3 in human cancer cells in vitamin D. In: Vitamin D: Basic and Clinical Aspects. Kumar R, ed. Boston: Martinus Nijhoff, 1984; 365–382.
20. Stumpf WE, Clark SA, Sar M, DeLuca HF. Topographical and developmental studies on target sites of 1, $25(OH)_2$-vitamin D_3 in skin. Cell Tissue Res 1984; 238:489–496.
21. Clemens TL, Adams JS, Horiuchi N, et al. Interaction of 1,25-dihydroxyvitamin D_3 with keratinocytes and fibroblasts from skin of normal subjects and a subject with vitamin D-dependent rickets, type II. J Clin Endocrinol Metab 1983; 56:824–830.
22. Smith EL, Walworth ND, Holick MF. Effect of 1,25-dihydroxyvitamin D_3 on the morphologic and biochemical differentiation of cultured human epidermal keratinocytes grown in serum-free conditions. J Invest Dermatol 1986; 86:709–714.
23. Bikle DD, Gee E, Pillai S. Regulation of keratinocyte growth, differentiation, and vitamin D metabolism by analogs of 1,25-dihydroxyvitamin D. J Invest Dermatol 1993; 101:713–718.
24. Mansur CP, Gordon PR, Ray S, Holick MF, Gilchrest BA. Vitamin D, its precursors, and metabolites do not affect melanization of cultured human melanocytes. J Invest Dermatol 1988; 91:16–21.
25. Milde P, Hauser U, Simon T, Mall G, Ernst V, Haussler MR, Frosch P, Rauterberg E. Expression of 1,25-dihydroxyvitamin D_3 receptors in normal and psoriatic skin. J Invest Dermatol 1991; 97:230–239.
26. Akerstrom VL, Walters M. Physiological effects of 1,25-dihydroxyvitamin D_3 in TM4 sertoli cell line. Am J Physiol 1992; 262:E884–E890.
27. Skowronski RJ, Peehl DM, Feldman D. Vitamin D and prostate cancer: 1,25-dihydroxyvitamin D_3 receptors and actions in human prostate cancer cell lines. Endocrinology 1993; 132:1952–1960.
28. Boland R, Norman A, Ritz E, Hausselbach W. Presence of 1,25-dihydroxyvitamin D_3 receptor in chick skeletal muscle myoblasts. Biochem Biophys Res Commun 1985;128:305–311.
29. Walters MR, Ilenchuk TT, Claycomb WC. 1,25-Dihydroxyvitamin D_3 stimulates $^{45}Ca^{2+}$ uptake by cultured adult rat ventricular cardiac muscle cells. J Biol Chem 1987; 262:2536–2541.
30. Weishaar RE, Simpson RU. The involvement of the endocrine system in regulating cardiovascular function: emphasis on vitamin D3. Endocr Rev 1989; 10:1–15.
31. Naveh-Many T, Silver J. Regulation of parathyroid hormone gene expression by hypocalcemia, hypercalcemia, and vitamin D in the rat. J Clin Invest 1990; 86:1313–1319.
32. Fukuda N, Tanaka H, Tominaga R, Fukagawa M, Kurokawa K, Seino Y. Decreased 1,25-dihydroxyvitamin D_3 receptor density is associated with a more severe form of parathyroid hyperplasia in chronic uremic patients. J Clin Invest 1993; 92:1436–1443.
33. Lemire JM, Adams JS, Sakai R, Jordan SC. 1α,25-Dihydroxyvitamin D_3 suppresses proliferation and immunoglobulin production by normal human peripheral blood mononuclear cells. J Clin Invest 1984; 74:657–661.
34. Branisteanu DD, Mathieu C, Cateels K, Bouillon R. Combination of vitamin D analogs and immuno suppressants. Clin Immunother 1996; 6:1–14.
35. Lorente F, Fontan G, Jara P, Casas C, Garcia-Rodriguez MC, Ojeda JA. Defective neutrophil motility in hypovitaminosis D rickets. Acta Paediatr Scand 1976; 65:695–699.
36. Holick MF. Noncalcemic actions of 1,25-dihydroxyvitamin D_3 and clinical applications. Bone 1995; 17(Suppl):107S–111S.
37. Koren R, Ravid A, Liberman UA, Hochberg Z, Weisman Y, Novogrodsky A. Defective binding and function of 1,25-dihydroxyvitamin D_3 receptors in peripheral mononuclear cells of patients with end-organ resistance to 1,25-dihydroxyvitamin D. J Clin Invest 1985; 76:2012–2015.
38. Binderup L. Immunological properties of vitamin D analogues and metabolites. Biochem Pharmacol 1992; 43:1885–1892.
39. Kitajima I, Maruyama I, Matsubara H, Osame M, Igata A. Immune dysfunction in hypophosphatemic vitamin D resistant rickets: immunoregulatory reaction of 1α-(OH) vitamin D_3. Clin Immunol Immunopathol 1989; 53:24–31.
40. Holick MF. Photobiology and noncalcemic actions of vitamin D. In: Principles of Bone Biology. Raisz LG, Rodan GA, Bilezikian JP, eds. San Diego: Academic Press, 1996; 447–460.
41. Weintroub S, Winter CC, Wahl SM, Wahl LM. Effect of vitamin D deficiency on macrophage and lymphocyte function in the rat. Calcif Tissue Int 1989; 44:125–130.
42. Suda T, Abe E, Miyaura C, Tanaka H, Shiina Y, Kuribayashi T. Vitamin D and its effects on myeloid leukemia cells. In: Vitamin D, Basic and Clinical Aspects. Kumar R, ed. Boston: Martinus Nijhoff, 1984; 365–382.
43. Amento EP. Vitamin D and the immune system. Steroids 1987; 49:55–72.

44. Fournier C, Gepner P, Sadouk MB, Charreire J. In vivo beneficial effects of cyclosporin A and 1,25-dihydrovitamin D_3 on the induction of experimental autoimmune thyroiditis. Immunol Immunopathol 1990; 54:53–63.
45. Lemire JM, Archer DC. 1,25-Dihydrovitamin D_3 prevents the in vivo induction of murine experimental autoimmune encephalomyelitis. J Clin Invest 1991; 87:1103–1107.
46. Chiocchia G, Boissier MC, Pamphile R, Fournier C. Enhancement of skin allograft survival in mice by association of 1-alpha-hydroxyvitamin D_3 to infratherapeutic does of cyclosporin A. In: Vitamin D. Gene Regulation, Structure-Function Analysis and Clinical Application. Norman AW, Bouillon R, Thomassett M, eds. Berlin: Walter de Gruyter, 1991; 514–515.
47. Mathieu C, Waer M, Laureys J, Rutgeerts O, Bouillon R. Prevention of autoimmune diabetes in NOD mice by 1,25 dihydroxyvitamin D_3. Diabetologia 1994; 37:552–558,.
48. Hayes CE, Cantorna MT, DeLuca HF. Vitamin D and multiple sclerosis. PSEBM 1997; 216:21–27.
49. Koeffler HP, Hirjik J, Iti L, the Southern California Leukemia Group. 1,25-Dihydroxyvitamin D_3: in vivo and in vitro effects on human preleukemic and leukemic cells. Cancer Treat Rep 1985; 69:1399–1407.
50. Arlet P, Nicodeme R, Adoue D, Larregain-Fournier D, Delsol G, Le Tallec Y. Clinical evidence for 1,25-dydroxycholecalciferol action in myelofibrosis. Lancet 1984; 1:1013,1014.
51. Bouillon R, Okamura WH, Norman AW. Structure-function relationships in the vitamin D endocrine system. Endocr Rev 1995; 16:200–257.
52. Perez A, Chen TC, Turner A, Raab R, Bhawan J, Poche P, Holick MF. Efficacy and safety of topical calcitriol (1,25-dihydroxyvitamin D_3) for the treatment of psoriasis. Br J Dermatol 1996; 134:238–246.
53. Kragballe K. Treatment of psoriasis by the topical application of the novel vitamin D_3 analogue MC 903. Arch Dermatol 1989; 125:1647–1652.
54. Kato T, Rokugo M, Terui T, Tagami H. Successful treatment of psoriasis with topical application of active vitamin D_3 analogue, $1\alpha,24$-dihydroxycholecalciferol. Br J Dermatol 1986; 115:431–433.
55. Perez A, Raab R, Chen TC, Turner A, Holick MF. Safety and efficacy of oral calcitriol (1,25-dihydroxyvitamin D_3) for the treatment of psoriasis. Br J Dermatol 1996; 134:1070–1078.
56. Huckins D, Felson DT, Holick MF. Treatment of psoriatic arthritis with oral 1,25-dihydroxyvitamin D_3: a pilot study. Arthritis Rheum 1990; 33:1723–1727.
57. Jimenez JJ, Yunis AA. Protection from chemotherapy-induced alopecia by 1,25-dihydroxyvitamin D_3. Cancer Res 1992; 52:5123–5125.
58. Holick MF, Chen ML, Kong XF, Sanan DK. Clinical uses for calciotropic hormones 1,25-dihydroxyvitamin D_3 and parathyroid hormone-related peptide in dermatology: a new perspective. J Invest Dermatol 1996; 1:1–19.
59. Suda T, Abe E, Miyaura C, et al. Vitamin D in the differentiation of myeloid leukemia cells. In: Vitamin D: Basic and Clinical Aspects. Kumar R, ed. The Hague: Nijhoff, 1984; 343–363.

14

Regulation of Parathyroid Hormone Gene Expression and Secretion by Vitamin D

Tally Naveh-Many and Justin Silver

1. INTRODUCTION

The discovery of the action of vitamin D on the parathyroids, with its therapeutic implications, is one of the success stories of clinical medicine. Vitamin D's classical sites of action were well known to be on the intestine (to increase calcium absorption) and on bone (to promote normal mineralization). With the discovery in 1970 by Fraser *(1)* that the kidney produced a more polar hydroxylated metabolite and the subsequent identification of this metabolite by Fraser and De Luca *(2)* as 1,25-dihydroxyvitamin D_3 [1,25(OH)$_2D_3$], it then became possible to study the biochemical and physiologic effects of vitamin D in greater depth. In the 1970s 1,25(OH)$_2D_3$ was shown to be effective for the treatment of renal bone disease, in particular the osteomalacic component, and in addition the features of secondary hyperparathyroidism. It was thought that the effect on the secondary hyperparathyroidism of 1,25(OH)$_2D_3$ was due to the increased intestinal absorption of calcium and the increased levels of serum calcium, leading to a decrease in parathyroid hormone (PTH) secretion. This is certainly correct, but it is only part of the story. It was later shown by a number of in vitro and in vivo studies that 1,25(OH)$_2D_3$ also has a direct action on the parathyroid. The use of 1,25(OH)$_2D_3$ is one of the cornerstones of secondary hyperparathyoidism management in all patients with chronic renal failure. This effect is discussed here.

2. PARATHYROID HORMONE BIOSYNTHESIS

PTH, a protein of 84 amino acids, is synthesized as a larger precursor, preproparathyroid hormone (prepro-PTH), which has a 25-residue "pre" or signal sequence, and a 6-residue "pro" sequence *(3,4)*. The signal sequence, along with the short "pro" sequence, functions to direct the protein into the secretory pathway. Like other signal sequences, the "pre" sequence binds to a signal recognition particle during protein synthesis. The signal recognition particle then delivers the nascent peptide chain to the rough endoplasmic reticulum, where it is threaded through a protein-lined aqueous pore *(5)*. During this transit, the signal sequence is cleaved off by a signal peptidase, and the "pre" sequence is rapidly degraded. Since this process of transport and cleavage occurs during protein synthesis, very little intact prepro-PTH is found within the parathyroid cell.

From: *Vitamin D: Physiology, Molecular Biology, and Clinical Applications*
Edited by: M. F. Holick © Humana Press Inc., Totowa, NJ

Mature PTH is the only form secreted from the parathyroid cell, and it has a molecular weight of approx 9.6 kDa. The amino acid sequence has been determined in several species, and there is a high degree of identity among species, particularly in the amino-terminal region of the molecule *(6)*.

3. THE PTH GENE

There is one gene encoding for PTH in all species studied including human. A related gene, which codes for parathyroid hormone-related peptide (PTHrP), is ancestrally related to PTH. In the human the PTH gene is on chromosome 11, and the PTHrP is on chromosome 12 *(7,8)*. Since the genes for PTH and PTHrP are located in similar positions on these sibling chromosomes, it is likely that they arose from a common precursor by chromosomal duplication. The human PTHrP gene, with three promoters, eight exons, and complex patterns of alternate splicing, is much more complicated than the PTH gene. Both PTH and PTHrP bind to the same receptor, the PTH/PTHrP receptor, to exert their biologic effects *(9)*. They both regulate calcium and phosphate metabolism under physiologic and pathologic conditions, and $1,25(OH)_2D_3$ has been shown to have an effect on both their genes *(10,11)*. However, it is the effect of $1,25(OH)_2D_3$ on the PTH gene that has physiologic relevance and that we review here.

Complementary DNAs encoding for human *(12,13)*, bovine *(14,15)*, rat *(16)*, pig *(16)*, chicken *(17,18)*, and dog *(19)* PTH have all been cloned. Corresponding genomic DNA has also been cloned from human *(13)*, bovine *(20)*, and rat *(21)*. The genes all have two introns or intervening sequences and three exons *(22)*. The primary RNA transcript consists of RNA transcribed from both the introns and exons, and then the RNA sequences derived from the introns are spliced out. The product of this RNA processing, which represents the exons, is the mature PTH mRNA, which will then be translated into prepro-PTH. The first intron separates the 5'-untranslated region of the mRNA from the rest of the gene, and the second intron separates most of the sequence encoding the precursor specific "prepro" region from exon 3, which encodes the mature PTH and the 3'-untranslated region (UTR). The three exons that result are thus roughly divided into functional domains. There is considerable identity among the mammalian PTH genes, which is reflected in an 85% identity between the human and bovine proteins, and 75% identity between the human and rat protein. There is less identity in the 3'-noncoding region. The two homologous TATA sequences flanking the human PTH gene direct the synthesis of two human PTH gene transcripts in both normal parathyroid glands and parathyroid adenomas *(23)*. The termination codon immediately following the codon for glutamine at position 84 of PTH indicates that there are no additional precursors of PTH with peptide extensions at the carboxyl position.

4. PROMOTER SEQUENCES

The regions upstream of the transcribed structural gene determine the tissue specificity and contain the regulatory sequences for the gene. For PTH this analysis has been hampered by the lack of a parathyroid cell line. Rupp et al. *(24)* analyzed the human PTH promoter region up to position −805 and identified a number of consensus sequences by computer analysis. These included a sequence resembling the canonical cyclic adenosine monophosphate (cAMP)-responsive element 5'-TGACGTCA-3' at position −81 with a single residue deviation. This element was fused to a reporter gene (CAT) and then transfected into different cell lines. Pharmacologic agents that increase cAMP led to increased expression of the CAT gene, suggesting a functional role for the cAMP-responsive element (CRE). The role of this possible CRE in the context of the PTH gene in the parathyroid remains to be established.

Demay et al. *(25)* identified DNA sequences in the human PTH gene that bind the $1,25(OH)_2D_3$ receptor. Nuclear extracts containing the $1,25(OH)_2D_3$ receptor were examined for binding to sequences in the 5'-flanking region of the human (h)PTH gene. A 25-bp oligonucleotide containing the sequences from −125 to −101 from the start of exon 1 bound nuclear proteins that were recognized by monoclonal antibodies against the $1,25(OH)_2D_3$ receptor. The sequences in this region contained a single copy of a motif (AGGTTCA) that is homologous to the motifs repeated in the upregulatory $1,25(OH)_2D_3$-response element of osteocalcin. When placed upstream to a heterologous viral promoter, the sequences contained in this 25-bp oligonucleotide mediated transcriptional repression in response to $1,25(OH)_2D_3$ in GH4C1 cells but not in ROS 17/2.8 cells. Therefore, this downregulatory element differs from upregulatory elements in both sequence composition and the requirement for particular cellular factors other than the vitamin D receptor (VDR) for repressing PTH transcription *(25)*. Further work is needed to demonstrate that this negative regulatory element functions in the context of the PTH gene's promoter and to establish whether other vitamin D response elements (VDREs) control PTH gene expression. Farrow et al. *(26,27)* have identified DNA sequences upstream of the bovine PTH gene that bind the $1,25(OH)_2D_3$ receptor. Liu et al. *(28)* have identified such sequences in the chicken PTH gene and demonstrated their functionality after transfection into the opossum kidney OK cell line.

5. REGULATION OF PTH GENE EXPRESSION

5.1. 1,25-Dihydroxyvitamin D

PTH is intimately involved in the homeostasis of normal serum concentrations of calcium and phosphate, which in turn regulate the synthesis and secretion of PTH. A major control mechanism in PTH synthesis occurs at the level of gene expression. $1,25(OH)_2D_3$ is also important in the maintenance of normal mineral metabolism, and there is a well-defined feedback loop between calcitriol and PTH.

PTH increases the renal synthesis of calcitriol. Calcitriol then increases blood calcium largely by increasing the efficiency of intestinal calcium absorption. Calcitriol also potently decreases the transcription of the PTH gene. This action was first demonstrated in vitro in bovine parathyroid cells in primary culture, in which calcitriol led to a marked decrease in PTH mRNA levels *(29,30)* and a consequent decrease in PTH secretion *(31–33)*. Earlier studies of the effect of $1,25(OH)_2D_3$ on PTH secretion were negative because insufficient time was allowed for the effect of $1,25(OH)_2D_3$ on PTH synthesis to become manifest *(34,35)*; studies at longer time intervals had shown an effect *(36)*. Cantley et al. *(31)* correlated the effect of $1,25(OH)_2D_3$ on PTH secretion and PTH mRNA levels in primary cultures of bovine parathyroid cells. In short-term incubations (30–120 min) $1,25(OH)_2D_3$ had no effect on PTH secretion, but in long-term incubations (24–96 h) there was a dose-dependent decrease in PTH mRNA levels and PTH secretion. This study and that of Chan et al. *(33)* confirm that the $1,25(OH)_2D_3$ effect on PTH gene expression is reflected in a decrease in PTH synthesis and subsequent secretion.

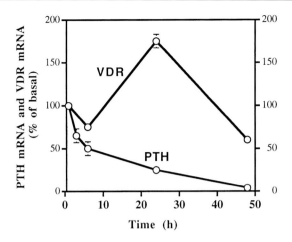

Fig. 1. Time course for the effect of 1,25(OH)$_2$D$_3$ on mRNA levels for PTH and the 1,25(OH)$_2$D$_3$ receptor (VDR) in rat thyroparathyroid glands. Rats were injected with either a single dose of 100 pmol 1,25(OH)$_2$D$_3$. The data represent the mean ± SE for four rats. (Reproduced with permission from ref. 44.)

The physiologic relevance of these findings was established by in vivo studies in rats *(37)*. Rats injected with amounts of calcitriol that did not increase serum calcium had marked decreases in PTH mRNA levels, reaching <4% of control at 48 h (Fig. 1). This effect was shown to be transcriptional in both in vivo studies in rats *(37)* and in vitro studies with primary cultures of bovine parathyroid cells *(38)*. When 684 bp of the 5'-flanking region of the human PTH gene were linked to a reporter gene and transfected into a rat pituitary cell line (GH4C1), gene expression was lowered by 1,25(OH)$_2$D$_3$ *(39)*. These studies suggest that 1,25(OH)$_2$D$_3$ decreases PTH transcription by acting on the 5'-flanking region of the PTH gene, probably at least partly through interactions with the vitamin D binding sequence noted above.

A further level at which 1,25(OH)$_2$D$_3$ might regulate the PTH gene would be at the 1,25(OH)$_2$D$_3$ receptor. 1,25(OH)$_2$D$_3$ acts on its target tissues by binding to the 1,25(OH)$_2$D$_3$ receptor, which regulates the transcription of genes with the appropriate recognition sequences. The concentration of the 1,25(OH)$_2$D$_3$ receptor in the 1,25(OH)$_2$D$_3$ target sites could allow a modulation of the 1,25(OH)$_2$D$_3$ effect, with an increase in receptor concentration leading to an amplification of its effect and a decrease in receptor concentration dampening the 1,25(OH)$_2$D$_3$ effect. Ligand- and cation-dependent upregulation of the 1,25(OH)$_2$D$_3$ receptor has been shown in vivo in rat intestine *(40,41)*, and in vitro in a number of systems *(42,43)*.

Naveh-Many et al. *(44)* injected 1,25(OH)$_2$D$_3$ into rats and measured the levels of the 1,25(OH)$_2$D$_3$ receptor mRNA (VDR mRNA) and PTH mRNA in the parathyrothyroid tissue. They showed that 1,25(OH)$_2$D$_3$ in physiologically relevant doses led to an increase in VDR mRNA levels in the parathyroid glands, in contrast to a decrease in PTH mRNA levels (Fig. 1). This increase in VDR mRNA occurred after a time lag of 6 h, and a dose response showed a peak at 25 pmol. Localization of the VDR mRNA to the parathyroids was demonstrated by *in situ* hybridization studies of the thyroparathyroid and duodenum (Fig. 2). VDR mRNA was localized to the parathyroids in the same concentration as in the duodenum, calcitriol's classic target organ. Weanling rats fed a

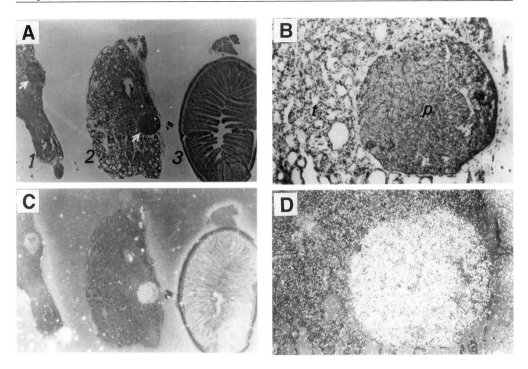

Fig. 2. *In situ* hybridization of parathyroid-thyroid and duodenum sections with 1,25(OH)$_2$D$_3$ receptor (VDR). (**A1**) Parathyroid-thyroid tissue from a control rat. (**A2**) Parathyroid-thyroid tissue from a 1,25(OH)$_2$D$_3$-treated rat (100 pmol at 24 h). (**A3**) Duodenum from the 1,25(OH)$_2$D$_3$-treated rat. The white arrow points at the parathyroid glands. (**B**) A higher power view of A2 showing the parathyroid gland (p) and thyroid follicle (t). Top figures were photographed under brightfield illumination, and bottom figures show darkfield illumination of the same sections. Hybridization was with an antisense VDR probe. After 4 d of autoradiographic exposure, sections were stained with Giemsa and photographed. Magnification in (B) is sevenfold the magnification of (A). (Reproduced with permission from ref. *44*.)

diet deficient in calcium were markedly hypocalcemic at 3 wk and had very high serum 1,25(OH)$_2$D$_3$ levels. Despite the chronically high serum 1,25(OH)$_2$D$_3$ levels, there was no increase in VDR mRNA levels, and furthermore PTH mRNA levels did not fall and were markedly increased. The low calcium may have prevented the increase in parathyroid VDR levels, which may partially explain the PTH mRNA suppression. Whatever the mechanism, the lack of suppression of PTH synthesis in the setting of hypocalcemia and increased serum 1,25(OH)$_2$D$_3$ is crucial physiologically, because it allows an increase in both PTH and 1,25(OH)$_2$D$_3$ at a time of chronic hypocalcemic stress. Russell et al. *(45)* studied the parathyroids of chicks with vitamin D deficiency and confirmed that 1,25(OH)$_2$D$_3$ regulates PTH and VDR gene expression in the avian parathyroid gland. The chicks in this study were fed a vitamin D-deficient diet from birth for 21 d and had established secondary hyperparathyroidism. These hypocalcemic chicks were then fed a diet with different calcium contents (0.5, 1.0, and 1.6%) for 6 d. The serum calciums were all still low (5, 6, and 7 mg/dL), with the expected inverse relationship between PTH mRNA and serum calcium. There was also a direct relationship between serum calcium and VDR mRNA levels. This result suggests either that VDR mRNA is not upregulated in the setting of secondary hyperparthroidism or that calcium directly regulates the VDR gene. Brown et al. *(46)* studied vitamin D-deficient rats and confirmed that calcitriol upregulated the parathyroid VDR mRNA and that in secondary hyperparathyroidism with hypocalcemia the PTH mRNA was upregulated without change in the VDR mRNA *(44)*.

All these studies show that $1,25(OH)_2D_3$ increases the expression of its receptor's gene in the parathyroid gland, which would result in increased VDR protein synthesis and increased binding of $1,25(OH)_2D_3$. This ligand-dependent receptor upregulation would lead to an amplified effect of $1,25(OH)_2D_3$ on the PTH gene and might help explain the dramatic effect of $1,25(OH)_2D_3$ on the PTH gene.

The use of calcitriol is limited by its hypercalcemic effect, and therefore a number of calcitriol analogs have been synthesized that are biologically active but are less hypercalcemic than calcitriol. These analogs usually involve modifications of the calcitriol side chain, such as 22-oxa-$1,25(OH)_2D_3$, which is the chemical modification in oxacacitriol *(47)*, or a cyclopropyl group at the end of the side chain in calcipotriol *(48,49)*. Brown et al. *(50)* showed that oxacalcitriol in vitro decreased PTH secretion from primary cultures of bovine parathyroid cells with a similar dose response to that of calcitriol. In vivo the injection of both vitamin D compounds led to a decrease in rat parathyroid PTH mRNA levels *(50)*. However, detailed in vivo dose–response studies showed that in vivo calcitriol is the most effective analog for decreasing PTH mRNA levels, even at doses that do not cause hypercalcemia *(51)*. Oxacalcitriol and calcipotriol are less effective for decreasing PTH RNA levels but have a wider dose range at which they do not cause hypercalcemia; this property might be useful clinically. The marked activity of calcitriol analogs in vitro compared with their modest hypercalcemic actions in vivo probably reflects their rapid clearance from the circulation *(52)*.

The ability of calcitriol to decrease PTH gene transcription is used therapeutically in the management of patients with chronic renal failure. They are treated with calcitriol to prevent the secondary hyperparathyroidism of chronic renal failure. The most effective dosing regime to suppress the elevated serum PTH levels without causing hypercalcemia has been shown to be possible when calcitriol is given in any of the current dose regimes, namely, single oral or intravenous doses three times a week or daily therapy *(53–55)*. The poor response in some patients may result from poor control of serum phosphate *(54,56)*, decreased vitamin D receptor concentration *(57)*, or tertiary hyperparathyroidism with monoclonal parathyroid tumors *(58)*.

Retinoid X receptors (RXRs) are involved in $1,25(OH)_2D_3$-mediated transcriptional events. MacDonald et al. *(59)* showed that all-*trans*- or 9-*cis* retinoic acid suppressed the release of PTH from bovine parathyroid cell cultures. Both retinoids were remarkably potent, with significant decreases evident at $10^{(-10)}$ *M* and a maximally suppressive effect (approx 65%) at $10^{(-7)}$ *M*. All-*trans*-retinol was considerably less potent in this system. The effect was not evident until 12 h, suggesting that retinoids did not affect the rapid secretion of pre-existing PTH stores. Prepro-PTH mRNA levels were also suppressed by retinoic acid, and the retinoid potencies were similar to those observed in the secretion studies. Combined treatment with $10^{(-6)}$ *M* retinoic acid and $10^{(-8)}$ *M* $1,25(OH)_2D_3$ more effectively decreased PTH secretion and prepro-PTH mRNA than did either compound alone. These data indicate that retinoic acid (1) elicits a bioresponse in bovine parathyroid cells; (2) attenuates PTH expression at the protein and mRNA levels; and (3) acts independently of $1,25(OH)_2D_3$ in the control of PTH expression. MacDonald et al. *(59)* suggested that, because all-*trans*- and 9-*cis* retinoic acid are equipotent, the suppressive effect may be mediated through the retinoic acid receptors and not through the RXRs, on condition that there are no enzymes that isomerize retinoic acid in the parathyroid. This is because 9-*cis* retinoic acid acts specifically on the RXR.

Liu et al. *(28)* characterized a response element (VDRE) in the 5'-flanking region of the chicken PTH gene that mediated negative regulation of gene transcription by $1,25(OH)_2D_3$ and bound the VDR. Using gel mobility shift assays, they showed that the chicken VDRE, incubated with partially purified nuclear extract from dog intestine, showed two bound complexes. These complexes were competed for by excess of unlabeled VDRE or monoclonal antibody specific for the VDR protein. In addition, they demonstrated similar protein–DNA complexes when the chicken VDRE was incubated with a mixture of purified preparations of recombinant VDR and RXR α-proteins *(28)*. By themselves, neither of the recombinant proteins were able to bind the VDRE significantly. This study suggests that RXR is necessary for binding of the VDR complex to the VDRE. Mackey et al. *(60)* studied the VDRE of the human PTH gene *(25)* with protein–DNA binding studies. They showed that with bovine parathyroid nuclear extracts, the VDR bound the VDRE independently of the RXR. In GH4C1 nuclear extracts, there were two VDR-containing complexes, one lacking and one containing RXR. The hPTH VDRE mediates transcriptional repression in GH4C1 cells but not in ROS 17/2.8 cells. In ROS 17/2.8 nuclear extracts, a single VDR-dependent complex was observed that contained RXR. By contrast, when the upregulatory rat osteocalcin VDRE was used as a probe, only VDR-RXR-containing complexes were generated using nuclear extracts from bovine parathyroids, and GH4C1 and ROS 17/2.8 cells.

From these studies, it is clear that retinoic acid acts to decrease PTH gene expression. The study of Liu et al. *(28)* using the chicken VDRE show that RXR is involved in this binding, but the study of Mackey et al. *(60)* suggests that this may not apply for the human VDRE. Therefore, the parathyroid nuclear proteins involved in the binding of the VDR to the downregulatory VDRE, need to be isolated to provide a more comprehensive understanding of how $1,25(OH)_2D_3$ decreases PTH gene transcription.

5.2. Calcium

The interrelationships of the effect of calcium and $1,25(OH)_2D_3$ on the PTH gene and parathyroid cell are particularly interesting. This is of particular relevance to the in vivo situation, in which a low serum calcium, for instance as a result of dietary calcium deficiency, leads to a marked increase in serum $1,25(OH)_2D_3$ levels, which would be expected to decrease PTH gene transcription. However, this dietary hypocalcemia is associated with a marked increase in PTH mRNA and serum PTH levels. The effect of calcium on the parathyroid, and in particular the interrelationship between hypocalcemia and the lack of effect of $1,25(OH)_2D_3$ is now reviewed.

5.2.1. IN VITRO STUDIES

A remarkable characteristic of the parathyroid is its sensitivity to small changes in serum calcium, which leads to large changes in PTH secretion. This calcium sensing is also expressed at the levels of PTH gene expression and parathyroid cell proliferation. In vitro and in vivo data agree that calcium regulates PTH mRNA levels, but the data differ in important ways. In vitro studies with bovine parathyroid cells in primary culture showed that calcium regulated PTH mRNA levels *(61,62)* with an effect mainly of high calcium to decrease PTH mRNA. These effects were most pronounced after more prolonged incubations, such as 72 h. The parathyroid calcium sensor is already markedly decreased in cells in culture at 24 h, so it is difficult to interpret the data at 72 h *(63,64)*.

5.2.2. IN VIVO STUDIES

Naveh-Many et al. *(65)* studied rats in vivo. They showed that small decreases in serum calcium from 2.6 to 2.1 mmol/L led to large increases in PTH mRNA levels, reaching threefold that of controls at 1 and 6 hours. A high serum calcium had no effect on PTH mRNA levels even at concentrations as high as 6.0 mmol/L. Interestingly, in these same thyroparathyroid tissue RNA extracts, calcium had no effect on the expression of the calcitonin gene. Thus, although a high calcium is a secretagogue for calcitonin, it does not regulate calcitonin gene expression. Yamamoto et al. *(66)* also studied the in vivo effect of calcium on PTH mRNA levels in rats. They showed that hypocalcemia induced by a calcitonin infusion for 48 h led to a sevenfold increase in PTH mRNA levels. Rats made hypercalcemic (2.9–3.4 m*M*) for 48 h had the same PTH mRNA levels as controls that had received no infusion (2.5 m*M*) but modestly lower than in those that had received a calcium-free infusion. In further studies Naveh-Many et al. *(67)* transplanted Walker carcinosarcoma 256 cells into rats, which led to serum calciums of 18 mg/dL at day 10. There was no change in PTH mRNA levels in these rats with marked chronic hypercalcemia *(67)*. The differences between in vivo and in vitro results probably reflect the instability of the in vitro system, but it is also impossible to eliminate the possibility that in vivo effects are influenced by indirect effects of a high or low serum calcium. Nevertheless, the physiologic conclusion is that common causes of hypercalcemia in vivo do not importantly decrease PTH mRNA levels; these results emphasize that the gland is geared to respond to hypocalcemia and not hypercalcemia.

5.2.3. MECHANISMS OF REGULATION OF PTH MRNA BY CALCIUM

The mechanism whereby calcium regulates PTH gene expression is particularly interesting. Changes in extracellular calcium are sensed by a calcium sensor, which then regulates PTH secretion *(68)*. It is not known what mechanism transduces the message of changes in extracellular calcium leading to changes in PTH mRNA. Okazaki et al. *(69)* have identified a negative calcium regulatory element (nCaRE) in the atrial natiuretic peptide gene, with a homologous sequence in the PTH gene. They identified a redox factor protein (ref1), known to activate several transcription factors via alterations of their redox state, which bound an nCaRE; the levels of ref1 mRNA and protein were elevated by an increase in extracellular calcium concentration *(70)*. They suggested that rcf1 had transcription repressor activity in addition to its function as a transcriptional auxiliary protein *(70)*. Because no parathyroid cell line is available, these studies were performed in nonparathyroid cell lines, so their relevance to physiologic PTH gene regulation remains to be established.

Moallem et al. *(71)* have shown that the effect of Ca^{2+} on PTH mRNA levels in vivo occurs with no detectable change in PTH gene transcription rate. These results suggest that the effect of calcium in vivo may well entail posttranscriptional mechanisms. Such mechanisms could well involve proteins that bind directly to PTH mRNA. To identify such proteins they used ultraviolet crosslinking and RNA gel mobility shift assays, which demonstrated that cytoplasmic proteins bind to the PTH mRNA 3'-UTR. RNA fragments spanning the full-length rat PTH cDNA incubated with thyroparathyroid cytoplasmic proteins revealed three bands that bound specifically to the mRNA. RNA corresponding to the 5'-UTR and the coding region of the gene did not bind any proteins. An RNA fragment representing the 3'-UTR bound parathyroid proteins with the same pattern as the full-length mRNA and therefore represents the protein binding sequence. Protein–RNA binding with parathyroid proteins from hypocalcemic rats was increased compared with control and decreased from hypophosphatemic rats (see later). These studies show that regions in the 3'-UTR of the PTH mRNA specifically bind cytosolic proteins, and they may be involved in the posttranscriptional increase in PTH gene expression induced by hypocalcemia.

In vitro studies by Hawa et al. (72) have also suggested a posttranscriptional effect of calcium on PTH gene expression. They incubated bovine parathyroid cells for 48 h in 0.4 mM calcium. This did not increase PTH mRNA levels compared with controls, but did increase the membrane-bound polysomal content of PTH mRNA by twofold. Actinomycin D reduced PTH mRNA levels by about 50% at 48 h in cells incubated in 0.4 and 1.0 mM calcium (72). However, actinomycin D did not prevent the rise in polysomal PTH mRNA induced by low calcium. Vadher et al. (73) have shown that the 3'-UTR is of major importance in mediating translational regulation of PTH synthesis by cytosolic regulatory proteins. 5'-UTR or 3'-UTR constructs of PTH mRNA fused to a luciferase reporter were synthesized and then translated with wheat germ lysate with or without parathyroid cell cytosol. The addition of cytosol inhibited translation from the 5'- and 3'-UTR PTH mRNA. These in vitro studies do not demonstrate the changes in PTH mRNA that are so marked in the in vivo studies, but they do show a mechanism of translational regulation of the PTH mRNA and a role for the UTR in this process. Together, these results suggest that calcium regulates PTH mRNA by posttranscriptional mechanisms, which might involve calcium sensitive proteins binding to the 5'- and 3'-UTRs. Transcriptional regulation of the PTH gene by calcium in parathyroid cells remains possible, but direct evidence for such regulation has not yet been obtained.

Pulse calcitriol decreases PTH gene transcription, but in chronic hypocalcemia there is a marked increase in $1,25(OH)_2D_3$ levels, which paradoxically do not decrease PTH mRNA levels. Calreticulin binds to the sequence KXGFFKR, found in steroid hormone receptors, resulting in altered transcription of steroid-responsive genes in vitro (74,75). Wheeler et al. (76) showed that calreticulin prevented $1,25(OH)_2D_3$ from increasing osteocalcin gene transcription in vitro. Sela et al. (77) have studied the role of calreticulin and the VDR in preventing the action of chronic high $1,25(OH)_2D_3$ levels on the PTH gene in vivo. Their results suggest that the response to $1,25(OH)_2D_3$ is associated with a decrease in calreticulin. The increase in VDR may amplify the effect of $1,25(OH)_2D_3$ on the PTH gene. In rats with chronic hypocalcemia, there may be enhanced levels of calreticulin, which would dominate over the effect of $1,25(OH)_2D_3$ on the PTH gene. These results raise the possibility that high circulating levels of $1,25(OH)_2D_3$ in diet-induced hypocalcemia do not decrease PTH gene transcription, because the increased levels of calreticulin in the parathyroid, together with decreased VDR, may prevent the binding of the VDR, or the VDR–RXR complex, to its response element in the PTH gene (77).

5.2.4. Secondary Hyperparathyroidism and Parathyroid Cell Proliferation

Chronic changes in the physiologic milieu often lead to changes in both parathyroid cell proliferation and parathyroid hormone gene regulation, as was discussed by Silver and Kronenberg (5). In such complicated settings, the regulation of PTH gene expression may well be controlled by mechanisms that differ from those in nonproliferating cells.

Secondary parathyroid hyperplasia is a complication of chronic renal disease *(78,79)* or vitamin D deficiency and may lead to disabling skeletal complications. The expression and regulation of the PTH gene has been studied in two models of secondary hyperparathyroidism: (1) rats with experimental uremia due to 5/6 nephrectomy; and (2) rats with nutritional secondary hyperparathyroidism due to diets deficient in vitamin D and/or calcium.

Higher serum creatinines and also appreciably higher levels of parathyroid gland PTH mRNA were found in 5/6 nephrectomy rats *(80)*. Their PTH mRNA levels decreased after single injections of $1,25(OH)_2D_3$, a response similar to that of normal rats *(80)*. Interestingly, the secondary hyperparathyroidism was characterized by an increase in parathyroid gland PTH mRNA but not in VDR mRNA *(80)*. This suggests that in 5/6 nephrectomy rats there was relatively less VDR mRNA per parathyroid cell, or a relative downregulation of the VDR, as has been reported in VDR binding studies *(81–84)*. Fukagawa et al. *(85)* also studied 5/6 nephrectomized rats and confirmed that calcitriol decreased PTH mRNA levels, as did the calcitriol analog 22-oxacalcitriol.

The second model of experimental secondary hyperparathyroidism studied was that due to dietary deficiency of vitamin D (–D) and/or calcium (–Ca), compared with normal vitamin D (ND) and normal calcium (NCa) *(86)*. These dietary regimes were selected to mimic the secondary hyperparathyroidism in which the stimuli for the production of hyperparathyroidism are the low serum levels of $1,25(OH)_2D_3$ and ionized calcium. Weanling rats were maintained on these diets for 3 wk and then studied. Rats on diets deficient in both vitamin D and calcium (–D-Ca) had a 10-fold increase in PTH mRNA compared with controls (ND-NCa), together with much lower serum calciums and also lower serum $1,25(OH)_2D_3$ levels. Calcium deficiency alone (–Ca-ND) led to a fivefold increase in PTH mRNA levels, whereas a diet deficient in vitamin D alone (–D-NCa) led to a twofold increase in PTH mRNA levels.

Since renal failure and prolonged changes in blood calcium and $1,25(OH)_2D_3$ can affect both parathyroid cell number and the activity of each parathyroid cell, the change in both these parameters must be assessed in each model, to understand the various mechanisms of secondary hyperparathyroidism. Parathyroid cell number was determined in thyroparathyroid tissue of normal rats and –D-Ca rats. To do this, the tissue was enzymatically digested into an isolated cell population, which was then passed through a flow cytometer fluorescence-activated cell sorter and separated by size into two peaks. The first peak of smaller cells contained parathyroid cells, as determined by the presence of PTH mRNA, and the second peak contained thyroid follicular cells and calcitonin-producing cells, which hybridized positively for thyroglobulin mRNA and calcitonin mRNA, but not PTH mRNA. There were 1.6-fold more cells in the –D-Ca rats than in the normal rats, compared with the 10-fold increase in PTH mRNA. Therefore, this model of secondary hyperparathyroidism is characterized by increased gene expression per cell, together with a smaller increase in cell number. Further studies by Naveh-Many et al. *(87)* have clearly demonstrated that hypocalcemia is a stimulus for parathyroid cell proliferation. They studied parathyroid cell proliferation by staining for proliferating cell nuclear antigen (PCNA) and found that a low calcium diet led to increased levels of PTH mRNA and a 10-fold increase in parathyroid cell proliferation *(87)*. The secondary hyperparathyroidism of 5/6 nephrectomized rats was characterized by an increase in both PTH mRNA levels and PCNA-positive parathyroid cells. Therefore hypocalcemia, and uremia induce parathyroid cell proliferation in vivo. The effect of $1,25(OH)_2D_3$ on parathyroid cell proliferation was also studied in this dietary model of secondary hyperparathyroidism. 1,25(OH)$_2$D$_3$ at a dose (25 pmol daily ×3) that lowered PTH mRNA levels had no effect on the number of PCNA-positive cells. However, in vitro (in bovine parathyroid cells using pharmacologic doses of 1,25(OH)$_2$D$_3$) *(88)* and in vivo (in rats with experimental uremia), 1,25(OH)$_2$D$_3$ decreased the amount of [^{3}H]thymidine incorporated into the parathyroid *(89)*. These findings emphasize the importance of a normal calcium in the prevention of parathyroid cell hyperplasia and indicate that the role of 1,25(OH)$_2$D$_3$ is not yet clear.

A further mechanism by which calcium might regulate parathyroid cell number is by inducing apoptosis. This has been studied in the parathyroids of hypocalcemic rats as well as in rats with experimental uremia fed different diets *(87)*. Apoptosis was determined by the TUNEL method, which detects nuclear DNA fragmentation *in situ*. In no situation were apoptotic cells detected in the parathyroids. However, apoptosis in the parathyroids needs to be explored in other models using different methodologies.

Wernerson and colleagues *(90,91)* have studied parathyroid cell number in dietary secondary hyperparathyroidism (ND-Ca) using stereoscopic electron microscopy and have shown that the cells are markedly hypertrophic without an increase in cell number. Thus, in models of secondary parathyroid enlargement such as this one, parathyroid cell hypertrophy can precede the development of parathyroid cell hyperplasia. These experimental findings are relevant to the management of patients with secondary hyperparathyroidism. Increased transcription of the PTH gene is readily reversible, but the reversibility of an increased number of parathyroid cells by accelerating cell death has not yet been demonstrated.

6. PHOSPHATE

Serum phosphate levels, like serum calcium levels, lead to large changes in serum 1,25(OH)$_2$D$_3$. A low serum phosphate increases serum 1,25(OH)$_2$D$_3$, and a high serum phosphate decreases 1,25(OH)$_2$D$_3$. However, in contrast to the effect of hypocalcemia (increased serum PTH despite high levels of 1,25(OH)$_2$D$_3$), in diet-induced hypophosphatemia the effect of both low phosphate and low 1,25(OH)$_2$D$_3$ is to decrease serum PTH. This implies a simple additive effect, but this may not be the case, because Kilav et al. *(92)* showed in vivo that the effect of hypophosphatemia is posttranscriptional. It would be expected that the transcriptional effect of the high 1,25(OH)$_2$D$_3$ would exert a dominant effect on the PTH gene. Therefore, there must be a mechanism at play preventing the chronically elevated serum 1,25(OH)$_2$D$_3$ from acting on the PTH gene.

PTH regulates serum phosphate concentration by its effect on the kidney, namely, decreased phosphate reabsorption. In moderate renal failure there is a decrease in phosphate clearance and an increase in serum phosphate, which becomes an important problem in severe renal failure. This has always been considered central to the pathogenesis of secondary hyperparathyroidism, but it has been difficult to separate the effects of hyperphosphatemia from those of the attendant hypocalcemia and decrease in serum 1,25(OH)$_2$D$_3$ levels. In the 1970s Slatopolsky and Bricker *(93)* showed in dogs with experimental chronic renal failure that dietary phosphate restriction prevented secondary hyperparathyroidism. Clinical studies *(94)* have demonstrated that phosphate restriction in patients with chronic renal insufficiency is effective in preventing the increase in serum PTH levels *(56,94–97)*. The mechanism of this effect was not clear, although at least part of it was considered to be due to changes in serum 1,25(OH)$_2$D$_3$ concentrations. In vitro *(98,99)* and in vivo *(94,100)*, phosphate directly regulated the production of 1,25(OH)$_2$D$_3$.

A raised serum phosphate decreases serum $1,25(OH)_2D_3$ levels, which then leads to decreased calcium absorption from the diet and eventually a low serum calcium. The raised phosphate complexes calcium, which is then deposited in bone and soft tissues, thereby decreasing serum calcium and also preventing the mobilization of calcium from bone. However, a number of careful clinical and experimental studies suggested that the effect of phosphate on serum PTH levels was independent of changes in both serum calcium and $1,25(OH)_2D_3$ levels. Lopez-Hilker et al. *(101)* have shown in dogs with experimental chronic renal failure that phosphate restriction corrected their secondary hyperparathyroidism independent of changes in serum calcium and $1,25(OH)_2D_3$ levels. They did this by placing the uremic dogs on diets deficient in both calcium and phosphate. This led to lower levels of serum phosphate and calcium, with no increase in the low levels of serum $1,25(OH)_2D_3$. Even so, there was a 70% decrease in PTH levels. Lopez-Hilker et al. *(101)* studied the effect of a low phosphate diet on serum PTH levels and suggested that phosphate had an effect on the parathyroid cell by a mechanism independent of its effect on serum $1,25(OH)_2D_3$ and calcium levels. Yi et al. *(102)* studied rats in early chronic renal failure. They showed that with a normal phosphate diet (0.6%) there was secondary hyperparathyroidism, with an increase in serum PTH and PTH mRNA levels with no change in serum calcium, phosphate or $1,25(OH)_2D_3$. However, rats fed a low phosphate diet (either 0.3% or 0.1%) had no evidence of secondary hyperparathyroidism *(102)*. At least in the rats fed the 0.3% phosphate diet, there was no increase in serum calcium or $1,25(OH)_2D_3$. Therefore, phosphate plays a central role in the pathogenesis of secondary hyperparathyroidism, both by its effect on serum $1,25(OH)_2D_3$ and calcium levels and also independently.

Kilav et al. *(103)* studied the expression of the PTH gene in hypophosphatemic rats and showed that phosphate regulated the gene independent of its effects upon serum calcium and $1,25(OH)_2D_3$. Weanling rats were fed diets with low, normal, or high phosphate content for 3 wk. The low-phosphate diet led to hypophosphatemia, hypercalcemia, and increased serum $1,25(OH)_2D_3$, together with decreased PTH mRNA levels (25% of controls) and a similar decrease in serum PTH levels. A high-phosphate diet led to increased PTH mRNA levels. *In situ* hybridization for PTH mRNA showed that hypophosphatemia decreased PTH mRNA in all the parathyroid cells. To separate the effect of low phosphate from changes in calcium and vitamin D, rats were fed diets to maintain them vitamin D deficient and normocalcemic despite the hypophosphatemia. Hypophosphatemic, normocalcemic rats with normal serum $1,25(OH)_2D_3$ levels still had decreased PTH mRNAs (Fig. 3). Nuclear transcript run-ons showed that the effect of low phosphate was posttranscriptional, unlike the transcriptional effect of $1,25(OH)_2D_3$ on the PTH gene. They therefore showed that dietary phosphate regulates the parathyroid by a mechanism that remains to be defined but is clearly independent of changes in serum calcium and $1,25(OH)_2D_3$ *(103)*.

There is now evidence showing a direct effect of phosphate on the parathyroid. Almaden et al. *(104)* showed that a high phosphate directly stimulated PTH secretion from whole rat parathyroid glands in culture and PTH mRNA levels in human parathyroid tissue in culture *(105)*. Slatopolsky et al. *(106)* and Nielsen et al. *(107)* showed similar results for PTH secretion from rat parathyroid glands maintained in primary culture. These results indicate that phosphate regulates the parathyroid directly. In these in vitro studies, parathyroid slices or tissue, rather than isolated cells, were incubated, suggesting either that the sensing mechanism for phosphate is damaged during the preparation of the isolated cells, or that the intact gland structure is important to the phosphate response.

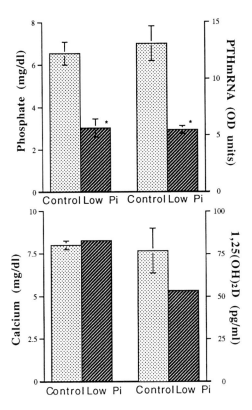

Fig. 3. Effect of dietary phosphate on PTH mRNA levels in second-generation vitamin D-deficient rats fed a vitamin D-deficient diet with both low phosphate and calcium contents for 1 d. Levels of serum phosphate and PTH mRNA (**top**) and serum calcium and 1,25(OH)$_2$D (**bottom**) for rats fed a vitamin D-deficient, normal phosphate, normal calcium diet (control group) or a vitamin D-deficient, low-phosphate, low-calcium diet for 1 d. The results are the mean ± SE for five rats in each group. *, $p \leq 0.01$. Data for 1,25(OH)$_2$D represent pooled serum from two or three rats with three samples for control rats and two samples for low-phosphate rats. (Reproduced with permission from ref. *103*.)

One of the effects on the parathyroid of either prolonged hypocalcemia or experimental chronic renal failure is to increase parathyroid cell proliferation *(87)*. Hypophosphatemic rats have a marked decrease in parathyroid cell proliferation. This effect of hypophosphatemia occurred in both normal and uremic rats. Rats fed high dietary phosphate had an increase in parathyroid cell proliferation, particularly those rats with experimental chronic renal failure *(87)*. Therefore phosphate regulates the parathyroid at a number of levels, namely, PTH gene expression, parathyroid cell proliferation, and serum PTH levels. However, the in vivo studies reported here utilized diets that led to very low serum phosphates, which may have no direct relevance to possible direct effects of high phosphate in renal failure. It is necessary to separate the nonspecific effects of very low phosphate from true physiologic regulation.

The effect of calcium on PTH secretion from dispersed bovine parathyroid cells occurs within seconds *(108,109)*. However, the effect of phosphate in vitro in parathyroid glands in culture takes about 4 h *(104,106,107)* before any change in PTH secretion occurs. The reason for this delay is not clear. It may reflect a delay in signal transduction or PTH synthesis. What is clear from the preliminary studies of Naveh-Many et al. *(110)* is that at least the effect on PTH gene expression correlates with a decrease in PTH mRNA

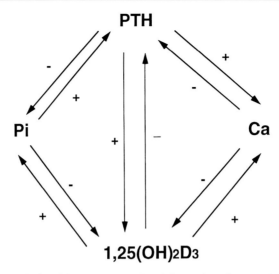

Fig. 4. Interrelationships among PTH, calcium, phosphate, and 1,25(OH)$_2$D$_3$.

protein binding. This compares with the effect of hypocalcemia, which increases this binding. These results indicate that the final pathway of low phosphate and low calcium on PTH mRNA shares a common mechanism (Fig. 4).

7. USE OF 1,25(OH)$_2$D$_3$ IN THE MANAGEMENT OF THE SECONDARY HYPERPARATHYROIDISM OF PATIENTS WITH RENAL DISEASE

A number of factors in renal failure lead to secondary hyperparathyroidism. There is a decrease in serum 1,25(OH)$_2$D$_3$ levels due to the decrease in functioning renal tissue *(111)*. Serum phosphate levels are increased because of the decreased clearance of phosphate by the failing kidney *(94,112)*. In addition, there are low levels of serum calcium because of the decreased levels of serum 1,25(OH)$_2$D$_3$, the complexing of calcium by the high serum phosphate, and decreased mobilization of calcium from bone by PTH due to the action of phosphate. The net result is a secondary hyperparathyroidism, with an increased synthesis and secretion of PTH by all the parathyroid cells, reflected by an increase in the size of the individual parathyroid cell, at least in experimental animals *(86)*, and an increase in the proliferation of the parathyroid cells *(87)*. This increased proliferation may eventually become monoclonal, resulting in the large parathyroids of tertiary hyperparathyoidism *(58)*, for which the only effective treatment is surgery.

Management of secondary hyperparathyoidism should start in mild to moderate renal failure. This includes maintenance of normal serum phosphate by judicious dietary restriction as well as the use of phosphate binders such as calcium carbonate and calcium acetate with meals. An important component of the therapeutic armamentarium is the use of 1,25(OH)$_2$D$_3$, or one of its analogs such as 1α-hydroxyvitamin D. As discussed in this chapter, 1,25(OH)$_2$D$_3$ decreases PTH gene transcription and the subsequent synthesis and secretion of PTH. 1,25(OH)$_2$D$_3$ may be given early in renal failure. However, care must be taken to ensure first that the serum phosphate is not too high and in particular that the calcium × phosphate product is not >60 (in mg/dL). With this therapy, 1,25(OH)$_2$D$_3$

or 1α-hydroxyvitamin D have been shown to be effective in either decreasing or preventing the expected increase in serum PTH levels, as well as the changes of renal bone disease in repeat bone biopsies *(113)*.

Daily calcium carbonate administration is also highly effective in treating secondary hyperparathyroidism in predialysis patients with mild to moderate renal failure. Therefore, vitamin D supplementation should be given to patients with increased plasma PTH levels and those with progressive hyperparathyroidism despite calcium supplementation. The goal of calcitriol administration is to achieve a plasma PTH level that is approximately two or three times the upper limit of normal in patients without renal disease. Appropriate suppression of PTH release should be associated with normalization of the plasma alkaline phosphatase concentration. Measurement of intact PTH is preferable to the midregion assay, which also detects inactive metabolites. Earlier studies had suggested that intravenous bolus doses given for instance, three times a week, for instance at each dialysis session, had a greater biologic effect than oral $1,25(OH)_2D_3$ *(53,114)*. However, these were uncontrolled studies. Subsequent controlled studies have shown that oral and intravenous pulse therapy is equally effective, and daily oral therapy is just as effective *(54,115,116)*. The most important parameter is that the patient receives the therapy and comply with the treatment schedule with due regard to the potential side effects such as excessive increases in serum calcium, as well as excess reduction in serum PTH, which may predispose to adynamic bone disease *(117)*.

However, regardless of the mode and dose of treatment, there are many patients who do not show a decrease in serum PTH levels, perhaps as many as 50% of dialysis patients *(54)*. Why does this occur? It certainly is reasonable to start treatment early before autonomy in PTH production develops, which may prevent autonomous function from developing *(113)*. Intolerance to higher doses of calcitriol because of higher serum calcium and phosphate levels was associated with a failure to respond to calcitriol *(54, 118)*, as well as larger gland size. Larger glands are more likely to have areas of nodular hyperplasia; nodules generally show a marked reduction in calcitriol receptors, which may contribute to the calcitriol resistance *(57)*. In addition, they are more likely to represent glands with monoclonal growths of parathyroid cells, which theoretically are less likely to respond to pharmacologic treatment *(58)*. In the future, there may be $1,25(OH)_2D_3$ analogs available with a more specific effect on the parathyroid than on the intestine in regard to increase of calcium and phosphate absorption *(51)*.

8. SUMMARY

The PTH gene is regulated by a number of factors. Calcitriol acts on the PTH gene to decrease its transcription; this action is used in the management of patients with chronic renal failure. Mechanisms of transcriptional downregulation are less well defined than mechanisms of transcriptional upregulation. It is clear that the effect of retinoic acid is additive to that of $1,25(OH)_2D_3$ in downregulating PTH gene expression, but the role of RXR in the binding of the VDR to the PTH gene's VDRE is controversial. A downregulatory VDRE has been defined in the PTH gene's promoter that requires other cellular factors for its function. What is clear is the powerful ability of $1,25(OH)_2D_3$ in vivo in experimental animals and also in patients to decrease PTH gene expression, which has been shown to be transcriptional. Why in in vivo situations of very high serum $1,25(OH)_2D_3$ levels there is no activity on the PTH gene is only beginning to be clarified. In the case of chronic hypocalcemia, studies suggest that calreticulin may play a role in preventing $1,25(OH)_2D_3$ effect on the PTH gene. Calcium and phosphate also determine parathyroid cell proliferation, mainly independently of $1,25(OH)_2D_3$ in vivo.

In diseases such as chronic renal failure, secondary hyperparathyroidism involves abnormalities in PTH secretion and synthesis. An understanding of how the parathyroid is regulated at each level will help us to devise rational therapy for the management of such conditions. The ability of $1,25(OH)_2D_3$ to decrease PTH gene transcription and serum PTH levels has led to the rational use of $1,25(OH)_2D_3$ in the management of patients with renal failure, an example of the effective application of advances in basic science and clinical investigation to patient well-being.

REFERENCES

1. Fraser DR. Regulation of the metabolism of vitamin D. Physiol Rev 1980; 60:551–613.
2. DeLuca HF. The metabolism and functions of vitamin D. Adv Exp Med Biol 1986; 196:361–375.
3. Potts JTJ, Kronenberg HM, Habener JF, Rich A. Biosynthesis of parathyroid hormone. Ann NY Acad Sci 1980; 343:38–55.
4. Habener JF. Regulation of parathyroid hormone secretion and biosynthesis. Annu Rev Physiol 1981; 43:211–223.
5. Silver J, Kronenberg HM. Parathyroid hormone—molecular biology and regulation. In: Principles of Bone Biology. Bilezikian JB, Raisz LG, Rodan GA, eds. San Diego: Academic Press, 1996; 325–338.
6. Habener JF, Potts JT Jr. Fundamental considerations in the physiology, biology, and biochemistry of parathyroid hormone. In: Metabolic Bone Disease, 2nd ed. Avioli LV, Krane SM, eds. Philadelphia: WB Saunders, 1990; 69–130.
7. Martin TJ, Allan EH, Caple IW, et al. Parathyroid hormone-related protein: isolation, molecular cloning, and mechanism of action. Recent Prog Horm Res 1989; 45:467–502.
8. Martin TJ, Moseley JM, Gillespeie MT. Parathyroid hormone-related protein: biochemistry and molecular biology. CRC Crit Rev Biochem 1991; 26:377–395.
9. Juppner H, Abou-Samra A-B, Freeman M, et al. A G protein-linked receptor for parathyroid hormone and parathyroid hormone-related peptide. Science 1991; 254:1024–1026.
10. Ikeda K, Lu C, Weir EC, Mangin M, Broadus AE. Transcriptional regulation of the parathyroid hormone-related peptide gene by glucocorticoids and vitamin D in a human C-cell line. J Biol Chem 1989; 264:15,743–15,746.
11. Falzon M. DNA sequences in the rat parathyroid hormone-related peptide gene responsible for 1,25-dihydroxyvitamin D_3-mediated transcriptional repression. Mol Endocrinol 1996; 10:672–681.
12. Hendy GN, Kronenberg HM, Potts JTJ, Rich A. Nucleotide sequence of cloned cDNAs encoding human preproparathyroid hormone. Proc Natl Acad Sci USA 1981; 78:7365–7369.
13. Vasicek TJ, McDevitt BE, Freeman MW, et al. Nucleotide sequence of the human parathyroid hormone gene. Proc Natl Acad Sci USA 1983; 80:2127–2131.
14. Kronenberg HM, McDevitt BE, Majzoub JA, et al. Cloning and nucleotide sequence of DNA coding for bovine preproparathyroid hormone. Proc Natl Acad Sci USA 1979; 76:4981–4985.
15. Weaver CA, Gordon DF, Kemper B. Nucleotide sequence of bovine parathyroid hormone messenger RNA. Mol Cell Endocrinol 1982; 28:411–424.
16. Schmelzer HJ, Gross G, Widera G, Mayer H. Nucleotide sequence of a full-length cDNA clone encoding preproparathyroid hormone from pig and rat. Nucleic Acids Res 1987; 15:6740.
17. Khosla S, Demay M, Pines M, Hurwitz S, Potts JTJ, Kronenberg HM. Nucleotide sequence of cloned cDNAs encoding chicken preproparathyroid hormone. J Bone Miner Res 1988; 3:689–698.
18. Russell J, Sherwood LM. Nucleotide sequence of the DNA complementary to avian (chicken) preproparathyroid hormone mRNA and the deduced sequence of the hormone precursor. Mol Endocrinol 1989; 3:325–331.
19. Rosol TJ, Steinmeyer CL, McCauley LK, Grone A, DeWille JW, Capen CC. Sequences of the cDNAs encoding cannine parathyroid hormone-related protein and parathyroid hormone. Gene 1995; 160:241–243.

20. Weaver CA, Gordon DF, Kissil MS, Mead DA, Kemper B. Isolation and complete nucleotide sequence of the gene for bovine parathyroid hormone. Gene 1984; 28:319–329.
21. Heinrich G, Kronenberg HM, Potts JTJ, Habener JF. Gene encoding parathyroid hormone. Nucleotide sequence of the rat gene and deduced amino acid sequence of rat preproparathyroid hormone. J Biol Chem 1984; 259:3320–3329.
22. Kronenberg HM, Igarashi T, Freeman MW, et al. Structure and expression of the human parathyroid hormone gene. Recent Prog Horm Res 1986; 42:641–663.
23. Igarashi T, Okazaki T, Potter H, Gaz R, Kronenberg HM. Cell-specific expression of the human parathyroid hormone gene in rat pituitary cells. Mol Cell Biol 1986; 6:1830–1833.
24. Rupp E, Mayer H, Wingender E. The promotor of the human parathyroid hormone gene contains a functional cyclic AMP-response element. Nucleic Acids Res 1990; 18:5677–5683.
25. Demay MB, Kiernan MS, DeLuca HF, Kronenberg HM. Sequences in the human parathyroid hormone gene that bind the 1,25-dihydroxyvitamin D_3 receptor and mediate transcriptional repression in response to 1,25-dihydroxyvitamin D_3. Proc Natl Acad Sci USA 1992; 89:8097–8101.
26. Farrow SM, Hawa NS, Karmali R, Hewison M, Walters JC, O'Riordan JL. Binding of the receptor for 1,25-dihydroxyvitamin D_3 to the 5'-flanking region of the bovine parathyroid hormone gene. J Endocrinol 1990; 126:355–359.
27. Hawa NS, O'Riordan JL, Farrow SM. Binding of 1,25-dihydroxyvitamin D_3 receptors to the 5'-flanking region of the bovine parathyroid hormone gene. J Endocrinol 1994; 142:53–60.
28. Liu SM, Koszewski N, Lupez M, Malluche HH, Olivera A, Russell J. Characterization of a response element in the 5'-flanking region of the avian (chicken) parathyroid hormone gene that mediates negative regulation of gene transcription by 1,25-dihydroxyvitamin D_3 and binds the vitamin D_3 receptor. Mol Endocrinol 1996; 10:206–215.
29. Silver J, Russell J, Sherwood LM. Regulation by vitamin D metabolites of messenger ribonucleic acid for preproparathyroid hormone in isolated bovine parathyroid cells. Proc Natl Acad Sci USA 1985; 82:4270–4273.
30. Russell J, Silver J, Sherwood LM. The effects of calcium and vitamin D metabolites on cytoplasmic mRNA coding for pre-proparathyroid hormone in isolated parathyroid cells. Trans Assoc Am Physicians 1984; 97:296–303.
31. Cantley LK, Russell J, Lettieri D, Sherwood LM. 1,25-Dihydroxyvitamin D_3 suppresses parathyroid hormone secretion from bovine parathyroid cells in tissue culture. Endocrinology 1985; 117: 2114–2119.
32. Karmali R, Farrow S, Hewison M, Barker S, O'Riordan JL. Effects of 1,25-dihydroxyvitamin D_3 and cortisol on bovine and human parathyroid cells. J Endocrinol 1989; 123:137–142.
33. Chan YL, McKay C, Dye E, Slatopolsky E. The effect of 1,25 dihydroxycholecalciferol on parathyroid hormone secretion by monolayer cultures of bovine parathyroid cells. Calcif Tissue Int 1986; 38:27–32.
34. Golden P, Greenwalt A, Martin K, et al. Lack of a direct effect of 1,25-dihydroxycholecalciferol on parathyroid hormone secretion by normal bovine parathyroid glands. Endocrinology 1980; 107: 602–607.
35. Chertow BS, Baker GR, Henry HL, Norman AW. Effects of vitamin D metabolites on bovine parathyroid hormone release in vitro. Am J Physiol 1980; 238:E384–E388.
36. Au WY. Inhibition by 1,25 dihydroxycholecalciferol of hormonal secretion of rat parathyroid gland in organ culture. Calcif Tissue Int 1984; 36:384–391.
37. Silver J, Naveh-Many T, Mayer H, Schmelzer HJ, Popovtzer MM. Regulation by vitamin D metabolites of parathyroid hormone gene transcription in vivo in the rat. J Clin Invest 1986; 78:1296–1301.
38. Russell J, Lettieri D, Sherwood LM. Suppression by 1,25(OH)$_2$D$_3$ of transcription of the pre-proparathyroid hormone gene. Endocrinology 1986; 119:2864–2866.
39. Okazaki T, Igarashi T, Kronenberg HM. 5'-Flanking region of the parathyroid hormone gene mediates negative regulation by 1,25-(OH)$_2$ vitamin D_3. J Biol Chem 1988; 263:2203–2208.
40. Costa EM, Feldman D. Homologous up-regulation of the 1,25(OH)$_2$ vitamin D_3 receptor in rats. Biochem Biophys Res Commun 1986; 137:742–747.
41. Favus MJ, Mangelsdorf DJ, Tembe V, Coe BJ, Haussler MR. Evidence for in vivo upregulation of the intestinal vitamin D receptor during dietary calcium restriction in the rat. J Clin Invest 1988; 82:218–224.
42. Huang Y, Lee S, Stolz R, et al. Effect of hormones and development on the expression of the rat 1,25-dihydroxyvitamin D_3 receptor gene. J Biol Chem 1989; 264:17,454–17,461.

43. Sandgren ME, DeLuca HF. Serum calcium and vitamin D regulate 1,25-dihydroxyvitamin D_3 receptor concentration in rat kidney in vivo. Proc Natl Acad Sci USA 1990; 87:4312–4314.
44. Naveh-Many T, Marx R, Keshet E, Pike JW, Silver J. Regulation of 1,25-dihydroxyvitamin D_3 receptor gene expression by 1,25-dihydroxyvitamin D_3 in the parathyroid in vivo. J Clin Invest 1990; 86:1968–1975.
45. Russell J, Bar A, Sherwood LM, Hurwitz S. Interaction between calcium and 1,25-dihydroxyvitamin D_3 in the regulation of preproparathyroid hormone and vitamin D receptor messenger ribonucleic acid in avian parathyroids. Endocrinology 1993; 132:2639–2644.
46. Brown AJ, Zhong M, Finch J, Ritter C, Slatopolsky E. The roles of calcium and 1,25-dihydroxyvitamin D_3 in the regulation of vitamin D receptor expression by rat parathyroid glands. Endocrinology 1995; 136:1419–1425.
47. Nishii Y, Abe J, Mori T, et al. The noncalcemic analogue of vitamin D, 22-oxacalcitriol, suppresses parathyroid hormone synthesis and secretion. Contrib Nephrol 1991; 91:123–128.
48. Kissmeyer AM, Binderup L. Calcipotriol (MC 903): pharmacokinetics in rats and biological activities of metabolites. A comparative study with 1,25(OH)2D3. Biochem Pharmacol 1991; 41:1601–1606.
49. Evans DB, Thavarajah M, Binderup L, Kanis JA. Actions of calcipotriol (MC 903), a novel vitamin D_3 analog, on human bone-derived cells: comparison with 1,25-dihydroxyvitamin D_3. J Bone Miner Res 1991; 6:1307–1315.
50. Brown AJ, Ritter CR, Finch JL, et al. The noncalcemic analogue of vitamin D, 22-oxacalcitriol, suppresses parathyroid hormone synthesis and secretion. J Clin Invest 1989; 84:728–732.
51. Naveh-Many T, Silver J. Effects of calcitriol, 22-oxacalcitriol and calcipotriol on serum calcium and parathyroid hormone gene expression. Endocrinology 1993; 133:2724–2728.
52. Bouillon R, Allewaert K, Xiang DZ, Tan BK, van-Baelen H. Vitamin D analogs with low affinity for the vitamin D binding protein: enhanced in vitro and decreased in vivo activity. J Bone Miner Res 1991; 6:1051–1057.
53. Slatopolsky E, Weerts C, Thielan J, Horst R, Harter H, Martin KJ. Marked suppression of secondary hyperparathyroidism by intravenous administration of 1,25-dihydroxy-cholecalciferol in uremic patients. J Clin Invest 1984; 74:2136–2143.
54. Quarles LD, Yohay DA, Carroll BA, et al. Prospective trial of pulse oral versus intravenous calcitriol treatment of hyperparathyroidism in ESRD. Kidney Int 1994; 45:1710–1721.
55. Caravaca F, Cubero JJ, Jimenez F, et al. Effect of the mode of calcitriol administration on PTH-ionized calcium relationship in uraemic patients with secondary hyperparathyroidism. Nephrol Dial Transplant 1995; 10:665–670.
56. Aparicio M, Combe C, Lafage MH, De Precigout V, Potaux L, Bouchet JL. In advanced renal failure, dietary phosphorus restriction reverses hyperparathyroidism independent of the levels of calcitriol. Nephron 1994; 63:122,123.
57. Fukuda N, Tanaka H, Tominaga Y, Fukagawa M, Kurokawa K, Seino Y. Decreased 1,25-dihydroxyvitamin D_3 receptor density is associated with a more severe form of parathyroid hyperplasia in chronic uremic patients. J Clin Invest 1993; 92:1436–1443.
58. Arnold A, Brown MF, Urena P, Gaz RD, Sarfati E, Drueke TB. Monoclonality of parathyroid tumors in chronic renal failure and in primary parathyroid hyperplasia. J Clin Invest 1995; 95:2047–2053.
59. MacDonald PN, Ritter C, Brown AJ, Slatopolsky E. Retinoic acid suppresses parathyroid hormone (PTH) secretion and PreproPTH mRNA levels in bovine parathyroid cell culture. J Clin Invest 1994; 93:725–730.
60. Mackey SL, Heymont JL, Kronenberg HM, Demay MB. Vitamin D binding to the negative human parathyroid hormone vitamin D response element does not require the retinoid X receptor. Mol Endocrinol 1996; 10:298–305.
61. Russell J, Lettieri D, Sherwood LM. Direct regulation by calcium of cytoplasmic messenger ribonucleic acid coding for pre-proparathyroid hormone in isolated bovine parathyroid cells. J Clin Invest 1983; 72:1851–1855.
62. Brookman JJ, Farrow SM, Nicholson L, O'Riordan JL, Hendy GN. Regulation by calcium of parathyroid hormone mRNA in cultured parathyroid tissue. J Bone Miner Res 1986; 1:529–537.
63. Mithal A, Kifor O, Kifor I, et al. The reduced responsiveness of cultured bovine parathyroid cells to extracellular Ca^{2+} is associated with marked reduction in the expression of extracellular Ca^{2+}-sensing receptor messenger ribonucleic acid and protein. Endocrinology 1995; 136:3087–3092.
64. Brown AJ, Zhong M, Ritter C, Brown EM, Slatopolsky E. Loss of calcium responsiveness in cultured bovine parathyroid cells is associated with decreased calcium receptor expression. Biochem Biophys Res Commun 1995; 212:861–867.

65. Naveh-Many T, Friedlaender MM, Mayer H, Silver J. Calcium regulates parathyroid hormone messenger ribonucleic acid (mRNA), but not calcitonin mRNA in vivo in the rat. Dominant role of 1,25-dihydroxyvitamin D. Endocrinology 1989; 125:275–280.
66. Yamamoto M, Igarashi T, Muramatsu M, Fukagawa M, Motokura T, Ogata E. Hypocalcemia increases and hypercalcemia decreases the steady-state level of parathyroid hormone messenger RNA in the rat. J Clin Invest 1989; 83:1053–1056.
67. Naveh-Many T, Raue F, Grauer A, Silver J. Regulation of calcitonin gene expression by hypocalcemia, hypercalcemia, and vitamin D in the rat. J Bone Miner Res 1992; 7:1233–1237.
68. Brown EM, Gamba G, Riccardi R, et al. Cloning and characterization of an extracellular Ca^{2+}-sensing receptor from bovine parathyroid. Nature 1993; 366:575–580.
69. Okazaki T, Ando K, Igarashi T, Ogata E, Fujita T. Conserved mechanism of negative gene regulation by extracellular calcium. Parathyroid hormone gene versus atrial natriuretic peptide. J Clin Invest 1992; 89:1268–1273.
70. Okazaki T, Chung U, Nishishita T, et al. A redox factor protein, ref1, is involved in negative gene regulation by extracellular calcium. J Biol Chem 1994; 269:27,855–27,862.
71. Moallem E, Kilav R, Silver J, Naveh-Many T. RNA–protein binding and post-transcriptional regulation of PTH gene expression by calcium and phosphate. J Biol Chem 1998; 273:5253–5259.
72. Hawa NS, O'Riordan JL, Farrow SM. Post-transcriptional regulation of bovine parathyroid hormone synthesis. J Mol Endocrinol 1993; 10:43–49.
73. Vadher S, Hawa NS, O'Riordan JLH, Farrow SM. Translation of parathyroid hormone gene expression and RNA: protein interactions. J Bone Miner Res 1996; 11:746–753.
74. Burns K, Duggan B, Atkinson EA, et al. Modulation of gene expression by calreticulin binding to the glucocorticoid receptor. Nature 1994; 367:476–480.
75. Dedhar S, Rennie PS, Shago M, et al. Inhibition of nuclear hormone receptor activity by calreticulin. Nature 1994; 367:480–483.
76. Wheeler DG, Horsford J, Michalak M, White JH, Hendy GN. Calreticulin inhibits vitamin D_3 signal transduction. Nucleic Acids Res 1995; 23:3268–3274.
77. Sela-Brown A, Russell J, Koszewski NJ, Michalak M, Naveh-Many T, Silver J. Calreticulin inhibits vitamin D's action on the PTH gene in vitro and may prevent vitamin D's effect in vivo in hypocalcemic rats. Mol Endocrinol 1998; 12:1193–1200.
78. Castleman B, Mallory TB. The pathology of the parathyroid gland in hyperparathyroidism. Am J Pathol 1932; 11:1–72.
79. Castleman B, Mallory TB. Parathyroid hyperplasia in chronic renal insufficiency. Am J Pathol 1937; 13:553–574.
80. Shvil Y, Naveh-Many T, Barach P, Silver J. Regulation of parathyroid cell gene expression in experimental uremia. J Am Soc Nephrol 1990; 1:99–104.
81. Brown AJ, Dusso A, Lopez-Hilker S, Lewis-Finch J, Grooms P, Slatopolsky E. 1,25-(OH)2D receptors are decreased in parathyroid glands from chronically uremic dogs. Kidney Int 1989; 35:19–23.
82. Korkor AB. Reduced binding of [^{3}H]1,25-dihydroxyvitamin D_3 in the parathyroid glands of patients with renal failure. N Engl J Med 1987; 316:1573–1577.
83. Merke J, Hugel U, Zlotkowski A, et al. Diminished parathyroid 1,25(OH)$_2$D$_3$ receptors in experimental uremia. Kidney Int 1987; 32:350–353.
84. Denda M, Finch J, Brown AJ, Nishii Y, Kubodera N, Slatopolsky E. 1,25-Dihydroxyvitamin D_3 and 22-oxacalcitriol prevent the decrease in vitamin D receptor content in the parathyroid glands of uremic rats. Kidney Int 1996; 50:34–39.
85. Fukagawa M, Kaname S, Igarashi T, Ogata E, Kurokawa K. Regulation of parathyroid hormone synthesis in chronic renal failure in rats. Kidney Int 1991; 39:874–881.
86. Naveh-Many T, Silver J. Regulation of parathyroid hormone gene expression by hypocalcemia, hypercalcemia, and vitamin D in the rat. J Clin Invest 1990; 86:1313–1319.
87. Naveh-Many T, Rahamimov R, Livni N, Silver J. Parathyroid cell proliferation in normal and chronic renal failure rats: the effects of calcium, phosphate and vitamin D. J Clin Invest 1995; 96:1786–1793.
88. Nygren P, Larsson R, Johansson H, Ljunghall S, Rastad J, Akerstrom G. 1,25(OH)2D3 inhibits hormone secretion and proliferation but not functional dedifferentiation of cultured bovine parathyroid cells. Calcif Tissue Int 1988; 43:213–218.
89. Szabo A, Merke J, Beier E, Mall G, Ritz E. 1,25(OH)$_2$ vitamin D_3 inhibits parathyroid cell proliferation in experimental uremia. Kidney Int 1989; 35:1049–1056.
90. Wernerson A, Svensson O, Reinholt FP. Parathyroid cell number and size in hypercalcemic rats: a stereologic study employing modern unbiased estimators. J Bone Miner Res 1989; 4:705–713.

91. Svensson O, Wernerson A, Reinholt FP. Effect of calcium depletion on the rat parathyroids. Bone Miner 1988; 3:259–269.
92. Kilav R, Silver J, Biber J, Murer H, Naveh-Many T. Coordinate regulation of the rat renal parathyroid hormone receptor mRNA and the Na-Pi cotransporter mRNA and protein. Am J Physiol 1995; 268:F1017–F1022.
93. Slatopolsky E, Bricker NS. The role of phosphorus restriction in the prevention of secondary hyperparathyroidism in chronic renal disease. Kidney Int 1973; 4:141–145.
94. Portale AA, Booth BE, Halloran BP, Morris RCJ. Effect of dietary phosphorus on circulating concentrations of 1,25-dihydroxyvitamin D and immunoreactive parathyroid hormone in children with moderate renal insufficiency. J Clin Invest 1984; 73:1580–1589.
95. Lucas PA, Brown RC, Woodhead JS, Coles GA. 1,25-Dihydroxycholecalciferol and parathyroid hormone in advanced chronic renal failure: effects of simultaneous protein and phosphorus restriction. Clin Nephrol 1986; 25:7–10.
96. Lafage MH, Combe C, Fournier A, Aparicio M. Ketodiet, physiological calcium intake and native vitamin D improve renal osteodystrophy. Kidney Int 1992; 42:1217–1225.
97. Combe C, Aparicio M. Phosphorus and protein restriction and parathyroid function in chronic renal failure. Kidney Int 1994; 46:1381–1386.
98. Tanaka Y, DeLuca HF. The control of vitamin D by inorganic phosphorus. Arch Biochem Biophys 1973; 154:566–570.
99. Condamine L, Vztovsnik F, Friedlander G, Menaa C, Garabedian M. Local action of phosphate depletion and insulin-like growth factor 1 on in vitro production of 1,25-dihydroxyvitamin D by cultured mammalian kidney cells. J Clin Invest 1994; 94:1673–1679.
100. Portale AA, Halloran BP, Curtis Morris J. Physiologic regulation of the serum concentration of 1,25-dihydroxyvitamin D by phosphorus in normal men. J Clin Invest 1989; 83:1494–1499.
101. Lopez-Hilker S, Dusso AS, Rapp NS, Martin KJ, Slatopolsky E. Phosphorus restriction reverses hyperparathyroidism in uremia independent of changes in calcium and calcitriol. Am J Physiol 1990; 259:F432–F437.
102. Yi H, Fukagawa M, Yamato H, Kumagai M, Watanabe T, Kurokawa K. Prevention of enhanced parathyroid hormone secretion, synthesis and hyperplasia by mild dietary phosphorus restriction in early chronic renal failure in rats: possible direct role of phosphorus. Nephron 1995; 70:242–248.
103. Kilav R, Silver J, Naveh-Many T. Parathyroid hormone gene expression in hypophosphatemic rats. J Clin Invest 1995; 96:327–333.
104. Almaden Y, Canalejo A, Hernandez A, et al. Direct effect of phosphorus on parathyroid hormone secretion from whole rat parathyroid glands in vitro. J Bone Miner Res 1996; 11:970–976.
105. Almaden Y, Hernandez A, Torregrosa V, Campistol J, Torres A, Rodriguez MS. High phosphorus directly stimulates PTH secretion by human parathyroid tissue. J Am Soc Nephrol 1995; 6:957 (abstract).
106. Slatopolsky E, Finch J, Denda M, et al. Phosphate restriction prevents parathyroid cell growth in uremic rats. High phosphate directly stimulates PTH secretion in vitro. J Clin Invest 1996; 97:2534–2540.
107. Nielsen PK, Feldt-Rasmusen U, Olgaard K. A direct effect of phosphate on PTH release from bovine parathyroid tissue slices but not from dispersed parathyroid cells. Nephrol Dial Transplant 1996; 11:1762–1768.
108. Nemeth EF, Scarpa A. Rapid mobilization of cellular Ca^{2+} in bovine parathyroid cells evoked by extracellular divalent cations. Evidence for a cell surface calcium receptor. J Biol Chem 1987; 262:5188–5196.
109. Brown EM. Extracellular $Ca2+$ sensing, regulation of parathyroid cell function, and role of $Ca2+$ and other ions as extracellular (first) messengers. Physiol Rev 1991; 71:371–411.
110. Naveh-Many T, Kilav R, Moallem E, Silver J. Post-transcriptional regulation of the PTH gene by calcium and phosphate in vivo. J Bone Miner Res 1994; 9:S338 (abstract).
111. Pitts TO, Piraino BH, Mitro R, et al. Hyperparathyroidism and 1,25-dihydroxyvitamin D deficiency in mild, moderate, and severe renal failure. J Clin Endocrinol Metab 1988; 67:876–881.
112. Portale AA, Booth BE, Tsai HC, Morris RCJ. Reduced plasma concentration of 1,25-dihydroxyvitamin D in children with moderate renal insufficiency. Kidney Int 1982; 21:627–632.
113. Hamdy NA, Kanis JA, Beneton MN, et al. Effect of alfacalcidol on natural course of renal bone disease in mild to moderate renal failure. Br Med J 1995; 310:358–363.

114. Andress DL, Norris KC, Coburn JW, Slatopolsky EA, Sherrard DJ. Intravenous calcitriol in the treatment of refractory osteitis fibrosa of chronic renal failure [see comments]. N Engl J Med 1989; 321:274–279.
115. Herrmann P, Ritz E, Schmidt-Gayk H, et al. Comparison of intermittent and continuous oral administration of calcitriol in dialysis patients: a randomized prospective trial. Nephron 1994; 67:48–53.
116. Indridason OS, Quarles LD. Oral versus intravenous calcitriol: is the route of administration really important? Curr Opin Nephrol Hypertens 1995; 4:307–312.
117. Goodman WG, Ramirez JA, Belin TR, et al. Development of adynamic bone disease in patients with secondary hyperparathyroidism after intermittent calcitriol therapy. Kidney Int 1994; 46:1160–1166.
118. Rodriguez M, Felsenfeld AJ, Williams C, Pederson JA, Llach F. The effect of long-term intravenous calcitriol administration on parathyroid function in hemodialysis patients. J Am Soc Nephrol 1991; 2:1014–1020.

15
Vitamin D Assays and Their Clinical Utility

Ronald L. Horst and Bruce W. Hollis

1. INTRODUCTION

More than 70 years ago, vitamin D was demonstrated to be an essential dietary constituent *(1)*. During the ensuing decades, its mode of action and physiologic function have received intense investigation. Research since 1968 has shown that vitamin D undergoes hydroxylations at the C25 *(2)* and C1*(3–5)*. This biotransformation to 1,25-dihydroxyvitamin D [$1,25(OH)_2D$] is necessary for the vitamin to become the most potent known stimulator of calcium and phosphorus absorption by the intestinal mucosa and calcium and phosphorus resorption from bone.

This simplistic picture outlined for vitamin D activation is complicated by the fact that vitamin D can be oxidatively metabolized to a variety of products. Most of these numerous metabolites have no identifiable biologic function, and many have been isolated from animals fed abnormally high amounts of vitamin D_3. Nevertheless, the evidence collected to date indicates that 25-hydroxylated vitamin D_3 metabolites are preferentially metabolized at the side chain. In particular, for vitamin D_3, carbon centers C23, C24, and C26 are readily susceptible to further oxidation. Figure 1 illustrates products of these oxidative pathways. As indicated, these oxidative pathways are shared by both 25-hydroxyvitamin D_3 [$25(OH)D_3$] and $1,25(OH)_2D_3$, and their importance is still a matter of controversy. For example, there is evidence that 24,25-dihydroxyvitamin D_3 [$24,25(OH)_2D_3$] may function to stimulate bone mineralization *(6,7)*, suppress parathyroid hormone secretion *(8)*, and maintain embryonic development *(9)*. For the most part, however, these side chain modifications are generally considered to be catabolic in nature.

Although these side chain oxidative pathways yield metabolites that are considered "nonfunctional," the presence of these compounds in the circulation can pose a problem in the analysis for $25(OH)D_3$ and $1,25(OH)_2D_3$ *(10)*. Further complicating the issue of understanding vitamin D activation, catabolism, and metabolite analysis is the presence of vitamin D_2. Vitamin D_2 has been shown to contribute significantly to the overall vitamin D status in humans and other mammals consuming supplemental vitamin D_2 *(11–13)*. Vitamin D_2 can also be metabolized in a similar fashion to produce several metabolites analogous to the vitamin D_3 endocrine system, including vitamin D_2's hormonally active form, 1,25-dihydroxyvitamin D_2 [$1,25(OH)_2D_2$] *(14)*. Simple inspection of the side chain, however, would imply that differences between metabolism of vitamins

From: *Vitamin D: Physiology, Molecular Biology, and Clinical Applications*
Edited by: M. F. Holick © Humana Press Inc., Totowa, NJ

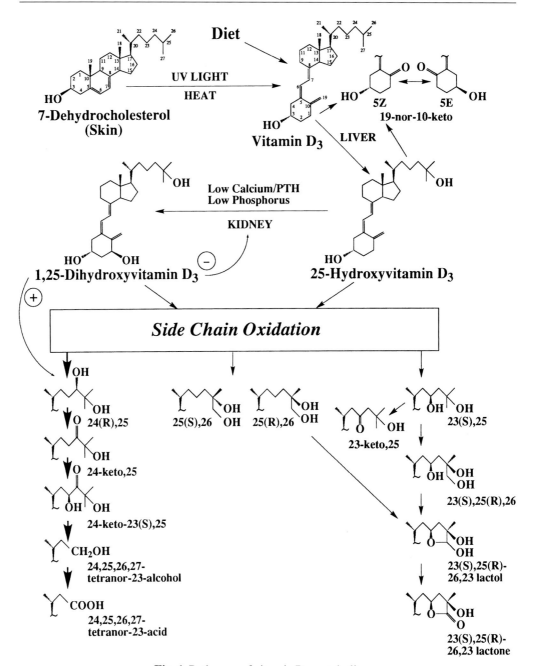

Fig. 1. Pathways of vitamin D_3 metabolism.

D_2 and D_3 may exist. The presence of unsaturation at carbon centers C22/C23, along with the additional methyl group at C24, would seem to preclude the existence of the same metabolic pathways for the two vitamins. Figure 2 outlines the known pathways of vitamin D_2 metabolism that have been shown to date.

For a period of approx 36 years (1922–1958), the only official method of determination of vitamin D in biologic fluids, tissues, and feedstuffs was the rat line test described by McCollum et al. (15). This test was based on the curative effects of the treatment, with

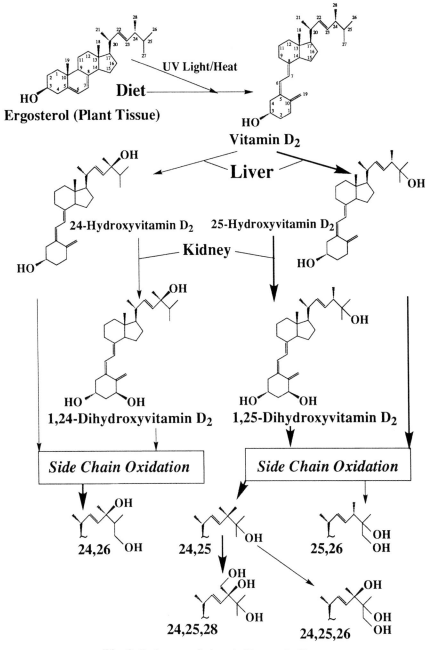

Fig. 2. Pathways of vitamin D_2 metabolism.

the supposition that after rickets was established in rats they were all cured of the lesion by the addition of vitamin D to the rachitogenic diet.

In 1958, a chemical method was adopted as the second official alternative to the rat line test. This method was utilized for pharmaceutical preparations containing large quantities (>250 μg) of vitamin D *(16)*. In this assay, the "purified" vitamin D preparation is reacted with antimony trichloride in ethylene dichloride. The resulting color is read spectrophotometrically and is linear over a wide range of vitamin D concentrations.

Table 1
Application of Assays for Plasma Concentrations of Vitamin D[a]

Extraction	Preliminary chromatography	Method of analysis	Plasma concentration (ng/mL)	Reference
Normal				
Chloroform/methanol	Silicic acid	CPBA	24–40	47
Methanol/dichloromethane	Lipidex-5000/HPLC	CPBA/	2.3 ± 1.1	34
Methanol/chloroform	Lipidex-5000	GLC/MS	5–11	29
Methanol/dichloromethane	Lipidex-5000	GLC/MS	3–17.7	31
Chloroform/methanol	TLC	GLC/MS	0–6.3	28
Methanol/chloroform	Preparative HPLC	HPLC/UV	2.2 ± 1.1	48
Methanol/hexane	Silica cartridges	HPLC/UV	9.1 ± 1.0	49
Methanol/dichloromethane	Lipidex-5000	HPLC/UV	3.2 ± 2.5	41
Sun exposure				
Methanol/dichloromethane	Lipidex 5000	HPLC/UV	27.1 ± 7.9	41
Methanol/dichloromethane	Lipidex 5000	CPBA	19.2–20.4	34
Vitamin D excess				
Ethyl acetate	C18-cartridge	HPLC/UV	2–283	125

[a]Results are values obtained from normal individuals as well as those receiving sun exposure and excess oral vitamin D. CPBA, competitive protein binding assay; GLC/MC, gas/liquid chromatography/mass spectrometry; HPLC, high-performance liquid chromatography; UV, ultraviolet; TLC, thin-layer chromatography.

Since 1958, knowledge of vitamin D physiology and metabolism has progressed rapidly. The introduction of high-pressure liquid chromatography (HPLC) and solid-phase extraction methods, as well as the discovery of specific binding proteins and antisera, was paramount in the quest for developing assays for vitamin D and vitamin D metabolites.

It is the purpose of this chapter to review the development of vitamin D and vitamin D metabolite assays and their applications in human and veterinary medicine. Each section provides some historical perspective on assays used in determining vitamin D and vitamin D metabolite activity, and also provides a detailed description of the authors' techniques of choice.

2. VITAMIN D ASSAY METHODOLOGY

The clinical application of an assay to measure vitamin D_2 or D_3 in biologic fluids is limited. Requirements for assessing nutritional vitamin D status are usually served well by plasma 25(OH)D analysis. The vitamin D assay, however, may have application in diseases of the liver or study of the effects of drugs that stimulate or inhibit liver hydroxylase activity. As a research tool, the vitamin D assay has been applied to various photobiology and metabolism experiments and to the quantitation of vitamin D in pharmaceutical preparations and foods.

Table 1 summarizes several of the different techniques and their applications in measuring vitamin D in normal plasma. It is readily apparent that a wide range of normal values exists for plasma vitamin D concentrations, depending on the assay used and the state of vitamin D nutrition. Generally, the HPLC methods give concentrations of 1–4 ng/mL in normal plasma, which are similar to those in gas/liquid chromatography/mass spectroscopy (GLC/MS) measurements.

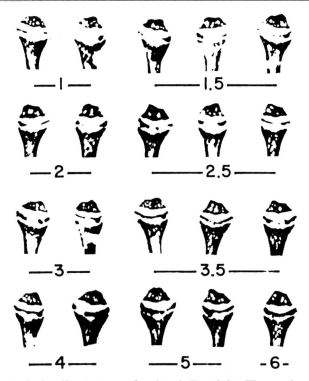

Fig. 3. Results from a typical rat line test assay for vitamin D activity. The numbers indicate the units of activity estimated from the degree of calcification (black area in the middle of the epiphysis plate) observed along the epipysial line. (Reproduced with permission from ref. *139*.)

2.1. Line Test

Shortly after the discovery of vitamin D, numerous biologic assays and critical evaluations of biologic assays were published. The most popular bioassay was the line test *(15)*, which is still used and is recognized as an official test for vitamin D activity. This test is very labor intensive and requires subjective judgments on the curative or preventive effects of the vitamin D preparation in question. The test is conducted in weanling rats fed a high-calcium, low-phosphorus vitamin D-deficient diet. Rickets generally develops after approx 3 wk. Animals are then subjected to either graded doses of vitamin D_3 or unknown test samples for 7 d. Following the test period, the animals are killed and the radii and ulna dissected. The bones are cleaned of muscle and connective tissue, sliced, and placed in a solution of silver nitrate. The assay is based on the deposition of silver in areas of the bone where calcium has been deposited. These areas turn black. The intensity and completeness of the line formed in the epiphyseal region are related to the amount of vitamin D in the samples (Fig. 3). Activity in unknown samples is compared with vitamin D_3 standards to determine concentration. This assay takes a lot of training and experience, but it can detect as little as 25 ng of vitamin D activity in samples. Because the test measures total vitamin D activity, it is of limited use in evaluating the concentration of individual metabolites present in a mixture of other metabolites with agonistic activity.

2.2. Chick Bioassays

The Association of Official Agricultural Chemists (AOAC) chick assay, as described by the AOAC, is used principally to assay for vitamin D_3 because it is 10–100 times more active in chicks than is vitamin D_2. The AOAC assay is conducted by placing newly hatched chicks on a vitamin D-deficient diet containing added levels of vitamin D_3 or the test compound. After 3 wk, the percent bone ash of dried, lipid-free tibias is determined. A rachitic bird typically has 25–27% bone ash, whereas a bird supplemented with 125 ng (50 IU) has 40–45% ash.

2.3. Intestinal Calcium Absorption

One of the biologic hallmarks of vitamin D presence is its capacity to stimulate intestinal calcium absorption. Consequently, one of the most useful tools in helping to understand the regulation of calcium absorption was the in vitro everted-gut-sac technique *(17)*. Modifications of the everted-gut-sac assay were found to be highly sensitive to vitamin D presence and have been frequently used for testing compounds with suspected vitamin D activity. A single iv dose of 12.5 ng (0.5 IU) of vitamin D to a vitamin D-deficient rat is capable of stimulating calcium transport. This assay has a working range of up to 2500 ng (100 IU) and requires that the test dose be administered ip or iv to vitamin-deficient rats at least 24 h prior to assessment of the calcium transport capacity of the duodenum.

Published in vitro variants have successfully utilized chicks in assessing calcium transport stimulation *(18,19)*. Like the everted-gut-sac assay, these tests are very expensive because the animals are generally fed purified diets for 2–3 wk. Also, like the everted-gut-sac assay, these procedures have been invaluable in assessing the ability of vitamin D metabolites and analogs to stimulate intestinal calcium transport. However, because the chicks discriminate against vitamin D_2, these assays are mainly used for testing the activity of vitamin D_3 preparations.

2.4. Stimulation of Intestinal Calcium Binding Proteins

During the search for mechanisms that augment intestinal calcium absorption, a vitamin D-dependent soluble protein isolated from the chick intestine was demonstrated *(20)*. This protein bound calcium with a high affinity when tested in vitro. Significantly, the calcium binding protein was not present in vitamin-deficient chick intestine, and its concentration peaked approx 72 h after vitamin D administration. The discovery of this protein has stimulated numerous investigations as to its possible role in mediating calcium transport by the intestine.

Because of the high correlation between intestinal calcium binding protein concentration and efficiency of intestinal calcium absorption, the calcium binding protein response to vitamin D and vitamin D metabolites was proposed as a bioassay for vitamin D activity *(21)*. In this assay, it was noted that 48 h after a single im injection of vitamin D_3 to rachitic chicks, a linear response in duodenal calcium binding protein content was obtained. The response was usable from 125 to 3100 ng (5–125 IU) and was related to the log dose of the test compound.

A recent in vitro intestinal organ culture system has been analyzed for sensitivity to vitamin D and vitamin D metabolites *(22,23)*. This system provides a uniquely relevant bioassay for the study of structure–activity relationships of vitamin D steroids, as the measurement of calcium binding protein biosynthesis is a primary physiologic response in a principal target organ. The assay offers the advantage of exquisite sensitivity [as little as 10 ng of vitamin D_3 and 10 pg of $1,25(OH)_2D_3$ induce detectable calcium binding protein] and a broad range of response [vitamin D_3 can be detected over a range of 10–10,000,000 ng and $1,25(OH)_2D_3$ can be detected over a range of 10–10,000,000 pg].

2.5. Colorimetric Assay

Prior to 1967, vitamin D was thought to act directly, independent of metabolism. Therefore, considerable effort was given to direct physical/chemical measurements of vitamin D in solutions and lipid extracts to avoid the tedious bioassays. One of the first attempts at a direct colorimetric method for the determination of vitamin D was described in 1936 by Brockmann and Chen *(24)*.

This assay involved the use of a cold, saturated solution of antimony trichloride in dry, alcohol-free chloroform. Both vitamins D_2 and D_3 gave an orange-yellow-colored solution with an absorption maximum at 500 nm. Tachysterol, cholesterol, sitosterol, ergosterol, 7-dehydrocholesterol, lumisterol, suprasterol I, and suprasterol II all gave much weaker colors and did not interfere unless present in concentrations over 30 times that of the vitamin D. Vitamin A also reacted with the antimony trichloride and gave an absorption maximum at 620 nm. Vitamin A interfered when present in concentrations six times that of the vitamin D. This reagent was therefore assessed as a dependable means for the colorimetric determination of vitamin D when other sterols and vitamin A were present in low concentrations.

2.6. Gas/Liquid Chromatography

The determination of vitamin D by GLC has been the subject of several investigations *(25–32)*. The major problem is that at GLC operating temperatures the vitamin D undergoes thermal cyclization into the pyro- and isopyrocalciferols and yields two peaks on the chromatogram. Several successful attempts have been made at derivatizing vitamin D to yield a single GLC peak. Examples are demonstrated in conversions to isovitamin D and isotachysterols, including their derivatives, such as heptafluorobutyrates. Although these techniques may offer value in the determination of vitamin D, they are hampered by factors such as reaction conditions, yields, and the heat stabilities of the compounds. In addition, the possibility of destroying the vitamin D compounds precludes the use of GLC as a preparative step in the ultimate purification of vitamin D metabolites and analogs.

2.7. Competitive Protein Binding Assays

The discovery of proteins capable of high-affinity specific binding (vitamin D binding proteins [DBPs]) of vitamin D and vitamin D metabolites has resulted in assessment of the use of these proteins in binding assays specific for vitamin D. Competitive protein binding assays (CPBAs) for vitamin D using ethanol extracts of plasma have been reported that did not include any prepurification and so resulted in erroneously high values reported for normal humans (Table 1). We now know that several vitamin D metabolites are cospecific or have enhanced specificity for the plasma DBP, and therefore extensive purification of the vitamin D is required before quantitation can be achieved using DBP assays. Two laboratories have recently described purification procedures for ultimate quantitation of vitamin D by CPBA *(33,34)*.

2.8. High-Pressure Liquid Chromatography

The technique of high-pressure liquid chromatography (HPLC) was developed in the late 1960s. Since that time, the development of high-performance columns and a better understanding of column technology have been motivating factors in the widespread application of HPLC. Shortly after its discovery, HPLC was applied to the separation of

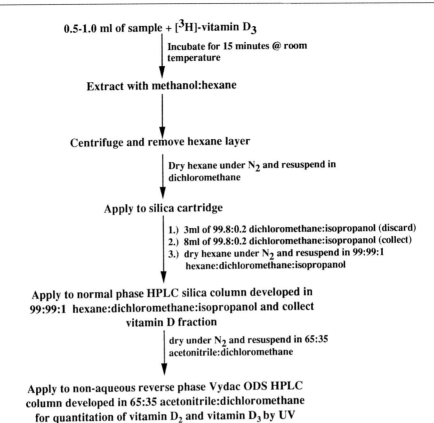

Fig. 4. Flow diagram for high-pressure liquid chromatographic (HPLC)/UV analysis of vitamins D_2 and D_3.

vitamin D metabolite mixtures, the separation of analogs during synthesis, and the isolation of vitamin D and vitamin D metabolites from biologic extracts. Recent improvements in detector stability at high sensitivity and automatic injection capabilities have led to frequent utilization of HPLC methodology in the analysis of vitamin D in biologic fluids and tissues *(10,34–45)*.

Vitamin D is difficult to quantitate in biologic fluids. It is the only antirachitic sterol that cannot be extracted from aqueous media utilizing solid-phase extraction techniques *(46)*. Therefore, unlike its more polar metabolites, vitamin D must be extracted from serum or plasma using liquid–liquid organic extraction techniques. Many of the initial studies used Bligh and Dyer-type total lipid extraction to extract vitamin D from serum samples *(37,47,48)*. However, these types of extractions remove an extraordinary amount of lipid from the plasma sample. A more selective organic extraction procedure incorporating methanol-hexane was therefore developed *(49)*. This extraction method, coupled with open cartridge silica chromatography, allowed for direct quantitation of vitamins D_2 and D_3 following nonaqueous reverse-phase HPLC. This procedure provides an accurate, convenient method to measure circulating vitamin D and is outlined in Fig. 4 and described below in detail as the authors' method of choice for quantitating vitamins D_2 and D_3 in biologic fluids.

2.9. Methodology of Choice

1. Sample extraction. From 0.5–1 mL of serum or plasma is placed into a 13 × 100-mm borosilicate glass culture tube containing 1000 cpm of [^{3}H]vitamin D_3 in 25 µL of ethanol to monitor recovery of the endogenous compound through the extraction and chromatographic procedures. Following a 15-min incubation with the tracer, two plasma volumes of HPLC-grade methanol are added to each sample. The sample is then vortex-mixed for 1 min, followed by the addition of 3 plasma volumes of HPLC-grade hexane. Each tube is capped and vortex-mixed for an additional min followed by centrifugation at 1,000g for 10 min. The hexane layer is removed and placed into another 13 × 100-mm culture tube. The aqueous layer is reextracted in the same fashion. The hexane layers are combined and dried in a heated waterbath at 55°C, under N_2. The lipid residue is then resuspended in 1 mL of HPLC-grade dichloromethane and capped.
2. Silica cartridge chromatography. Silica Bond-Elut cartridges (500 mg) and Vac-Elut cartridge racks were obtained from Varian Instruments (Harbor City, CA). The silica cartridges are washed in order with 5 mL of HPLC-grade methanol, 5 mL of HPLC-grade isopropanol, and 10 mL of HPLC-grade dichloromethane. The sample, in 1 mL of dichloromethane, is then applied to the cartridge and eluted through the cartridge under vacuum into waste. This initial step is followed by 3 mL of 0.2% isopropanol in dichloromethane (discard) and 8 mL of the same solvent applied and collected as vitamin D. This 8-mL fraction contains both vitamins D_2 and D_3 and is dried in a heated waterbath at 55°C under N_2.
3. Preparative normal-phase HPLC. This normal-phase HPLC step is performed using a 0.4 × 25-cm Zorbax-Sil column. However, any equivalent column could be utilized. The mobile phase is comprised of hexane/dichloromethane/isopropanol (99:99:1 v/v) at a flow rate of 2 mL/min. The sample residue from the silica cartridge is dissolved in 150 µL of the mobile phase and injected onto the HPLC column that has been previously calibrated with 10 ng of standard vitamin D_3. The vitamin D fraction is collected in a 12 × 75-mm glass culture tube and dried under N_2 at 55°C.
4. Quantitative reverse-phase HPLC. The final quantitative step was performed using non-aqueous reverse-phase HPLC. The column used was a 0.4 × 25-cm Vydac TP-201-54, 5 µm, wide-pore, non-endcapped ODS silica material. The mobile phase is comprised of acetonitrile/dichloromethane (65:35, v/v) at a flow rate of 1.2 mL/min. This system provides clear resolution of vitamins D_2 and D_3 (Fig. 5A). The system is calibrated with varying amounts of vitamins D_2 and D_3 (1–100 ng). The sample residue from normal-phase HPLC is dissolved in 15 µL dichloromethane followed by the addition of 135 µL of the acetonitrile and injected into the HPLC. Elution and final quantitation of vitamins D_2 and D_3 is by direct monitoring at 265 nm. The vitamin D_3 portion is collected, dried under N_2, and subjected to liquid scintillation counting in order to determine the final endogenous recovery of vitamin D_3 from the sample. Calculations are then performed and the results reported in ng/mL vitamin D_2 and/or D_3.

3. 25(OH)D ASSAY METHODOLOGY

The most abundant metabolite in plasma in the normal state or during vitamin D excess is 25(OH)D. C25 hydroxylation of vitamin D occurs in the liver microsomes and is the first step in the activation of vitamin D. Normally, 25(OH)D_3 is the major circulating form of 25(OH)D in plasma; however, 25(OH)D_2 is the major form in patients undergoing vitamin D_2 therapy for such diseases as pseudohypoparathyroidism or hypophosphatemia rickets. The 25(OH)D circulates at concentrations of 15–30 ng/mL. Table 2 summarizes the normal plasma concentration of 25(OH)D in humans, as reported by

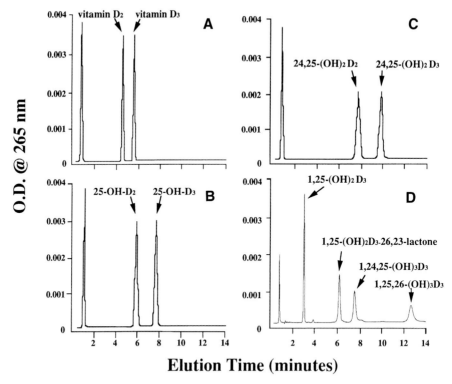

Fig. 5. High-pressure liquid chromatographic (HPLC) profiles of vitamins D_2 and D_3 (**A**), $25(OH)D_2$ and $25(OH)D_3$ (**B**), $24,25(OH)_2D_2$ and $24,25(OH)_2D_3$ (**C**), and the trihydroxyvitamin D_3s and $1,25(OH)_2D_3$-26,23-lactone (**D**). Column conditions are as described in the text.

several laboratories using different analytical procedures. The relatively high circulating concentrations of plasma 25(OH)D make this metabolite a useful clinical indicator of vitamin D status in states of nutritional excess or deficiency, as well as during disease or treatment states such as enterohepatic disorders, anticonvulsant drug therapy, gastrointestinal diseases or surgery, or liver and renal diseases. An excellent review of the clinical usefulness of the 25(OH)D assay has been published *(50)*.

3.1. Competitive Protein Binding Assays

Like vitamin D, the tissue and plasma proteins can be used for the analysis of 25(OH)D in biologic fluids *(50)*. These techniques offer the advantages of low sample volume and minimal sample processing to achieve the final results. The binding protein from either mammalian kidney or plasma is cospecific for $25(OH)D_2$ and $25(OH)D_3$, and therefore a mixture can be measured in a single determination. Avian sera or tissue extracts, however, preferentially bind the vitamin D_3 metabolites *(51)* and should therefore be avoided in the development of a CPBA.

The amount of sample purification required for ultimate quantitation of 25(OH)D has received considerable attention. One of the first published CPBAs for 25(OH)D advocated the use of a nonchromatographic method *(52)*. Ethanol-precipitated plasma is used directly without further preparation. In samples from normal subjects, the results of this assay agree reasonably well with assays using preparative chromatography. A major shortcoming of this assay, however, is that metabolites such as $24,25(OH)_2D_3$, $25,26(OH)_2D_3$

Table 2
Application of Assays for Plasma Concentrations of 25-Hydroxyvitamin D[a]

Extraction	Chromatography	Analysis	Concentration (ng/mL)	Reference
Normal				
Ethanol	None	CPBA	20–100	52
Methanol/chloroform	Silicic acid	CPBA	15.2–39.5	126
Ether	HPLC	CPBA	43.2 ± 9.5	56
Methanol/chloroform	Sephadex LH-20	GC/MS	45.0	127
Methanol/chloroform	None	CPBA	21.6 ± 10.1	128
Methanol/dichloromethane	HPLC	HPLC	34.0 ± 9.9	65
Ethanol	None	CPBA	5.0–72.3	129
Acetonitrile	C18/Silica Sep Pak	GLC/MS	2.3–32.6	71
Acetonitrile	None	RIA	9.9–41.5	73
Methanol/chloroform	Sephadex LH-20	HPLC/UV	31.9 ± 1.7	55
Ether	Silicic acid	CPBA	27.3 ± 11.8	130
Sun exposure				
Methanol/dichloromethane	Sephadex LH-20	HPLC/UV	55.5 ± 3.8	41
Vitamin D excess				
Ethanol	None	CPBA	119–1100	129
Ethanol	None	CPBA	500–3200	52
Acetonitrile	C18/Silica Sep Pak	GLC/MS	89–656	71
Methanol/chloroform	Silicic acid	CPBA	564 ± 11.2	126
Ethyl acetate	C18 cartridge	HPLC/UV	293 ± 174	125

[a] Results are values obtained from normal individuals as well as those receiving sun exposure and excess oral vital D. For abbreviations, see footnote to Table 1.

(which are cospecific for DBP) *(53)*, and 25(OH)D$_3$-26,23-lactone (which has enhanced affinity for the DBP) *(54)* could offer significant interference in the direct 25(OH)D assays, especially in situations of vitamin D excess.

3.2. HPLC and GLC Methods

CPBA for 25(OH)D dominated the literature until 1977, when the first valid direct quantitative HPLC assay was introduced *(55)*. Following this initial report, a number of assays have been reported dealing with the HPLC analysis of plasma 25(OH)D *(10,41, 42,44,56–70)*. The metabolite 25(OH)D circulates in the ng/mL range and thus could be directly quantitated by ultraviolet (UV) detection following its separation by normal-phase HPLC. Also, HPLC detection provided the advantage of being able to quantitate 25(OH)D$_2$ and 25(OH)D$_3$ individually (Fig. 5B). The disadvantages of HPLC quantitation methods are their requirement for expensive equipment, large sample size, and cumbersome nature, along with the technical expertise required to perform this type of analysis. However, HPLC analysis for 25(OH)D is still frequently used in research environments, including our own, and has provided a great deal of significant information.

Recently, a novel approach to the analysis of 25(OH)D has been suggested. The method involves low-pressure liquid chromatography (LPLC) and HPLC preparative chromatography followed by isomerization of the 25(OH)D isolate to its isotachysterol derivative *(42)*. Following derivatization, the 25-isotachysterol is rechromatographed by HPLC and the UV monitored at 290 nm. Reading the UV at the higher absorbance setting gives the advantage of less lipid interference, whereas the isomerization of the compounds to their isotachysterol derivatives results in a doubling of the peak heights.

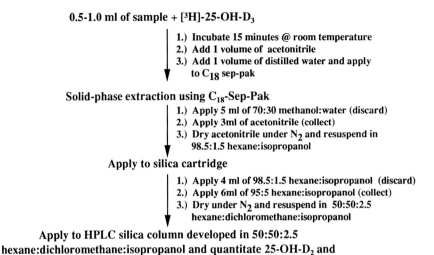

Fig. 6. Flow diagram for high-pressure liquid chromatographic (HPLC)/UV analysis of 25(OH)D_2 and 25(OH)D_3.

In contrast to the large number of HPLC methods for 25(OH)D analysis, there have been only few reports concerning analysis of this metabolite by GLC/MS *(31,70,71)*. This is probably a result of the need for extensive equipment and technical expertise to perform the assay. When applied to the analysis of 25(OH)D in human plasma, this procedure gives results similar to those obtained by HPLC or CPBA. The methods reported, however, were not extended to the simultaneous analysis of 25(OH)D_2 and 25(OH)D_3.

3.3. Radioimmunoassay

As the clinical demand for circulating 25(OH)D analysis increased, it was clear that simpler, rapid, yet valid assay procedures would be required. All valid assays required liquid–liquid organic extraction, some sort of chromatographic prepurification, and evaporation of the organic solvents. The first valid radioimmunoassay (RIA) for assessing circulating 25(OH)D was introduced *(72)*. This RIA eliminated the need for sample prepurification prior to assay and had no requirement for organic solvent evaporation. However, it was still based on the use of [^{3}H]25(OH)D_3 as a tracer. This final shortcoming was solved in 1993 when a [^{125}I] tracer was developed and incorporated into the RIA for 25(OH)D *(73)*. This assay has become the method of choice for assessing 25(OH)D status and has recently become the first test for vitamin D approved for clinical diagnosis by the Food and Drug Administration (FDA). This RIA, along with an HPLC-based procedure routinely used in our laboratory for research purposes, is outlined in Figs. 6 and 7 and described in detail below.

3.4. Methodologies of Choice

3.4.1. HPLC METHODOLOGY

1. Sample extraction. Serum or plasma (0.5 mL) is placed into a 12 × 75-mm borosilicate glass culture tube containing 1000 cpm of [^{3}H]25(OH)D_3 in 25 µL of ethanol to monitor recovery of endogenous compound through the extraction and chromatographic proce-

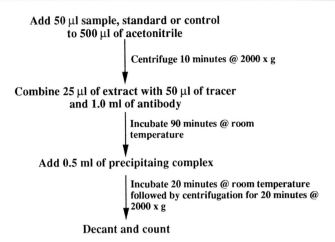

Fig. 7. Flow diagram for the quantitation of 25(OH)D by radioimmunoassay (RIA).

dures. Following a 15-min incubation with the tracer, 1 plasma volume of HPLC-grade acetonitrile is added to each sample. The sample is then vortex-mixed for 1 min, followed by centrifugation at 1000g for 10 min. The supernatant is removed into another 12 × 75-mm culture tube, and 1 plasma volume of distilled water is added.

2. Solid-phase extraction chromatography. C18 silica Sep-Pak cartridges (500 mg) and Sep-Pak racks were obtained from Waters Associates (Milford, MA). The C18 cartridges are washed in order with 5 mL of HPLC-grade isopropanol and 5 mL of HPLC-grade methanol. The sample is then applied to the cartridge and eluted through the cartridge under vacuum into waste. This initial step is followed by 5 mL of 30% water in methanol (discard); then 3 mL of acetonitrile is added and collected as 25(OH)D. This acetonitrile fraction is dried in a heated waterbath at 55°C, under N_2. The lipid residue is then resuspended in 1 mL of 1.5% isopropanol in hexane and capped. The C18 cartridges can be cleaned and regenerated by washing with 2 mL of methanol and reused many times.

3. Silica cartridge chromatography. Silica Bond-Elut cartridges (500 mg) and Vac-Elut cartridge racks were obtained from Varian Instruments. The silica cartridges are washed in order with 5 mL of HPLC-grade methanol, 5 mL of HPLC-grade isopropanol, and 5 mL of HPLC-grade hexane. The sample, in 1 mL of 1.5% isopropanol in hexane, is then applied to the cartridge and eluted through the cartridge under vacuum into waste. This initial elution is followed by 4 mL of 1.5% isopropanol in hexane (discard) and 6 mL of 5% isopropanol in hexane. This 6-mL fraction contains 25(OH)D_2 and 25(OH)D_3 and is subsequently dried in a heated waterbath at 55°C, under N_2.

4. Quantitative normal-phase HPLC. The final quantitative step is performed with normal-phase HPLC, using a 0.4 × 25-cm Zorbax-Sil column packed with 5-μm spherical silica. The mobile phase is comprised of hexane/dichloromethane/isopropanol (50:50:2.5, v/v) at a flow rate of 2 mL/min. The sample residue from the silica cartridge is dissolved in 150 μL of mobile phase and injected onto the HPLC column previously calibrated with varying amounts of 25(OH)D_2 and 25(OH)D_3 (1–100 ng). This HPLC system provides clear resolution of 25(OH)D_2 and 25(OH)D_3 (Fig. 5B). Elution and final quantitation of 25(OH)D_2 and 25(OH)D_3 is by direct UV monitoring at 265 nm. The 25(OH)D_3 portion is collected, dried under N_2, and subjected to liquid scintillation counting to determine the final endogenous recovery of 25(OH)D_2 and 25-OHD_3 from the sample. Calculations are then performed and the results reported in ng/mL 25(OH)D_2 and/or 25(OH)D_3.

3.4.2. RADIOIMMUNOASSAY

1. Preparation of assay calibrators. One of the goals of this RIA procedure was to eliminate the need for individual sample recovery. Another goal was to obtain FDA approval for clinical use of this procedure in the United States. Both of these goals required placing 25(OH)D_3 in a human serum-based set of assay calibrators. To prepare these calibrators, human serum was "stripped" free of vitamin D metabolites by treatment with activated charcoal. Absence of endogenous 25(OH)D in the stripped sera was confirmed by its absence using HPLC as described above. Subsequently, crystalline 25(OH)D_3 dissolved in absolute ethanol was added to the stripped sera to yield calibrators at concentrations of 0, 5, 12, 40, and 100 ng/mL.
2. Sample and calibrator extraction. 25(OH)D is extracted from calibrators and samples as follows. Acetonitrile (0.5 mL) is placed into a 12 × 75-mm borosilicate glass tube after which 50 µL of sample or calibrator is dropped through the acetonitrile. After vortex-mixing, the tubes are centrifuged (1,000g, 4°C, 5 min) and 25 µL of supernatant is transferred to 12 × 75-mm borosilicate glass tubes and placed on ice.
3. Radioimmunoassay. The assay tubes are 12 × 75-mm borosilicate glass tubes containing 25 µL of acetonitrile-extracted calibrators or samples. To each tube add [^{125}I]25(OH)D derivative (50,000 cpm in 50 µL of 1:1 ethanol, 0.01 M phosphate buffer, pH 7.4) that was synthesized as previously described *(73)*. Then add to each tube 1.0 mL of primary antibody diluted 1:15,000-fold in sodium phosphate buffer (50 mM, pH 7.4 containing 0.1% swine-skin gelatin). Nonspecific binding is estimated using the above buffer minus the antibody. Vortex-mix the contents of the tubes and incubate them for 90 min at 20–25°C. Then add 0.5 mL of a second-antibody precipitating complex to each tube, vortex-mix, incubate at 20–25°C for 20 min, and centrifuge (20°C, 2000g, 20 min). Discard the supernate and count the tubes in a gamma well counting system. 25(OH)D values are calculated directly from the standard curve by the counting system using a smooth-spline or log-linear method of calculation (Fig. 7).

3.4.2.1. Comments. This [^{125}I] tracer-based RIA is similar to an RIA we introduced several years ago that used [^{3}H]25(OH)D_3 as a tracer *(72)*. In both cases, antisera were raised against the synthetic vitamin D analog 23,24,25,26,27-pentanor-C22-carboxylic acid. The synthesis of this analog and its [^{125}I]-labeled counterpart has been described in detail *(72,73)*. Coupling this compound to bovine serum albumin allowed us to generate antibodies that crossreacted equally with most vitamin D_2 and D_3 metabolites. Because our analog retained the intact structure of vitamin D only up to C22, the structural differences between vitamins D_2 and D_3 were not involved in the antibody recognition, and antibodies directed against this analog could not discriminate with respect to side-chain metabolism of vitamin D. The antibody, however, was specific for the open B-ring *cis*-triene structure containing a 3β-hydroxyl group that is inherent in all vitamin D compounds.

Many vitamin D metabolites other than 25(OH)D are present in the circulation. However, they contribute only a small percentage (6–7%) to the overall assessment of nutritional vitamin D status compared with 25(OH)D *(74)*. This fact is supported by comparison of the present 25(OH)D RIA with the UV-quantitative HPLC assay for 25(OH)D described earlier in this chapter on a variety of human serum samples. Furthermore, the present [^{125}I]-based RIA was shown to identify vitamin D deficiency in biliary atresia patients, as well as vitamin D toxicity in hypoparathyroid patients who were receiving massive vitamin D therapy for the maintenance of plasma calcium *(73)*.

Table 3
Application of Assays for Plasma Concentrations of 24,25-Dihydroxyvitamin D[a]

Extraction	Chromatography	Analysis	Concentration (ng/mL)	Reference
Normal				
Acetonitrile	C18/HPLC	CPBA	2.9 ± 1.9	81
Methanol/dichloromethane	Sephadex LH-20/HPLC	CPBA	1.9 ± 0.83	83
Methanol	C18/NH$_2$	CPBA	2.6 ± 0.3	64
Methanol/dichloromethane	Sephadex LH-20	CPBA	8.24 ± 0.34	88
Methanol/ethyl acetate	Sephadex LH-20/HPLC	CPBA	2.3 ± 1.4	89
Dichloromethane	Sephadex LH-20/HPLC	CPBA	2.5 ± 0.75	131
Methanol/dichloromethane	Sephadex LH-20	CPBA	3.7 ± 0.2	77
Methanol/dichloromethane	Sephadex LH-20/HPLC	CPBA	3.5 ± 0.2	41
Methanol/dichloromethane	Sephadex LH-20/HPLC	CPBA	2.4 ± 1.1	90
Methanol/dichloromethane	Sephadex LH-20/HPLC	RIA	0.1–4.0	132
Acetonitrile	C18-OH/Silica	CPBA	3.1 ± 0.7	122
Ether-methanol/dichloromethane	Sephadex LH-20/HPLC x2	CPBA	2.8 ± 0.7	78
Chronic renal failure				
Methanol/chloroform	Sephadex LH-20/HPLC	CPBA	0.5 ± 0.16	133
Anephric				
Methanol/dichloromethane	Sephadex LH-20	CPBA	3.0 ± 0.7	77
Methanol/dichloromethane	Sephadex LH-20/HPLC	CPBA	1.9 ± 1.3	41
Ether-methanol/dichloromethane	Sephadex LH-20/HPLC x2	CPBA	<0.2	78

[a] Results are values obtained from normal individuals, patients with chronic renal failure, and anephric patients. RIA, radioimmunoassay. For other abbreviations, see Table 1 footnote.

4. 24,25-DIHYDROXYVITAMIN D ASSAY METHODOLOGY

The compound 24,25-dihydroxyvitamin D_3 [24,25(OH)$_2$D$_3$] was discovered in 1970, but its structure was erroneously reported as 21,25-dihydroxyvitamin D_3 *(75)*. The actual structure was later demonstrated to be 24,25(OH)$_2$D$_3$ *(76)*. 24,25(OH)$_2$D circulates in plasma at concentrations of 1–3 ng/mL (Table 3) and in normal subjects seems to be dependent on the concentration of circulating 25(OH)D. There is considerable controversy as to whether 24,25(OH)$_2$D has a physiologic role; evidence for and against such a role has been reported *(6–9)*. Therefore, effort has been given to the quantitation of this metabolite in biologic fluids.

Like 25(OH)D, 24,25(OH)$_2$D can be quantitated using diluted plasma as a source of the DBP. Table 3 summarizes some of the results for plasma 24,25(OH)$_2$D estimations currently published. The reported presence of 24,25(OH)$_2$D in anephrics consuming normal vitamin D by some *(77)* and its lack of detection in similar sera by others *(78)* seems to be totally a result of differences in preparation of samples for analysis. Generally, the purer the preparations, the lower the concentrations of 24,25(OH)$_2$D. Indeed, Horst et al. *(78)* have shown that anephrics consuming normal dietary levels of vitamin D produce little, if any, 24,25(OH)$_2$D.

Several laboratories have published 24,25(OH)$_2$D assays with various modifications *(53,56,64,71,79–90)*. These assays usually incorporated the use of HPLC to resolve unwanted metabolites and related compounds from 24,25(OH)$_2$D. Generally, the analyses involved a Sephadex LH-20 preparative step followed by final purification using HPLC silica gel columns developed in hexane/isopropanol. Using this approach, the plasma 24,25(OH)$_2$D of anephrics ranged from 0.5–1.0 ng/mL, whereas normal values

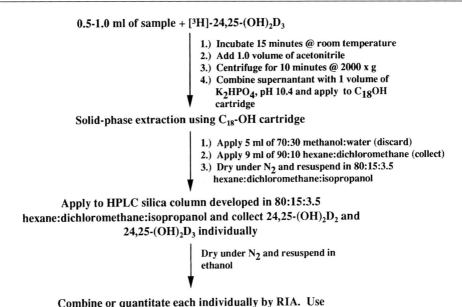

Fig. 8. Flow diagram for high-pressure liquid chromatographic (HPLC)/radioimmunoassay (RIA) for quantitation of 24,25(OH)$_2$D.

ranged from 2.0–3.0 ng/mL (Table 3). Later, Horst et al. *(78)* demonstrated that 25-hydroxyvitamin D$_3$-26,23-lactone [25(OH)D$_3$-26,23-lactone] and 25,26-dihydroxyvitamin D$_2$ comigrated with 24,25(OH)$_2$D on HPLC silica gel columns developed in hexane/isopropanol. They suggested that the 24,25(OH)$_2$D be further purified by HPLC on a silica gel column developed in dichloromethane/isopropanol. Using this solvent system, all the aforementioned metabolites could be resolved and quantitated individually by CPBAs. This additional purification step resulted in the demonstration that anephric humans consuming normal quantities of vitamin D have nondetectable 24,25(OH)$_2$D.

The lack of vitamin D$_2$ metabolite standards has resulted in some erroneous assumptions in previously described assays. The compound 24,25(OH)$_2$D$_2$ is now known to be partially resolved from 24,25(OH)$_2$D$_3$ on Sephadex LH-20 *(10)*. Also, this compound is two to three times less competitive than 24,25(OH)$_2$D$_3$ in the CPBA *(78)*. When 24,25(OH)$_2$D$_2$ is the major circulating form, underestimation of its concentration would certainly occur if proper precautions were not taken to avoid discriminating against its collection from Sephadex LH-20 or HPLC columns and, if correction is not made, to compensate for its inferior competition in the CPBAs. We therefore have chosen to use the RIA for final quantitation of 24,25(OH)$_2$D because the antibody used is cospecific for the vitamin D$_2$ and D$_3$ forms *(73)*, and thus only one compound, 24,25(OH)$_2$D$_3$, is required to construct the standard curve. A flow diagram of the preferred methodology appears in Fig. 8 and is described below in detail.

4.1. Method of Choice

1. Sample extraction. Serum or plasma (0.5–1.0 mL) is placed into a 12 × 75-mm borosilicate glass culture tube containing 1000 cpm of [^{3}H]24,25(OH)$_2$D$_3$ in 25 µL of ethanol to

monitor recovery of endogenous compound through the extraction and chromatographic procedures. Following a 15-min incubation with the tracer, 1 plasma volume of HPLC-grade acetonitrile is added to each sample. The sample is then vortex-mixed for 1 min followed by centrifugation at 1000g for 10 min. The supernatant is removed into another 12 × 75-mm culture tube, and 1 plasma volume of distilled water is added.

2. Solid-phase extraction chromatography. C18-OH silica bond elute cartridges (500 mg) and Vac-Elut cartridge racks were obtained from Varian Instruments. The C18-OH cartridges were modified by removing the upper polypropylene frit and replacing it with a stainless steel frit also purchased from Varian. This modification resulted in significantly better recoveries. The C18-OH cartridges are washed in order with 5 mL of HPLC-grade dichloromethane, 5 mL of HPLC-grade isopropanol, and 5 mL of HPLC-grade methanol. The sample is then applied to the cartridge and eluted under vacuum into waste. This initial step is followed by 5 mL of 30% water in methanol, which is discarded. The 24,25(OH)$_2$D is then eluted [along with 25(OH)D] by adding 5 mL of 10% dichloromethane in hexane. This final fraction is dried in a heated waterbath at 55°C, under N$_2$. The lipid residue is resuspended in 150 µL HPLC mobile phase (described below).

3. Preparative normal-phase HPLC. This normal-phase HPLC step is performed using a 0.4 × 25-cm Zorbax-Sil column packed with 5 µm silica. The mobile phase is comprised of hexane/dichloromethane/isopropanol (80:15:3.5 v/v) at a flow rate of 2 mL/min. The sample residue from the C18-silica cartridge is injected onto the HPLC column that had been previously calibrated with 10 ng of 24,25(OH)$_2$D$_2$ and 24,25(OH)$_2$D$_3$. The fractions containing 24,25(OH)$_2$D$_2$ and 24,25(OH)$_2$D$_3$ (Fig. 5C) can be collected individually or together in 12 × 75-mm glass tubes and dried under N$_2$ at 55°C. The residue is then redissolved in 500 µL absolute ethanol and capped. Two 25-µL aliquots of 24,25(OH)$_2$D is used for quantitation by RIA (*see below*). The reminder of the 500 µL not used in the RIA is used for estimating recovery.

4. Radioimmunoassay. The assay tubes are 12 × 75-mm borosilicate glass tubes containing 25 µL of the HPLC-purified extracts in ethanol. The standards for this RIA, which are 24,25(OH)$_2$D$_3$, are placed in 12 × 75-mm tubes in 25 µL ethanol at concentrations from 0–200 pg/tube. To each tube add [^{125}I] tracer (50,000 cpm in 50 µL 1:1 ethanol, 0.01 M phosphate buffer, pH 7.4). Then add to each tube 1.0 mL of primary antibody diluted 1:15,000-fold in sodium phosphate buffer (50 mM, pH 7.4 containing 0.1% swine-skin gelatin). Nonspecific binding is estimated using the above buffer minus the antibody. Vortex-mix the contents of the tubes and incubate for 90 min at 20–25°C. Then add 0.5 mL of a second-antibody precipitating complex to each tube, vortex-mix, incubate at 20–25°C for 20 min and centrifuge (20°C, 2000g for 20 min). Discard the supernate and count the tubes in a gamma-well counting system. The 24,25(OH)$_2$D values are calculated from the standard curve in pg/tube. To convert this value to ng/mL, correct for dilution and final recovery of [^{3}H]24,25(OH)$_2$D$_3$ (Fig. 8).

5. 1,25-DIHYDROXYVITAMIN D ASSAY METHODOLOGY

One of the most significant recent discoveries in vitamin D research has been the isolation and identification of 1,25(OH)$_2$D$_3$ *(3–5)*. This metabolite is biologically the most active form of vitamin D$_3$ in stimulating intestinal calcium transport and bone calcium resorption. 1,25(OH)$_2$D circulates at very low concentrations (20–50 pg/mL) in normal sera (Table 4). Therefore, detection requires significantly more sensitive assays than those applied to other dihydroxyvitamin D metabolites.

Table 4
Application of Assays for Plasma Concentrations of 1,25-Dihydroxyvitamin D[a]

Extraction	Chromatography	Analysis	Concentration (ng/mL)	Reference
Normal				
Dichloromethane	Sephadex LH-20/HPLC	RIA	41 ± 2.5	109
Methanol/dichloromethane	Sephadex LH-20/HPLC	RRA	31.9 ± 9	41
$CHCl_3$/methanol	Sephadex LH-20/HPLC	GLC/MS	55 ± 10	93
Extrelute/dichloromethane	HPLC	RRA	36	98
Benzene	HPLC	Bioassay	33 ± 6	134
Clin-elute dichloromethane/acetone	HPLC	RRA	55 ± 14	99
Clin-elute benzene/ether	—	CRA	34 ± 11	135
Methanol/chloroform	Silici acid/Sephadex LH-20/celite	RRA	39 ± 8	136
Methanol/dichloromethane/	Sephadex LH-20/HPLC	RRA	29 ± 2	96
Acetonitrile	C18-OH/Silica	RRA	37.4 ± 2.2	13
Acetonitrile	C18-OH	RRA	28.2 ± 11.3	97
Acetonitrile	C18-OH/Silica	RIA	32.2 ± 8.5	110
Chronic renal failure				
Acetonitrile	C18-OH	RRA	10.9 ± 5.2	97
Acetone/dichloromethane	Sephadex LH-20/HPLC	RIA	17.2 ± 10.3	137
Extrelut-1/diisopropyl ether	HPLC	RIA	3–11	138
Acetonitrile	C18-OH/Silica	RRA	10.6 ± 1.5	13
Clin-elute benzene/ether	—	CRA	11.5 ± 6.7	135
Primary hyperparathyroid				
Acetone/MeCl2	Sephadex LH-20/HPLC	RIA	53.2 ± 19.5	137
Acetonitrile	C18-OH/Silica	RRA	68.9 ± 5.0	13
Clin-elute benzene/ether	—	CRA	50.2 ± 15.3	135

[a] Results are values obtained from normal individuals and those with chronic renal failure or primary hyperparathyroidism. RIA, radioimmunoassay; RRA, radioreceptor assay. For other abbreviations, see Table 1 footnote.

5.1. Bioassays

To date, only one bioassay has been applied to the measurement of 1,25(OH)$_2$D in sera *(91)*. This assay was based on the release of previously incorporated ^{45}C from 19-d-old fetal rat radii and ulnae and can detect as little as 2–3 pg of either 1,25(OH)$_2$D$_2$ or 1,25(OH)$_2$D$_3$. The 1,25(OH)$_2$D must be purified by HPLC prior to quantitation in this assay. Disadvantages of this system are the need to develop skill in the tissue culture technique and the possibly greater variability in this assay as opposed to the protein binding assay. The advantages are that this assay seems to be more sensitive and is based on a biologic end-organ response; it would therefore be useful in validating results from other methods.

5.2. Cytoreceptor Assay

Manolagus and Deftos *(92)* have demonstrated a novel technique for the measurement of 1,25(OH)$_2$D in plasma. The technique is basically a radioreceptor assay (RRA), except that the receptor is maintained intracellularly. This assay takes advantage of the fact that the intact cell membrane in combination with the DBPs on the outside of the cell and the 1,25(OH)$_2$D receptor inside the cells perform in a biologic manner the separation of 1,25(OH)$_2$D from other metabolites of vitamin D present in extracts of sera. This procedure obviates the need for extensive preparative chromatography and is therefore more rapid and less intense than other RRAs. However, the procedure requires expertise and facilities for handling live cells, which has precluded its success as a marketable assay.

5.3. GLC/MS Assays

An isotope dilution method mass fragmentography assay for the analysis of $1,25(OH)_2D_3$ has been described for human sera *(93)*. This represents the first attempt at such an approach and, because of the complexity of the approach and the expensive equipment required, it has received little interest. In addition, this procedure is specific for $1,25(OH)_2D_3$ and, therefore, high concentrations of $1,25(OH)_2D_2$ would preclude analysis by this method. The use of isotope dilution mass fragmentography would, however, seem to be useful in the verification of results obtained by competitive protein binding analysis.

5.4. Radioreceptor Assays

The experiments by Brumbaugh and Haussler *(94,95)* demonstrating a high-affinity, low-capacity receptor in chick intestine preparations were landmarks in the development of sensitive RRAs for $1,25(OH)_2D$. Since these original efforts, several RRAs have been described for measuring $1,25(OH)_2D$ in sera *(13,41,96–100)*.

The first successful $1,25(OH)_2D$ RRA, as reported by Brumbaugh et al. *(101)*, involved lipid extraction of the sera followed by purification on three LPLC columns. The $1,25(OH)_2D$ isolated from the final column was assayed by RRA. As demonstrated by these authors, the need for the successive chromatographic steps was caused by the presence of interfering substances that were not removed until the last chromatographic step was performed. This original attempt demonstrated the need for careful verification of the assay technique that, in this case, involved demonstrating that $1,25(OH)_2D$ is not present in anephric human plasma or plasma from vitamin D-deficient animals. The binding assay was performed using a mixture of freshly prepared intestinal cytosol and nuclear chromatin.

In 1976, Eisman et al. *(96)* reported several modifications to the original Brumbaugh et al. *(101)* method. They were able to demonstrate the advantages of greater receptor stability achieved by washing the chick intestinal mucosa prior to homogenation, capacity to store lyophilized intestinal cytosol at low temperatures for up to 3 mo, purification of the $1,25(OH)_2D$ by HPLC following prepurification on Sephadex LH-20, and use of polyethylene glycol (PEG)-6000 for the separation of bound from free steroid.

The use of HPLC results in a more rapid preparation of $1,25(OH)_2D$ analysis but also allows for the simultaneous purification of other metabolites in a single purification step. HPLC purification of the $1,25(OH)_2D$ Sephadex LH-20 isolates also results in non-detectable $1,25(OH)_2D$ in anephric plasma taken from patients not consuming dihydrotachysterol. Dihydrotachysterol therapy seems to result in the production of a metabolite (possibly 25-hydroxydihydrotachysterol) that comigrates with $1,25(OH)_2D_3$ on HPLC and also competes in the CPBA *(102–104)*.

Following these original advances in the RRA analysis of $1,25(OH)_2D$, most efforts have dealt with more sensitive and faster purification methods. One of the most significant contributions resulting in improved assay sensitivity has been the introduction of high specific activity $[^3H]1,25(OH)_2D_3$. Prior to 1978, the highest specific activity material available was 10–20 Ci/mmol. In 1978, Yamada et al. *(105)* demonstrated the synthesis of 80–90 Ci/mmol $23,24[^3H]25(OH)D_3$, which could easily be converted to $23,24[^3H]1,25(OH)_2D_3$ in vitro. Shortly after, Napoli et al. *(106)* synthesized 180 Ci/mmol

26,27[^{3}H]1,25(OH)$_2$D$_3$, which is the highest specific-activity radioactive vitamin D sterols available.

For many years, 1,25(OH)$_2$D analyses were conducted using chick receptor protein as the source of binding protein and HPLC for sample purification. Because of the extremely technical nature of these assays and the cost of HPLC systems, only a few laboratories could afford to measure circulating 1,25(OH)$_2$D. Furthermore, because these early techniques were so cumbersome, commercial laboratories did not offer 1,25(OH)$_2$D determinations as a clinical service. This changed in 1984 with the introduction of a new concept for the determination of circulating 1,25(OH)$_2$D *(13)*.

This new RRA utilized solid-phase extraction of 1,25(OH)$_2$D from serum along with silica cartridge purification of 1,25(OH)$_2$D. As a result, the need for HPLC sample prepurification was eliminated. Also, this assay utilized the vitamin D receptor (VDR) isolated from calf thymus, which, if lyophilized, proved to be quite stable at room temperature and thus had to be prepared only periodically. Furthermore, the volume requirement was reduced to 1 mL of serum or plasma. This assay opened the way for any laboratory to measure circulating 1,25(OH)$_2$D. The procedure also resulted in the production of the first commercial kit for 1,25(OH)$_2$D measurement. This RRA was further simplified in 1986 by decreasing the required chromatographic purification steps *(97)* (Fig. 9). During the next decade, no new advances were reported with respect to the quantitation of circulating 1,25(OH)$_2$D.

As good as the calf thymus RRA for 1,25(OH)$_2$D was, it still possessed two serious shortcomings. First, VDR still had to be isolated from thymus glands, which was still a difficult technique. Second, because the VDR is so specific for its ligand, only [^{3}H]1,25(OH)$_2$D$_3$ could be used as a tracer as opposed to an [^{125}I]-based tracer. This is a major handicap, especially for the commercial laboratory, where λ-emitting tracers are preferred to β-emitting tracers.

5.5. Radioimmunoassays

Many RIAs have been published and validated for the quantification of 1,25(OH)$_2$D, but all have included HPLC steps for sample prepurification *(107–109)*. Development of an RIA for quantification of circulating 1,25(OH)$_2$D has been hampered from the beginning by the relatively poor specificity of the antibodies that have been generated. To date, the best antibodies toward 1,25(OH)$_2$D have, at best, a crossreactivity with the non-1-hydroxylated metabolites of vitamin D of approximately 1%. In comparison, the VDR used in the RRA has a cross-reactivity of approximately 0.01% to these more abundant metabolites *(13)*. Given that these non-1-hydroxylated metabolites circulate at concentrations >1000-fold that of 1,25(OH)$_2$D, the magnitude of the problem becomes clear. However, the VDR is so specific that any attempt to introduce a radionuclide such as [^{125}I] into 1,25(OH)$_2$D$_3$ erodes the binding between this steroid hormone and the VDR. Therefore, if one wishes to develop [^{125}I]-based assays to quantify 1,25(OH)$_2$D, RIA is the only choice.

Recently two new RIA procedures have been introduced that incorporate an [^{125}I] tracer as well as standards in an equivalent serum matrix, so individual sample recoveries and HPLC are no longer required. One assay is described *(110)* and the other is an unpublished assay marketed by IDS (Tyne and Wear, UK). The antibody chosen for use in the development for both of these RIAs has been characterized previously *(111)*. This antibody crossreacts by 1–2% with the more abundant vitamin D metabolites, and 1,25(OH)$_2$D$_2$ is only 60–70% as competitive as 1,25(OH)$_2$D$_3$ *(111)*. The substantial

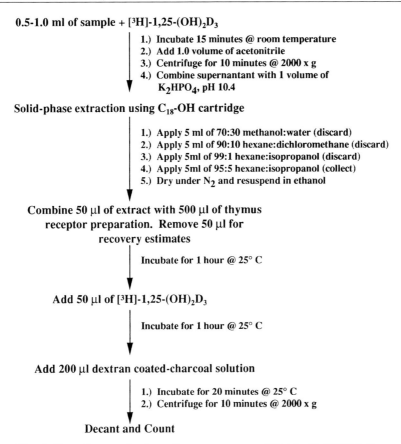

Fig. 9. Flow diagram for the radioreceptor assay (RRA) of 1,25(OH)$_2$D.

crossreactivity with non-1-hydroxylated metabolites meant that some prepurification method for 1,25(OH)$_2$D was needed prior to RIA.

In the method described by Hollis et al. *(110)*, a simplified chromatographic procedure that had been previously incorporated into the RRA for 1,25(OH)$_2$D was used to avoid HPLC. When this purification procedure was applied to the RIA, the "apparent" circulating 1,25(OH)$_2$D concentrations were approx 50% greater than the RRA results. Further investigation revealed that the vitamin D metabolite most responsible for this overestimation was 25,26(OH)$_2$D. Efforts to resolve this metabolite by chromatographic means short of HPLC failed. Therefore, extracted samples were pretreated with sodium periodate. Instituting this pretreatment step converted all of the 24,25(OH)$_2$D$_3$ and 25,26(OH)$_2$D$_3$ into their respective aldehyde and ketone forms, which are easily removed by the current chromatographic scheme. An extra silica column is incorporated into this procedure to ensure that all contaminating substances were removed. Use of this purification procedure reduced contamination by non-1-hydroxylated vitamin D compounds to insignificant levels. Failure to use the sodium periodate pretreatment step results in a 40% overestimation of the actual concentration of circulating 1,25(OH)$_2$D in normal human subjects. This new RIA was validated against the standard RRA with excellent results *(110)*.

The methodology of the IDS kit followed a different approach and used selective immunoextraction of 1,25(OH)$_2$D from serum or plasma with a specific monoclonal antibody bound to a solid support. This antibody is directed toward the 1-hydroxylated A-ring of 1,25(OH)$_2$D$_3$ *(112)*. This immunoextraction procedure, however, is also highly

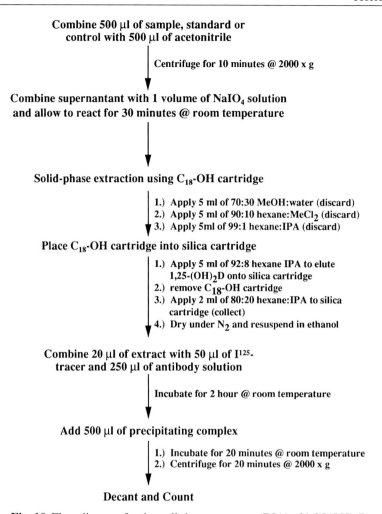

Fig. 10. Flow diagram for the radioimmunoassay (RIA) of 1,25(OH)$_2$D.

specific for other 1-hydroxylated forms of vitamin D *(113)* and therefore has the disadvantage of removing other 1-hydroxylated metabolites, which are known to circulate and include 1,24,25-trihydroxyvitamin D$_3$ [1,24,25(OH)$_3$D$_3$], 1,25,26-trihydroxyvitamin D$_3$ [1,25,26(OH)$_3$D$_3$], 1,25-dihydroxyvitamin D$_3$-26,23-lactone [1,25(OH)$_2$D$_3$-26,23-lactone], 24-oxo-1,25-dihydroxyvitamin D$_3$ [1,25(OH)$_2$-24-oxo-D$_3$], calcitropic acid, and possibly various water-soluble, side-chain conjugates. Of particular significance is the presence of 1,25(OH)$_2$D$_3$-26,23-lactone, which is known to circulate at high concentrations *(114)* under physiologic conditions and has been demonstrated to affect the analysis of 1,25(OH)$_2$D significantly using the immunoextraction protocol *(113)*.

We describe in detail below the preferred RRA and RIA methodologies for the analysis of 1,25(OH)$_2$D. Flow diagrams for each method are shown in Figs. 9 and 10.

5.6. RRA Method of Choice

1. Preparation of calf thymus VDR was as described by Reinhardt et al. *(13)*, and all steps are carried out at 4°C. Briefly, thymus tissue is minced with a meat grinder and homogenized

(20% w/v) in a buffer containing 50 mM K_2HPO_4, 5 mM dithiothreitol, 1 mM EDTA, and 400 mM KCl, pH 7.5. The tissue is homogenized using five 30-s bursts of a Polytron PT-20 tissue disrupter at a maximum power setting. The homogenate is then centrifuged for 15 min at 20,000g to remove large particles. The resulting supernatant is then centrifuged at 100,000g for 1 h and the cytosol is collected minus the floating lipid layer. The cytosol is then fractionated by the slow addition of solid $(NH_4)_2SO_4$ to 35% saturation. The cytosol $(NH_4)_2SO_4$ mixture is stirred for 30 min while maintaining the temperature at 4°C. The mixture is then aliquoted into 15-mL centrifuge tubes and centrifuged at 20,000g for 20 min. The supernatant is discarded and tubes allowed to drain for 5 min. The pelleted receptor is lyophilized and stored under inert gas at −70°C. Receptor prepared in this manner is stable for up to 60 h at room temperature. As an alternative, calf thymus VDR preparations can also be purchased from commercial sources (INCSTAR, Stillwater, MN).

2. Sample extraction. Serum or plasma (1.0 mL) is placed into a 12 × 75-mm borosilicate glass culture tube containing 700 cpm of [^{3}H]1,25(OH)$_2$D$_3$ in 25 µL of ethanol to monitor recovery of endogenous compound through the extraction and chromatographic procedures. Following a 15-min incubation with the tracer, 1 mL of HPLC-grade acetonitrile is added to each tube. The sample is then vortex-mixed for 1 min, followed by centrifugation at 1000g for 10 min. The supernatant is removed and placed in another 12 × 75-mm culture tube; 1 mL of 0.4 M K_2HPO_4, pH 10.6, is added, followed by vortex-mixing.

3. Solid-phase extraction and purification chromatography. C18-OH silica Bond-Elut cartridges (500 mg) and Vac-Elut cartridge racks were obtained from Varian Instruments. The C18-OH cartridges are washed in order with 5 mL of HPLC-grade dichloromethane, 5 mL of HPLC-grade isopropanol, and 5 mL of HPLC-grade methanol. The sample is then applied to the cartridge and eluted through the cartridge under vacuum into waste. This initial step is followed by 5 mL of 30% water in methanol (discard), 5 mL of 10% dichloromethane in hexane (discard), 5 mL of 1% isopropanol in hexane (discard), and 5 mL of 3% isopropanol in hexane, and is collected as 1,25(OH)$_2$D. This final fraction is dried in a heated waterbath at 55°C, under N_2. The residue is then resuspended in 200 µL absolute ethanol and capped.

4. Radioreceptor assay. Prior to assay, the VDR-containing pellet is reconstituted to its original volume with assay buffer. The assay buffer contains 50 mM K_2HPO_4, 5 mM dithiothreitol, 1.0 mM EDTA, and 150 mM KCl at pH 7.5. The receptor pellet is redissolved by gentle stirring on ice using a magnetic stir bar. The receptor solution is allowed to mix for 20–30 min. Typically, a small portion of the pellet resists solubilization and is removed by centrifugation at 3000g for 10 min. The receptor solution is then diluted 1:3–1:9 with assay buffer and kept on ice until use. The correct dilution of receptor used in the assay is determined for each new batch of receptor. At the appropriate dilution for assay use, specific binding in the absence of cold 1,25(OH)$_2$D is 1600–2000 cpm; nonspecific binding is 200–300 cpm. These results assume a specific activity of 130 Ci/mmol for [^{3}H]1,25(OH)$_2$D$_3$ and a 40% counting efficiency for tritium.

The assay tubes are 12 × 75-mm borosilicate glass tubes containing 50 µL of C18-OH purified extracts in ethanol. The standards for this RRA, which are 1,25(OH)$_2$D$_3$, are placed in 12 × 75-mm tubes in 50 µL ethanol at concentrations from 1–15 pg/tube. Nonspecific binding is estimated by adding 1 ng/tube of 1,25(OH)$_2$D$_3$. To each tube, add 0.5 mL of reconstituted thymus cytosol, vortex-mix, and incubate for 1 h at 15–20°C. Following this initial incubation, each tube receives [^{3}H]1,25(OH)$_2$D$_3$ (5000 cpm in 25 µL ethanol), and the incubation proceeds for an additional hour at 15–20°C. Finally, place the assay tubes in an ice bath and add 0.2 mL of 0.1 M borate buffer containing 1.0% Norit-A charcoal, 0.1% dextran T-70, vortex-mix, incubate 20 min, and centrifuge (4°C, 2000g, 10 min). Remove the supernatant and place in vials, add scintillation fluid, and monitor for radioactive content in a scintillation counter. 1,25(OH)$_2$D values are calculated from the standard curve in pg/tube. To convert this value to pg/mL, correct for dilution used, as well as final recovery of [^{3}H]1,25(OH)$_2$D$_3$ added at the beginning of the sample extraction procedure.

5.7. RIA Method of Choice

1. Preparation of assay calibrators. As was described for the 25(OH)D RIA, one of the goals of this RIA procedure was to eliminate the need for individual sample recovery. To prepare these calibrators, human serum was "stripped" free of vitamin D metabolites. Absence of endogenous 1,25(OH)$_2$D in the stripped sera was confirmed by RRA for 1,25(OH)$_2$D, as described above. Subsequently, crystalline 1,25(OH)$_2$D$_3$ dissolved in absolute ethanol was added to the stripped sera to yield calibrators at concentrations of 0. 5, 15, 30, 60, and 160 pg/mL.

2. Sample and calibrator extraction and pretreatment. 1,25(OH)$_2$D is extracted from calibrators and samples as follows. Serum or plasma (0.5 mL) is placed into a 12 × 75-mm borosilicate glass culture tube and 0.5 mL of HPLC-grade acetonitrile is added and vortex-mixed for 1 min followed by centrifugation at 1000g for 10 min. The supernatant is removed into another 12 × 75-mm culture tube, and 0.5 mL of a 25-mg/mL solution of sodium m-periodate is added. The samples are now incubated for 30–60 min at room temperature.

3. Solid-phase extraction and silica purification chromatography. C18-OH silica, silica Bond-Elut cartridges (500 mg), and Vac-Elut cartridge racks were obtained from Varian Instruments. The supernate-sodium periodate mix was applied to a C18-OH cartridge that had been prewashed successively with 5 mL of isopropanol and 5 mL of methanol. Next, the cartridge was successively washed with 5 mL of 30% water in methanol (discard), 5 mL of 10% dichloromethane in hexane (discard), and 5 mL of 1% isopropanol in hexane (discard). The C18-OH cartridge is now placed into a silica cartridge previously washed successively with 5 mL of 1% methanol, 5 mL of isopropanol, and 5 mL of 20% isopropanol in hexane, and 1,25(OH)$_2$D is eluted onto the silica cartridge with 5 mL of 8% isopropanol in hexane (discard). The C18-OH cartridge is removed from the silica cartridge, and the silica cartridge is eluted with 2 mL of 8% isopropanol in hexane (discard) and 5 mL of 20% isopropanol in hexane, which is collected as 1,25(OH)$_2$D. Each C18-OH cartridge can be regenerated for reuse by washing with 2 mL of methanol. The silica cartridges can be reused without any further washing steps. This final fraction is dried in a heated waterbath at 55°C, under N$_2$. The residue is then resuspended in 50 µL absolute ethanol and capped.

4. Radioimmunoassay. The assay tubes are 12 × 75-mm borosilicate glass tubes containing 20 µL of the ethanol-reconstituted extracted calibrators or samples. To each tube add [^{125}I]1,25(OH)$_2$D derivative (50,000 cpm in 50 µL 1:1 ethanol, 0.01 M phosphate buffer, pH 7.4) that was synthesized as previously described *(110)*. Then add to each tube 0.25 mL of primary antibody diluted 1:200,000-fold in sodium phosphate buffer (50 mM, pH 6.2) containing 0.1% swine-skin gelatin and 0.35% polyvinyl alcohol (MW 13,000–23,000). Nonspecific binding is estimated by using the above buffer without the antibody. Vortex-mix the contents of the tubes and incubate them for 2 h at 20–25°C for 20 min and centrifuge (20°C, 2000g, 20 min). Discard the supernate and count the tubes in a gamma-well counting system. 1,25(OH)$_2$D values are calculated directly from standard curve.

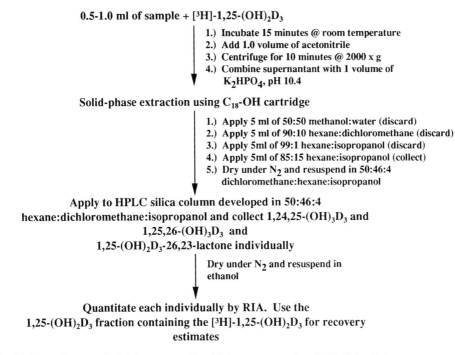

Fig. 11. Flow diagram for high-pressure liquid chromatographic (HPLC)/radioimmunoassay (RIA) for 1,24,25(OH)$_3$D$_3$, 1,25,26(OH)$_3$D$_3$, and 1,25(OH)$_2$D$_2$-26,23-lactone.

6. TRIHYDROXYVITAMIN D$_3$ METABOLITES AND 1,25-DIHYDROXYVITAMIN D$_3$-26,23-LACTONE

The metabolites 1,24,25(OH)$_3$D$_3$, 1,25,26(OH)$_3$D$_3$, and 1,25(OH)$_2$D$_3$-26,23-lactone are known to circulate under physiologic conditions *(114–116)* and are significantly elevated following treatment with 1,25(OH)$_2$D$_3$ *(117,118)*. A few assay methods have been published with ultimate quantitation achieved by using RRA following separation by HPLC *(119,120)*.

A flow diagram of the authors' method of choice for assaying these metabolites is presented in Fig. 11 and described in detail below.

6.1. Method of Choice

1. Sample extraction. Serum or plasma (1.0 mL) is placed into a 12 × 75-mm borosilicate glass culture tube (Fig. 10). Because trihydroxyvitamin D$_3$ metabolites are generally unavailable, the samples are treated with 700 cpm of [^{3}H]1,25(OH)$_2$D$_3$ in 25 µL of ethanol to monitor recoveries through the extraction and chromatographic procedures. Following a 15-min incubation with the tracer, 1 mL of HPLC-grade acetonitrile is added to each sample. The sample is then vortex-mixed for 1 min, followed by centrifugation at 1000*g* for 10 min. The supernatant is removed into another 12 × 75-mm culture tube and 1 mL of 0.4 *M* K$_2$HPO$_4$, pH 10.6, is added followed by vortex-mixing.
2. Solid-phase extraction and silica purification chromatography. The supernatant is applied to a C18-OH cartridge, which was prewashed successively with 5 mL of isopropanol and 5 mL of methanol. Next, the cartridge is successively washed with 5 mL of 50% water in methanol (discard), 5 mL of 10% dichloromethane in hexane (discard), 5 mL of 1%

isopropanol in hexane (discard), and finally with 5 mL of 15% isopropanol in hexane to elute the 1,25(OH)$_2$D and the polar metabolites. (Note that the water concentration in this step is increased to 50% to retain the very polar vitamin D sterols on the column and that the isopropanol concentration in the final step is elevated to 15% to elute these metabolites.) The final fraction is dried in a heated waterbath at 55°C, under N$_2$. The residue is then resuspended in 150 µL of mobile phase (described below) in preparation for HPLC.

3. The trihydroxyvitamin D$_3$ metabolites and 1,25(OH)$_2$D$_3$-26,23-lactone can be separated by HPLC using an Econosphere 3-µm silica column (Altech, Deerfield, IL) developed in a mobile phase of 50:46:4 hexane/dichloromethane/isopropanol (Fig. 5D). Each metabolite is collected individually for quantitation by RIA as described in the 1,25-dihydroxyvitamin D section. The antisera used for quantitating 1,25(OH)$_2$D$_3$ do not discriminate against the trihydroxyvitamin D$_3$ metabolites or 1,25(OH)$_2$D$_3$-26,23-lactone. Therefore, 1,25(OH)$_2$D$_3$ can be used for constructing standards curves. The metabolite values are calculated from the standard curve in pg/tube. To convert this value to pg/mL, correct for dilution and final recovery of [^{3}H]1,25(OH)$_2$D$_3$.

7. MULTIPLE ASSAYS

The term *multiple* implies that more than one determination can be made from a single sample. Several multiple assays have been published in the vitamin D literature. The most popular determinations are those of 25(OH)D, 24,25(OH)$_2$D$_3$, and 1,25(OH)$_2$D. One of the first multiple assays, which included the determination of the active vitamin D metabolite 1,25(OH)$_2$D, was reported by Hughes et al. *(121)*. Since that time, HPLC and GLC/MS have been applied to the resolution and quantitation of multiple forms of vitamin D sterols in biologic fluids *(10,41,48,64,69,71,86,98,102,122)*.

Horst et al. *(10)* demonstrated the most comprehensive multiple assay procedure described to date. Their protocol included the capacity to determine individually the vitamin D$_2$ and D$_3$ forms of vitamin D, 25(OH)D, 24,25(OH)$_2$D, 25,26(OH)$_2$D, and 1,25(OH)$_2$D. The assay also provides for the analysis of 25(OH)D$_3$-26,23-lactone and trihydroxyvitamin Ds. In their procedure, plasma lipids are extracted with peroxide-free diethylether, followed by extraction with methanol/dichloromethane to facilitate removal of 25(OH)D$_3$-26,23-lactone. The resulting lipid extract was chromatographed on a Sephadex LH-20 column. This column resolved the vitamin D metabolites into four fractions: vitamin D, 25(OH)D, dihydroxyvitamin D, and trihydroxyvitamin D metabolites. Each fraction was handled individually and was usually subjected to one or more additional purification steps before final quantitation.

A multiple assay can also be constructed using protocols described in this chapter. For example, by combing the procedures for assaying 24,25(OH)$_2$D and 1,25(OH)$_2$D, one can construct an assay that can measure 25(OH)D, 24,25(OH)$_2$D, and 1,25(OH)$_2$D.

8. SUMMARY

Recent advances in chromatographic, CPBA, and RIA techniques have resulted in the ability to quantitate very small amounts of vitamin D and vitamin D metabolites. Emphasis, however, has focused on only two of these metabolites, namely, 25(OH)D and 1,25(OH)2D, for clinically relevant information. For example, subnormal circulating levels of 25(OH)D usually results from inadequate vitamin D intake and/or insufficient sunlight exposure. This combination of events can put elderly patients at risk of developing vitamin D deficiency and associated complications *(123)*. This in turn has been shown to result in an increased risk of hip fractures in the elderly due to osteomalacia *(124)*. Other conditions that contribute to nutritional vitamin D deficiency include nephrotic syndrome, chronic renal disease, cirrhosis, and malabsorption syndromes such as biliary atresia. Vitamin D intoxication, although rare, still occurs and is most accurately diagnosed by determining circulating 25(OH)D. Thus, from a clinical standpoint, the determination of circulating 25(OH)D is the most frequently requested antirachitic sterol measurement. Circulating 1,25(OH)$_2$D on the other hand, is diagnostic for several clinical conditions, including vitamin-dependent rickets types I and II, hypercalcemia associated with sarcoidosis, and other hypercalcemic disorders causing increased 1,25(OH)$_2$D levels. These other disorders include tuberculosis, fungal infections, Hodgkin's disease, lymphoma, and Wegener's granulomatosis. In all other clinical conditions involving the vitamin D endocrine system, including hypoparathyroidism, hyperparathyroidism, and chronic renal failure, the assay of 1,25(OH)$_2$D is a confirmatory test. It is also important to remember that circulating 1,25(OH)$_2$D provides essentially no information with respect to the patient's nutritional vitamin D status. Thus, circulating 1,25(OH)$_2$D should never be used as an indicator for hypo- or hypervitaminosis D when nutritional factors are suspected.

NOTE

Names are necessary to report factually on available data; however, the USDA neither guarantees nor warrants the standard of the product, and the use of the name by the USDA implies no approval of the product to the exclusion of others that may also be suitable.

REFERENCES

1. Mellanby E. An experimental investigation on rickets. Lancet I 1919; 4985:407–412.
2. Blunt JW, DeLuca HF, Schnoes HK. 25-Hydroxycholecalciferol: a biologically active metabolite of vitamin D$_3$. Biochemistry 1968; 7:3317–3322.
3. Lawson DEM, Fraser DR, Kodicek E, Morris HR, Williams DH. Identification of 1,25-dihydroxycholecalciferol, a new kidney hormone controlling calcium metabolism. Nature 1971; 230:228–230.
4. Holick MF, DeLuca HF. A new chromatographic system for vitamin D$_3$ and its metabolites: resolution of a new vitamin D$_3$ metabolite. J Lipid Res 1971; 12:460–465.
5. Norman AW, Myrtle JF, Midgett RJ, Nowicki HG. 1,25-Dihydroxycholecalciferol: identification of the proposed active form of vitamin D$_3$ in the intestine. Science 1971; 173:51–54.
6. Bordier P, Rasmussen H, Marie P, Miravet L, Gueris J, Ryckwaert A. Vitamin D metabolism and bone mineralization in man. J Clin Endocrinol Metab 1978; 46:284–294.
7. Ornoy A, Goodwin D, Noff D, Edelstein S. 24,25-Dihydroxyvitamin D is a metabolite of vitamin D essential for bone formation. Nature 1978; 276:517–519.
8. Canterbury JM, Lerman S, Claflin AJ, Henry H, Norman A, Reiss E. Inhibition of parathyroid hormone secretion by 25-hydroxycholecalciferol and 24,25-dihydroxycholecal ciferol in the dog. J Clin Invest 1977; 78:1375–1383.
9. Henry HL, Norman AW. Vitamin D: two dihydroxylated metabolites are required for normal chicken egg hatchability. Science 1978; 201:835–847.
10. Horst RL, Littledike ET, Riley JL, Napoli JL. Quantitation of vitamin D and its metabolites and their plasma concentrations in five species of animals. Anal Biochem 1981; 116:189–203.
11. Hollis BW, Pittard WB. Relative concentrations of 25-hydroxyvitamin D$_2$/D$_3$ and 1,25-dihydroxyvitamin D$_2$/D$_3$ in maternal plasma at delivery. Nutr Res 1984; 4:27–32.
12. Hartwell D, Hassager C, Christiansen C. Effect of vitamin D$_2$ and vitamin D$_3$ on the serum concentrations of 1,25(OH)$_2$D$_2$ and 1,25(OH)$_2$D$_3$ in normal subjects. Acta Endocrinol 1987; 115:378–384.
13. Reinhardt TA, Horst RL, Orf JW, Hollis BW. A microassay for 1,25-dihydroxyvitamin D not requiring high performance liquid chromatography: application to clinical studies. J Clin Endocrinol Metabol 1984; 58:91–98.
14. Jones G, Kung M, Kano K. The isolation and identification of two new metabolites of 25-hydroxyvitamin D$_3$ produced in the kidney. J Biol Chem 1983; 258:12,920–12,928.

15. McCollum EV, Simmonds N, Shipley PG, Park EA. Studies on experimental rickets. XVI. A delicate biological test for calcium depositing substances. J Biol Chem 1922; 54:41–50.
16. United States Pharmacopeia. Rockville, MD: Mack Printing, 1950; XVI:910–933.
17. Schachter D, Rosen SM. Active transport of Ca^{45} by the small intestine and its dependence of vitamin D. Am J Physiol 1959; 196:357–365.
18. Hibberd KA, Norman AW. Comparative biological effects of vitamin D_2 and D_3 and dihydrotachysterol-2 and dihydrotachysterol-3 in chick. Biochem Pharmacol 1964; 18:2347–2355.
19. Coates ME, Holdsworth ES. Vitamin D_3 and absorption of calcium in the chick. Br J Nutr 1961; 15:131–147.
20. Wasserman RH, Corradino RA, Taylor AN. Vitamin D-dependent calcium-binding protein: purification and some properties. J Biol Chem 1968; 243:3978–3986.
21. Bar A, Wasserman H. Duodenal calcium binding protein in the chick: a new bioassay for vitamin D. J Nutr 1974; 104:1202–1207.
22. Corradino RA. Induction of calcium-binding protein in embryonic chick duodenum in vitro: direct assessment of biopotency of vitamin D steroids. In: Vitamin D: Basic and Applied Aspects. Kumar R, ed. Hingham, MA: Martinus Nijhoff 1984; 325–341.
23. Corradino RA. Calcium-binding protein of intestine: induction by biologically significant cholecalciferol-like steroid in vitro. J Steroid Biochem 1978; 9:1183–1187.
24. Brockmann H, Chen YH. Über eine Methode zur quantitativen Bestimmung von Vitamin D. Hoppe-Seylers Z Physiol Chem 1936; 241:129–133.
25. Tsukida K, Saiki K. Determination of vitamins D by gas-liquid chromatography. II. Rapid assay for vitamin D_2 in the presence of vitamins A and E. J Vitaminol 1972; 18:165–171.
26. Tsukida K, Saiki K. Determination of vitamins D by gas-liquid chromatography: I. Differentiation and assay of vitamins D_2 and D_3. J Vitaminol 1970; 16:293–296.
27. Sklan D, Budowski P, Katz M. Determination of 25-hydroxycholecalciferol by combined thin-layer and gas chromatography. Anal Biochem 1973; 56:606–609.
28. Bjorkhem I, Larsson A. A specific assay of vitamin D_3 in human serum. Clin Chim Acta 1978; 88:559–568.
29. De Leenheer AP, Cruyl AAM. Vitamin D_3 in plasma quantitation by mass fragmentography. Anal Biochem 1978; 91:293–303.
30. Avioli LV, Lee SW. Detection of nanogram quantities of vitamin D by gas liquid chromatography. Anal Biochem 1966; 16:193–199.
31. Seamark DA, Trafford DJH, Makin HLJ. The estimation of vitamin D and some metabolites in human plasma by mass fragmentography. Clin Chim Acta 1980; 106:51–62.
32. Wilson PW, Lawson DEM, Kodicek E. Gas-liquid chromatography of ergocalciferol and cholecalciferol in nanogram quantities. J Chromatog 1969; 39:75–77.
33. Horst RL, Reinhardt TA, Beitz DC, Littledike ET. A sensitive competitive protein binding assay for vitamin D in plasma. Steroids 1981; 37:581–592.
34. Hollis BW, Roos BA, Lambert PW. Vitamin D in plasma quantitation by a nonequilibrium ligand binding assay. Steroids 1981; 37:609–620.
35. Cruyl AAM, De Leenheer AP. Assay of vitamin D_3 (cholecalciferol) in plasma using gas chromatography/selected ion monitoring (GC/SIM). In: Vitamin D: Biochemical, Chemical and Clinical Aspects Related to Calcium Metabolism. Norman AW, Schaefer K, Coburn JW, et al., eds. Berlin: Walter de Gruyter, 1977; 455–457.
36. Henderson SK, Wickroski AF. Reverse phase high pressure liquid chromatographic determination in vitamin D in fortified milk. J Assoc Offic Anal Chem 1978; 61:1130–1134.
37. de Vries EJ, Zeeman J, Esser RJE, Borsje B, Mulder FJ. Analysis of fat-soluble vitamins. XXI. High pressure liquid chromatographic assay methods for vitamin D in vitamin D concentrations. J Assoc Offic Anal Chem 1979; 62:129–135.
38. Koshy KT, VanDerSlik AL. High-pressure liquid chromatographic method for the determination of 25-hydroxycholecalciferol in cow plasma. Anal Biochem 1976; 74:282–291.
39. Hofsass H, Grant A, Alicino NJ, Greenbaum SB. High pressure liquid chromatographic determination of vitamin D_3 in resins, oils, dry concentrates, and dry concentrates containing vitamin A. J Assoc Offic Anal Chem 1976; 59:251–260.
40. Hofsass H, Alicino JH, Hirsch AL, Ameika L, Smith LD. Comparison of high pressure liquid chromatographic and chemical methods for vitamin D_3 concentrates. II. Collaborative study. J Assoc Offic Anal Chem 1978; 61:735–745.

41. Shepard RM, Horst RL, Hamstra AJ, DeLuca HF. Determination of vitamin D and its metabolites in plasma from normal and anephric man. Biochem J 1979; 182:55–69.
42. Seamark DA, Trafford DJH, Hiscocks PG, Makin HLJ. High-performance liquid chromatography of vitamin D: enhanced ultraviolet absorbance by prior conversion to isotachysterol derivatives. J Chromatogr 1980; 197:271–273.
43. Adams JS, Clemens TL, Holick MF. Silica Sep-Pak preparative chromatography for vitamin D and its metabolites. J Chromatogr 1981; 226:198–201.
44. Aksnes L. A simplified high-performance liquid chromatographic method for determination of vitamin D_3, 25-hydroxyvitamin D_2 and 25-hydroxyvitamin D_3 in human serum. Scand J Clin Lab Invest 1992; 52:177–182.
45. Jones G, DeLuca HF. High-pressure liquid chromatography: separation of the metabolites of vitamins D_2 and D_3 on small-particle silica columns. J Lipid Res 1975; 16:448–453.
46. Rhodes CJ, Claridge PA, Trafford DJH, Makin HLJ. An evaluation of the use of Sep-Pak C_{18} cartridges for the extraction of vitamin D_3 and some of its metabolites from plasma and urine. J Steroid Biochem 1983; 19:1349–1354.
47. Belsey R, DeLuca HF, Potts JT Jr. Competitive binding assay for vitamin D and 25-OH vitamin D. J Clin Endocrinol Metabol 1971; 33:554–557.
48. Jones G. Assay of vitamins D_2 and D_3, and 25-hydroxyvitamins D_2 and D_3 in human plasma by high-performance liquid chromatography. Clin Chem 1978; 24:287–298.
49. Liel Y, Ulmer E, Shary J, Hollis BW, Bell NH. Low circulating vitamin D in obesity. Calcif Tiss Int 1988; 43:199–201.
50. Haddad Jr JG. Competitive protein-binding radioassays for 25(OH)D: clinical applications. In: Vitamin D: Molecular Biology and Clinical Nutrition, vol. 2. Norman AW, ed. New York: Marcel Dekker, 1980; 579–602.
51. Hay AWM, Watson G. Vitamin D_2 in vertebrate evolution. Comp Biochem Physiol 1977; 56B: 375–380.
52. Belsey RE, DeLuca HF, Potts JT Jr. A rapid assay for 25-OH vitamin D_3 without preparative chromatography. J Clin Endocrinol Metab 1974; 38:1046–1051.
53. Horst RL, Shepard RM, Jorgensen NA, DeLuca HF. The determination of 24,25-dihydroxyvitamin D and 25,26-dihydroxyvitamin D in plasma from normal and nephrecto mized man. J Lab Clin Med 1979; 93:277–285.
54. Horst RL, Littledike ET. 25-OHD_3-26,23 lactone: demonstration of kidney-dependent synthesis in the pig and rat. Biochem Biophys Res Commun 1980; 93:149–154.
55. Eisman JA, Shepard RM, DeLuca HF. Determination of 25-hydroxyvitamin D_2 and 25-hydroxyvitamin D_3 in human plasma using high pressure liquid chromatography. Anal Biochem 1977; 80: 298–305.
56. Bishop JE, Norman AW, Coburn JW, Roberts PA, Henry HL. Studies on the metabolism of calciferol. XVI. Determination of the concentration of 25-hydroxyvitamin D, 24,25-dihydroxyvitamin D, and 1,25-dihydroxyvitamin D in a single two-milliliter plasma sample. Miner Electrolyte Metab 1980; 3:181–189.
57. Bruton J, Wray HL, Dawson E, Butler V. 25-Hydroxyvitamin D_2 and 25-hydroxyvitamin D_3 as measured by liquid chromatography and by competitive protein binding. Clin Chem 1985; 31:783–784.
58. Dabek JT, Harkonen M, Wahlroos O, Aldercreutz H. Assay for plasma 25-hydroxyvitamin D_2 and 25-hydroxyvitamin D_3 by "high-performance" liquid chromatography. Clin Chem 1981; 27: 1346–1351.
59. Gilbertson TJ, Stryd RP. High performance liquid chromatographic assay for 25-hydroxyvitamin D_3 in serum. Clin Chem 1977; 23:1700–1704.
60. Horst RL, Shepard RM, Jorgensen NA, DeLuca HF. The determination of the vitamin D metabolites on a single plasma sample: changes during parturition in dairy cows. Arch Biochem Biophys 1979; 192:512–523.
61. Horst RL, Reinhardt TA, Hollis BW. Improved methodology for the analysis of plasma vitamin D metabolites. Kidney Int 1990; 38:S28–35.
62. Jones G. Application of high pressure liquid chromatography for assay of vitamin D metabolites. In: Vitamin D: Biochemical, Chemical and Clinical Aspects Related to Calcium Metabolism. Norman AW, Schaefer K, Coburn JW, et al., eds. Berlin: Walter de Gruyter, 1977; 491–500.
63. Matsuyama N, Okano T, Takada K, et al. Assay of 25-hydroxyvitamin D_3 in human plasma by high performance liquid chromatography. J Nutr Sci Vitaminol 1979; 25:469–478.

64. McGraw CA, Hug G. Simultaneous measurement of 25-hydroxy-, 24,25-dihydroxy-, and 1,25-dihydroxyvitamin D without use of HPLC. Med Lab Sci 1990; 47:17–25.
65. Okano T, Mizuno N, Shida S, et al. A method for simultaneous determination of 25-hydroxyvitamin D_2 and 25-hydroxyvitamin D_3 in human plasma by using two steps of high performance liquid chromatography. J Nutr Sci Vitaminol 1981; 27:43–54.
66. Osredkar J, Vrhovec I. A method for determination of 25-hydroxyvitamin D_3 and 1,25-dihydroxyvitamin D_3 in human serum by using high-performance liquid chromatography with gradient elution and radioimmunologic detection. J Liquid Chromatogr 1989; 12:1897–1907.
67. Stryd RP, Gilbertson TJ. Some problems in development of high performance liquid chromatographic assay to measure 25-hydroxyvitamin D_2 and 25-hydroxyvitamin D_3 simultaneously in human serum. Clin Chem 1978; 24:927–930.
68. Turnbull H, Trafford DJH, Makin HLJ. A rapid and simple method for the measurement of plasma 25-hydroxyvitamin D_2 and 25-hydroxyvitamin D_3 using Sep-Pak C_{18} cartridges and a single high performance liquid chromatographic step. Clin Chim Acta 1982; 120:65–76.
69. Lambert PW, Syverson BS, Arnaud CD, Spelsberg TC. Isolation and quantitation of endogenous vitamin D and its physiologically important metabolites in human plasma by high pressure liquid chromatography. J Steroid Biochem 1977; 8:929–937.
70. Trafford DJH, Seamark DA, Turnbull H, Makin HLJ. High-performance liquid chromatography of 25-hydroxyvitamin D_2 and 25-hydroxyvitamin D_3 in human plasma. Use of isotachysterols and a comparison with gas chromatography-mass spectrometry. J Chromatogr 1981; 226:351–361.
71. Coldwell RD, Trafford DJH, Varley MJ, Kirk DN, Makin HLJ. Measurement of 25-hydroxyvitamin D_2, 25-hydroxyvitamin D_3, 24,25-dihydroxyvitamin D_2 and 25,26-dihydroxyvitamin D_2 in a single plasma sample by mass fragmentography. Clin Chim Acta 1989; 180:157–168.
72. Hollis BW, Napoli JL. Improved radioimmunoassay for vitamin D and its use in assessing vitamin D status. Clin Chem 1985; 31:1815–1819.
73. Hollis BW, Kamerud JQ, Selvaag SR, Lorenz JD, Napoli JL. Determination of vitamin D status by radioimmunoassay with an ^{125}I-labeled tracer. Clin Chem 1993; 39:529–533.
74. Hollis BW, Pittard III WB. Evaluation of the total fetomaternal vitamin D relationships at term: evidence for racial differences. J Clin Endocrinol Metab 1984; 59:652–657.
75. Suda T, DeLuca HF, Schnoes HK, Ponchon G, Tanaka Y, Holick MF. 21,25-Dihydroxycholecalciferol: a metabolite of vitamin D_3 preferentially active on bone. Biochemistry 1970; 9:2917–2922.
76. Holick MF, Schnoes HK, DeLuca HF, Gray RW, Boyle IT, Suda T. Isolation and identification of 24,25-dihydroxycholecalciferol, a metabolite of vitamin D_3 made in the kidney. Biochemistry 1972; 11:4251–4255.
77. Haddad JG Jr, Min C, Mendolsohn M, Slatopolsky E, Hahn TJ. Competitive protein-binding radioassay of 24,25-dihydroxyvitamin D in sera from normal and anephric subjects. Arch Biochem Biophys 1977; 182:390–395.
78. Horst RL, Littledike ET, Gray RW, Napoli JL. Impaired 24,25-dihydroxyvitamin D production in anephric human and pig. J Clin Invest 1981; 67:274–280.
79. Weisman Y, Occhipinti M, Knox G, Reiter E, Root A. Concentrations of 24,25-dihydroxyvitamin D and 25-hydroxyvitamin D in paired maternal-cord sera. Am J Obstet Gynecol 1978; 130:704–707.
80. Weisman Y, Reiter E, Root A. Measurement of 24,25-dihydroxyvitamin D in sera of neonates and children. J Pediatr 1977; 91:904–908.
81. Traba ML, Babe M, Piedra C, Marin A. 24,25-Dihydroxyvitamin D_3 in serum: sample purification with Sep-Pak C-18 cartridges and liquid chromatography before protein-binding assay. Clin Chem 1983; 29:1806,1807.
82. Taylor CM, Wallace JE, Cundy T, Kanis JA. Effect of vitamin D_2 on the assay of 24,25-dihydroxyvitamin D_3 in man. Miner Electrolyte Metab 1982; 7:15–20.
83. Shimotsuji T, Hiejima T, Seino Y, et al. A specific competitive protein binding assay for serum 24,25-dihydroxyvitamin D in normal children and patients with nephrotic syndrome. Clin Chim Acta 1980; 106:145–154.
84. Reinhardt TA, Horst RL. Simplified assays for the determination of 25-OHD, 24,25-$(OH)_2$D and 1,25-$(OH)_2$D. In: Vitamin D: Molecular, Cellular and Clinical Endocrinology. Norman AW, Schaefer K, Grigoleit H-G, Herrath D-V, eds. Berlin: Walter de Gruyter, 1988; 720–726.
85. O'Riordan JLH, Graham RF, Dolev E. An assay for 24,25-dihydroxycholecalciferol and 25,26-dihydroxycholecalciferol in sera. In: Vitamin D: Biochemical, Chemical and Clinical Aspects Related to Calcium Metabolism. Norman AW, Schaefer K, Coburn JW, et al., eds. Berlin: Walter de Gruyter, 1977; 519–521.

86. Norman AW, Henry H, Bishop JE, Coburn JW. Simultaneous measurement of 25-hydroxyvitamin D, 24,25-dihydroxyvitamin D and 1,25-dihydroxyvitamin D in plasma samples by steroid competition assay. Clin Res 1978; 26:423A.
87. Mason RS, Lissner D, Reek C, Posen S. Assay of 1,25-dihydroxycholecalciferol and 24,25-dihydroxycholecalciferol in human serum: some technical considerations. In: Vitamin D: Basic Research and its Clinical Application. Norman AW, Schaefer K, Herrath D-v, et al., eds. Berlin: Walter de Gruyter, 1979; 243–246.
88. Kremer R, Guillemant S. A simple and specific competitive protein binding assay for 24,25-dihydroxyvitamin D in human serum. Clin Chim Acta 1978; 86:187–194.
89. Imawari M. A simple and sensitive assay for 25-hydroxyvitamin D, 24,25-dihydroxyvitamin D, and 1,25-dihydroxyvitamin D in human serum. Clin Chim Acta 1982; 124:63–73.
90. Dreyer CE, Goodman DBP. A simple direct spectrophotometric assay for 24,25-dihydroxyvitamin D_3. Anal Biochem 1981; 114:37–41.
91. Stern PH, Hamstra AJ, DeLuca HF, Bell NH. A bioassay capable of measuring 1 picogram of 1,25-dihydroxyvitamin D_3. J Clin Endocrinol Metab 1978; 46:891–896.
92. Manolagas SC, Deftos LJ. Cytoreceptor assay for 1,25-dihydroxyvitamin D_3: a novel radiometric method based on binding of the hormone to intracellular receptors in vitro. Lancet 1980; 2:401,402.
93. Bjorkhem I, Holmberg I, Kristiansen T, Pedersen JI. Assay of 1,25-dihydroxyvitamin D_3 by isotope dilution mass fragmentography. Clin Chem 1979; 25:584–588.
94. Brumbaugh PF, Haussler MR. 1α,25-Dihydroxycholecalciferol receptors in intestine. II. Temperature-dependent transfer of the hormone to chromatin via a specific cytosol receptor. J Biol Chem 1974; 249:1258–1262.
95. Brumbaugh PF, Haussler MR. 1α,25-Dihydroxycholecalciferol receptors in intestine. I. Association of 1α,25-dihydroxycholecalciferol with intestinal mucosa. J Biol Chem 1974; 249:1251–1257.
96. Eisman JA, Hamstra AJ, Kream BE, DeLuca HF. A sensitive, precise, and convenient method for determination of 1,25-dihydroxyvitamin D in human plasma. Arch Biochem Biophys 1976; 176:235–243.
97. Hollis BW. Assay of circulating 1,25-dihydroxyvitamin D involving a novel single-cartridge extraction and purification procedure. Clin Chem 1986; 32:2060–2063.
98. Mason RS, Lissner D, Grunstein HS, Posen S. A simplified assay for dihydroxylated vitamin D metabolites in human serum: application to hyper- and hypovitaminosis D. Clin Chem 1980; 26:444–450.
99. Dokoh S, Pike JW, Chandler JS, Mancini JM, Haussler MR. An improved radioreceptor assay for 1,25-dihydroxyvitamin D in human plasma. Anal Biochem 1981; 116:211–222.
100. Jongen MJM, van der Vijgh WJF, Willems HJJ, Netelenbos JC. Analysis for 1,25-dihydroxyvitamin D in human plasma, after a liquid-chromatographic purification procedure, with a modified competitive protein binding assay. Clin Chem 1981; 27:444–450.
101. Brumbaugh PF, Haussler DH, Bressler R, Haussler MR. Radioreceptor assay for 1α,25-dihydroxyvitamin D_3. Science 1974; 183:1089–1091.
102. Jongen MJM, van der Vijgh WJF, Lips P, Netelenbos JC. Measurement of vitamin D metabolites in anephric subjects. Nephron 1984; 36:230–234.
103. Gray RW, Adams ND, Lemann J Jr. The measurement of 1,25-dihydroxyvitamin D by competitive protein binding assay in the plasma of anephric patients: the effects of dihydrotachysterol. In: Vitamin D: Basic Research and its Clinical Application. Norman AW, Schaefer K, Herrath D-V, et al., eds. Berlin: Walter de Gruyter, 1979; 839–841.
104. Gray RW, Adams ND, Lemann J. The effects of dihydrotachysterol therapy on the measurement of plasma 1,25-$(OH)_2$-vitamin D in humans. J Lab Clin Med 1979; 93:1031–1034.
105. Yamada S, Schnoes HK, DeLuca HF. Synthesis of 25-hydroxy[23,24-^{3}H]vitamin D_3. Anal Biochem 1978; 85:34–41.
106. Napoli JL, Mellon WS, Fivizzani MA, Schnoes HK, DeLuca HF. Direct chemical synthesis of 1α,25-dihydroxy[26,27-^{3}H]vitamin D_3 with high specific activity: its use in receptor studies. Biochemistry 1980; 19:2515–2521.
107. Bouillon R, De Moor P, Baggiolini EG, Uskokovic MR. A radioimmunoassay for 1,25-dihydroxycholecalciferol. Clin Chem 1980; 26:562–567.
108. Gray TK, McAdoo T. Radioimmunoassay for 1,25-dihydroxycholecalciferol. Clin Chem 1983; 29:196–200.
109. Clemens TL, Hendy GN, Papapoulos SE, Fraher LF, Care AD, O'Riordan JLH. Measurement of 1,25-dihydroxycholecalciferol in man by radioimmunoassay. Clin Endocrinol 1979; 11:225–234.

110. Hollis BW, Kamerud JQ, Kurkowski A, Beaulieu J, Napoli JL. Quantification of circulating 1,25-dihydroxyvitamin D by radioimmunoassay with an ^{125}I-labeled tracer. Clin Chem 1996; 42:586–592.
111. Fraher LJ, Adami S, Clemens TL, Jones G, O'Riordan JLH. Radioimmunoassay of 1,25-dihydroxyvitamin D_2: studies on the metabolism of vitamin D_2 in man. Clin Endocrinol 1983; 18:151–165.
112. Mawer EB, Berry JL, Bessone J, Shany S, Smith H, White A. Selection of high-affinity and high-specificity monoclonal antibodies for $1\alpha,25$-dihydroxyvitamin D. Steroids 1985; 46:741–754.
113. Hollis BW. 1,25-Dihydroxyvitamin D_3-26,23-lactone interferes in determination of 1,25-dihydroxyvitamin D by RIA after immunoextraction. Clin Chem 1995; 41:1313,1314.
114. Ishizuka S, Ohba T, Norman AW. $1,25$-$(OH)_2D_3$-26,23-lactone is a major metabolite of $1,25$-$(OH)_2D_3$ under physiological conditions. In: Vitamin D: Molecular, Cellular, and Clinical Endocrinology. Norman AW, Schaefer K, Grigoleit H-G, Herrath D-v, eds. Berlin: Walter de Gruyter, 1988; 143,144.
115. Reinhardt TA, Napoli JL, Beitz DC, Littledike ET, Horst RL. A new in vivo metabolite of vitamin D_3: 1,25,26-trihydroxyvitamin D_3. Biochem Biophys Res Comm 1981; 99:302–307.
116. Reinhardt TA, Napoli JL, Beitz DC, Littledike ET, Horst RL. 1,24,25-Trihydroxyvitamin D_3: a circulating metabolite in vitamin D_3-treated bovine. Arch Biochem Biophys 1982; 213:163–168.
117. Ohnuma N, Bannai K, Yamaguchi H, Hashimoto Y, Norman AW. Isolation of a new metabolite of vitamin D produced in vivo, $1\alpha,25$-dihydroxyvitamin D_3-26,23-lactone. Arch Biochem Biophys 1980; 204:387–391.
118. Horst RL, Wovkulich PM, Baggiolini EG, Uskokovic MR, Engstrom GW, Napoli JL. (23S)-1,23,25-Trihydroxyvitamin D_3: its biologic activity and role in 1,25-dihydroxyvitamin D_3 26,23-lactone biosynthesis. Biochemistry 1984; 23:3973–3979.
119. Ishizuka S, Naruchi T, Hashimoto Y, Orimo H. Radioreceptor assay for $1\alpha,24(R)25$-trihydroxyvitamin D_3 in human serum. J Nutr Sci Vitaminol 1981; 27:71–75.
120. Horst RL, Hove K, Littledike ET, Reinhardt TA, Uskokovic MR, Partridge JJ. Plasma concentrations of 1,25-dihydroxyvitamin D, $1,24R,25$-trihydroxyvitamin D_3, and 1,25,26-trihydroxyvitamin D_3 after their administration to dairy cows. J Dairy Sci 1983; 66:1455–1460.
121. Hughes MR, Baylink DJ, Jones PG, Haussler MR. Radioligand receptor assay for 25-hydroxyvitamin D_2/D_3 and $1\alpha,25$-dihydroxyvitamin D_2/D_3: application to hypervitaminosis D. J Clin Invest 1976; 58:61–70.
122. Wei S, Tanaka H, Kubo T, Ichikawa M, Seino Y. A multiple assay for vitamin D metabolites without high-performance liquid chromatography. Anal Biochem 1994; 222:359–365.
123. Gloth III, FM, Tobin JD, Sherman SS, Hollis VW. Is the recommended daily allowance for vitamin D too low for the homebound elderly? J Am Geriatr Soc 1991; 39:137–141.
124. Chapuy MC, Arlot ME, DuBoeuf F, Brun J, Arnaud S. Meunier P. Vitamin D_3 and calcium to prevent hip fractures in elderly women. N Engl J Med 1992; 327:1637–1642.
125. Jacobus CH, Holick MF, Shao Q, et al. Hypervitaminosis D associated with drinking milk. N Engl J Med 1992; 326:1173–1177.
126. Aksnes L. An improved competitive protein binding assay for 25-hydroxyvitamin D. Scand J Clin Lab Invest 1978; 38:677–686.
127. Bjorkhem I, Holmberg I. A novel specific assay of 25-hydroxyvitamin D_3. Clin Chim Acta 1976; 68:215–221.
128. Shimotsuji T, Seino Y, Yabuuchi H. A competitive protein binding assay for plasma 25-hydroxyvitamin D_3 in normal children. Tohoku J Expt Med 1976; 118:233–240.
129. Morris JF, Peacock M. Assay of plasma 25-hydroxyvitamin D. Clin Chim Acta 1976; 72:383–392.
130. Haddad JG, Chyu KJ. Competitive protein-binding radioassay for 25-hydroxycholecalciferol. J Clin Endocrinol Metabol 1971; 33:992–925.
131. Guillemant S, Kremer R, Eurin J, Guillemant J, Prier A, Camus JP. Radio competitive protein binding assays for 25-hydroxyvitamin D, 24,25-dihydroxyvitamin D, and 1,25-dihydroxyvitamin D in human serum. In: Vitamin D: Basic Research and its Clinical Application. Norman AW, Schaefer K, Herrath D-v, et al., eds. Berlin: Walter de Gruyter, 1979; 247–250.
132. Hummer L, Christiansen C. A sensitive and selective radioimmunoassay for serum 24,25-dihydroxycholecalciferol in man. Clin Endocrinol 1984; 21:71–79.
133. Weisman Y, Eisenberg Z, Leib L, Harell A, Shasha SM, Edelstein S. Serum concentrations of 24,25-dihydroxy vitamin D in different degrees of chronic renal failure. Br Med J 1980; 281:712,713.
134. Stern PH, Phillips TE, Navreas T. Bioassay of 1,25-dihydroxyvitamin D in human plasma purified by partition, alkaline extraction, and high pressure chromatography. Anal Biochem 1980; 102:22–30.

135. Manolagas SC, Culler FL, Howard JE, Brickman AS, Deftos LJ. The cytoreceptor assay for 1,25-dihydroxyvitamin D and its application to clinical studies. J Clin Endocrinol Metabol 1983; 56: 751–759.
136. Brumbaugh R, Haussler DH, Bursac KM, Haussler MR. Filter assay for 1α,25-dihydroxyvitamin D_3. Utilization of the hormone's target tissue chromatin receptor. Biochemistry 1974; 13:4091–4097.
137. Peacock M, Taylor GA, Brown W. Plasma 1,25$(OH)_2$ vitamin D measured by radioimmunoassay and cytosol radioreceptor assay in normal subjects and patients with primary hyperparathyroidism and renal failure. Clin Chim Acta 1980; 101:93–102.
138. Scharla S, Schmidt-Gayk H, Reichel H, Mayer E. A sensitive and simplified radioimmunoassay for 1,25-dihydroxyvitamin D_3. Clin Chim Acta 1984; 142:325–338.
139. Official Methods of Analysis, 12th ed. Washington, DC: Association of Official Analytical Chemists, 1975; 885.

2. CLINICAL FEATURES OF VITAMIN D DEFICIENCY

The clinical presentation of vitamin D deficiency differs depending on the age of the patient at onset of the disease.

2.1. Neonates

Vitamin D deficiency during pregnancy seldom causes maternal osteomalacia *(4–6)*, but clinical consequences of this deficiency are observed in the neonates. Although findings of bone deformities suggestive of rickets are exceptional *(6–9)*, vitamin D-deficient newborns, either full-term or premature, may present with clinical signs of hypocalcemia, fine tremor of the fingers and chin, episodes of tachypnea followed by apnea and cyanosis, vomiting, or even convulsions *(10,11)*. Two types of neonatal hypocalcemia are usually distinguished. Premature babies present with an "early type," often asymptomatic. In this type, biologic and electrocardiographic changes, mainly a prolonged QT interval, are found from the first 24–48 h of life and last 5–10 d *(11)*. The mechanism leading to early hypocalcemia is not totally understood. It does not merely result from maternal vitamin D deficiency or from an immaturity of the enzyme systems involved in vitamin D metabolism *(12,13)*. Vitamin D deficiency may facilitate its occurrence, however *(14)*. The second type is defined as "late" neonatal hypocalcemia. It mostly occurs in full-term babies and is associated with convulsions or motor and sensory symptoms of tetany. Some cases clearly result from congenital disorders of the parathyroid function, but, as in the early type, maternal vitamin D deficiency is likely to be involved in a significant number of cases, as suggested by the observation of seasonal fluctuations in the incidence of "early" and "late" neonatal hypocalcemia in regions with suboptimal amounts of sunlight or excessive atmospheric pollution *(14)*. Neonatal hypocalcemia is also found in "at-risk" populations of neonates born to mothers who did not expose themselves to solar radiation for climatic, religious, or socioeconomic reasons *(15,16)*, and even more when mothers are vegetarian, possibly because of the reduced vitamin D content of these diets *(17)*. Administration of vitamin D supplements to such pregnant women significantly reduces the percentage of neonates with signs of hypocalcemia *(14–16)*. In a total population of 13,377 neonates, for instance, this percentage decreased from 0.55 to 0.21% following vitamin D supplementation of the pregnant mothers *(14)*.

Congestive heart failure is another short-term consequence of intrauterine and neonatal vitamin D deficiency. It is exceptional but requires immediate care and treatment *(18,19)*.

Maternal vitamin D deficiency also has long-term consequences. It may be associated with slightly lower weight gain and linear growth during the first year of life *(20)*. Most of all, it increases the risk for hypoplasia of the dental enamel of primary teeth *(16,21)*. A study of 61 infants in their third year of life showed a defect of dental enamel in 48% of infants born to mothers with a low vitamin D status and in only 7% of those born to vitamin D-supplemented mothers *(21)*.

Finally, maternal vitamin D deficiency may have long-term consequences for the mother. Numbers of pregnancies, moderate calcium deficiency during pregnancy, and breast-feeding do not appear to be long-term determinants of bone density in women *(22)*. However, the influence of vitamin D deficiency on the mother's bone mass and on her risk of developing osteoporosis after menopause has not yet been investigated.

Table 1
Clinical and Radiologic Signs of Vitamin D Deficiency:
Frequency (%) According to Age in 46 Infants

	Age (mo)			
	2–3	4–6	7–12	13–18
Growth				
Ponderal defect (>2 SD)	87	62	36	0
Statural defect (>2 SD)	0	0	0	27
Rickets				
Hypotonia	93	100	100	93
Bone pain	—	75	73	75
Rachitic rosary	40	87	91	67
Swelling of wrists, knees, and ankles	60	62	64	50
Craniotabes	53	62	36	25
Craniostenosis	0	0	0	0
Fractures[a]	20	12	9	8
Other signs				
Convulsions	20	12	27	25
Infection of the respiratory tract	74	74	63	59
Ricketic lung[a]	67	50	73	58

[a] Signs observed on thoracic and skeletal radiographs.

2.2. Infants

Infants are particularly prone to develop severe vitamin D deficiency and present typical skeletal signs like rickets, poor mineralization of the skeleton, bone deformities, and slow growth *(3)*. Fortification of infant formulas and/or routine supplementation of the infants with vitamin D have markedly decreased the incidence of this disease during the first 2–4 yr of life. Rickets can still be seen among unsupplemented breast-fed infants *(23)* or in infants fed macrobiotic *(24)* or strictly vegetarian diets *(25)*.

Neurologic symptoms linked to hypocalcemia are observed in 10–30% of the infants with vitamin D deficiency, especially at the onset of the disease (Tables 1 and 2). They may be severe enough to require immediate treatment, such as convulsions, fits, laryngismus stridulus, or mild paresthesis. Carpopedal spasms and Trousseau's sign are characteristic. They may be spontaneous or occur after a voluntary contraction or overbreathing. Chvostek's sign is another classical sign, although it is found in as many as 5–20% of healthy individuals *(26)*. Other signs of tetany include intestinal manifestations, spasms of the diaphragm, and spasms of voluntary and smooth respiratory muscles.

Muscular hypotonia, skeletal changes, and bone pain are the main features of vitamin D deficiency during infancy (Table 1). Early skeletal signs of rickets are palpable costochondral beading, termed rachitic rosary, and epiphyseal enlargement at the wrists, knees, and ankles. Although a slight costochondral beading may be seen in lean infants *(3)*, these signs are considered to be the most reliable signs of rickets, especially when

Table 2
Age, Fractures, and Extraskeletal Signs (%)
According to the Severity of Vitamin D Deficiency

	Stage	
	Early	Advanced
Age (mo)		
2–3	40	60
4–6	63	37
7–12	73	27
13–18	50	50
Fractures on X-rays	8	24
Extraskeletal signs		
Convulsions	36	5
Delay in standing or walking	64	76
Rhinopharyngitis	32	5
Bronchopulmonary infection	24	76
Palpable liver and spleen	24	24
Hemoglobin <10 g/100 mL	28	5

present simultaneously. Other signs are not specific and may lead to overdiagnosis. Thus, delay in closure of the anterior fontanelle and delay in walking may result from malnutrition. Softening of the skull (craniotabes) may be physiologic before the age of 3 mo. Its presence in older infants is associated with rickets in only 5–25% of the cases in the tropics *(3)*. It is not constant in children with vitamin D deficiency, 25–36% of the children after 6 months of age in our experience (Table 1), and appears to be more frequent in countries where babies remain in a supine position for long periods than in countries where babies are carried in a sitting position *(3)*.

Signs of advanced rickets can be observed at any age (Table 2). They include characteristic deformities that affect different parts of the skeleton depending on the age of the child at the onset of the disease *(2)*. Deformities of the head, with frontal bossing and parietal or occipital flattening of the skull, are mostly found in young infants. Older infants show deformities of the chest (pigeon chest, Harrison's groove), which interfere with ventilation, and deformities of the spinal column and pelvis (kyphosis, scoliosis, and coxa vara), which contribute to the waddling gait. They also show bending of the forearms, which is related to the sitting position, and bowing of the legs, knock knees, genu valgum, or genu varum. Green-stick fractures may occur in the long bones. They may remain clinically unrecognized because of the severity of bone pain and tenderness of the limbs even in the absence of fracture.

Other clinical signs in the younger infants are failure to thrive, especially during the first year of life (Table 1). Older infants may show some statural defect, in part related to the skeletal deformities, delay in standing or walking, and muscular hypotonia with pot-belly, in association with their bone deformations. These defects result from softening of the bones, relaxation of ligaments, poor development and weakness of limb, thoracic, and abdominal muscles. Several other associated signs, such as respiratory infections, bronchitis, and bronchopneumonia, and anemia with iron deficiency are typical rachitic complications. They are mostly found in infants with advanced rickets

(Table 2). Palpable liver and spleen mainly result from muscular hypotonia, but they may reflect true hepatosplenomegaly in children with severe anemia. Older infants also show delayed eruption of the teeth and enamel hypoplasia as a consequence of hypocalcemia.

2.3. Children and Adolescents

Although rare, cases of vitamin D deficiency rickets have been reported in "at-risk" populations of school children and adolescents in the tropics *(3)*, in Europe *(27–34)*, and in northern China, where the incidence of late-onset rickets was recently reported to be as high as 5–15% among adolescents *(35)*. Most cases are found in geographic areas with suboptimal solar irradiation or atmospheric pollution, particularly in communities with limited sun exposure for climatic, religious, or socioeconomical reasons and in subjects with heavy skin pigmentation.

Vitamin D deficiency at this age usually manifests by moderate deformations of the joints and bone pains at the spinal column and limbs, but more severe deformations, i.e., genu varum, genu valgum, and coxa vara, can be observed. Vitamin D deficiency should therefore be excluded by measuring the serum 25-hydroxyvitamin D before considering surgical correction of such long bone deformities during childhood and adolescence.

3. RADIOLOGIC SIGNS

Distinctive radiologic features of vitamin D deficiency during growth result from the alteration of cell growth and activity in the hypertrophic zone of the growth plate cartilage, leading to disorganization of the chondrocyte columns, and from deficient mineralization of the zone of provisional calcification (*see* Chapter 1). They are best observed in the fastest growing parts of the skeleton, the costochondral junctions of the middle ribs, the distal femur, both ends of the tibia, the proximal humerus and the distal ulna and radius *(2)*. The first signs are found at the costochondral junctions and the distal growth plates of the ulna and radius, which appear cupped and widened. Later on, this widening progress and beak-like prolongations of the ossification line on each side may be seen. The metaphyseal zone of provisional calcification becomes irregular, and fraying and poor definition of the trabecular bone in the metaphyses can be observed. This metaphyseal zone of calcification is hardly visible in the most advanced cases. A simultaneous defect in the ossification of the epiphyses results in late-appearing, irregular, and indistinct epiphyseal centers and a widening of the space separating the metaphyseal line from the epiphyseal nucleus.

Diaphyses of the long bones show rarefaction of the shafts. Cortices are thin and translucent, with occasional lamellar deposits of bone under the periosteum. Other signs of secondary hyperparathyroidism may be seen. Deformities of the diaphyses produce bowing of the legs, and sometimes of the clavicles, and coxa vara. The cranial vault may be thin or thickened, particularly in the occipital and frontal regions. Some children may have craniostenosis. Fractures may be observed in the legs, femoral neck, forearms, and ribs. Cortical radiolucent lines perpendicular or oblique to the bone shaft may be found on the femoral and humeral necks, pubic rami, ribs, and axillary borders of the scapula in adolescents and adults. These pseudofractures (Looser-Milkman zones) are mostly symmetric and bilateral. Unlike true fractures, they are not painful.

Typical signs of rickets at the metaphyseoepiphyseal zones of the long bones and osteopenia with thinning of the cortices suggest vitamin D deficiency. However, they are

Table 3
Serum Data in 46 Infants According to the Severity of Vitamin D Deficiency (mean ± 1 SD)

Parameter	Stage		Normal values[b]
	Early	Advanced[a]	
Serum calcium (mg/dL)	8.05 ± 0.7	7.50 ± 0.6	10.0 ± 0.4 ($n = 162$)
Incidence of low serum calcium (%) (<8.0 mg/dL)	32	71	—
Serum phosphorus (mg/dL)	5.41 ± 0.22	4.70 ± 0.17	Age-dependent
Alkaline phosphatase activity (IU/L)	346 ± 36	455 ± 59	199 ± 47 ($n = 127$)
25(OH)D (ng/mL)	3.2 ± 1.2	1.3 ± 0.8	31 ± 11[c] ($n = 158$)
1,25(OH)$_2$D (pg/mL)	66 ± 42	45 ± 27	84 ± 33 ($n = 71$)

[a] Infants with clinical and radiologic signs of advanced rickets

[b] Values found with the same techniques in 162 healthy 8–11-mo-old infants (69). Number of infants in parentheses.

[c] All infants were supplemented with vitamin D.

indistinguishable from those encountered in rickets secondary to calcium deficiency, to acquired or inherited disorders of vitamin D metabolism (see Chapter 18), or to hereditary resistance to 1,25-dihydroxyvitamin D (see Chapter 20) *(1,2)*. Determination of the precise etiology in these cases relies on familial history and laboratory data. Biologic investigation is also required to exclude other rachitic syndromes, which share several clinical and radiologic features with vitamin D deficiency rickets, like those resulting from hereditary or acquired hypophosphatemia, hypophosphatasia, renal diseases, and gastrointestinal malabsorption *(2)*. But differential diagnosis with metaphyseal chondrodysplasia (Schmid type) is usually easy, as metaphyses and diaphyses are well mineralized and no Looser's zones or signs of secondary hyperparathyroidism are found in this disease *(2)*.

4. BIOCHEMICAL SIGNS

4.1. Typical Biology in Patients with Clinical and/or Radiologic Signs of Vitamin D Deficiency

Basal exploration of calcium and phosphorus homeostasis shows decreased serum levels of total and ionized calcium in most patients. Hypocalcemia becomes more severe as the vitamin D deficiency progresses (Table 3), as opposed to the clinical manifestations of this disorder, which are mainly found during the early phase of the disease (Table 2). A few patients may have normocalcemia at this early phase or because of recent, unnoticed administration of vitamin D. However, all have hypocalciuria, reflecting inadequate intestinal absorption of calcium and the elevated requirements of the undermineralized skeleton. Serum levels of phosphorus are normal or low in the more advanced forms (Table 3). They mainly result from decreased tubular reabsorption of this ion.

These biologic disorders are associated with signs of secondary hyperparathyroidism, as evidenced by elevated levels of parathyroid hormone (PTH). Urinary and nephrogenic

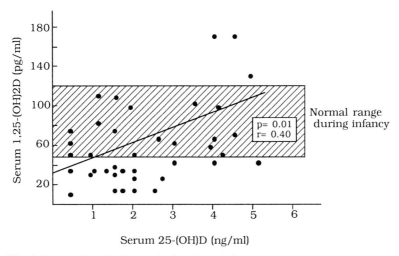

Fig. 1. Serum vitamin D metabolites in 39 infants with vitamin D deficiency.

cyclic adenosine monophosphate (cAMP) excretions are also elevated, but their levels are no longer determined since the development of immunoradiometric methods to assay the intact PTH molecule. Secondary hyperparathyroidism aggravates, or is responsible for, decreased reabsorption by the proximal tubule of phosphates, bicarbonates, and amino acids. Slight hyperchloremic acidosis and isolated hyperaminoaciduria are observed in patients with vitamin D deficiency *(36)*.

Secondary hyperparathyroidism also appears to be responsible for the increased rate of bone turnover suggested by the elevated alkaline phosphatase activity in the serum. This activity increases with the severity of the vitamin D deficiency (Table 3). Osteocalcin is another marker of osteoblast activity. Its serum levels tend to be lower than normal in children with vitamin D deficiency rickets *(37,38)*, as in children with calcium deficiency rickets *(39)*. In practice, the assay of osteocalcin in the serum of vitamin D-deficient children is of no help for the clinician. as published levels are not consistently altered before treatment. Moreover, considerable changes in these levels follow minimal vitamin D administration, for instance, the 10 times increase over basal values observed 3 wk after the onset of vitamin D treatment *(37)*.

The association of hypocalcemia and secondary hyperparathyroidism strongly suggests vitamin D deficiency in children with clinical and radiologic signs of rickets. However, identical data are obtained in patients with calcium deficiency rickets *(39–41)*, and in those with a hereditary or acquired defect in the renal production of $1,25(OH)_2D$, or with a hereditary resistance to this active metabolite of vitamin D *(42)*. Confirmation of vitamin D deficiency thus necessitates the assay of vitamin D metabolites. Typically, serum levels of the major circulating form of vitamin D, 25-hydroxyvitamin D [25(OH)D], are low or undetectable before vitamin D treatment in patients with vitamin D deficiency *(43–47)* (Table 3), and normal in children with calcium deficiency *(39–41)* or pseudo-deficiency rickets *(42)*.

Assay of the active form, 1,25-dihydroxyvitamin D, is much less informative. Circulating levels of this metabolite do decrease with the severity of vitamin D deficiency (Table 3) and are positively correlated with the levels of its precursor, 25(OH)D (Fig. 1). However, absolute values of serum $1,25(OH)_2D$ are either low, normal, or even elevated in vitamin D-deficient patients before treatment (Figs. 1 and 2) *(43–48)*. Such high serum

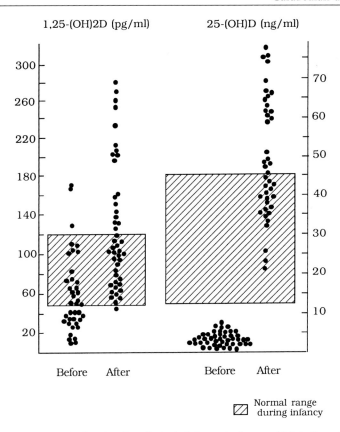

Fig. 2. Serum vitamin D metabolites in vitamin D deficiency before and 21 d after administration of 600,000 IU vitamin D_3.

1,25(OH)$_2$D levels, despite the deficiency of substrate, probably result from an activation of the renal 25(OH)D-1-hydroxylase by low serum calcium, low serum phosphate, and elevated PTH levels, three major stimulators of this enzyme activity *(49)* and determinants of the circulating 1,25(OH)$_2$D levels *(50,51)*. Another difficulty arises from the age-dependent variations in 1,25(OH)$_2$D levels. Interpretation of the data thus requires precise knowledge of reference values for the age (Table 4).

This assay may help to confirm or exclude the diagnosis of vitamin D deficiency in the following cases: (1) the child has received vitamin D before blood collection and its 25(OH)D levels are therefore in the normal range; or (2) the child has low 25(OH)D levels indicating a low vitamin D status, but its disease results from an impaired production of 1,25(OH)$_2$D (pseudodeficiency rickets type I, pseudohypoparathyroidism, renal disease) or from a resistance to 1,25(OH)$_2$D (pseudodeficiency rickets type II). If rickets results from vitamin D deficiency, the kidney's ability to produce 1,25(OH)$_2$D is high and only limited by substrate deficiency. Vitamin D administration is thus rapidly followed, within the first 2–4 wk, by a marked increase in 1,25(OH)$_2$D, up to levels in the normal range or above (Fig. 2 and Table 5), and by the normalization of serum calcium and phosphorus *(45–48)*. Children with resistance to 1,25(OH)$_2$D show a similar, or even greater, increase in 1,25(OH)$_2$D, but with no effect on their low serum calcium and phosphate. Children with impaired production of 1,25(OH)$_2$D show no increase in 1,25(OH)$_2$D, or only a slight increase, and no normalization of serum calcium and phosphorus. The

Table 4
Serum 1,25(OH)$_2$D Levels in Healthy Subjects
According to Age (Mean and Extreme Range)[a]

Populations	1,25(OH)$_2$D (pg/mL)	1,25(OH)$_2$D (pmol/L)
Neonates (n = 84)	50 (20–80)	120 (50–190)
Infants (n = 71)	80 (50–120)	190 (120–290)
Children (n = 80)	50 (20–70)	120 (50–170)
Adolescents (n = 53)	50 (30–80)	120 (70–190)
Adults (n = 40)	40 (20–60)	100 (50–140)
Pregnant women (n = 84)	70 (40–90)	170 (100–220)

[a] Data taken from personal results in healthy populations living in France.

Table 5
Alkaline Phosphatase Activity and Serum Vitamin D
Metabolites During Treatment of Vitamin D Deficiency[a]

	Before	+ 11–30 d	+ 45 d
Alkaline phosphatase (IU/L)			
Study 1	400 ± 47	280 ± 80	220 ± 60
Study 2	460 ± 100	—	—
Study 3	420 ± 80	320 ± 100	—
25(OH)D (ng/mL)			
Study 1	2 ± 1	55 ± 20	53 ± 9
Study 2	5 ± 2	19 ± 8	—
Study 3	4 ± 2	9 ± 7	—
1,25(OH)$_2$D (pg/mL)			
Study 1	54 ± 35	113 ± 70	106 ± 68
Study 2	186 ± 141	230 ± 127	—
Study 3	152 ± 109	325 ± 117	—

[a] Study 1, 46 infants (2–17 mo) treated with one oral dose of 600,000 IU vitamin D$_2$ (unpublished results); study 2, five infants and children (4 mo–4 yr) treated with one oral dose of 10 µg/kg 25(OH)D$_3$ (46); study 3, eight infants and children (7 mo–12 yr) treated with daily oral doses of 50 µg/d of vitamin D$_2$ (46).

kinetics study of the serum calcium, phosphate, and 1,25(OH)$_2$D responses to vitamin D administration may thus be used to confirm vitamin D deficiency in complicated situations. Serum 1,25(OH)$_2$D may remain elevated up to 1–2 mo after the onset of vitamin D administration to vitamin D-deficient children, long after normalization of the serum calcium, phosphate, and even PTH levels (Table 5). Thus findings of elevated alkaline phosphatase and 1,25(OH)$_2$D values in the serum of a child with clinical and radiologic signs of rickets but no other biologic disorder suggest that the origin of the rickets is of a past vitamin D deficiency.

Neonates

Some neonates with clinical signs of hypocalcemia show severe hypocalcemia, hyperphosphatemia, and low serum PTH levels during their first weeks of life *(52,53)*. This association suggests transient congenital hypoparathyroidism *(54)*. Its relation to maternal vitamin D deficiency is not clear, but should be investigated more extensively since this condition may be aggravated by vitamin D deficiency *(54)* and has been found to be associated with low 25(OH)D levels *(53)*.

4.3. Subclinical Vitamin D Deficiency

Finally, asymptomatic biologic signs of vitamin D deficiency have been observed in a substantial proportion of apparently healthy populations living in countries with suboptimal amounts of sunlight during the winter-spring period. This has been shown for Asian immigrants *(7,17)*, as well as for nonimmigrant pregnant women and neonates *(16,53)*, and elderly persons *(55–58)*. A few long-term consequences of subclinical vitamin D deficiency have been considered, including slow growth during infancy *(20)* and increased risk for enamel defect of primary teeth, when the deficiency occurs during fetal life *(16)*, and increased risk for hip fractures and bone loss, when the deficiency occurs in elderly women (*see* Chapter 17) *(56,57)*. Other long-term consequences need to be investigated, for example, on the acquisition of peak bone mass, on cell differentiation, and on the immune defense system. To do so, true subclinical deficiency has to be distinguished from hypovitaminosis D with no pathologic consequence. Biologic criteria for subclinical vitamin D deficiency have been proposed in neonates *(53)*, i.e., the association of low 25(OH)D concentrations (<12 ng/mL, 30 nmol/L) and high PTH concentrations (>60 pg/mL). A similar approach should be taken to define criteria for asymptomatic vitamin D deficiency in older children, adolescents, and adults.

5. TREATMENT

Different therapeutic strategies have been proposed *(45–48,59)*. As a rule, the cure of simple rickets requires a total of 5–15 mg of vitamin D_2 or D_3 given orally. These doses can be given either as a single-day therapy (Table 5) *(59)* or as daily doses of 2000–4000 IU/d (50–100 µg/d) for 3–6 mo. All these strategies result in a similar increase in 25(OH)D concentrations (Table 5), resulting in correction of the calcium and phosphorus disorders within 6–10 d, normalization of PTH levels within 1–2 mo, normalization of alkaline phosphatase activity, and healing of radiologic signs of rickets within 3–6 mo, depending on the severity of the deficiency. A single large doses (5 mg), possibly repeated 3 mo later, may be preferred to eliminate lack of compliance. Data on 25(OH)D levels during treatment, like those presented in Table 5, clearly show that the small daily doses can cure rickets but do not restore adequate stores of vitamin D as rapidly as large single doses. The oral route of administration is advisable, except in cases with severe fat malabsorption, because of its better bioavailability than subcutaneous or intramuscular administrations *(60)*.

In any case, treatment should also provide sufficient calcium for correction of the demineralization defect and to avoid the complication of hypocalcemia. Children with low serum calcium, <8.0 mg/dL, should receive calcium infusion, from a few hours before the first administration of vitamin D up to normalization of serum calcium, to avoid the occurrence of clinical signs of hypocalcemia. Other children should have daily intakes of calcium of at least 1 g/d during the first months of treatment, via either dietary intake or oral calcium supplements.

6. PREVENTION OF VITAMIN D DEFICIENCY DURING PREGNANCY AND CHILDHOOD

Regular sun exposure is the most physiologic way to prevent vitamin D deficiency. Geographic, climatic, or pollution conditions may limit the intensity of the ultraviolet light adequate for skin production of vitamin D in the solar light, especially during the winter–spring season *(61)* (*see* Chapter 2). In addition, the surface of the skin exposed, duration of the exposure, or ability of the skin to produce vitamin D may be insufficient in some sociocultural, ethnic, and age groups *(4,5,7,17, 27,32,33,54–57,61)*. Enrichment of infant food with vitamin D is an appropriate way to compensate for the lack of sun exposure during pregnancy and early life. But children with a higher risk of rickets (because they recently emigrated to industrial areas, have restricted outdoor activities, are poorly fed, escape regular medical surveys, were born to vitamin D-deficient mothers, and so forth) are also often breast-fed and therefore cannot benefit from the enriched food. Daily vitamin D supplementation has been extensively advocated in countries where infant milk is not supplemented with vitamin D. In this case, oral doses of 1000–1500 IU/d, all year round up to the age of 2 yr and during the winter up to 5 yr, have been shown to prevent vitamin D deficiency adequately with no sign of vitamin D intoxication *(62)*. This prevention scheme may therefore be proposed to "at-risk" infants who are exclusively breast-fed or who are too old to take formula. Prevention with lower daily doses (400–500 IU/d) may be recommended to all the other breast-fed infants *(63)*, and to "at-risk" neonates and young infants receiving formula *(53)*. Comparison of the effects of maternal and infant supplementation during lactation suggests that prevention of vitamin D deficiency is best achieved by direct vitamin D administration to the child *(63)*.

Intermittent high oral doses of vitamin D may be required for populations of infants who escape systematic survey and when a lack of compliancy is suspected *(64,65)*. If very high doses (15 mg) appear clearly excessive on the basis of their effect on serum levels of 25(OH)D *(64,65)*, lower doses, (5 mg every 6 mo, or better, 2.5 mg every 3 mo) may provide adequate protection during the first 2 years of life *(65)*. A similar preventive regime can be proposed to "at-risk" older children and adolescents.

Therefore, prevention of vitamin D deficiency during early life also implies the prevention of maternal deficiency. Different strategies have been proposed with comparable preventive effects and little overloading risk, as suggested by maternal and neonatal 25(OH)D levels: daily oral doses of vitamin D, 1000 IU/d during the last trimester of pregnancy *(15,20,66,67)* or 400 IU/d from the wk 12 of pregnancy *(16)*, or a single oral dose of 2.5 mg during the sixth or seventh month of pregnancy *(53,68)*.

REFERENCES

1. Glorieux FH, ed. Rickets. Nestlé Nutrition Workshop Series, vol 21. New York: Raven, 1991; 285.
2. Pitt MJ. Rickets and osteomalacia are still around. Radiol Clin North Am 1991; 29:97–118.
3. Bhattacharyya AK. Nutritional rickets in the tropics. In: Nutritional Triggers for Health and in Disease. Simopoulos AP, ed. World Rev Nutr Diet 1992; 67:140–197.
4. Felton DJC, Stone WD. Osteomalacia in Asian immigrants during pregnancy. Br Med J 1966; 1: 1521,1522.
5. Rab SM, Baseer A. Occult osteomalacia amongst healthy and pregnant women in Pakistan. Lancet 1976; 2:1211–1213.
6. Park W, Paust H, Kaufmann HJ, Offermann G. Osteomalacia of the mother. Rickets of the newborn. Eur J Pediatr 1987; 146:292,293.
7. Preece M, McIntosh W, Tomlinson S, Ford J, Dunnigan M, O'Riordan J. Vitamin D deficiency among Asian immigrants to Britain. Lancet 1973; i:907–910.
8. Ford JA, Davidson DC, Mc Intosh WB, Fyfe WM, Dunnigan MG. Neonatal rickets in Asian immigrant population. Br Med J 1973; 3:211,212.
9. Moncrieff M, Fadahursi TO. Congenital rickets due to maternal vitamin D deficiency. Arch Dis Child 1974; 49:810,811.

10. Rosen JF, Roginsky M, Nathenson G, Finberg L. 25-hydroxyvitamin D. Plasma levels in mothers and their premature infants with neonatal hypocalcemia. Am J Dis Child 1974; 127:220–223.
11. Rösli A, Fanconi A. Neonatal hypocalcemia: "Early type" in low birth weight newborns. Helv Pediatr Acta 1973, 28:443–457.
12. Hillman LS, Rojanasathit S, Slatopolsky E, et al. Serial measurements of serum calcium, magnesium, parathyroid hormone, calcitonin and 25-hydroxyvitamin D in premature and term infants during the first week of life. Pediatr Res 1977; 11:739–744.
13. Glorieux FH, Salle BL, Delvin E, et al. Vitamin D metabolism in preterm infants: serum calcitriol values during the first 5 days of life. J Pediatr 1981; 99:640–643.
14. Mallet E, Henocq A, De Menibus CH. Effects of vitamin D supplementation in pregnant women on the frequency of neonatal hypocalcemia. In: Vitamin D, A Pluripotent Steroid Hormone: Structural Studies, Molecular Endocrinology and Clinical Applications. Norman AW, Bouillon R, Thomasset M, eds. Berlin: Walter de Gruyter, 1994; 869–870.
15. Brooke OG, Brown IRF, Bone CDM, Carter ND, Cleeve HJW, Maxwell JD, Robinson VP, Winder SM. Vitamin D supplements in pregnant Asian women: effects on calcium status and fetal growth. Br Med J 1980; 280:751–754.
16. Cockburn F, Belton NR, Purvis RJ, Giles MM, Brown JK, Turner TL, Wilkinson EM, Forfar JO, Barrie WJM, McKay GS, Pocock SJ. Maternal vitamin D intake and mineral metabolism in mothers and their newborn infants. Br Med J 1980; 281:11–14.
17. Brooke OG, Brown IRF, Cleeve HJW. Observations on the vitamin D state of pregnant asian women in London. Br Med J 1981; 88:18–26.
18. Gillore A, Groneck P, Kaiser J, Schmitz-Stolbrink A. Congestive heart failure in an infant due to vitamin D deficiency rickets. Monatsschr Kinderheilkd 1989; 137:108–110.
19. Brunvand L, Haga P, Tangsrud SE, Haug E. Congestive heart failure caused by vitamin D deficiency? Act Paediatr 1995; 84:106–108.
20. Brooke OG, Butters F, Wood C. Intrauterine vitamin D nutrition and post natal growth in Asian infants. Br Med J 1981; 283:1024.
21. Purvis RJ, Barrie WJM, Mac Kay GS, Wilkinson EM, Cockburn F, Belton NR, Forfar JO. Enamel hypoplasia of the teeth associated with neonatal tetany: a manifestation of maternal vitamin D deficiency. Lancet 1973; 2:811–814.
22. Kritz-Silverstein D, Barrett-Connor E, Hollenbach KA. Pregnancy and lactation as determinants of bone mineral density in postmenopausal women. Am J Epidemiol 1992; 136:1052–1059.
23. Edidin D, Levitsky L, Schey W, Dumbovic N, Campos A. Resurgences of nutritional rickets associated with breast-feeding and special dietary practices. Pediatrics 1980; 65:232–235.
24. Dagnelie PC, Vergote F, Van Staveren WA, van den Berg H, Dingjan PG, Hautvast JGAJ. High prevalence of rickets in infants on macrobiotic diets. Am J Clin Nutr 1990; 51:202–208.
25. Key LL. Nutritional rickets. Trends Endocrinol Metab 1991; 2:81–85.
26. Hoffman E. The Chovstek sign; a clinical study. Am J Surg 1958; 96:33–37.
27. Ford JA, Colhoun EM, McIntosh WB, Dunnigan MG. Rickets and osteomalacia in the Glasgow Pakistani community 1961–1971. Br Med J 1972;2:677–680.
28. Moncrieff MW, Lunt HRW, Arthur LJH. Nutritional rickets at puberty. Arch Dis Child 1973; 48:221–24.
29. Cooke WT, Asquith P, Ruck N, Melikian V, Swan CHJ. Rickets, growth, and alkaline phosphatase in urban adolescents. Br Med J, 1974; 2:293–297.
30. Corbeel L, Bouillon R, Deschamps L, Geussens H. Vitamin D sensitive rickets in puberty. Acta Paediatr Belg 1976; 29:103–108.
31. Ford JA, McIntosh WB, Butterfield R, et al. Clinical and subclinical vitamin D deficiency in Bradford children. Arch Dis Child 1976; 51:939–943.
32. Freycon MT, Pouyau G, Abeille A, Frederich A, Durr F, Loras B, Freycon F. Rachitisme carentiel chez le grand enfant. A propos de deux observations. Pediatrie 1983; 38:485–490.
33. O'Hare AE, Uttley WS, Belton NR et al. Persisting vitamin D deficiency in the Asian adolescent. Arch Dis Child 1984; 59:766–770.
34. Bénichou JJ, Sallière D, Labrune B. Rachitisme carentiel chez une adolescente. Arch Fr Pediatr 1985; 42:443–445.
35. Zhou H. Rickets in China. In: Rickets. Nestlé Nutrition Workshop Series, vol 21. Glorieux FH, ed. New York: Raven, 1991; 253–261.

36. Fraser D, Kooh SW, Scriver CR. Hyperparathyroidism as the cause of hyperaminoaciduria and phosphaturia in human vitamin D deficiency. Pediatr Res 1967; 1:425–436.
37. Greig F, Casas J, Castells S. Plasma osteocalcin before and during treatment of various forms of rickets. J Pediatr 1989;114:820–823.
38. Kruse K, Kracht U. Evaluation of serum osteocalcin as an index of altered bone metabolism. Eur J Pediatr 1986; 145:27–33.
39. Oginni LM, Worsfold M, Sharp CA, Oyelami OA, Powell DE, Davie MW. Plasma osteocalcin in healthy Nigerian children and in children with calcium-deficiency rickets. Calcif Tissue Int 1996; 59: 424–427.
40. Kooh SW, Fraser D, Reilly BJ, Hamilton JR, Gall DG, Bell L. Rickets due to calcium deficiency. N Engl J Med 1977; 297:1264–1266.
41. Pettifor JM, Ross FP, Travers R, Glorieux FH, De Luca HF. Dietary calcium deficiency: a syndrome associated with bone deformities and elevated serum 1,25-dihydroxyvitamin D concentrations. Metab Bone Dis Rel Res 1981; 2:301–305.
42. Thakker RV, O'Riordan JLH. Inherited forms of rickets and osteomalacia. Bailleres Clin Endocrinol Metab 1988; 2:157–191.
43. Stanbury SW, Taylor CM, Lumb GA, Mawer EB, Berry J, Hahn J, Wallace J. Formation of vitamin D metabolites following correction of human vitamin D deficiency. Observations in patients with nutritional osteomalacia. Miner Electrolyte Metab 1981; 5:212–227.
44. Rosen JF, Chesney RW. Circulating calcitriol concentrations in health and disease. J Pediatr 1983; 103:1–16.
45. Chesney RW, Zimmerman J, Hamstra A, De Luca HF, Mazess RB. Vitamin D metabolite concentrations in vitamin D deficiency: are calcitriol levels normal? Am J Dis Child 1981; 135:1025.
46. Garabédian M, Vainsel M, Mallet E, Guillozo H, Toppet M, Grimberg R, Nguyen TM, Balsan S. Circulating vitamin D metabolite concentrations in children with nutritional rickets. J Pediatr 1983; 103:381–386.
47. Markestad T, Halvorsen S, Seeger Halvorsen K, Asknes L, Aarskog D. Plasma concentrations of vitamin D metabolites before and during treatment of vitamin D deficiency rickets in children. Acta Paediatr Scand 1984; 73:225–231.
48. Venkataraman P, Tsang RC, Buckley DD, Ho M, Steichen JJ. Elevation of serum 1,25-dihydroxyvitamin D in response to physiologic doses of vitamin D in vitamin D deficient infants. J Pediatr 1983; 103:416–419.
49. Kumar R. The metabolism and mechanism of action of 1,25-dihydroxyvitamin D3. Kidney Int 1986; 30:793–803.
50. Dawson-Hugues B, Harris S, Dallal GE. Serum ionised calcium, as well as phosphorus and parathyroid hormone, is associated with the plasma 1,25-dihydroxyvitamin D3 concentration in normal postmenopausal women. J Bone Miner Res 1991; 6:461–468.
51. Ho SC, Mac Donald D, Chan C, Fan YK, Chan SSG, Swaminathan R. Determinants of serum 1,25-dihydroxyvitamin D concentration in healthy premenopausal subjects. Clin Chim Acta 1994; 230: 21–33.
52. David L, Anast CS. Calcium metabolism in newborn infants. The interrelationship of parathyroid function and calcium, magnesium, and phosphorus metabolism in normal, "sick" and hypocalcemic newborns. J Clin Invest 1974; 54:287–296.
53. Zeghoud F, Vervel C, Guillozo H, Walrant O, Boutigon H, Garabédian M. Subclinical vitamin D deficiency in neonates: definition and response to vitamin D supplements. Am J Clin Nutr 1997; 65:771–778.
54. Fanconi A, Prader A. Transient congenital idiopathic hypoparathyroidism. Helv Paediat Acta 1967; 22:342–359.
55. Sherman SS, Hollis BW, Tobin JD. Vitamin D status and related parameters in a healthy population: the effects of age, sex and season. J Clin Endocrinol Metab 1990; 71:405–413.
56. Chapuy MC, Arlot ME, Duboeuf F, Brun J, Crouzet B, Arnaud S, Delmas PD, Meunier PJ. Vitamin D3 and calcium to prevent hip fractures in elderly women. N Engl J Med 1992; 327:1637–1642.
57. Ooms ME, Roos JC, Bezemer PD, Van Der Vijgh WJF, Bouter LM, Lips P. Prevention of bone loss by vitamin D supplementation in elderly women: a randomized double-blind trial. J Clin Endocrinol Metab 1995; 80:1052–1058.
58. Gloth FM III, Gundborg CM, Hollis BW, Haddad JG, Tobin JD. Vitamin D deficiency in homebound elderly persons. JAMA 1995; 274:1683–1686.

59. Shah BR, Finberg L. Single-day therapy for nutritional vitamin D deficiency rickets: a preferred method. J Pediatr 1994; 125:487–490.
60. Whyte MP, Haddad JG, Walters DD, Stamp TCB. Vitamin D bioavailability: serum 25-hydroxyvitamin D levels in man after oral, subcutaneous, intramuscular, and intraveinous vitamin D administration. J Clin Endocrinol Metab 1979; 48:906–911.
61. Webb AR, Holick MF. The role of sunlight in the cutaneous production of vitamin D_3. Annu Rev Nutr 1988; 8:375–399.
62. Garabédian M, Zeghoud F, Rossignol C. Vitamin D status of infant living in France: results of a multicentre survey on 411 healthy infants 8–10 mo old age (1988–1990). In: Journées Parisiennes de Pédiatrie. Paris: Flammarion Médecine-Sciences, 1991; 51–57 (in French).
63. Ala-Houhala M. 25-Hydroxyvitamin D levels during breast feeding with or without maternal or infantile supplementation of vitamin D. J Pediatr Gastroenterol Nutr 1985; 4:220–226.
64. Markestad TM, Hesse V, Siebenhuner M, et al. Intermittent high-dose vitamin D prophylaxis during infancy: effect on vitamin D metabolites, calcium and phosphorus. Am J Clin Nutr 1987; 46:652–658.
65. Zeghoud F, Ben-Mekhbi H, Djeghri N, Garabédian M. Vitamin D prophylaxis during infancy: comparison of the long-term effects of three intermittent doses (15,5, or 2.5 mg) on 25-hydroxyvitamin D concentrations; Am J Clin Nutr 1994; 60:393–396.
66. Delvin EE, Salle BL, Glorieux FH, Adeleine P, David LS. Vitamin D supplementation during pregnancy: effect on neonatal calcium homeostasis. J Pediatr 1986; 109:328–334.
67. Mallet E, Gugi B, Brunelle Ph, Hénocq A, Basuyau JP, Lemeur H. Vitamin D supplementation in pregnancy: a controlled trial of two methods. Obstet Gynecol 1986; 68:300–304.
68. Zeghoud F, Garabedian M, Jardel A, Bernard N, Melchior J. Administration of a single dose of vitamin D_3 (100 000 IU) to pregnant women in winter. Effects on the serum calcium of their newborns. J Gynecol Biol Reprod 1988; 17:1099–1105 (in French).
69. Suarez F, Zeghoud F, Rossignol C, Walrant O, Garabédian M. Association between vitamin D receptor gene polymorphism and sex-dependent growth during the first two years of life. J Clin Endocrinol Metab 1997; 82:2966–2970.

17 Vitamin D and Bone Health in Adults and the Elderly

Clifford J. Rosen

1. INTRODUCTION

Vitamin D is one of the principle hormonal regulators of calcium homeostasis in the body. Besides being critically important for calcium and phosphate absorption in the intestine, vitamin D is essential for normal mineralization of bone and has major regulatory effects on bone cells in the bone remodeling unit. In addition to distant skeletal and intestinal effects, the active form of vitamin D, 1,25-dihydroxyvitamin D_3 [1,25(OH)$_2$D$_3$], also regulates its own synthesis in the kidney and parathyroid hormone (PTH) secretion in the parathyroid gland. These multisystem effects firmly establish the importance of this hormone in the maintenance of skeletal health. Moreover, perturbations in vitamin D synthesis, secretion, or action have been implicated as potential pathogenetic factors in the development of osteoporosis. For these reasons, there has been sustained interest in vitamin D and 1,25(OH)$_2$D$_3$ as therapeutic agents in several metabolic bone disorders. However, the relationship between active vitamin D and its metabolites and calcified tissue components is complex and redundant. Hence a thorough understanding of the role vitamin D plays in the bone remodeling process (either directly or indirectly) is extremely important. In turn, complete delineation of the physiologic role of vitamin D in mineral homeostasis illustrates why vitamin D deficiency, especially in the elderly, is now being recognized as a major public health issue.

This chapter focuses on the role of vitamin D in skeletal remodeling, its relationship to osteoporosis/osteomalacia, and the utility of vitamin D analogs as therapeutic modalities for various metabolic bone diseases.

2. BONE REMODELING AND ITS RELATIONSHIP TO VITAMIN D HOMEOSTASIS

2.1. Physiology of Remodeling: The Bone Multicellular Unit

The adult skeleton is not static but rather undergoes dynamic renewal in a process commonly known as remodeling or turnover. During this carefully orchestrated process, bone dissolution or resorption is followed by new bone formation *(1)*. At the end of this process, the quantum of bone, which is resorbed, is matched by newly deposited bone

From: *Vitamin D: Physiology, Molecular Biology, and Clinical Applications*
Edited by: M. F. Holick © Humana Press Inc., Totowa, NJ

matrix. In general, the remodeling cycle takes between 100 and 130 d, and this process permits mechanical competence to be maintained and calcium homeostasis to be preserved.

Remodeling is controlled by various hormonal and skeletal factors that influence the behavior of the basic multicellular unit (BMU), the basic functional unit of bone turnover. The BMU is composed of three types of cells, the osteoclast, the osteoblast (OB), and the osteocyte. Each has a defined role in remodeling and each has unique regulatory circuits that control activity during active skeletal turnover.

OBs are bone-forming cells that are derived from progenitor cells of mesenchymal origin *(2)*. These mononucleated cells serve to synthesize and secrete collagen and proteoglycan complexes that constitute osteoid and also play a major role in matrix mineralization *(3)*. OBs may also regulate ionic channels and movement of cations/anions in and out of bone fluid *(2,3)*. In addition, mature OBs secrete several protein products that may be critical to the process of mature bone formation; these include osteo calcin (bone gla protein) and alkaline phosphatase *(4)*. OBs form a layer of cells that cover bone surfaces and lay down matrix in humans at a rate of approx 100 μm^3 of matrix volume/d *(5)*.

Fully differentiated and premature OBs are regulated by two major calciotropic hormones [PTH and $1,25(OH)_2D_3$], systemic mediators (growth hormone, insulin, thyroxine, glucocorticoids and sex steroids), and skeletal growth factors (e.g., insulin-like growth factor-I, prostaglandins, transforming growth factor-β, several cytokines). Each of these is involved in a cascade of events that initiate and propagate bone formation. Several key skeletal growth factors are also regulated by calciotropic hormones.

Osteoblasts are terminally differentiated *(2)*. Once these cells lay down new bone, the OB can undergo apoptosis or bury itself within skeletal matrix and become another cell, the osteocyte. The osteocyte maintains cytoplasmic and vascular connections with osteoblasts on the surface of bone *(6)*. Although it appears inert by microscopic analysis, the osteocyte serves two roles: (1) as a transducer for mechanical loading; and (2) to modulate minute to minute exchange of minerals within the matrix *(6,7)*.

Osteoclasts are large multinucleated giant cells that arise from hematopoietic stem cells in the bone marrow *(8)*. These cells are responsible for bone resorption through four major mechanisms: carbonic anhydrase production, proton secretion by the osteoclast, calcium-dependent adenosine triphosphatase activity, and the sodium/potassium pump *(9)*. As might be predicted, osteoclastic activity is also tightly regulated by $1,25(OH)_2D_3$, PTH, and calcitonin. The recruitment and differentiation of premature osteoclasts from stem cells are also under the control of hematopoietic growth factors including granulocyte/macrophage colony-stimulating factor (GM-CSF) and several interleukins *(10)*. These peptides are released by osteoblasts during initial activation of the remodeling cycle. In contrast to the deliberate and orchestrated process of bone formation, which is completed in 2.5–3 mo, bone resorption can be accomplished in 2–5 d *(1)*.

The remodeling cycle begins when osteoblasts are activated; this in turn leads to the release of osteoclast-activating substances that accelerate recruitment and differentiation of osteoclast precursor cells. During bone resorption, growth factors that activate and recruit surface osteoblasts are released from the skeletal matrix. This permits coupling between these two distinct processes. The cycle is finished when bone formation and matrix maturation is complete. Calciotropic hormones act at every point in the remodeling cycle through all three cell types. In fact, $1,25(OH)_2D_3$ is a major regulator of bone turnover, and its effects are profound in both physiologic and pathologic states.

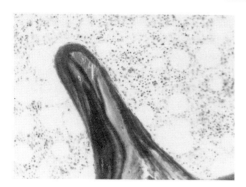 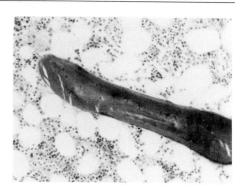

Fig. 1. Human trabecular bone (×125). (Left) Osteomalacia, showing impaired mineralization. Trabecular surface is covered with thick osteoid, which is the dark rim around the mineralized bone. (Right) Osteoporosis, showing thin and eroded trabecular surface with little or no osteoid.

2.2. Pathophysiologic States of Remodeling

Osteoporosis is defined as a reduction in bone mass *(11)*. This results from uncoupling in the remodeling unit whereby the rate of resorption exceeds the rate of formation. In most instances, accelerated remodeling leads to increased recruitment and differentiation of osteoclastic precursors that cannot be totally matched by upregulation of osteoblastic activities. Estrogen withdrawal during menopause or increased PTH secretion resulting from reduced calcium intake and/or absorption are the two most common etiologic factors in idiopathic osteoporosis *(12)*. However, thyroid hormone excess, exogenous glucocorticoid administration, or anticonvulsant treatments can also produce significant mismatches between resorption and formation *(13)*. These factors contribute to rapid bone loss and an enhanced risk for osteoporotic fractures of the spine, hip, and wrist.

Osteomalacia results from impaired mineralization of newly synthesized osteoid *(14)* (Fig. 1). This can lead to pseudofractures and bone pain or can complicate coexisting osteoporosis. Inherited disorders of vitamin D metabolism, systemic acidosis, specific types of secondary hyperparathyroidism with renal insufficiency, and calcium/vitamin D malabsorption can all contribute to a reduction in the mineral component of the bone matrix.

In both osteoporosis and osteomalacia, abnormalities in bone cells or products of these cells lead to classic phenotypic disorders of the skeleton. Most changes in osteoblast function can be at least partially related to alterations in calciotropic hormone regulation or their skeletal second messengers. In particular, vitamin D plays a major pathogenetic role because of its powerful direct and indirect influences on the remodeling unit.

2.3. Role of $1,25(OH)_2D_3$ in Bone Remodeling

2.3.1. VITAMIN D AND THE OSTEOBLAST

The effects of $1,25(OH)_2D_3$ on bone are an integral component of vitamin D's role in calcium homeostasis, maintaining serum calcium levels for deposition of mineral while simultaneously supporting the body's responsiveness to a broad spectrum of physiologic demands. Vitamin D can accomplish this through its actions on the osteoclast (*see* Section 2.3.2.), the osteoblast, or nonskeletal target tissues (*see* Section 3.). The effects of vitamin D on the osteoblast are direct, but upregulation of osteoblastic activity by $1,25(OH)_2D_3$ also leads to increased synthesis of cytokines active on bone that promote osteoclast differentiation and function.

In general, young osteoblasts express the vitamin D receptor (VDR) and in response to $1,25(OH)_2D_3$ progress from immature proliferating cells to nondividing terminally differentiated cells that synthesize matrix proteins and mineralize bone *(15)*. However, vitamin D exhibits selective effects on the osteoblast depending on its stage of proliferation, the in vitro model system employed, and the role of the target osteoblast in remodeling. The addition of $1,25(OH)_2D_3$ to mature bone cell cultures results in upregulation of osteocalcin and osteopontin and downregulation of type I collagen synthesis *(16)*. Cell number is increased and cell morphology in mature osteoblasts is also altered by $1,25(OH)_2D_3$, as is the expression of alkaline phosphatase, an enzyme critical to mineralization of the protein matrix *(17)*. On the other hand, in young proliferating and neonatal osteoblasts, $1,25(OH)_2D_3$ blocks differentiation marker expression *(18)*. Thus, depending on the cell stage, $1,25(OH)_2D_3$ can have a major effect on osteoblast destiny. Why the same vitamin D response element (VDRE) on a particular gene can upregulate or downregulate phenotypic expression markers remains to be determined.

Osteocalcin (bone gla protein) is expressed by differentiated osteoblasts once mineralization begins *(19)*. Although its precise function is unknown, it probably serves to regulate the size of the mineralization crystal as well as performing some role in the recruitment and coupling of osteoclasts *(20)*. One of the major regulators of osteocalcin synthesis is $1,25(OH)_2D_3$ *(21)*. A VDRE is present in the promotor region of the osteocalcin gene, which allows binding of the VDR to *cis*-acting elements upstream from the transcription start site *(22)*. Although $1,25(OH)_2D_3$ may not stimulate mineralization directly, it is likely that vitamin D plays a major role in crystalline growth and maturation by working in an indirect manner through regulation of osteocalcin expression and calcium absorption in the gut.

2.3.2. VITAMIN D AND THE OSTEOCLAST

$1,25(OH)_2D_3$ stimulates bone resorption both in vivo and in vitro *(23)*. Indeed, generation of osteoclasts from isolated macrophages or bone marrow cultures requires the presence of $1,25(OH)_2D_3$ *(24)*. However, the effects of vitamin D on osteoclastogenesis are complex. For example, even though osteoclast formation is promoted by $1,25(OH)_2D_3$, osteoblasts are a necessary component to drive that differentiation *(25)*. Moreover, mature osteoclasts do not express the VDR *(26)*.

Not unlike its effect on the osteoblast, $1,25(OH)_2D_3$ acts as a potent maturational agent affecting both transformed and normal cell types. Its target is primarily the early osteoclast precursor, the macrophage, and in these cells $1,25(OH)_2D_3$ induces multinucleation, the expression of osteoclast-specific enzymes (tartrate-resistant acid phosphatase, carbonic anhydrase), and the induction of the M-CSF receptor *(27)*. Of note, however, is the fact that mature osteoclasts require accessory cells such as stromal cells or osteoblasts, to respond completely to $1,25(OH)_2D_3$ and express the fully differentiated osteoclast phenotype *(28)*. What the $1,25(OH)_2D_3$-inducible factor is that links the osteoclast to the osteoblast is unknown at this time but may turn out to be of significance in understanding how vitamin D regulates the entire remodeling sequence.

2.3.3. VITAMIN D AND THE OSTEOCYTE

Osteocytes are difficult, if not impossible, to study in vitro. Hence there are virtually no experiments examining the role of $1,25(OH)_2D_3$ in the life cycle of the osteocyte. In vivo effects of active vitamin D on the osteocyte are also extremely hard to quantify by histomorphometry, and the role of 1,25(OH)$_2$D$_3$ in osteocyte-mediated processes, such as load modulation, remain to be explored.

2.3.4. SUMMARY

Vitamin D plays a major role as a differentiation agent for osteoclasts and osteoblasts. It modulates bone formation and matrix mineralization and also stimulates bone resorption. Indirectly, 1,25(OH)$_2$D$_3$ promotes osteocalcin synthesis and increases calcium absorption in the gut. Hence, this vitamin/hormone is vitally important for the skeleton. On the other hand, as will be seen from clinical trials, 1,25(OH)$_2$D$_3$ can inhibit mineralization, enhance resorption, and increase bone loss. These differences reinforce the complexity of vitamin D in skeletal metabolism and lead to caution concerning the use of vitamin D analogs in the treatment of low bone mass.

3. EFFECTS OF VITAMIN D ON NONSKELETAL TARGET TISSUE AND ITS RELATIONSHIP TO BONE REMODELING

3.1. Effects of Vitamin D on the Intestine

Vitamin D promotes calcium absorption in the proximal duodenum and phosphate absorption in the jejunum and ileum *(29)*. The transport of these ions is critical to the process of new bone formation, and the action of 1,25(OH)$_2$D$_3$ on bone mineralization is primarily related to its activity in the intestine. The significance of this fact is noted by the almost universal expression of the VDR along the length of the intestine, with the highest concentration of VDR noted proximally *(30)*. Intestinal calcium absorption can occur via three mechanisms: (1) passive transport of calcium; (2) active transcellular passage by a vitamin D-mediated process; and (3) rapid transmembrane flux stimulated by vitamin D in a nongenomic manner *(31)*. More than 80% of calcium transport out of the gut occurs via vitamin D-mediated mechanisms, with the transcellular process dominant *(32)*. This process takes several hours and is related to the induction of vitamin D-induced calcium binding proteins (calbindins) *(33)*. This family of proteins serves to hand calcium off to several carriers in an apical to basolateral direction in the intestine. The relative contribution of the rapid nongenomic mechanism to overall efficiency of calcium absorption remains to be determined.

3.2. Relationship of Vitamin D to Skin

Most of the active vitamin D in humans, which is necessary for intestinal and skeletal activity, is derived from percutaneous synthesis after exposure to ultraviolet radiation (UV) *(34)*. Solar high-energy ultraviolet photons (UVB: 290–315 nm) are absorbed by a precursor of cholesterol, 7-dehydrocholesterol, in the skin. This leads to a transformation of previtamin D$_3$ to vitamin D$_3$, which is then transported in the circulation to the liver by vitamin D binding proteins *(35)*. Since >60% of circulating (OH)D$_3$ is derived from percutaneous nonenzymatic production, it is certain that this form of endogenous production has an indirect, but important, role at the level of the skeletal remodeling unit *(36)*.

Several factors regulate percutaneous synthesis of vitamin D. These include:

1. The amount of melanin present in skin (an increased melanin content means that longer solar exposure is required for equivalent synthesis of vitamin D);
2. Exogenous sunscreen use (use of a sun protection factor >8 will reduce percutaneous production of vitamin D by nearly 100%);

3. Social factors (living indoors, e.g., a nursing home residency, protective clothing) that contribute to reduce cutaneous synthesis;
4. Season of the year and latitude (winter sunlight in northern latitudes results in a reduction in the zenith angle of the sun; this makes the UVB photons strike the atmosphere at a more oblique angle, thereby increasing the absorption by the ozone layer of the atmosphere); and
5. Aging [with advanced age there is a marked reduction in skin concentration of 7-dehydrocholesterol (provitamin D_3); this leads to 60–70% less percutaneous synthesis of vitamin D_3] *(36,37)*.

3.3. Vitamin D Action in the Kidney

Vitamin D plays a critical role in the kidney, not so much in terms of regulating calcium and phosphate excretion (which it does, but only to a small degree and primarily through its feedback on PTH secretion) but in terms of control over active $1,25(OH)_2D_3$ synthesis. VDR protein is present in the kidney, as are the vitamin D-dependent proteins calmodulin and calbindin *(38)*. However, the primary action of $1,25(OH)_2D_3$ is to regulate its own synthesis and that of the inactive $24,25(OH)_2D_3$. Adequate levels of $1,25(OH)_2D_3$ inhibit 1α-hydroxylase activity while inducing 24-hydroxylase gene expression, thereby pushing 25-hydroxyvitamin D into the inactivation pathways and $24,25(OH)_2D_3$ production *(39)*. $1,25(OH)_2D_3$ can also (1) directly inhibit PTH secretion, which is the major stimulator of 1α-hydroxylase activity; and (2) indirectly suppress PTH, via calcium-promoting actions on bone and intestine *(40)*. Thus, although $1,25(OH)_2D_3$ may have a minor role in calcium reabsorption in the kidney, its major renal effect is to control its own synthesis and that of its inactive metabolite, $24,25(OH)_2D_3$. These regulatory circuits are critical for maintaining mineral homeostasis, especially at the skeletal remodeling unit.

4. PATHOPHYSIOLOGY AND TREATMENT OF OSTEOPOROSIS IN RELATION TO VITAMIN D

As noted earlier, osteoporosis is defined as a reduction in bone mass due to uncoupling of the bone remodeling unit. Imbalances between resorption and formation eventually lead to bone loss, skeletal fragility, and osteoporotic fractures. In fact, low bone mineral density (BMD) has emerged as the most powerful risk factor for this disease. However, it has become very clear in the last few years that bone loss due to uncoupling of the remodeling unit is not only a function of the rate of bone loss during a given period (e.g., menopause or aging) but also the rate of acquisition of peak bone mass (between the ages of 15 and 20 yr). During the critical phase of acquisition and consolidation immediately after linear growth ceases, the remodeling unit favors bone formation over resorption, resulting in an increase in bone mass. There are three periods in a woman's life (peak bone mass, menopause, and after age 65) when the bone remodeling unit can be influenced either positively or negatively and in such a way as to dramatically affect a woman's future risk of osteoporotic fracture.

From previous discussions it is clear that vitamin D occupies a critical role during acquisition as well as during the process of age-related bone loss. In both situations, calcium balance is critical to the maintenance of bone balance. Several longitudinal studies have highlighted the importance of calcium intake during adolescence and the pathophysiologic nature of reduced calcium intake during aging *(41,42,43)*. Less clear is the role of calcium and vitamin D in the development of type I or postmenopausal osteoporosis. This section highlights studies that identify the role of vitamin D in the pathogenesis of low bone mass. In addition, the potential therapeutic role of vitamin D in osteoporosis is examined.

4.1. Bone Acquisition and Vitamin D

Several epidemiologic studies have confirmed that peak bone mass accounts for most of the variance in BMD at any subsequent age *(44,45)*. Genetic factors are responsible for 60–70% of peak bone mass, even though at this time those determinants are not known *(46)*. However, from calcium intervention studies in teens, it is clear that calcium supplementation and other environmental factors can modulate genetic influences *(41–43)*. Thus, vitamin D may be important in this regard. No studies to date have yet identified alterations in the PTH/vitamin D axis in adolescents who have low calcium intakes. Indeed, the PTH increase, with a subsequent rise in 1α-hydroxylase activity [and hence an increase in $1,25(OH)_2D_3$], is brisk and prompt following calcium deprivation in younger individuals *(47)*. In fact, the overall metabolism of vitamin D during this critical phase of adolescent bone "life" is probably normal if the pathophysiologic syndromes of glucocorticoid exposure or gastrointestinal disturbances are not considered.

However, the recent identification of polymorphisms in the VDR gene has suggested that vitamin D may still be important in bone mass acquisition. Morrison et al. *(48)* were the first to identify that a polymorphism in an intron close to the 3' untranslated region of the VDR gene was associated with bone density among Australian twins. In that original study and in subsequent examinations, the absence of a cut site for one restriction enzyme (*Bsm*I) was denoted as having the *BB* genotype; the presence of his restriction site was denoted as *bb (48)*. Heterozygotes were considered *Bb*. Two other restriction sites are also present near this polymorphism in the VDR gene, but for the most part, they are linked to the presence or absence of the B restriction site.

In the original twin study, as well as in several population studies, the *BB* genotype has been associated with low bone mass, and several investigators have claimed that this polymorphism was responsible for 50–60% of the variance in BMD *(48–50)*. Other groups have been unable to show that VDR polymorphisms could predict BMD in their populations of older postmenopausal women, although still others have shown that the relationship is stronger in younger women, in whom environmental factors have less influence on bone turnover *(50–53)*. One set of investigators has demonstrated that the effects of the three vitamin D genotypes on bone density are most important during states of relative calcium deficiency *(53)*.

However, the significance of VDR alleles in the acquisition of peak bone mass remains to be determined for several reasons. First, polymorphisms in introns are generally not associated with functional or structural abnormalities of the protein product coded for by the nucleotide sequence. In the case of the putative VDR polymorphisms, the number and function of VDR receptors does not appear different among the various genotypes. Still, alterations in nucleotide sequences are important for mapping and linkage studies, so it is possible that the various VDR alleles are linked to other genes (or other parts of the VDR gene) that are important in bone acquisition. Second, bone mass is a quantitative trait (like final adult height), which means that multiple genes almost certainly control its expression. It is highly unlikely, therefore, that a single polymorphism in an untranslated region of the gene could account for 60% of the variance in BMD across all ages. Finally, although vitamin D is central to calcium balance, it remains to be demonstrated how it is possible for alterations in the VDR protein to lead to major changes in BMD *without* demonstrable changes in calcium metabolism. In sum, further studies are needed before a definite role for genotypic differences in the VDR are associated with the acquisition phase of bone mass.

4.2. Perimenopausal and Postmenopausal Bone Loss: The Role of Vitamin D

Bone loss during and immediately after menopause may be dramatic and certainly is a major pathogenetic factor in type I osteoporosis. The uncoupling of the remodeling unit is a function of estrogen deprivation, which leads to markedly increased bone resorption and an increase in bone formation that cannot match the rate of resorption *(54)*. The role of vitamin D and its metabolites in this disorder has been studied. In general, levels of 25(OH)D and $1,25(OH)_2D_3$ have been either normal or decreased in women with postmenopausal osteoporosis and vertebral fractures *(55)*. One study identified an inverse relationship between $1,25(OH)_2D_3$ levels and radial bone density *(56)*. However, most studies have only been able to show that age is associated with low bone mass and declining levels of vitamin D and its metabolites *(57)*. Moreover, postmenopausal women between the ages of 50 and 65 respond normally to PTH by increasing their serum $1,25(OH)_2D_3$ levels *(58)*. This is concordant with other reports showing that PTH levels generally are suppressed in postmenopausal women with osteoporosis *(59)*. Therefore, it does not seem likely that the PTH/vitamin D axis is a major pathophysiologic component of type I postmenopausal osteoporosis. However, this has not precluded treating such women with vitamin D in an attempt to enhance bone formation and suppress bone resorption.

Numerous studies have been performed in postmenopausal women with osteoporotic fractures to determine whether vitamin D or its metabolites might prevent bone loss and fractures. More than 20 years ago, Buring et al. *(60)* treated women with back pain and spinal osteoporosis with vitamin D_2 (35,000 IU/d) plus 1 g of calcium/d *(60)*. Although there was no control group, radial BMD did not change. Nordin et al. *(61)* gave 10,000–50,000 IU of caciferol in a nonrandomized trial of women with vertebral fractures and found that metacarpal cortical area bone loss was accelerated by vitamin D treatment. In addition, among seven different treatment groups, only the vitamin D group showed an increase in fractures. In 1982, Riggs et al. *(62)* published a retrospective analysis of patients treated with 50,000 IU of vitamin D up to two times weekly along with either calcium, hormone replacement therapy, or fluoride. There was no evidence from that study that vitamin D was effective in postmenopausal women with osteoporosis. Christiansen et al. *(63)* used low-dose cholecalciferol (400 IU/d) or placebo for 2 yr in early postmenopausal women and also found no differences in radial BMD between the two groups. Thus, physiologic or pharmacologic vitamin D alone or in combination with other agents does not have a role in the prevention of fractures or postmenopausal bone loss among women within 15 yr of menopause. In older women, the story may be entirely different, however.

Another strategy employed by investigators for the treatment of type I osteoporosis has been the use of $1,25(OH)_2D_3$, a much more potent analog of calciferol. Administration of $1,25(OH)_2D_3$ in doses as low as 0.25 µg/d increases calcium absorption in the gut and markedly enhances serum osteocalcin levels *(64)*. However, this should not be interpreted as meaning that there is direct stimulation of bone formation, since $1,25(OH)_2D_3$ directly stimulates transcriptional regulation of the osteocalcin gene *(16)*. Moreover, the effects of varying doses of $1,25(OH)_2D_3$ on other indices of bone turnover are variable and can include suppression of alkaline phosphatase activity as well as increased urinary hydroxyproline excretion at higher doses of the medication *(65)*. Still there have been multiple trials with $1,25(OH)_2D_3$ to determine its effectiveness in the management of postmenopausal osteoporosis.

At least 11 studies in the last 15 yr have used $1,25(OH)_2D_3$ in various doses for women with vertebral fractures. Some of the trials included much older postmenopausal women, but many of the women were younger than 70 yr of age. In general, the early studies between 1981 and 1987 showed variable results but employed doses of <0.50 μg/d *(66–70)*. In addition, at least one study suggested that vertebral fracture rates could be greater in those patients receiving calcitriol *(70)*.

Three major studies between 1988 and 1990 were conducted in the United States to determine the fracture efficacy of $1,25(OH)_2D_3$ in postmenopausal women. Although the studies differed somewhat and were conducted at different sites, each started with a dose of 0.50 μg/d of calcitriol *(71–74)*. Postmenopausal vertebral fracture patients were randomized to placebo or $1,25(OH)_2D_3$, and the dose of calcitriol was subsequently adjusted until hypercalcuria or hypercalcemia occurred. At the end of 2 yr, one group [that of Aloia et al. *(71)*] had women on doses of calcitriol averaging 0.80 μg/d *(71)*. These women ($n = 27$) had an increase in lumbar BMD and distal radial bone mineral content (BMC) compared with placebo. However, some of those women also developed hypercalcuria and hypercalcemia. In addition, fracture rates were not statistically different. Gallagher et al. *(72)* followed the same protocol, but the average dose of $1,25(OH)_2D_3$ was slightly lower (0.62 μg/d) because of dosage adjustments. Spine BMD increased 2% in the calcitriol group and decreased >2% in placebo women. Fracture rates did not differ between groups. Ott and Chestnut *(73)* studied 72 women with postmenopausal osteoporosis; once again, the doses varied. In this study, the mean $1,25(OH)_2D_3$ dose was 0.43 μg/d. Calcium intake was adjusted downward from a high value at the beginning of the study of 1000 mg/d. Distal radial BMC in the calcitriol group actually decreased to a significant degree, even though bone biopsy parameters did not change. There were no differences in fracture rates between the two groups. Subgroup analysis of the women receiving the highest doses of $1,25(OH)_2D_3$ did demonstrate an increase in BMD consistent with earlier studies *(74)*. However, the numbers were small.

The largest study to date was performed by Tilyard et al. *(75)* in New Zealand. Women with vertebral fractures ($n = 622$) were randomly assigned to 0.5 μg/d calcitriol or calcium 1 g/d and were followed for 3 yr *(75)*. The only end point was vertebral fracture. Two-thirds of the women completed the study, and there were more fractures in the calcium-treated women than in the calcitriol-treated subjects. In fact, the calcium-only women had a threefold increase in new vertebral fractures. However, the design was not blinded, >30% of the women did not complete the study, and it is unclear what the baseline 25(OH)D levels were in these women. Since New Zealand is located at a low latitude, it seems likely that many of these subjects were at least subclinically vitamin D deficient (although it is unclear from the paper what that percentage of people was!). Thus, even though the numbers were large, the trial itself left many questions unanswered.

In sum, it remains to be seen whether $1,25(OH)_2D_3$ can be used as a therapy for postmenopausal osteoporotic women who are vitamin D replete. At higher doses, its use is associated with significant side effects (hypercalcuria and hypercalcemia), and concern about increased bone resorption and fracture rates after long-term therapy will probably limit its potential utility as a primary therapy for this disease. However, as noted above, the use of vitamin D in elderly patients may represent an entirely different scenario.

A final note should be made about 1α-hydroxycholecalciferol (alphacalcidol), a synthetic vitamin D compound hydroxylated in position 1 but not position 25. It is rapidly 25-hydroxylated in the liver after oral administration, making its pharmacologic profile similar to that of 1,25(OH)$_2$D$_3$ *(76)*. Since its introduction in the mid-1970s, approx 10 studies have been conducted to determine potential fracture and bone mass efficacy for this compound. In most of those studies, BMD in the spine increased significantly, but fracture efficacy was not established *(77–81)*. Interestingly, most of these studies also showed that bone resorption indices were suppressed by administration of this metabolite. In the most recent study to date of postmenopausal women with osteoporosis, vertebral BMD increased in the alphacalidol group, whereas placebo women showed a decrease; there were statistically fewer fractures of the spine for the women treated with 1α-hydroxycholecalciferol *(82)*. Once again, these studies are inconclusive, but there may be some suggestion that 1α-hydroxycholecalciferol could be more beneficial with fewer side effects than calcitriol. Further studies are required.

4.3. Age-Associated Bone Loss and Vitamin D

Type II osteoporosis has been classically defined by the secondary hyperparathyroidism that results in the elderly from reduced calcium intake, decreased vitamin D synthesis and ingestion, and the marked rise in PTH necessary to preserve serum calcium. Several longitudinal, placebo-controlled trials have demonstrated that calcium supplementation can prevent age-related bone loss and suppress, at least partially, the secondary hyperparathyroidism associated with this process *(83–85)*. However, the pathophysiologic role of vitamin D in this process has only recently been appreciated. Although patients with vertebral fractures do not have lower serum 25(OH)D levels than controls, hip fracture patients do *(86)*. This is almost certainly a function of age and relates to several important variables that control vitamin D levels in the body. These include (1) lack of sunlight exposure due to sunscreen, clothing, or placement in a nursing home; (2) aging itself; (3) changes in seasons of the year and time of day; and (4) an increase in latitude *(36,37)*.

As noted previously, serum levels of 25(OH)D are primarily dependent on sun exposure and to a lesser degree on dietary intake. Unfortunately, very few foods (certain fish and cod liver oil) other than milk contain vitamin D, and even in milk, the amount of vitamin D supplemented can vary from the advertised 400 IU/qt *(87)*. Hence for the elderly who consume little milk, their principle source of vitamin D comes from the sun. However, as women age, their capacity to synthesize vitamin D in the skin declines, their exposure lessens, and for those living in northern latitudes, the lower zenith of the sun's angle effectively prevents adequate vitamin D stores to be maintained during winter months. The seasonal effects of inadequate sun exposure are manifested by a significant decline in spine and hip BMD (1–3%), a 20–25% decline in serum 25(OH)D, and a marked rise in serum PTH *(88)*. Moreover, winter for northern New England women who consume low calcium is associated with marked increases in serum osteocalcin, and urinary *N*-telopeptide, further evidence that bone turnover is accelerated by the seasonal decline in vitamin D *(89)*.

It is uncertain how much of a role slightly low levels of serum 25(OH)D play in the pathogenesis of age-related bone loss and hip fractures. However, it is clear that nursing home patients are at highest risk for hip fractures and that between 50 and 80% of these residents also have serum 25(OH)D levels <20 ng/mL *(90)*. Moreover, other events occur during aging in humans that might contribute to the pathogenesis of hip fractures. For example, the number of VDRs in the intestine decline with age *(91)*. Elderly women are less capable of synthesizing $1,25(OH)_2D_3$ in response to increased PTH secretion, suggesting a partial age-associated defect in 1α-hydroxylase activity *(58)*. Finally, polymorphisms in the VDR receptor could theoretically play a role during states of relative calcium deficiency, preventing full expression of the VDR protein necessary for activation of calcium transport *(53)*.

Based on much of the evidence noted above, clinical trials with vitamin D in elders have yielded interesting and exciting results. Heikinheimo et al. *(92)* studied 800 elders living in Finland. One-third of the people were in nursing homes, and two-thirds were independent. Elderly subjects (mean age 86 yr) were randomized to receive 150,000 IU/d of vitamin D_2 per year or nothing. Serum 25(OH)D levels normalized in the treated group only, and fracture rates were reduced by 25% after 3 yr of vitamin D therapy. Since the study was conducted at a northerly latitude where vitamin D deficiency is relatively common, the authors concluded that vitamin D was efficacious in preventing fractures. Chapuy et al. *(93)* randomized more than 3000 elderly nursing home residents (mean age 84 yr) to calcium, 1.2 g/d and 800 IU of vitamin D_3 or double placebo for 3 yr *(93)*. This study was unique for its numbers and for the fact that all three primary end points showed significant changes with treatment after only 18 mo; these were (1) a nearly 30% reduction in hip fractures; (2) preservation of femoral bone density (vs 4.5% loss in placebo); and (3) a nearly 50% suppression in PTH in the treated group (using analysis from a subset of each group). It is important to note that in most of the patients in whom 25(OH)D was measured, their levels (both groups) were quite low (13–16 ng/mL). Hence an argument could be made that this trial was effective because it treated subclinical vitamin D deficiency syndrome. Still, the results from both studies are meaningful for clinicians since nursing home residents often have low levels of serum 25(OH)D and are at high risk for hip fractures, the most costly end point of osteoporosis.

The argument continues regarding whether it is calcium or vitamin D that is essential for preventing secondary hyperparathyroidism and subsequent bone loss. Vitamin D increases the efficiency of calcium absorption in the gut from 15–20% (without vitamin D) to 70% *(94)*. It seems logical that if calcium intake is adequate in elders (approx 1500 mg/d) the modest declines in vitamin D that accompany seasonal changes are not likely to have a major impact on the skeleton, especially since during the summer, serum 25(OH)D concentrations, PTH levels, and bone density return toward their baseline levels *(88)*. However, in elders with poor calcium intake (<800 mg/d), low serum levels of vitamin D may be extremely important, not only for the skeleton but also for muscle function.

Unfortunately, the RDA for vitamin D has not been established with scientific certainty, in part because there have been no dose-ranging studies in elders. Therefore, it is unclear what dose of vitamin D is really necessary to prevent bone loss and secondary hyperparathyroidism. Currently, most investigators recommend sunlight exposure if at the proper latitude, a vitamin D supplement, or a pharmaceutical vitamin D preparation.

In summary, vitamin D production and intake decline with age. This can complicate calcium economy, especially among women with poor dietary intake of calcium, by producing a form of secondary hyperparathyroidism that leads to bone loss and eventual fracture. Adequate calcium and vitamin D are essential components of any therapeutic regimen in elders, in whom fracture prevention is a key concern.

5. OSTEOMALACIA AND VITAMIN D

Osteomalacia was originally defined in the late 1800s as a general softening of bone that led to crippling deformities *(14)*. This was well before the discovery of vitamin D, but soon thereafter, bone softening was linked to vitamin D deficiency and treatment of osteomalacia meant administration of vitamin D. Even today, vitamin D deficiency from various causes is the principle cause of osteomalacia, although there are clearly other pathophysiologic disorders that can lead to bone softening and increased unmineralized bone matrix. Regardless of the etiology, the osteomalacia syndrome has characteristic clinical, biochemical, radiologic, and histologic features that can be related to vitamin D.

5.1. Clinical Presentation of Osteomalacia

The clinical manifestations of this syndrome vary according to the age of the individual. Defective mineralization in childhood leads to rickets and if untreated produces severe deformities and gross radiologic abnormalities. This discussion focuses on adult-onset osteomalacia, which usually presents subtly, although occasionally patients with nontropical sprue or severe malabsorption will demonstrate the full clinical phenotype.

Classic symptoms of osteomalacia include bone pain, muscle weakness and skeletal tenderness *(95)*. Severe cases may be associated with symptoms of tetany resulting from hypocalcemia. Onset early in life leads to short stature and can include kyphosis, coxa vara, and pigeon breast *(96)*. Late-onset disease often does not manifest phenotypically characteristic symptoms but can mimic or exist with type I or type II osteoporosis. Proximal muscle weakness is especially common in older individuals with vitamin D deficiency and osteomalacia *(97)*. Radiologically, there is generalized thinning of cortical and trabecular bone (because of increased PTH), often referred to as *demineralization*, which can be detected by densitometry or X-ray *(98)*. Bone shape may also be altered. The best known radiologic feature of osteomalacia is a lucent band adjacent to the periosteum that represents an unhealed insufficiency type stress fracture, also called a Looser zone *(99)*. These are most commonly seen in the ribs, pelvis, and scapulae but can also be detected in the femoral neck and the shaft of long bones. Besides stress fractures, other types of skeletal failure occur. These include rib and sternal fractures as well as hip fractures. Biochemically, patients have characteristic increases in PTH, low urinary calcium excretion, and increased indices of bone turnover including alkaline phosphatase and collagen crosslink excretion *(100)*. Serum levels of vitamin D metabolites are classically altered according to the biochemical defect.

5.2. Alterations in Vitamin D Metabolism

Vitamin D metabolism can be affected at one of six levels and can lead to the same clinical syndrome of osteomalacia. Alterations in vitamin D metabolites that characterize that particular abnormality help in identifying the defect responsible for undermineralization of the skeleton. Once again, comments are generally confined to adult-onset syndromes.

The first and most common defect in vitamin D metabolism responsible for osteomalacia is *extrinsic vitamin D depletion*. As noted earlier, this conditon may be suspected in elderly individuals due to latitude, lack of sun exposure, and poor dietary intake of vitamin D *(36,37)*. Without histomorphometric evidence, the manifestations are often attributed to the osteoporotic syndrome. However, it is not uncommon for osteomalacia to exhibit variable presentations depending on the individual. Hence, osteomalacic lesions can develop in Paget's disease (because of high turnover) or during pregnancy (resulting from increased needs) or with lactation. Still, the two most susceptible groups are the elderly and Asian/African migrants who translocate to northern Europe and who may have both genetic susceptibility and poor sunlight exposure combined with unique dietary habits *(101,102)*. Characteristically, serum 25(OH)D levels are very low while 1,25(OH)$_2$D$_3$ concentrations are either normal or only slightly decreased.

Intrinsic vitamin D depletion can also result in osteomalacia and is more often recognized because of the underlying disease state even though it is less common worldwide than extrinsic deficiencies. The most common gastrointestinal disorders associated with this condition include adult celiac disease (nontropical sprue), gastric bypass surgery, postgastrectomy states, chronic pancreatitis, short bowel, inflammatory bowel disease, and biliary cirrhosis *(103)*. Not infrequently, both extrinsic and intrinsic deficiencies coexist because of the nature of the gastrointestinal disorder. Classically 25(OH)D levels are very low while 1,25(OH)$_2$D$_3$ serum levels are normal *(104)*.

Adult celiac disease is probably the most unrecognized disorder leading to osteomalacia *(105)*. Up to half the patients with this disease (even with variable presentations) present with vertebral fractures and osteoporotic symptoms. Bone density and urinary calcium excretion are both very low, PTH levels are high, and serum 25(OH)D concentrations are extremely low *(106)*. In mid-adult life, treatment with vitamin D, 25(OH)D, or calcitriol can suppress PTH levels and increase bone mass transiently *(107)*. However, complete reversal of the osteoporotic syndrome (low bone mass and vertebral fractures) in nontropical sprue does not appear to be possible even with vitamin D therapy and a gluten free diet.

Impaired 25-hydroxylation of vitamin D is another cause of osteomalacia. In this condition, 25-hydroxylase activity is reduced because of hepatobiliary disease *(108)*. Often all three vitamin D defects coexist, making it somewhat difficult if not impossible to identify the major component. Clearly though, alcoholic liver disease, primary biliary cirrhosis, and hemochromatosis can be associated with significant aberrancies in 25-hydroxylase activity *(109)*. Once again, serum levels of 25(OH)D are low, whereas 1,25(OH)$_2$D$_3$ levels are normal or low. Treatment with 25(OH)D or calcitriol is necessary to correct the underlying disorder.

Increased catabolism of 25(OH)D is also a cause of vitamin D-deficient osteomalacia. In this syndrome, conversion of 25(OH)D to inactive metabolites in the liver is accelerated due to several different medications that induce P450 cytochrome oxidase activity. These include anticonvulsant medications, rifampin, and the barbiturates *(110)*. In general, these agents tend to increase vitamin D requirements modestly in adults and moderately in children. Adequate sunlight and dietary intake usually prevent the florid manifestations of osteomalacia. However, in elderly individuals, or people who are institutionalized and have virtually no sunlight exposure, this enzymatic alteration could have clinical significance. Not unlike the other causes of vitamin D-induced osteomalacia mentioned so far, serum 25(OH)D levels are low and 1,25(OH)$_2$D$_3$ levels are usually normal or increased. Treatment with prophylactic vitamin D, or active calcitriol once the condition is noted, usually corrects the defect.

The last two levels in which vitamin D alterations can lead to osteomalacia *include abnormal 1-hydroxylation of 25(OH)D*, almost always as a function of kidney failure, and defects in the VDR, which lead to hereditary forms of vitamin D-dependent and vitamin D-resistant rickets. The latter are discussed in another chapter. The former, renal osteodystrophy, is commonly associated with osteoporosis, osteomalacia, and disorders of calcium homeostasis.

The kidney plays a major role in controlling calcium balance both through its reabsorption of calcium and phosphorus in the tubules and in the synthesis of $1,25(OH)_2D_3$ *(111)*. As noted previously, calcitriol has major effects on the intestine and the skeleton and is clearly the most potent metabolite of vitamin D. Several factors contribute to functional vitamin D deficiency in renal failure. First, 1α-hydroxylase activity is markedly reduced *(112)*. This leads to impaired calcium absorption and secondary hyperparathyroidism. Second, impaired phosphate excretion leads to high serum levels of phosphorus, which in turn suppresses any residual 1α-hydroxylase activity *(113)*. Third, alterations in the number of VDRs present in the parathyroids may also affect calcium homeostasis *(114)*. Finally, there may be peripheral resistance to PTH action in bone, which contributes to progressive hypocalcemia *(115)*. All these factors worsen secondary hyperparathyroidism and the bone disease that follows. However, osteomalacia may also coexist in this syndrome, either as a result of aluminum-induced low bone turnover, or secondary to chronic hypocalcemia, hypophosphatemia, and vitamin D deficiency related not only to impaired 1α-hydroxylation but also to extrinsic or even intrinsic disorders.

The clinical manifestations of renal osteodystrophy cannot be separated by pathogenetic factors. However, characteristically, bone pain, pseudofractures, growth retardation, muscle weakness, and soft tissue calcifications can be detected *(116)*. In addition, full-blown osteitis fibrosa cystica may be seen in which subperiosteal erosions of the phalanges are detected by X-ray. Bone cysts and slipped epiphysis can also be detected with this disorder *(117)*.

There are several therapeutic approaches to renal osteodystrophy that include vitamin D treatment. Calcitriol, calciferol, and calcidiol have all been employed to prevent secondary hyperparathyroidism in combination with calcium, phosphate binders, and dialysis. Several studies have documented clinical and radiologic improvement with $1,25(OH)_2D_3$ treatment for patients with renal osteodystrophy and osteitis fibrosa cystica *(118,119)*. Muscle strength improves, bone pain declines, and growth in children is restored, as are some of the histologic features of osteitis *(120,121)*. Doses of calcitriol have ranged from 0.25 to 1.5 µg/d and can be administered either intravenously or orally. Hypercalcemia is the major side effect from this form of therapy.

6. CONCLUSIONS

Vitamin D is a critical component of skeletal health across all ages. It may be particularly important during acquisition of bone mass and late in life when calcium balance is so important to the integrity of the remodeling unit. Therapeutic approaches to vitamin D deficiency states have generally led to positive results in terms of bone mass and fracture efficacy. Use of pharmacologic doses of vitamin D metabolites to enhance bone density has generally produced mixed results. Further studies are currently under way with analogs of vitamin D metabolites that might lessen the potential side effects of daily treatment while maximizing impact on bone. The future role of vitamin D in the treatment of osteoporosis remains to be determined.

ACKNOWLEDGMENTS

This work was supported in part by NIH grant RO1 AG10192.

REFERENCES

1. Rosen CJ. Biochemical markers of bone turnover. In: Osteoporosis: Diagnostic and Therapeutic Principles. Rosen CJ, ed. Totowa, NJ: Humana, 1996; 129–141.
2. Lian JB, Stein GS. Osteoblast biology. In: Osteoporosis. Marcus R, Feldman D, Kelsey J, eds. San Diego: Academic, 1996; 23–59.
3. Marks SC, Popoff SN. Bone cell biology: the regulation of development structure and function in the skeleton. Am J Anat 1988; 183:1–44.
4. Lian JB, Stewart C, Puchacz E, Mackowiak S, Shalhoub V, Collart D, Sambetti G, Stein GS. Structure of the rat osteocalcin gene and regulation of vitamin D-dependent expression. Proc Natl Acad Sci USA 1989; 86:1143–1147.
5. Glowacki J. The cellular and biochemical aspects of bone remodeling. In: Osteoporosis. Rosen CJ, ed. Totowa, NJ: Humana, 1996; 3–15.
6. Menton DN, Simmons DJ, Chang SL, Orr BY. From bone lining cell to osteocyte—an SEM study. Anat Rec 1984; 209:29–39.
7. Holtrop ME. The ultrastructure of bone. Ann Clin Lab Sci 1975; 5:264–271.
8. Teitelbaum SL, Tondravi MM, Ross FP. Osteoclast biology. In: Osteoporosis. Marcus R, Feldman D, Kelsey J, eds. San Diego: Academic, 1996; 61–94.
9. Blair HC, Teitelbaum SL, Grosso LE, Lacey DL, Tan H, McCort DW, Jeffrey JJ. Extracellular matrix degradation at acid pH: avian osteoclast acid collagenase isolation and characterization. Biochem J 1993; 290:873–884.
10. Kurihara N, Chenu C, Miller M, Civin C, Roodman GD. Identification of committed mononuclear precursors for osteoclast like cells formed in long term human marrow cultures. Endocrinology 1991; 126:2733–2741.
11. Miller PD. Guidelines for the clinical utilization of bone mass measurement in the adult population. Calcif Tiss Int 1995; 57:251,252.
12. Riggs BL. Overview of osteoporosis. West J Med 1991; 154:63–77.
13. Slemenda CW. Adult bone loss. In: Osteoporosis. Marcus R, ed. Boston: Blackwell Scientific, 1994; 107–124.
14. Parfitt AM. Osteomalacia and related disorders. In: Metabolic Bone Diseases and Clinically Related Disorders, vol. 2. Avioli L, Krane S, eds. Philadelphia: Saunders, 1990; 329–381.
15. Chen TL, Li JM, Van Ye T, Cone CM, Feldman D. Hormonal responses to 1,25 dihydroxyvitamin D_3 in cultured mouse osteoblast like cells—modulation by changes in receptor level. J Cell Physiol 1986; 126:21–28.
16. Noda M, Vogel RL, Craig AM, Prahl J, DeLuca HF, Denhardt DT. Identification of a DNA sequence responsible for binding of the 1,25 dihydroxyvitamin D_3 receptor and 1,25dihydroxyvitamin D_3 enhancement of mouse secreted phosphoprotein 1(osteopontin)gene expression. Proc Natl Acad Sci USA 1990; 87:9995–9999.
17. Canalis E, Lian JB. 1,25 Dihydroxyvitamin D effects on collagen and DNA synthesis in periosteum and periosteum free cultures. Bone 1985; 6:457–460.
18. Henthorn PS, Raducha M, Fedde KN, Lafferty MA, Whyte MP. Different missense mutations at the tissue specific alkaline phosphatase gene locus in autosomal recessively inherited forms of mild and severe hypophosphatasia. Proc Natl Acad Sci USA 1992; 89:9924–9928.
19. Stein GS, Lian JB. Molecular mechanisms mediating developmental and hormone regulated expression of genes in osteoblasts. In: Cellular and Molecular Biology of Bone. Noda M, ed. San Diego: Academic, 1993; 47–95.
20. Desbois C, Bradley A, Karsenty G. Study of osteocalcin function in mouse by targeted disruption. J Bone Miner Res 1994; 9:S376.
21. Towler DA, Bennett CD, Rodan GA. Activity of the rat osteocalcin basal promoter in osteoblastic cells is dependent upon homeodomain and CP1 binding motifs. Mol Endocrinol 1994; 8:614–624.
22. Hoffmann HM, Catron KM, van Wijnen AJ, McCabe LR, Lian JB, Stein GS, Stein JL. Transcriptional control of the tissue specific developmentally regulated osteocalcin gene requires a binding motif for the MSX family of homoeodomain proteins. Proc Natl Acad Sci USA 1994; 91:1287–1291.
23. Raisz LG, Trummel CL, Holick MF, DeLuca HF. 1,25 Dihydroxycholecalciferol: a potent stimulator of bone resorption in tissue culture. Science 1972; 175:768,769.
24. Suda T, Tkahashi N, Martin TJ. Modulation of osteoclast differentiation. Endocr Rev 1992; 13: 66–80.

18 Inherited Defects of Vitamin D Metabolism

Marie Demay

1. INTRODUCTION

The absence of biologic effects of vitamin D, resulting from deficient synthesis, dietary intake, or lack of activation of vitamin D (or to resistance to the biologic effects of the active metabolite), presents primarily with signs and symptoms that reflect impaired intestinal calcium absorption. These include signs and symptoms of neuromuscular irritability, including tetany and seizures, which are a direct result of the hypocalcemia. Long-standing deficiency of, or resistance to vitamin D metabolites leads to impaired bone mineralization as a result of calcium and phosphate deficiency. In the growing skeleton, growth plate abnormalities known as rickets are also observed (*see* Chapter 18). Secondary hyperparathyroidism is also observed, due to the hypocalcemia and the lack of antiproliferative and antitranscriptional effects of 1,25-dihydroxyvitamin D on the parathyroid glands (*see* Chapter 13).

With the institution of solar radiation and fish oil therapy for the treatment of rickets, it became clear that there was a rare subset of individuals who were resistant to this treatment. Measurements of the circulating vitamin D metabolites in this group of patients revealed that they fell into two categories: those who had adequate levels of 25-hydroxyvitamin D but low or undetectable levels of 1,25-dihydroxyvitamin D, and those in whom 1,25-dihydroxyvitamin D levels were elevated. These two groups were classified as having vitamin D-dependent rickets types I and II (VDDR I and II), respectively. More recently, two kindreds have been described in whom the biochemical defect points to impaired 25-hydroxylation of vitamin D *(1,2)*.

2. VITAMIN D-DEPENDENT RICKETS TYPE I

The metabolic pathway by which vitamin D, derived from cutaneous production or dietary sources, is converted to the active hormone has been discussed extensively in Chapter 2. Following 25-hydroxylation in the liver, vitamin D is converted to its active metabolite, 1,25-dihydroxyvitamin D, in the proximal tubule of the kidney. This is the only clinically significant source of the hydroxylase in normal humans; however, 25-hydroxyvitamin D 1α-hydroxylase activity has also been detected in granulomata *(3)*, in the decidual cells of the placenta *(4)*, and in keratinocytes *(5)*. VDDR I is a rare inborn error of metabolism, inherited in an autosomal recessive fashion. It is most commonly found

From: *Vitamin D: Physiology, Molecular Biology, and Clinical Applications*
Edited by: M. F. Holick © Humana Press Inc., Totowa, NJ

Table 1
Biochemical Data from Nine Patients Affected
with VDDR I Treated with Various Doses of Vitamin D[a]

Patient	Age (yr)	Total Calcium (mg/dL) [9–10.5]	Phosphorus (mg/dL) [3.8–5.0]	Alkaline phosphatase (IU/L) [100–300]	iPTH (mEq/L) [20–150]	25(OH)D (µg/L) [18–36]	1,25(OH)$_2$D (ng/L) [33–35]	Dose of vitamin D$_2$ (IU)
1	4	9.2	3.0	528	295	148	11	10,000
2	14	6.1	3.6	1160	292	108	9	5000
3	14	8.5	1.6	1100	372	140	6	75,000
4	12	6.3	4.2	595	258	135	6	60,000
5	15	6.4	4.4	378	398	105	12	100,000
6	21	8.6	2.6	395	375	262	12	50,000
7	1	7.6	2.2	1620	213	30	3	0
8	1	6.2	3.0	5900	1440	30	2	0
9	10	6.8	3.6	1268	445	37	8	9000–12,000 (age 1–8)

[a] Data in brackets represent normal ranges. Modified with permission from ref. *14*.

in the French Canadian population *(6)*, and the genetic locus maps to chromosome 12q14 *(7)*. The recent cloning of the vitamin D 1α-hydroxylase cDNA *(5,8–10)* has permitted confirmation of the hypothesis that VDDR I results from mutation of this gene *(5)*.

2.1. Clinical Presentation and Management

Since VDDR I is an inborn error of metabolism, it presents within the first few months of life. Because the main effect of 1,25-dihydroxyvitamin D is to maintain mineral ion homeostasis by promoting intestinal calcium absorption, the first clinical signs are those of acute or chronic hypocalcemia. Infants may present with hypocalcemic seizures as early as 4 wk postpartum or may present later (usually before the age of 2 years) with growth retardation, deformity, and bone pain secondary to rickets and osteomalacia. These abnormalities are clearly visible radiologically. In the *in utero* environment, the fetus' calcium homeostasis is thought to be relatively normal, due to maternal effects; therefore, teeth that calcify *in utero* are normal, but those that calcify postnatally often have marked enamel hypoplasia *(11)*. Because hypocalcemia elicits the normal physiologic response of secondary hyperparathyroidism, affected individuals may also present with aminoaciduria *(12)* in addition to hypocalcemia and hypophosphatemia. In patients with this clinical presentation, serum levels of vitamin D and its metabolites establish the diagnosis of VDDR I. Levels of vitamin D and 25-hydroxyvitamin D are normal or elevated, whereas levels of the active metabolite, 1,25-dihydroxyvitamin D, are low *(13,14)*. This observation was the basis for the hypothesis that impaired 1α-hydroxylation is the underlying pathophysiologic defect. Although therapeutic responses have been observed using pharmacologic doses of vitamin D (40–54.5 µg/kg/d) *(15)* and 25-hydroxyvitamin D (3–18 µg/kg/d) *(12)*, these doses are two orders of magnitude higher than those required to cure classical VDDR. Patients treated with these metabolites are often given insufficient doses or fail to respond to non-1α-hydroxylated metabolites, as evidenced by lack of remission of the clinical disorder (Table 1). Furthermore, although this treatment leads to markedly elevated levels of 25-hydroxyvitamin D, levels of 1,25-dihydroxyvitamin D remain low. The availability of 1α-hydroxylated metabolites of vitamin D has led to a complete remission of the clinical disorder (Fig. 1). Furthermore, the recommended doses of these 1α-hydroxylated metabolites are similar to those used to

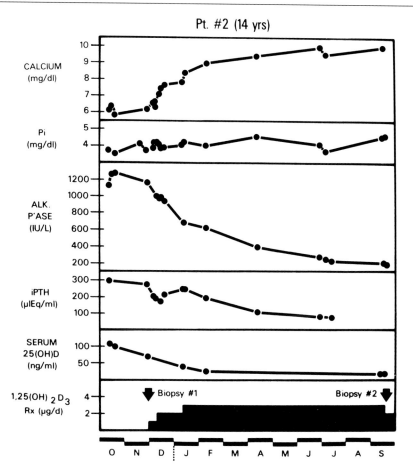

Fig. 1. Biochemical response to calcitriol therapy (of patient 2 in Table 1). Serum calcium (Ca), phosphorus (Pi), alkaline phosphatase (Alk Ptse), immunoreactive parathyroid hormone (iPTH) and 25-hydroxyvitamin D levels are shown, 1,25-Dihydroxyvitamin D was increased from 1 to 3 µg/d until a therapeutic response was observed. After the second bone biopsy revealed healing of osteomalacic lesions, the 1,25-dihydroxyvitamin D dosage was decreased to 2 µg/d. (Reprinted with permission from ref. *14*.)

treat patients with secondary or acquired 1α-hydroxylation defects (such as hypoparathyroidism) and result in normal serum levels of 1,25-dihydroxyvitamin D. These observations support the finding that VDDR I is a result of impaired or absent 25-hydroxyvitamin D 1α-hydroxylase activity.

The recommended treatment for this disorder is lifelong therapy with either 1α-hydroxyvitamin D (80–100 ng/kg/d) *(15)* or 1,25-dihydroxyvitamin D (8–400 ng/kg/d) *(14)*. This treatment results in a marked increase in intestinal calcium absorption (Fig. 2), accompanied by an increase in serum calcium within 24 h and radiologic healing of osteomalacic lesions within 2–3 mo *(15)*. Histomorphometric documentation of healing of osteomalacic lesions has been obtained in two affected siblings after 9 and 10 mo of treatment with 1,25-dihydroxyvitamin D *(14)*.

All patients receiving replacement therapy for vitamin D deficiency, including those with VDDR I, require careful monitoring to ensure adequate replacement and to minimize

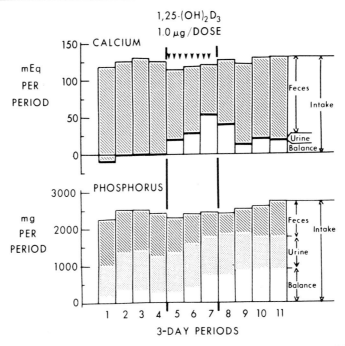

Fig. 2. Calcium balance studies in a patient with VDDR I. The patient was treated with 1.0 µg of 1,25-hydroxyvitamin D daily in periods 5, 6, and 7. (Reprinted with permission from ref. *12*.)

the risk of toxicity. The parameters to be monitored include serum calcium, which should normalize within 1 wk of institution of adequate therapy. At least initially, to promote healing of the osteomalacic lesions, calcium supplements should be included in the regimen. Serum levels of alkaline phosphatase may increase initially as the skeleton remineralizes, but should normalize within approx 6 mo. Secondary hyperparathyroidism regresses with adequate treatment; however, the time required for involution often reflects the duration and severity of untreated 1,25-dihydroxyvitamin D deficiency (Fig. 2). The main complications of therapy, hypercalcemia and nephrolithiasis, can be avoided by careful monitoring. Serum calcium levels should be maintained in the low-normal range and urinary calcium excretion below the lithogenic threshold (<4 mg/kg/d). Healing of the osteomalacic and rachitic lesions should be documented radiologically within 6–9 mo of institution of therapy. In growing children, monitoring to ensure adequate dosage and to avoid complications should be performed every 6–8 wk. The required replacement doses of 1α-hydroxylated metabolites may change rapidly in those patients with a growing skeleton. The growth rate of these children also needs to be followed carefully. In those patients with a mature skeleton, monitoring can be performed less frequently (every 3–6 mo) since their 1,25-dihydroxyvitamin D requirements are more stable. However, any patient with intercurrent intestinal disease or pregnancy requires closer monitoring. Placental 1α-hydroxylase activity has been shown to be present in the decidua *(4,16)*, which is maternally derived; therefore, the physiologic increase in 1,25-dihydroxyvitamin D levels found in normal pregnancy is absent in affected individuals. During pregnancy, therefore, the dose of 1α-hydroxylated vitamin D metabolites may need to be increased by 50–100%, with careful monitoring of serum and urinary parameters.

Careful analysis of the biochemical parameters of affected individuals led to a therapy capable of reversing the clinical and metabolic abnormalities in affected individuals long before the molecular basis of VDDR I was identified. The dramatic response to physiologic doses of 1α-hydroxylated vitamin D metabolites was the basis for the hypothesis that the disorder wass caused by an inherited mutation in the 25-hydroxyvitamin D 1α-hydroxylase or in a factor required for its enzymatic activity. Although prenatal diagnostic screening has not been performed, early postnatal diagnosis of the disorder can prevent the development of skeletal and dental complications and help affected individuals achieve their normal growth potential.

3. VITAMIN D-DEPENDENT RICKETS TYPE II

The therapeutic efficacy of 1α-hydroxylated vitamin D metabolites in the treatment of VDDR I led to the identification of a clinical syndrome that did not respond to physiologic doses of these metabolites. Analysis of the biochemical parameters revealed that these patients, in fact, had normal or elevated levels of 1,25-dihydroxyvitamin D prior to treatment. This led to the hypothesis that affected individuals were resistant to the biologic effects of 1,25-dihydroxyvitamin D. The cloning of the vitamin D receptor (VDR) *(17)* enabled the identification of the molecular basis of this disease as mutations in the VDR (see Chapter 22). Reports have begun to emerge, however, of normal VDR cDNAs in kindreds with clinical and biochemical parameters suggestive of VDDR II *(18,19)*.

3.1. Clinical Presentation and Management

Like VDDR I, VDDR II is an inborn error of metabolism, inherited in an autosomal recessive fashion. VDDR II presents in infancy or childhood with rickets, osteomalacia, and signs of hypocalcemia including tetany and seizures *(20)*. Affected individuals may have enamel hypoplasia and oligodentia *(21)*. Interestingly, a few cases have not presented until late adolescence *(22)*. Like VDDR I, affected patients have hypocalcemia, accompanied by secondary hyperparathyroidism, which results in hypophosphatemia and occasionally aminoaciduria. Normal levels of vitamin D and 25-hydroxyvitamin D are present, whereas levels of 1,25-dihydroxyvitamin D are elevated because of stimulation of the renal 1α-hydroxylase by parathyroid hormone (PTH). One unique feature of VDDR II is the alopecia totalis observed in some kindreds (Fig. 3), which is thought to be associated with the finding that VDRs are normally found in the external root sheath of hair follicles *(23)*. It was suggested that patients presenting with alopecia had a more severe resistance to 1,25-dihydroxyvitamin D *(20,24)*. However, spontaneous remissions have been described in patients with alopecia *(25)*, and severely affected individuals with normal hair growth have also been described *(26,27)*. Interestingly, the VDR cDNA in one patient with clinical and biochemical parameters suggestive of VDDR II, including alopecia, has been shown to be normal by sequence analysis and in vitro transcriptional activation studies *(18)*.

Histomorphometric analysis of bone biopsies from patients with VDDR II has demonstrated features consistent with osteomalacia *(22,28–30)*. In patients with long-standing secondary hyperparathyroidism, marrow fibrosis may also be seen *(22,29)*. Intestinal resistance to 1,25-dihydroxyvitamin D has been documented by marked impairment of intestinal calcium absorption and high fecal calcium losses in affected individuals, despite elevated serum levels of 1,25-dihydroxyvitamin D *(28,29,31)*; however, an improvement in intestinal calcium absorption has been observed concomitant with therapeutic responses to vitamin D metabolites.

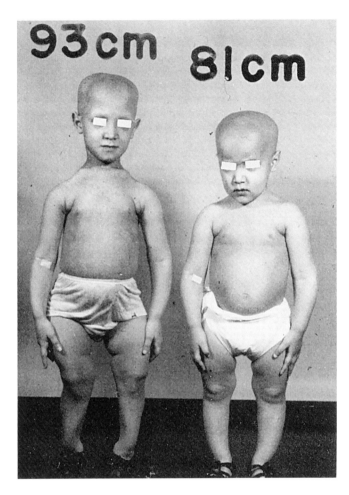

Fig. 3. Two sisters affected by VDDR II (ages 7 and 3) demonstrate short stature, bowing of the legs and alopecia. (Reprinted with permission from ref. *28*.)

Because the physiologic abnormality underlying VDDR II is a receptor defect, no treatment has been uniformly successful. Affected individuals have been shown to have variable therapeutic responses to pharmacologic doses of vitamin D metabolites and oral calcium supplements (1–6 g/d). Although 1α-hydroxylated metabolites have been the favored treatment in recent years, the intact metabolic pathway of vitamin D activation suggests that vitamin D or 25-hydroxyvitamin D may be as effective. The recommended therapeutic doses vary from 0.1–50 mg/d of vitamin D, 0.05–1.5 mg/d of 25-hydroxyvitamin D, and 5–20 µg/d of 1,25-dihydroxyvitamin D *(21)*, all of which result in markedly elevated levels of 1,25-dihydroxyvitamin D. In spite of these pharmacologic doses of vitamin D metabolites, in some severely affected individuals hypocalcemia persists, as do the rachitic and osteomalacic lesions. Because the main physiologic effect of 1,25-dihydroxyvitamin D is thought to be the promotion of intestinal calcium absorption, some patients have been treated with parenteral calcium infusions in an attempt to circumvent the intestinal resistance to 1,25-dihydroxyvitamin D. This has led to near normalization of the associated biochemical abnormalities as well as clinical and radiologic

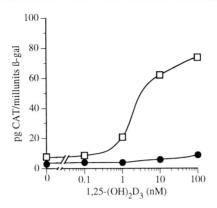

Fig. 4. Transactivation assay in COS-7 cells. COS-7 cells were transiently transfected with the Gly46Asp mutant (●) or wild-type (□) receptor cDNA construct and a VDRE-CAT reporter plasmid. The cells were then treated with 1,25(OH)$_2$D$_3$ to activate expression of the CAT gene. The CAT protein was only induced by the wild-type VDR. The Gly46Asp mutation eliminated the transactivation capacity of the VDR.

at amino acid 46 being changed to an aspartic acid (Gly46Asp) (Fig. 2). In contrast to the other DBD mutations described above, which were all located in highly conserved amino acids, the mutation at Gly46 occurs in an amino acid that is not conserved in all members of the steroid–thyroid–retinoid receptor superfamily. However, Gly46 is conserved among receptors that form heterodimers with RXR proteins. The recreated mutant VDR expressed in COS-7 cells exhibited normal [^{3}H]1,25(OH)$_2$D$_3$ binding and showed a reduced affinity for DNA. Although this mutation occurred in a nonconserved amino acid, the mutant VDR was transcriptionally inactive in reporter gene assays (Fig. 4). Using PCR and a restriction fragment length polymorphism (RFLP) generated by the mutation, Lin et al. *(85)* demonstrated that the patient was homozygous for the mutation in exon 2 and that the patient's father was a carrier of the mutant allele (Fig. 5).

A previously unreported patient with HVDRR from a Moroccan family was examined for mutations in the VDR gene by Wiese et al. *(75)*. At the cellular level, this patient exhibited the receptor-negative phenotype, suggesting that the defect was located in the VDR LDB. However, Wiese et al. *(75)* found a mutation in the VDR DBD. A unique C to T base change replaced the codon for arginine (CGA) with an opal termination codon (TGA) (Arg73stop) (Figs. 2 and 3). The Arg73stop mutation occurs in the middle of the second zinc finger, deleting the entire LDB and monoclonal antibody binding sites. The 72-amino acid polypeptide predicted from the sequence could not be demonstrated in the cultured cells. Interestingly, the Arg73stop mutation (CGA to TGA) occurs in the same codon that gives rise to the G family mutation (CGA to CAA) (Arg73Gln) *(60)* but at a different nucleotide base (Fig. 3).

5.2. Ligand Binding Domain Mutations

To investigate the molecular basis of the abnormality in the VDR in the receptor-negative class of defects, investigators also used PCR to amplify the exons of the VDR gene to facilitate their studies. The initial molecular analysis of the receptor-negative phenotype was performed by Ritchie et al. *(64)* in three related families (C, E, and H families) *(13,20,21,24)*. A single unique base change was identified at nucleotide 970 in exon 8 *(64)*. This single base change replaced a tyrosine codon (TAC) with an ochre

Since the discovery of HVDRR as a genetic disorder, the resistance to $1,25(OH)_2D_3$ seen in HVDRR patients has been suspected to be caused by mutations in the VDR. However, although the VDR is a principle player in the $1,25(OH)_2D_3$ action pathway, target organ resistance to $1,25(OH)_2D_3$ may be caused by mutations in other interacting proteins that participate in the transactivation process. Some of the likely candidates include RXR, or other transcription factors, and coactivators or corepressors. Defects in interacting proteins may inhibit VDR binding to DNA or disrupt the contact between the VDR and the interacting protein.

Hewison et al. *(90)* have described a case of HVDRR that may be caused by a defect in an interacting protein. The patient, a young girl, exhibited all the hallmarks of the disease, including alopecia. Examination of the patient's fibroblasts showed that they expressed a normal size VDR transcript and VDR that had a normal binding affinity for $[^3H]1,25(OH)_2D_3$. However, no 24-hydroxylase activity could be detected after treating the fibroblasts with up to 1 μM $1,25(OH)_2D_3$. Although the cells were clearly resistant to $1,25(OH)_2D_3$, the authors could not find a mutation in the coding region of the VDR gene. The patient's VDR cDNA was reassembled from mRNA from the resistant cells by reverse transcription and PCR. The VDR expressed from the patient's VDR cDNA exhibited a normal transactivation response to $1,25(OH)_2D_3$ in VDRE-CAT reporter assays in CV-1 cells, which proved that the receptor was normal and that the tissue resistance was not due to a defect in the VDR. Their data suggest that hormone resistance and HVDRR may be caused by mutations in an essential protein that participates in the $1,25(OH)_2D_3$ action pathway.

6. THERAPY

Most patients with HVDRR do not respond to treatment with vitamin D analogs, even at supraphysiologic doses. Therapies using combinations of calcium and active vitamin D metabolites have been tried in attempts to cure the signs and symptoms of the disease. In a few of the earlier reports, HVDRR patients without alopecia responded clinically, and X-rays showed improvement in rickets following administration of pharmacologic doses of vitamin D ranging from 5000 to 40,000 IU/d *(1,2,5)*, 20 to 200 µg of 25(OH)D/d, and 17 to 20 µg of $1,25(OH)_2D_3$/d *(2)*. The molecular basis of HVDRR in these earlier cases has not been described. These patients may have had minor aberrations in the VDR LDB that resulted in a decreased binding affinity for $1,25(OH)_2D_3$, which could be overcome by treatment with high doses of the hormone. Indeed, in two patients without alopecia, mutations were found in the VDR LBD *(72,74,89)*. In one case, an Arg274Leu mutation was identified *(74)*. However, this patient did not respond to treatment with massive doses of $1,25(OH)_2D_3$, and his fibroblasts were also unresponsive to hormone treatment. Interestingly, the recreated Arg274Leu mutant VDR did exhibit transcriptional activity when high doses of hormone were given *(74)*. In the second case, a His305Gln mutation was identified *(89)*. This patient responded to treatment with high doses of $1,25(OH)_2D_3$ (30 µg/d) *(88)*. These studies suggest that HVDRR patients without alopecia may have defects in the VDR LBD, and the defect may sometimes be overcome by treatment with high doses of vitamin D metabolites.

In general, HVDRR patients who do not develop alopecia are more responsive to treatment with vitamin D preparations, whereas those who have alopecia are generally less responsive *(91)*. However, a small number of patients with alopecia have been successfully treated with vitamin D metabolites including vitamin D or 1α(OH)D *(9,12)*, 25(OH)D as well as 1α(OH)D *(14)*, and 1α(OH)D or 1,25(OH)$_2$D$_3$ *(6,23,26,34,38)*. Interestingly, in one case in which vitamin D and 1,25(OH)$_2$D$_3$ therapies were ineffective, the patient responded to oral phosphorous therapy *(4)*.

Calcium administration has also been an effective therapy for treating HVDRR patients. Sakati et al. *(30)* used a high-dose oral calcium therapy of 3–4 g of elemental calcium orally per day to treat a patient who failed to respond to calciferols. The patient showed clinical improvement within the 4 mo of therapy. Long-term intravenous calcium infusions have also been used to treat HVDRR patients. Balsan et al. *(92)* showed the beneficial effects of intravenous calcium infusions in a child with HVDRR and alopecia who did not respond to prior treatment with large doses of vitamin D derivatives or oral calcium supplements. The child received high doses of calcium intravenously during the nocturnal hours over a 9-mo period. Relief from bone pain was observed within the first 2 wk of intravenous therapy, and within 7 mo the child gained weight and height. Intravenous calcium infusions bypassed the calcium absorption defect in the intestine caused by the disease. Several other studies have used intravenous calcium infusion to treat children with HVDRR *(35,93,94)*. High-dose oral calcium therapy after radiologic healing of the rickets by intravenous calcium infusion has been shown to be an effective means of maintaining normal serum calcium concentrations *(94)*. Children with HVDRR are now routinely started on this two-step protocol at 2 yr of age *(94)*.

Spontaneous improvement in the disease has been observed in a few cases of HVDRR *(20,21,23)*. Sometimes the improvement occurs after long-term relatively ineffective treatment with vitamin D metabolites and mineral replacements; in other cases the improvement was noted after the treatment was stopped *(23)*. Interestingly, fibroblasts taken from HVDRR patients after spontaneous healing occurred continued to exhibit resistance to 1,25(OH)$_2$D$_3$ *(23)*. Spontaneous improvement has occurred in patients exhibiting the receptor-negative phenotype *(20,21)* as the result of a Try295stop mutation *(64,68)* as well as in the receptor-positive phenotype *(23)* due to an Arg73Gln mutation *(60)*. It is interesting to note that in all the patients who showed spontaneous improvement, the alopecia remained unchanged *(20,21.23)*.

7. CONCLUDING REMARKS

HVDRR is a rare recessive genetic disorder caused by mutations in the VDR that results in end-organ resistance to 1,25(OH)$_2$D$_3$ action. The major effect of the defective VDR on the vitamin D endocrine system is to decrease intestinal calcium and phosphate absorption, which results in decreased bone mineralization and rickets. Since 1978, more than 40 families exhibiting signs and/or symptoms of HVDRR have been studied. In all cases, the assignment of HVDRR has been based on resistance to vitamin D in combination with high circulating levels of 1,25(OH)$_2$D$_3$. A number of cases have been analyzed for 1,25(OH)$_2$D$_3$ binding and bioactivity, which showed that the disease was caused by heterogeneous defects in the VDR resulting from mutations in the VDR gene. A number of cases of HVDRR have not yet been examined for mutations in the VDR gene. Since some of these cases presented late in life, they may have been the result of nonhereditary resistance to 1,25(OH)$_2$D$_3$.

Analysis of the HVDRR syndrome provides many interesting insights into vitamin D physiology and the role of the VDR in mediating 1,25(OH)$_2$D$_3$ action. VDRs have been found in many tissues in the body, widening the scope of potential vitamin D target cells. In addition to maintaining calcium homeostasis, $1,25(OH)_2D_3$ regulates a number of other biologic processes *(61,95–99)*. Important biologic actions for vitamin D have been postulated in many of these sites, particularly in the immune and endocrine systems. VDRs have been found in endocrine glands such as pituitary, pancreas, parathyroid, gonads, and placenta, and $1,25(OH)_2D_3$ has been shown to regulate hormone synthesis and secretion from these glands *(61,95–99)*. VDRs have also been found in hematolymphopoietic cells, and $1,25(OH)_2D_3$ has been shown to regulate cell differentiation and the production of interleukins and cytokines *(100)*. However, despite the many processes shown to be regulated by $1,25(OH)_2D_3$, children with HVDRR only exhibit symptoms that relate to their calcium deficiency and alopecia. Hochberg et al. *(101)* examined hormone secretion in patients with HVDRR and found no abnormalities in insulin, thyroid-stimulating hormone, prolactin, growth hormone, and testosterone levels in serum. Even et al. *(102)* showed that urinary cyclic adenosine monophosphate and renal excretion of potassium, phosphorous, and bicarbonate were normal in HVDRR patients treated with PTH. However, PTH failed to decrease urinary calcium and sodium excretion in these patient to the extent found in the control patients. This suggests that $1,25(OH)_2D_3$ may selectively modulate the renal response to PTH and facilitate the PTH-induced reabsorption of calcium and sodium *(102)*. Although minor aberrations have been noted in the fungicidal activity of neutrophils from HVDRR patients *(103)*, the patients do not exhibit any clinically apparent immunologic defects.

The improvement of rickets by chronic intravenous calcium infusion raises interesting questions about the role of vitamin D in bone homeostasis. First, correction of hypocalcemia and secondary hyperparathyroidism leads to healing of the rickets, as assessed by X-ray and bone biopsy. Thus, although there are many well-defined actions of vitamin D on osteoblasts, the calcium infusion data suggest that $1,25(OH)_2D_3$ action on osteoblasts is not essential to form normal bone. The implication is that $1,25(OH)_2D_3$ action is mainly on the intestine to provide calcium and phosphate for bone formation. The same conclusion was reached by Underwood and DeLuca *(104)*, who showed that the development of rickets could be prevented in totally vitamin D-deficient rats by calcium and phosphate infusions in the absence of vitamin D.

Second, although $1,25(OH)_2D_3$ is an inhibitor of PTH production in some settings, in the HVDRR children, normalizing serum calcium by intravenous infusion is enough to suppress their PTH overproduction. In addition, intravenous calcium therapy without phosphate is sufficient to correct all the metabolic abnormalities in children with HVDRR. This suggests that the hypophosphatemia in these patients is the result of secondary hyperparathyroidism.

Third, a number of interesting facts concerning alopecia and HVDRR are worth noting. Since VDRs have been found in hair follicles *(105,106)*, $1,25(OH)_2D_3$ action through the VDR appears to be essential in the differentiation of this structure during embryogenesis. Also, Marx et al. *(91)* has analyzed a number of HVDRR patients and shown that there is some correlation between the severity of rickets and the presence of alopecia. Patients with alopecia tend to be less responsive to calciferols than those without alopecia. The alopecia or some degree of hair loss appears to be associated with DBD mutations or premature stop mutations, which usually result in complete hormone resistance. Patients with LBD mutations in general do not develop alopecia. Alopecia remains unchanged in patients who undergo intravenous calcium infusion or who show spontaneous improvement. It is of interest to note that alopecia has not been found in other conditions related to vitamin D, including VDDR I and vitamin D deficiency states. Thus, in families with a prior history of the disease, the absence of body hair in newborns provides initial evidence for HVDRR.

At this time, 11 point mutations have been found in the VDR DBD, 1 in the hinge region, and 4 in the LDB. Deletion of exons 7–9 of the VDR gene has been described in one family. Mutations in the VDR DBD prevent the receptor from activating gene transcription even though $1,25(OH)_2D_3$ binding is normal. Conversely, missense mutations in the LDB cause a less profound defect; in one case, $1,25(OH)_2D_3$ responsiveness was restored to an individual by therapy with high doses of hormone. On the other hand, mutations that introduce premature termination codons, which truncate the VDR, lead to complete hormone resistance.

A prenatal diagnosis of HVDRR is now possible in pregnant women from high risk families. Cultured cells from chorionic villus samples or amniotic fluid have been used to ascertain whether the fetus has HVDRR using $[^3H]1,25(OH)_2D_3$ binding, induction of 24-hydroxylase activity and RFLP analyses *(107,108)*.

A final point is the interesting dilemma regarding the spontaneous improvement in some HVDRR children as they get older. A hypothesis to explain normalization of the $1,25(OH)_2D_3$ endocrine system in the face of inactive VDRs is that some other regulatory factor can substitute for the defective system. It has recently been shown that the VDRE in the osteocalcin gene confers responsiveness to retinoic acid in addition to vitamin D. The RAR binds to the VDRE and is capable of transactivating the osteocalcin gene. A hypothesis to explain the spontaneous improvement of some HVDRR children, despite the continued presence of defective VDR, is that the RAR can ultimately substitute for the nonfunctional VDR and activate the appropriate target genes to reverse the hypocalcemia and restore the bones to normal. This speculative hypothesis remains to be validated.

Biochemical and genetic analysis of the VDR in the HVDRR syndrome has yielded important insights into the structure and function of the receptor in mediating $1,25(OH)_2D_3$ action. Similarly, studies of the children with HVDRR continue to provide further insight into the biologic role of $1,25(OH)_2D_3$ in vivo. A concerted investigative approach of HVDRR at the clinical, cellular, and molecular levels has proved exceedingly valuable in understanding the mechanism of action of $1,25(OH)_2D_3$ and in improving the diagnostic and clinical management of this rare genetic disease.

REFERENCES

1. Brooks MH, Bell NH, Love L, Stern PH, Orfei E, Queener SF, Hamstra AJ, DeLuca HF. Vitamin-D-dependent rickets type II. Resistance of target organs to 1,25-dihydroxyvitamin D. N Engl J Med 1978; 298:996–999.
2. Marx SJ, Spiegel AM, Brown EM, Gardner DG, Downs RW Jr, Attie M, Hamstra AJ, DeLuca HF. A familial syndrome of decrease in sensitivity to 1,25-dihydroxyvitamin D. J Clin Endocrinol Metab 1978; 47:1303–1310.
3. Balsan S, Garabedian M, Lieberherr M, Gueris J, Ulmann A. Serum 1,25-dihydroxyvitamin D concentrations in two different types of pseudo-deficiency rickets. In: Vitamin D: Basic Research and Its Clinical Application. Fourth Workshop on Vitamin D. Norman AW, Bouillon R, Thomasset M, eds. New York: Walter de Gruyter, 1979; 1143–1149.
4. Rosen JF, Fleischman AR, Finberg L, Hamstra A, DeLuca HF. Rickets with alopecia: an inborn error of vitamin D metabolism. J Pediatr 1979; 94:729–735.
5. Zerwekh JE, Glass K, Jowsey J, Pak CY. An unique form of osteomalacia associated with end organ refractoriness to 1,25-dihydroxyvitamin D and apparent defective synthesis of 25-hydroxyvitamin D. J Clin Endocrinol Metab 1979; 49:171–175.

6. Fujita T, Nomura M, Okajima S, Furuya H. Adult-onset vitamin D-resistant osteomalacia with the unresponsiveness to parathyroid hormone. J Clin Endocrinol Metab 1980; 50:927–931.
7. Liberman UA, Samuel R, Halabe A, Kauli R, Edelstein S, Weisman Y, Papapoulos SE, Clemens TL, Fraher LJ, O'Riordan JL. End-organ resistance to 1,25-dihydroxycholecalciferol. Lancet 1980; 1: 504–506.
8. Sockalosky JJ, Ulstrom RA, DeLuca HF, Brown DM. Vitamin D-resistant rickets: end-organ unresponsiveness to 1,25(OH)$_2$D$_3$. J Pediatr 1980; 96:701–703.
9. Tsuchiya Y, Matsuo N, Cho H, Kumagai M, Yasaka A, Suda T, Orimo H, Shiraki M. An unusual form of vitamin D-dependent rickets in a child: alopecia and marked end-organ hyposensitivity to biologically active vitamin D. J Clin Endocrinol Metab 1980; 51:685–690.
10. Beer S, Tieder M, Kohelet D, Liberman OA, Vure E, Bar-Joseph G, Gabizon D, Borochowitz ZU, Varon W, Modai D. Vitamin D resistant rickets with alopecia: a form of end organ resistance to 1,25-dihydroxyvitamin D. Clin Endocrinol 1981; 14:395–402.
11. Eil C, Liberman UA, Rosen JF, Marx SJ. A cellular defect in hereditary vitamin-D-dependent rickets type II: defective nuclear uptake of 1,25-dihydroxyvitamin D in cultured skin fibroblasts. N Engl J Med 1981; 304:1588–1591.
12. Kudoh T, Kumagai T, Uetsuji N, Tsugawa S, Oyanagi K, Chiba Y, Minami R, Nakao T. Vitamin D dependent rickets: decreased sensitivity to 1,25-dihydroxyvitamin D. Eur J Pediatr 1981; 137: 307–311.
13. Feldman D, Chen T, Cone C, Hirst M, Shani S, Benderli A, Hochberg Z. Vitamin D resistant rickets with alopecia: cultured skin fibroblasts exhibit defective cytoplasmic receptors and unresponsiveness to 1,25(OH)$_2$D$_3$. J Clin Endocrinol Metab 1982; 55:1020–1022.
14. Balsan S, Garabedian M, Liberman UA, Eil C, Bourdeau A, Guillozo H, Grimberg R, Le Deunff MJ, Lieberherr M, Guimbaud P, Broyer M, Marx SJ. Rickets and alopecia with resistance to 1,25-dihydroxyvitamin D: two different clinical courses with two different cellular defects. J Clin Endocrinol Metab 1983; 57:803–811.
15. Clemens TL, Adams JS, Horiuchi N, Gilchrest BA, Cho H, Tsuchiya Y, Matsuo N, Suda T, Holick MF. Interaction of 1,25-dihydroxyvitamin-D$_3$ with keratinocytes and fibroblasts from skin of normal subjects and a subject with vitamin-D-dependent rickets, type II: a model for study of the mode of action of 1,25-dihydroxyvitamin D$_3$. J Clin Endocrinol Metab 1983; 56:824–830.
16. Griffin JE, Zerwekh JE. Impaired stimulation of 25-hydroxyvitamin D-24-hydroxylase in fibroblasts from a patient with vitamin D-dependent rickets, type II. A form of receptor-positive resistance to 1,25-dihydroxyvitamin D$_3$. J Clin Invest 1983; 72:1190–1199.
17. Liberman UA, Eil C, Marx SJ. Resistance to 1,25(OH)$_2$D$_3$: association with heterogeneous defects in cultured skin fibroblasts. J Clin Invest 1983; 71:192–200.
18. Liberman UA, Eil C, Holst P, Rosen JF, Marx SJ. Hereditary resistance to 1,25-dihydroxyvitamin D: defective function of receptors for 1,25-dihydroxyvitamin D in cells cultured from bone. J Clin Endocrinol Metab 1983; 57:958–962.
19. Adams JS, Gacad MA, Singer FR. Specific internalization and action of 1,25-dihydroxyvitamin D$_3$ in cultured dermal fibroblasts from patients with X-linked hypophosphatemia. J Clin Endocrinol Metab 1984; 59:556–560.
20. Chen TL, Hirst MA, Cone CM, Hochberg Z, Tietze HU, Feldman D. 1,25-Dihydroxyvitamin D resistance, rickets, and alopecia: analysis of receptors and bioresponse in cultured fibroblasts from patients and parents. J Clin Endocrinol Metab 1984; 59:383–388.
21. Hochberg Z, Benderli A, Levy J, Vardi P, Weisman Y, Chen T, Feldman D. 1,25-Dihydroxyvitamin D resistance, rickets, and alopecia. Am J Med 1984; 77:805–811.
22. Gamblin GT, Liberman UA, Eil C, Downs RWJ, Degrange DA, Marx SJ. Vitamin D dependent rickets type II: defective induction of 25-hydroxyvitamin D$_3$-24-hydroxylase by 1,25-dihydroxyvitamin D$_3$ in cultured skin fibroblasts. J Clin Invest 1985; 75:954–960.
23. Hirst MA, Hochman HI, Feldman D. Vitamin D resistance and alopecia: a kindred with normal 1,25-dihydroxyvitamin D binding, but decreased receptor affinity for deoxyribonucleic acid. J Clin Endocrinol Metab 1985; 60:490–495.
24. Hochberg Z, Gilhar A, Haim S, Friedman-Birnbaum R, Levy J, Benderly A. Calcitriol-resistant rickets with alopecia. Arch Dermatol 1985; 121:646,647.
25. Koren R, Ravid A, Liberman UA, Hochberg Z, Weisman Y, Novogrodsky A. Defective binding and function of 1,25-dihydroxyvitamin D$_3$ receptors in peripheral mononuclear cells of patients with end-organ resistance to 1,25-dihydroxyvitamin D. J Clin Invest 1985; 76:2012–2015.

26. Castells S, Greig F, Fusi MA, Finberg L, Yasumura S, Liberman UA, Eil C, Marx SJ. Severely deficient binding of 1,25-dihydroxyvitamin D to its receptors in a patient responsive to high doses of this hormone. J Clin Endocrinol Metab 1986; 63:252–256.
27. Fraher LJ, Karmali R, Hinde FR, Hendy GN, Jani H, Nicholson L, Grant D, O'Riordan JL. Vitamin D-dependent rickets type II: extreme end organ resistance to 1,25-dihydroxy vitamin D_3 in a patient without alopecia. Eur J Pediatr 1986; 145:389–395.
28. Liberman UA, Eil C, Marx SJ. Receptor-positive hereditary resistance to 1,25-dihydroxyvitamin D: chromatography of receptor complexes on deoxyribonucleic acid-cellulose shows two classes of mutation. J Clin Endocrinol Metab 1986; 62:122–126.
29. Liberman UA, Eil C, Marx SJ. Clinical features of hereditary resistance to 1,25-dihydroxyvitamin D (hereditary hypocalcemic vitamin D resistant rickets type II). Adv Exp Med Biol 1986; 196:391–406.
30. Sakati N, Woodhouse NJY, Niles N, Harfi H, de Grange DA, Marx S. Hereditary resistance to 1,25-dihydroxyvitamin D: clinical and radiological improvement during high-dose oral calcium therapy. Horm Res 1986; 24:280–287.
31. Takeda E, Kuroda Y, Saijo T, Toshima K, Naito E, Kobashi H, Iwakuni Y, Miyao M. Rapid diagnosis of vitamin D-dependent rickets type II by use of phytohemagglutinin-stimulated lymphocytes. Clin Chim Acta 1986; 155:245–250.
32. Laufer D, Benderly A, Hochberg Z. Dental pathology in calcitirol resistant rickets. J Oral Med 1987; 42:272–275.
33. Nagler A, Merchav S, Fabian I, Tatarsky I, Hochberg Z. Myeloid progenitors from the bone marrow of patients with vitamin D resistant rickets (type II) fail to respond to $1,25(OH)_2D_3$. Br J Haematol 1987; 67:267–271.
34. Takeda E, Kuroda Y, Saijo T, Naito E, Kobashi H, Yokota I, Miyao M. 1 Alpha-hydroxyvitamin D_3 treatment of three patients with 1,25-dihydroxyvitamin D-receptor-defect rickets and alopecia. Pediatrics 1987; 80:97–101.
35. Bliziotes M, Yergey AL, Nanes MS, Muenzer J, Begley MG, Viera NE, Kher KK, Brandi ML, Marx SJ. Absent intestinal response to calciferols in hereditary resistance to 1,25-dihydroxyvitamin D: documentation and effective therapy with high dose intravenous calcium infusions. J Clin Endocrinol Metab 1988; 66:294–300.
36. Barsony J, McKoy W, DeGrange DA, Liberman UA, Marx SJ. Selective expression of a normal action of the 1,25-dihydroxyvitamin D_3 receptor in human skin fibroblasts with hereditary severe defects in multiple actions of that receptor. J Clin Invest 1989; 83:2093–2101.
37. Malloy PJ, Hochberg Z, Pike JW, Feldman D. Abnormal binding of vitamin D receptors to deoxyribonucleic acid in a kindred with vitamin D-dependent rickets, type II. J Clin Endocrinol Metab 1989; 68:263–269.
38. Takeda E, Yokota I, Kawakami I, Hashimoto T, Kuroda Y, Arase S. Two siblings with vitamin-D-dependent rickets type II: no recurrence of rickets for 14 years after cessation of therapy. Eur J Pediatr 1989; 149:54–57.
39. Koeffler HP, Bishop JE, Reichel H, Singer F, Nagler A, Tobler A, Walka M, Norman AW. Lymphocyte cell lines from vitamin D-dependent rickets type II show functional defects in the 1 alpha,25-dihydroxyvitamin D_3 receptor. Mol Cell Endocrinol 1990; 70:1–11.
40. Takeda E, Yokota I, Ito M, Kobashi H, Saijo T, Kuroda Y. 25-Hydroxyvitamin D-24-hydroxylase in phytohemagglutinin-stimulated lymphocytes: intermediate bioresponse to 1,25-dihydroxyvitamin D_3 of cells from parents of patients with vitamin D-dependent rickets type II. J Clin Endocrinol Metab 1990; 70:1068–1074.
41. Yokota I, Takeda E, Ito M, Kobashi H, Saijo T, Kuroda Y. Clinical and biochemical findings in parents of children with vitamin D-dependent rickets Type II. J Inherit Metab Dis 1991; 14:231–240.
42. Simonin G, Chabrol B, Moulene E, Bollini G, Strouc S, Mattei JF, Giraud F. Vitamin D-resistant rickets type II: apropos of 2 cases. Pediatrie 1992; 47:817–820.
43. Lin JP, Uttley WS. Intra-atrial calcium infusions, growth, and development in end organ resistance to vitamin D. Arch Dis Child 1993; 69:689–692.
43a. Kitanaka S, Takeyama K, Murayama A, Sato T, Okumura K, Nogami M, Hasegawa Y, Niimi H, Yanagisawa J, Tanaka T, et al. Inactivating mutations in the 25-hydroxyvitamin D_3 1alpha-hydroxylase gene in patients with pseudovitamin D-deficiency rickets. N Engl J Med 1998; 338:653–661.
43b. Fu GK, Lin D, Zhang MY, Bikle DD, Shakleton CH, Miller WL, Portale AA. Cloning of human 25-hydroxyvitamin D-1alpha-hydroxylase and mutations causing vitamin D-dependent rickets type 1. Mol Endocrinol 1997; 11:1961–1970.

44. Pike JW, Sleator NM. Hormone-dependent phosphorylation of the 1,25-dihydroxyvitamin D_3 receptor in mouse fibroblasts. Biochem Biophys Res Commun 1985; 131:378–385.
45. Evans RM. The steroid and thyroid hormone receptor superfamily. Science 1988; 240:889–895.
46. McDonnell DP, Scott RA, Kerner SA, O'Malley BW, Pike JW. Functional domains of the human vitamin D_3 receptor regulate osteocalcin gene expression. Mol Endocrinol 1989; 3:635–644.
47. Carson-Jurica MA, Schrader WT, O'Malley BW. Steroid receptor family: structure and functions. Endocr Rev 1990; 11:201–220.
48. Rastinejad F, Perlmann T, Evans RM, Sigler PB. Structural determinants of nuclear receptor assembly on DNA direct repeats. Nature 1995; 375:203–211.
49. Zilliacus J, Wright AP, Carlstedt-Duke J, Gustafsson JA. Structural determinants of DNA-binding specificity by steroid receptors. Mol Endocrinol 1995; 9:389–400.
50. Hsieh JC, Jurutka PW, Galligan MA, Terpening CM, Haussler CA, Samuels DS, Shimizu Y, Shimizu N, Haussler MR. Human vitamin D receptor is selectively phosphorylated by protein kinase C on serine 51, a residue crucial to its trans-activation function. Proc Natl Acad Sci USA 1991; 88:9315–9319.
51. Hsieh JC, Jurutka PW, Nakajima S, Galligan MA, Haussler CA, Shimizu Y, Shimizu N, Whitfield GK, Haussler MR. Phosphorylation of the human vitamin D receptor by protein kinase C. Biochemical and functional evaluation of the serine 51 recognition site. J Biol Chem 1993; 268:15,118–15,126.
52. Nakajima S, Hsieh JC, MacDonald PN, Galligan MA, Haussler CA, Whitfield GK, Haussler MR. The C-terminal region of the vitamin D receptor is essential to form a complex with a receptor auxiliary factor required for high affinity binding to the vitamin D-responsive element. Mol Endocrinol 1994; 8:159–172.
53. Wagner RL, Apriletti JW, McGrath ME, West BL, Baxter JD, Fletterick RJ. A structural role for hormone in the thyroid hormone receptor. Nature 1995; 378:690–697.
54. Renaud JP, Rochel N, Ruff M, Vivat V, Chambon P, Gronemeyer H, Moras D. Crystal structure of the RAR-gamma ligand-binding domain bound to all-trans retinoic acid. Nature 1995; 378:681–689.
55. Bourguet W, Ruff M, Chambon P, Gronemeyer H, Moras D. Crystal structure of the ligand-binding domain of the human nuclear receptor RXR-alpha. Nature 1995; 375:377–382.
56. Forman BM, Yang CR, Au M, Casanova J, Ghysdael J, Samuels HH. A domain containing leucine-zipper-like motifs mediate novel in vivo interactions between the thyroid hormone and retinoic acid receptors. Mol Endocrinol 1989; 3:1610–1626.
57. Faraco JH, Morrison NA, Baker A, Shine J, Frossard PM. ApaI dimorphism at the human vitamin D receptor gene locus. Nucleic Acids Res 1989; 17:2150.
58. Szpirer J, Szpirer C, Riviere M, Levan G, Marynen P, Cassiman JJ, Wiese R, DeLuca HF. The Sp1 transcription factor gene (SP1) and the 1,25-dihydroxyvitamin D_3 receptor gene (VDR) are colocalized on human chromosome arm 12q and rat chromosome 7. Genomics 1991; 11:168–173.
59. Labuda M, Fujiwara TM, Ross MV, Morgan K, Garcia-Heras J, Ledbetter DH, Hughes MR, Glorieux FH. Two hereditary defects related to vitamin D metabolism map to the same region of human chromosome 12q13-14. J Bone Miner Res 1992; 7:1447–1453.
60. Hughes MR, Malloy PJ, Kieback DG, Kesterson RA, Pike JW, Feldman D, O'Malley BW. Point mutations in the human vitamin D receptor gene associated with hypocalcemic rickets. Science 1988; 242:1702–1705.
61. Pike JW. Vitamin D_3 receptors: structure and function in transcription. Annu Rev Nutr 1991; 11: 189–216.
62. Pike JW. Molecular mechanisms of cellular response to the vitamin D_3 hormone. In: Disorders of Bone and Mineral Metabolism. Coe FL, Favus MJ, eds. New York: Raven, 1992; 163–193.
62a. Miyamoto K, Kesterson RA, Yamamoto H, Taketani Y, Nishiwaki E, Tatsumi S, Inoue Y, Morita K, Takeda E, Pike JW. Structural organization of the human vitamin D receptor chromosome gene and its promoter. Mol Endocrinol 1997; 11:1165–1179.
63. Baker AR, McDonnell DP, Hughes M, Crisp TM, Mangelsdorf DJ, Haussler MR, Pike JW, Shine J, O'Malley BW. Cloning and expression of full-length cDNA encoding human vitamin D receptor. Proc Natl Acad Sci USA 1988; 85:3294–3298.
64. Ritchie HH, Hughes MR, Thompson ET, Malloy PJ, Hochberg Z, Feldman D, Pike JW, O'Malley BW. An ochre mutation in the vitamin D receptor gene causes hereditary 1,25-dihydroxyvitamin D_3-resistant rickets in three families. Proc Natl Acad Sci USA 1989; 86:9783–9787.
65. Sone T, Scott RA, Hughes MR, Malloy PJ, Feldman D, O'Malley BW, Pike JW. Mutant vitamin D receptors which confer hereditary resistance to 1,25-dihydroxyvitamin D_3 in humans are transcriptionally inactive in vitro. J Biol Chem 1989; 264:20,230–20,234.

66. Feldman D, Malloy PJ. Hereditary 1,25-dihydroxyvitamin D resistant rickets: molecular basis and implications for the role of $1,25(OH)_2D_3$ in normal physiology. Mol Cell Endocrinol 1990; 72:C57–62.
67. Hughes MR, Malloy PJ, O'Malley BW, Pike JW, Feldman D. Genetic defects of the 1,25-dihydroxyvitamin D_3 receptor. J Recept Res 1991; 11:699–716.
68. Malloy PJ, Hochberg Z, Tiosano D, Pike JW, Hughes MR, Feldman D. The molecular basis of hereditary 1,25-dihydroxyvitamin D_3 resistant rickets in seven related families. J Clin Invest 1990; 86:2071–2079.
69. Malloy PJ, Weisman Y, Feldman D. Hereditary 1 alpha,25-dihydroxyvitamin D-resistant rickets resulting from a mutation in the vitamin D receptor deoxyribonucleic acid-binding domain. J Clin Endocrinol Metab 1994; 78:313–316.
70. Sone T, Marx SJ, Liberman UA, Pike JW. A unique point mutation in the human vitamin D receptor chromosomal gene confers hereditary resistance to 1,25-dihydroxyvitamin D_3. Mol Endocrinol 1990; 4:623–631.
71. Saijo T, Ito M, Takeda E, Huq AH, Naito E, Yokota I, Sone T, Pike JW, Kuroda Y. A unique mutation in the vitamin D receptor gene in three Japanese patients with vitamin D-dependent rickets type II: utility of single-strand conformation polymorphism analysis for heterozygous carrier detection. Am J Hum Genet 1991; 49:668–673.
72. Rut AR, Hewison M, Rowe P, Hughes M, Grant D, O'Riordan JLH. A novel mutation in the steroid binding region of the vitamin D receptor (VDR) gene in hereditary vitamin D resistant rickets (HVDRR). In: Vitamin D: Gene Regulation, Structure-Function Analysis, and Clinical Application. Eighth Workshop on Vitamin D. Norman AW, Bouillon R, Thomasset M, eds. New York: Walter de Gruyter, 1991; 94–95.
73. Yagi H, Ozono K, Miyake H, Nagashima K, Kuroume T, Pike JW. A new point mutation in the deoxyribonucleic acid-binding domain of the vitamin D receptor in a kindred with hereditary 1,25-dihydroxyvitamin D-resistant rickets. J Clin Endocrinol Metab 1993; 76:509–512.
74. Kristjansson K, Rut AR, Hewison M, O'Riordan JL, Hughes MR. Two mutations in the hormone binding domain of the vitamin D receptor cause tissue resistance to 1,25 dihydroxyvitamin D_3. J Clin Invest 1993; 92:12–16.
75. Wiese RJ, Goto H, Prahl JM, Marx SJ, Thomas M, al-Aqeel A, DeLuca HF. Vitamin D-dependency rickets type II: truncated vitamin D receptor in three kindreds. Mol Cell Endocrinol 1993; 90:197–201.
76. Rut AR, Hewison M, Kristjansson K, Luisi B, Hughes MR, O'Riordan JL. Two mutations causing vitamin D resistant rickets: modelling on the basis of steroid hormone receptor DNA-binding domain crystal structures. Clin Endocrinol 1994; 41:581–590.
77. Feldman D, Chen T, Hirst M, Colston K, Karasek M, Cone C. Demonstration of 1,25-dihydroxyvitamin D_3 receptors in human skin biopsies. J Clin Endocrinol Metab 1980; 51:1463–1465.
78. Pike JW, Donaldson CA, Marion SL, Haussler MR. Development of hybridomas secreting monoclonal antibodies to the chicken intestinal 1 alpha,25-dihydroxyvitamin D_3 receptor. Proc Natl Acad Sci USA 1982; 79:7719–7723.
79. Pike JW, Marion SL, Donaldson CA, Haussler MR. Serum and monoclonal antibodies against the chick intestinal receptor for 1,25-dihydroxyvitamin D_3. Generation by a preparation enriched in a 64,000-dalton protein. J Biol Chem 1983; 258:1289–1296.
80. Pike JW. Monoclonal antibodies to chick intestinal receptors for 1,25-dihydroxyvitamin D_3. Interaction and effects of binding on receptor function. J Biol Chem 1984; 259:1167–1173.
81. Dokoh S, Haussler MR, Pike JW. Development of a radioligand immunoassay for 1,25-dihydroxycholecalciferol receptors utilizing monoclonal antibody. Biochem J 1984; 221:129–136.
82. Pike JW, Dokoh S, Haussler MR, Liberman UA, Marx SJ, Eil C. Vitamin D_3-resistant fibroblasts have immunoassayable 1,25-dihydroxyvitamin D_3 receptors. Science 1984; 224:879–881.
83. McDonnell DP, Mangelsdorf DJ, Pike JW, Haussler MR, O'Malley BW. Molecular cloning of complementary DNA encoding the avian receptor for vitamin D. Science 1987; 235:1214–1217.
84. Saiki RK, Gelfand DH, Stoffel S, Scharf SJ, Higuchi R, Horn GT, Mullis KB, Erlich HA. Primer-directed enzymatic amplification of DNA with a thermostable DNA polymerase. Science 1988; 239:487–491.
85. Lin NU-T, Malloy PJ, Sakati N, Al-Ashwal A, Feldman D. A novel mutation in the deoxyribnucleic acid-binding domain of the vitamin D receptor gene causes hereditary 1,25-dihydroxyvitamin D resistant rickets. J Clin Endocrinol Metab 1996; 81:2564–2569.
86. Malloy PJ, Hughes MR, Pike JW, Feldman D. Vitamin D receptor mutations and hereditary 1,25-dihydroxyvitamin D resistant rickets. In: Vitamin D: Gene Regulation, Structure-Function Analy-

sis, and Clinical Application. Eighth Workshop on Vitamin D. Norman AW, Bouillon R, Thomasset M, eds. New York: Walter de Gruyter, 1991; 116–124.
87. Thompson E, Kristjansson K, Hughes M. Molecular scanning methods for mutation detection: application to the 1,25-dihydroxyvitamin D receptor. Abstracts of the Eighth Workshop on Vitamin D, Paris, France, 1991, p. 6.
88. Van Maldergem L, Bachy A, Feldman D, Bouillon R, Maassen J, Dreyer M, Rey R, Holm C, Gillerot Y. Syndrome of lipoatrophic diabetes, vitamin D resistant rickets, and persistent müllerian ducts in a Turkish boy born to consanguineous parents. Am J Med Genet 1996; 64:506–513.
89. Malloy PJ, Eccleshall TR, Gross C, Van Maldergem L, Bouillon R, Feldman D. Hereditary vitamin D resistant rickets caused by a novel mutation in the vitamin D receptor that results in decreased affinity for hormone and cellular hyporesponsiveness. J Clin Invest 1996.
90. Hewison M, Rut AR, Kristjansson K, Walker RE, Dillon MJ, Hughes MR, O'Riordan JL. Tissue resistance to 1,25-dihydroxyvitamin D without a mutation of the vitamin D receptor gene. Clin Endocrinol 1993; 39:663–670.
91. Marx SJ, Bliziotes MM, Nanes M. Analysis of the relation between alopecia and resistance to 1,25-dihydroxyvitamin D. Clin Endocrinol 1986; 25:373–381.
92. Balsan S, Garabedian M, Larchet M, Gorski AM, Cournot G, Tau C, Bourdeau A, Silve C, Ricour C. Long-term nocturnal calcium infusions can cure rickets and promote normal mineralization in hereditary resistance to 1,25-dihydroxyvitamin D. J Clin Invest 1986; 77:1661–1667.
93. Weisman Y, Bab I, Gazit D, Spirer Z, Jaffe M, Hochberg Z. Long-term intracaval calcium infusion therapy in end-organ resistance to 1,25-dihydroxyvitamin D. Am J Med 1987; 83:984–990.
94. Hochberg Z, Tiosano D, Even L. Calcium therapy for calcitriol-resistant rickets. J Pediatr 1992; 121:803–808.
95. Reichel H, Koeffler HP, Norman AW. The role of the vitamin D endocrine system in health and disease. N Engl J Med 1989; 320:980–991.
96. Walters MR. Newly identified actions of the vitamin D endocrine system. Endocr Rev 1992; 13:719–764.
97. Bikle DD. Clinical counterpoint: vitamin D: new actions, new analogs, new therapeutic potential. Endocr Rev 1992; 13:765–784.
98. Darwish H, DeLuca HF. Vitamin D-regulated gene expression. Crit Rev Eukaryot Gene Expr 1993; 3:89–116.
99. MacDonald PN, Dowd DR, Haussler MR. New insight into the structure and functions of the vitamin D receptor. Semin Nephrol 1994; 14:101–118.
100. Manolagas SC, Yu XP, Girasole G, Bellido T. Vitamin D and the hematolymphopoietic tissue: a 1994 update. Semin Nephrol 1994; 14:129–143.
101. Hochberg Z, Borochowitz Z, Benderli A, Vardi P, Oren S, Spirer Z, Heyman I, Weisman Y. Does 1,25-dihydroxyvitamin D participate in the regulation of hormone release from endocrine glands? J Clin Endocrinol Metab 1985; 60:57–61.
102. Even L, Weisman Y, Goldray D, Hochberg Z. Selective modulation by vitamin D of renal response to parathyroid hormone: a study in calcitriol-resistant rickets. J Clin Endocrinol Metab 1996; 81:2836–2840.
103. Etzioni A, Hochberg Z, Pollak S, Meshulam T, Zakut V, Tzehoval E, Keisari Y, Aviram I, Spirer Z, Benderly A, Weisman Y. Defective leukocyte fungicidal activity in end-organ resistance to 1,25-dihydroxyvitamin D. Pediatr Res 1989; 25:276–279.
104. Underwood JL, DeLuca HF. Vitamin D is not directly necessary for bone growth and mineralization. Am J Physiol 1984; 246:E493–498.
105. Stumpf WE, Sar M, Reid FA, Tanaka Y, DeLuca HF. Target cells for 1,25-dihydroxyvitamin D_3 in intestinal tract, stomach, kidney, skin, pituitary, and parathyroid. Science 1979; 206:1188–1190.
106. Colston K, Hirst M, Feldman D. Organ distribution of the cytoplasmic 1,25-dihydroxycholecalciferol receptor in various mouse tissues. Endocrinology 1980; 107:1916–1922.
107. Weisman Y, Jaccard N, Legum C, Spirer Z, Yedwab G, Even L, Edelstein S, Kaye AM, Hochberg Z. Prenatal diagnosis of vitamin D-dependent rickets, type II: response to 1,25-dihydroxyvitamin D in amniotic fluid cells and fetal tissues. J Clin Endocrinol Metab 1990; 71:937–943.
108. Weisman Y, Malloy PJ, Krishnan AV, Jaccard N, Feldman D, Hochberg Z. Prenatal diagnosis of calcitriol resistant rickets (CRR) by 1,25$(OH)_2D_3$ binding, 24-hydroxylase induction and RFLP analysis. Presented at the Ninth Workshop on Vitamin D, Orlando, FL, 1994.

20 Extrarenal Production of 1,25-Dihydroxyvitamin D and Clinical Implications

John S. Adams

1. INTRODUCTION

This chapter addresses the pathophysiology and cellular biochemistry of dysregulated vitamin D metabolism that occurs in some patients with granuloma-forming and malignant lymphoproliferative disorders. The principal focus of discussion is the human granuloma-forming disease sarcoidosis. Of all of the human conditions associated with the extrarenal overproduction of an active vitamin D metabolite with consequent endogenous vitamin D intoxication, sarcoidosis is the most commonly recognized and most carefully studied. Hence the first part of the chapter reviews what is known about the mechanics and regulation of the vitamin D-metabolizing enzymes present in human inflammatory cells. This is followed by a discussion of the diagnosis, treatment, and prevention of hypercalcemia and hypercalciuria in patients suffering from endogenous vitamin D intoxication. The concluding portion of the chapter addresses the issue of *why* active vitamin D metabolites are made in these diseases. There is now consensus agreement among investigators in the vitamin D and immunology fields that macrophage-derived vitamin D metabolites can play an important role in the modulation of the local human immune response in these diseases. The reader is referred to other chapters in the text that discuss in detail some of the noncalcemic (Chapter 13) and nongenomic (Chapter 12) actions of vitamin D metabolites and analogs (Chapter 25) that are particularly relevant to the issue of the immunomodulatory effects of vitamin Ds in human disease (i.e., cancer and psoriasis).

2. VITAMIN D AND GRANULOMA-FORMING DISEASE

2.1. Sarcoidosis: A Historical Overview

Sarcoidosis is the human disease most commonly complicated by endogenous vitamin D intoxication *(1)*. A pathophysiologic relationship between vitamin D and sarcoidosis was first recognized by Harrell and Fisher in 1939 *(2)*; they observed a steep rise in the serum calcium concentration in sarcoidosis patients following ingestion of vitamin D-enriched cod liver oil. The next major contribution to our knowledge on the subject came almost 20 years later. In 1956 Henneman et al. *(3)* demonstrated that the hypercalcemic

From: *Vitamin D: Physiology, Molecular Biology, and Clinical Applications*
Edited by: M. F. Holick © Humana Press Inc., Totowa, NJ

syndrome of sarcoidosis, characterized by increased intestinal calcium absorption and bone resorption, was remarkably similar to that of exogenous vitamin D intoxication. Taylor and coworkers *(4)* performed the first, large-scale seasonal evaluation of serum calcium levels in patients with sarcoidosis in 1963. They found that there was significant increase in the mean serum calcium concentration in 345 patients with sarcoidosis from winter to summer but no such change in over 12,000 control subjects. This was the first evidence of an association between enhanced cutaneous vitamin D synthesis during the summer months and the blood level of calcium in patients with sarcoidosis. This observation was prospectively confirmed by Dent et al. *(5)* and Hendrix *(6)*, who achieved, respectively, resolution of hypercalcemia and hypercalciuria in two patients with sarcoidosis by institution of vitamin D-deficient diets and environmental sunlight deprivation and an ability to increase the serum calcium concentration in patients with active sarcoidosis on exposure to whole-body ultraviolet (UV) radiation.

In 1964 Bell et al. *(7)* were the first to propose that development of a clinical abnormality in calcium balance in patients with active sarcoidosis resulted from an increase in target organ responsiveness to vitamin D, a view that persisted for nearly two decades. However, after the discovery of 1,25-dihydroxyvitamin D [$1,25(OH)_2D$] as the active vitamin D hormone *(8,9)* and the development of sensitive and specific assays for the hormone in blood *(10–14)*, investigators were quick to determine that the hypercalcemia of sarcoidosis was the result of an increase in the circulating concentrations of a $1,25(OH)_2D$-like metabolite that interacted with the vitamin D receptor (VDR) *(15–17)*. The fact that the vitamin D hormone was made outside the kidney in hypercalcemic patients with sarcoidosis was first discovered by Barbour and colleagues in 1981 *(18)*, who described high concentrations of a vitamin D metabolite, detected as $1,25(OH)_2D$ in a VDR binding assay, in the circulation of a hypercalcemic, anephric patient with active sarcoidosis. Two years later, Adams et al. *(19)* determined the macrophage to be the extrarenal source of this active vitamin D metabolite, with unequivocal structural characterization of the metabolite as $1,25(OH)_2D$ being obtained by these same investigators in 1985 *(20)*.

2.2. Sarcoidosis: Pathophysiology of Disordered Calcium Balance

In a worldwide review of serum calcium concentrations in 3676 patients with sarcoidosis, James et al. *(21)* reported an 11% incidence of hypercalcemia (serum calcium ≥10.5 mg/dL). Studdy et al. *(22)* studied 547 patients with biopsy-proven sarcoidosis in Great Britain and found hypercalcemia to be 38% more frequent in men than women and more common among white than among individuals of West Indian descent. The frequency of hypercalcemia among patients with sarcoidosis tends to be consistently higher in North America than in Northern Europe *(1)*; this is perhaps due to the lower latitude and more direct sunlight exposure in the United States.

Although the fractional urinary calcium excretion may be decreased in patients with renal insufficiency *(23)*, the principal source of calcium that accumulates in the blood of patients with sarcoidosis is the skeleton. This fact is perhaps most strongly supported by the work of Rizatto et al. *(24)*. Compared with age- and sex-matched control subjects, these workers documented a significant decrease over time in bone mineral density in a group of patients with chronic active sarcoidosis in whom antiinflammatory agents were not used in management. These findings also support the long-standing observation that hypercalcemia persists in patients with active sarcoidosis in the absence of ingested calcium *(25)*. The proximal cause of bone loss is increased osteoclast-mediated bone resorption *(26)*. This process does not require the presence of extensive granulomata in the bone *(27)*, suggesting that a circulating stimulator of bone resorption exists in this disease. One such stimulator of bone resorption is 1,25(OH)$_2$D.

The synthesis of 1,25(OH)$_2$D by the renal vitamin D-1-hydroxylase (1-hydroxylase) has been extensively reviewed in other chapters. The enzyme is normally strictly regulated with levels of 1,25(OH)$_2$D being some 1000-fold less plentiful in the circulation than that of the principal substrate for the enzyme, 25-hydroxyvitamin D [25(OH)D]. Hormone synthesis in the kidney is stimulated by an increase in the serum parathyroid hormone (PTH) concentration, a decrease in the serum phosphate concentration, and a decrease in the activity of the competing vitamin D 24-hydroxylase. Conversely, 1,25(OH)$_2$D synthesis by the renal enzyme is inhibited by a decrease in circulating PTH, an increase in the serum phosphate, and an increase in the activity of the vitamin D 24-hydroxylase. Numerous pieces of clinical evidence indicate that endogenous 1,25(OH)$_2$D production in hypercalcemic/hypercalciuric patients with sarcoidosis is dysregulated [i.e., not bound by the same set of endocrine factors known to regulate 1,25(OH)$_2$D synthesis in the kidney]. That evidence is summarized below.

Hypercalcemic patients with sarcoidosis possess a frankly high or inappropriately elevated serum 1,25(OH)$_2$D concentration, although their serum PTH level is suppressed and their serum calcium and phosphate concentration is relatively elevated *(28,29)*. If 1,25(OH)$_2$D synthesis were under the regulation of PTH, phosphate, and 1,25(OH)$_2$D itself, then 1,25(OH)$_2$D concentrations in such patients should be low, not elevated. Unlike the situation in normal individuals, the serum 1,25(OH)$_2$D concentration in patients with active sarcoidosis is highly sensitive to even a small increase in available substrate *(30)*. Clinically this is manifest by the appearance of hypercalciuria and/or hypercalcemia in sarcoidosis patients in the summer months *(31)* or following holidays to geographic locations at lower latitudes than those at which the patient normally resides *(32)*. This link between an increase in cutaneous vitamin D synthesis and the development of calcium imbalance can be (1) replicated by the oral administration of vitamin D$_3$ *(17)* to such patients, and (2) substantiated on a biochemical basis by demonstration of an abnormal, positive correlation between the serum 25(OH)D and 1,25(OH)$_2$D concentrations in patients with active sarcoidosis *(29)*. In addition, the serum calcium and 1,25(OH)$_2$D concentrations are positively correlated to indices of disease activity in patients with sarcoidosis *(23,33,34)*. Finally, the rate of endogenous 1,25(OH)$_2$D production, which is significantly increased in patients with sarcoidosis *(35)*, is sensitive to inhibition by drugs that do not influence the renal vitamin D-1-hydroxylase at the same doses. The best example of this are the antiinflammatory glucocorticoids. These drugs have been used clinically for a long time as effective combatants of sarcoidosis-associated hypercalcemia *(25,36)* and more recently have been shown to lower elevated 1,25(OH)$_2$D levels dramatically *(29)*. Chloroquine and its hydroxylated analog hydroxychloroquine, are other examples of pharmaceutical agents that appear to act preferentially on the extrarenal vitamin D-1-hydroxylation reaction and not the renal 1-hydroxylase *(37–39)*.

2.3. Sarcoidosis: Correlates In Vitro of Clinical Features of Dysregulated Vitamin D Homeostasis

Investigators have now generated a substantial body of experimental data from inflammatory cells harvested directly from patients with sarcoidosis to indicate that the dysregulated vitamin D hormone synthesis in sarcoidosis is not due (1) to expression of

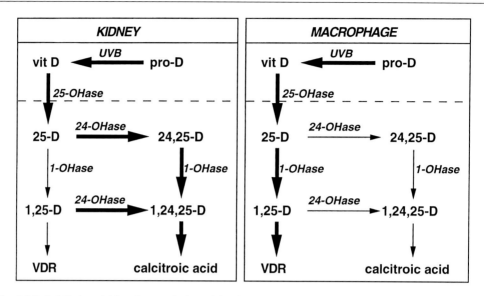

Fig. 1. Model distinguishing the regulation of the vitamin D-1-hydroxylase (1-OHase) in the proximal renal tubular epithelial cell of the kidney (**left**) and in the granuloma-forming disease-activated macrophage (**right**). Before gaining exposure to the 1-OHase (above the dotted lines) (1) 7-dehydroxycholesterol (proD) is converted to vitamin D in the skin under the influence of sunlight (UVB) and (2) available substrate vitamin D is converted to 25-hydroxyvitamin D (25-D) by the high-capacity hepatic 25-hydroxylase (25-OHase) system. In the kidney the enzymatic conversion of substrate 25-D to the product 1,25-dihydroxyvitamin D (1,25-D) via the 1-OHase is subject to negative feedback control with downregulation of enzyme activity under the influence of (1) a calcium-mediated decrease in circulating parathyroid hormone (PTH), (2) a 1,25-D-mediated increase in the serum phosphate (P) level; and (3) a 1,25-D-mediated increase in vitamin D 24-hyroxylase (24-OHase) activity. Conversely, there is little or no traffic through the macrophage 24-OHase pathway with diversion of most 25-D down the 1-OHase pathway.

an enzyme different from the renal 1-hydroxylase but (2) to expression of the authentic 1-hydroxylase in a macrophage, not a kidney cell *(40)*.

2.3.1. SIMILARITIES OF THE MACROPHAGE AND RENAL VITAMIN D 1-HYDROXYLASE

Similar to the bona fide 1-hydroxylase of renal origin, the macrophage enzyme is a cytochrome P-450-associated mixed function oxidase confined to mitochondria *(41)* (Fig. 1). Like the renal 1-hydroxylase reconstituted from mitochondrial extracts, a flavoprotein, ferredoxin reductase, an electron source, and molecular oxygen (O_2) are all required for electron transfer to the cytochrome P-450 and for the insertion of an oxygen atom in the substrate *(41)*. We also now know that the macrophage 1-hydroxylase is inhibited by the napthoquinones, molecules that compete with reductase for donated electrons, and by the imidazoles such as ketoconazole, which compete with the enzyme for receipt of O_2 *(42)*. Similar to the renal 1-hydroxylase, the macrophage 1-hydroxylase (1) requires a secosterol (vitamin D sterol molecule with an open B-ring) as substrate *(40)*, and (2) has a particular affinity for secosterols bearing a carbon-25 hydroxy group; the calculated K_m (affinity) of the 1-hydroxylase in pulmonary alveolar macrophages derived directly from patients with active sarcoidosis is in the range of 50–100 nM for the two preferred substrates for this enzyme, 25(OH)D and 24,25-dihydroxyvitamin D [24,25(OH)$_2$D] *(40–43)*. Absolute confirmation that the renal and macrophage enzymes are identical still awaits successful cloning and expression of the macrophage cDNA of the 1-hydroxylase. If they are identical, then the dramatic differences in expression of the renal and macrophage 1-hydroxylase in vivo must result from the cell in which the enzyme is expressed and the regulatory pathways that impinge on the mitochondrial 1-hydroxylase in these two cell types.

2.3.2. DIFFERENCES IN REGULATION OF THE MACROPHAGE AND THE RENAL 1-HYDROXYLASE

In contrast to the renal enzyme, the macrophage 1-hyroxylase is not affected by PTH or phosphate *(42,44)* (Fig. 1, right). There is no evidence that PTH receptors, even if present in the macrophage membrane, are responsive to PTH in terms of stimulating the protein kinase signaling pathways that are associated with stimulation of the renal 1-hydroxylase *(42)*. Similarly, the macrophage enzyme appears to be uninfluenced by changes in the extracellular phosphate concentration *(42)*. Conversely, exposure of activated macrophages expressing the 1-hydroxylase to calcium or calcium ionophore stimulates the synthesis of 1,25(OH)$_2$D *(40)*, whereas increasing the extracellular calcium concentration will have an inhibitory effect on the renal 1-hydroxylase *(45)*. These observations appear to confirm the fact that the three most important extracellular signaling systems for the renal 1-hydroxylase, calcium, phosphate, and PTH, are not heeded by the macrophage enzyme. This also provides an explanation for why 1,25(OH)$_2$D production by the macrophage in diseases like sarcoidosis is not subject to negative feedback control by a drop in the serum PTH concentration and an increase in the circulating calcium and phosphate level. Furthermore, with the possible exception of insulin-like growth factor-1 *(46)*, there is no evidence that the macrophage 1-hydroxylation reaction is influenced by any of the other endocrine factors, including estrogen, prolactin, and growth hormone, purported to increase production of 1,25(OH)$_2$D in the kidney *(47–50)*.

The other major contributor to the circulating 1,25(OH)$_2$D level is the activity of the vitamin D-24-hydroxylase (Fig. 1, left). Like the 1-hydroxylase, 24-hydroxylase is a heme-binding mitochondrial enzyme requiring reduced nicotinamide adenine dinucleotide phosphate (NADPH), molecular oxygen, and magnesium ions *(51)*. The cDNA for the enzyme, now referred to as P450cc24, has been recently cloned *(52–55)*. Expression of P450cc24 is stimulated in kidney cells by 1,25(OH)$_2$D *(56,57)*. PTH appears to exert an inhibitory effect on P450cc24 gene transcription *(55)* and 24,25(OH)$_2$D synthesis *(58,59)*. There is a dual impact of the 24-hydroxylase on vitamin D and calcium balance. The first point of impact is on regulation of substrate 25(OH)D available to the 1-hydroxylase. Like the 1-hydroxylase, the 24-hydroxylase exhibits a preference for 25-hydroxylated secosterol substrates *(60)*; although its affinity for 25(OH)D is reported to be somewhat less than that of renal 1-hydroxylase, its capacity for substrate is substantially greater *(51)*. Hence, when upregulated under the influence of circulating or locally produced 1,25(OH)$_2$D or diminished serum PTH levels, the 24-hydroxylase has the capacity to compete with the 1-hydroxylase for substrate 25(OH)D. The second point of impact of the 24-hydroxylase on the circulating 1,25(OH)$_2$D concentration is at the level of catabolism of the 1,25(OH)$_2$D hormone. The affinity of the 24-hydroxylase for 1,25(OH)$_2$D is as great as it is for 25(OH)D. Considering the fact that the 24-hydroxylase is the initial step in the conversion of 1,25(OH)$_2$D to nonbiologically active, excretable metabolites, upregulation of this enzyme will contribute to a lowering of 1,25(OH)$_2$D hormone levels. Because the macrophage lacks detectable 24-hydroxylase activity (Fig. 1, right) *(40)*, the production as the amount of NO generated inside the macrophage continues to increase *(80)*. This suggests that there is some kind of a built-in limit (i.e., negative feedback) on NO-stimulated macrophage 1,25(OH)$_2$D production. This inhibitory effect of NO on the macrophage 1-hydroxylase is almost certainly due to competition of NO with O$_2$ for binding to the heme center of the enzyme; a similar effect has been very recently demonstrated for a number of heme-containing enzymes *(81–85)*, including those involved in steroid hormone metabolism *(86)*. Thus, in summary, it appears that NO exerts a biphasic effect on the macrophage 1-hydroxylase. The stimulatory effect on the 1-hydroxylase is mediated by relatively low intracellular NO levels and is initiated by the donation of an electron (*e*-) from NO to oxidized NADP. This results in the formation of NADPH, which in turn supplies the electron transport chain linked to the 1-hydroxylase. The inhibitory effect of relatively high NO levels in the cell, on the other hand, results from competition of NO with O$_2$ for binding to the heme center of the cytochrome P-450-linked hydroxylase in a fashion analogous to the inhibition by carbon monoxide of P-450-linked oxidative enzymes.

2.3.1.3. Other Factors. There is now evidence *(72)* that macrophages may be producing an autocrine factor that sustains expression of the 1-hydroxylase, at least in vitro, and that its ability to promote 1,25(OH)$_2$D production may be mediated through the generation of NO. The increase in NO production under the influence of this factor is mediated by an increase in transcription of the iNOS gene. Although not yet characterized, this autoregulator appears to be a non-endotoxin-related, small molecular weight protein (3000–10,000 kDa) that is exported from the macrophage before interacting with the macrophage cell membrane and stimulating an increase in intracellular NO synthesis.

Another potential, but not yet proved, autoregulator of macrophage 1,25(OH)$_2$D synthesis is the stress-induced hsp70 family of proteins *(87)*. In macrophage-like cells hsp70 expression is known to be induced by both physical and chemical (i.e., IFN) stimuli, and its expression is dramatically enhanced by the vitamin D hormone 1,25(OH)$_2$D *(88)*. Proteins in the hsp70 family have recently been shown to be high-capacity intracellular binding proteins for 25-hydroxylated vitamin D metabolites *(89)*. It is possible that up-regulated hsp70 expression in disease-activated macrophages is another way to increase 1,25(OH)$_2$D production by these cells. Because of its capacity to bind 25(OH)D, hsp70 may concentrate (on a relatively low-affinity, high-capacity binding protein) substrate for the 1-hydroxylase. In fact, by virtue of their organelle-targeting sequences *(90)*, hsp70s may be critical in the directed translocation of 25(OH)D to the inner mitochondrial membrane where the 1-hydroxylase actually resides.

2.4. Other Human Disorders Associated with the Extrarenal Overproduction of Vitamin D

2.4.1. Granuloma-Forming Diseases

Aside from sarcoidosis, tuberculosis is the most common human granuloma-forming disease reported to be complicated by an alteration in calcium balance. Hypercalcemia has been recognized as a complication of infection with *Mycobacterium tuberculosis* for over 80 years now *(91)*. The disturbance in calcium balance in tuberculosis, like sarcoidosis, is caused by the extrarenal over-production of an active vitamin D metabolite *(92,93)*. As is the case with sarcoidosis, the circulating vitamin D metabolite causing hypercalcemia appears to be 1,25(OH)$_2$D synthesized by disease-activated macrophages *(94–97)*. Significantly increased production of the metabolite can occur with only small

Table 1
Diseases Associated with Hypercalcemia
and the Extrarenal Overproduction
of an Active Vitamin D Metabolite

Noninfectious granuloma-forming diseases
    Sarcoidosis
    Berylliosis
    Silicone granulomatosis
    Infantile fat necrosis
Infectious granuloma-forming diseases
    Tuberculosis
    Leprosy
    Candidiasis
    Coccidioidomycosis
    Crytococcosis
Lymphoproliferative diseases
    Lymphoma
    Lymphomatoid granulomatosis
    Eosinophilic granuloma
    Wegener's granulomatosis

changes in the serum concentration of substrate 25-hydroxyvitamin D [25(OH)D] *(98)*. The prevalence of hypercalcemia in tuberculosis patients has been reported to be as high as 26% *(91)* and may be even higher in the era of the acquired immunodeficiency syndrome (AIDS) owing to the more frequent occurrence of disseminated disease in immunocompromised patients. As in sarcoidosis, dysregulated vitamin D and calcium balance can be corrected by the administration of glucocorticoid *(99,100)*.

As depicted in Table 1, hypercalciuria and/or overt hypercalcemia can complicate a number of other infectious granuloma-forming diseases, mostly characterized by widespread granuloma formation and macrophage proliferation in infected tissue. Included among these are leprosy *(101,102)*, disseminated candidiasis *(103)*, histoplasmosis *(104)*, coccidioidomycosis *(105)*, and crytococcosis *(79)*. Hypercalcemia in most of these conditions has been documented to be associated with inappropriately elevated serum concentrations of $1,25(OH)_2D$. Similar to the expectation with tuberculosis, it is likely that these abnormalities in vitamin D metabolism and action will increase in frequency as the number of immunocompromised patients (i.e. those with AIDS) increases worldwide *(106)*. The syndrome of extrarenal overproduction of $1,25(OH)_2D$ has also been documented in adult patients with widespread granuloma-forming disease of noninfectious origin other than sarcoidosis (Table 1). This includes silicone-induced granulomata *(107)*, berylliosis *(108)*, and massive subcutaneous fat necrosis in newborns *(109)*.

2.4.2. OTHER DISORDERS

A vitamin D-mediated disturbance in calcium metabolism is not confined to patients with granuloma-forming diseases. It can also be observed in patients with lymphoproliferative disorders *(108,110–116)* (Table 1). Most recent reports *(117,118)* indicate that the extrarenal overproduction of $1,25(OH)_2D$ is the most common cause of hypercalciuria and hypercalcemia in patients with non-Hodgkin's and Hodgkin's lymphoma. This is especially true of patients with B-cell neoplasms, whether or not the tumor is associated with AIDS *(113)*. In fact, in the study of Seymour et al. *(118)*, 71% of normocalcemic patients with non-Hodgkin's lymphoma had hypercalciuria, and most of these patients had circulating 1,25(OH)$_2$D levels in the normal range or frankly elevated. As is the situation with patients with granuloma-forming disease and elevated circulating 1,25(OH)$_2$D levels, the serum concentrations of PTH are suppressed and PTH-related peptide is normal (i.e., not elevated) in lymphoma patients; this is, of course, indicative of the state of dysregulated overproduction of the active vitamin D hormone. Results of clinical studies of hypercalcemic patients with lymphoma before and after curative antitumor therapy (either chemotherapeutic or surgical) *(113,117–119)* are most compatible with the tumor being either (1) an immediate source of an active vitamin D metabolite *(120)* or (2) the source of a soluble factor that stimulates the production of 1,25(OH)$_2$D in neighboring inflammatory cells. Although less well documented, there is also evidence that extrarenal 1,25(OH)$_2$D production can complicate the course of solid neoplasms *(121,122)*.

3. MANAGEMENT OF ENDOGENOUS VITAMIN D INTOXICATION

3.1. Making the Diagnosis

The state of "endogenous" vitamin D intoxication is confirmed when (1) hypercalciuria and/or hypercalcemia occurs in a patient with an inappropriately elevated serum 1,25(OH)$_2$D; (2) exogenous intoxication with an active vitamin D metabolite or vitamin D prohormone [i.e., vitamin D, 25(OH)D, 1α-hydroxyvitamin D, or dihydrotachysterol] is ruled out; (3) the serum PTH level is appropriately suppressed; and (4) the calcium-sensing receptor in the plasma membrane of the host's parathyroid cell is normally operative *(79)*. These criteria clearly distinguish the patient with endogenous vitamin D intoxication from the individual with primary hyperparathyroidism and elevated 1,25(OH)$_2$D levels; the other major exception here is the patient with absorptive hypercalciuria who possesses, as a primary or secondary abnormality, an inappropriately elevated circulating 1,25(OH)$_2$D concentration *(123)*.

3.2. Screening and Prevention Strategies

The best way to treat vitamin D-mediated abnormalities in calcium balance is to prevent their occurrence. The first step is to identify patients at risk, mainly those with granuloma-forming disease and with malignant lymphoproliferative disorders, especially B-cell and Hodgkin's lymphoma. In these groups of patients production of the offending vitamin D metabolite is directly related to two distinct factors, the amount of substrate 25(OH)D available to the macrophage 1-hydroxylase (Fig. 1) and the severity and activity of the underlying disease. For example, in the case of sarcoidosis those patients at risk would have: (1) widespread, active disease; (2) a previous history of hypercalciuria or hypercalcemia; (3) a diet or treatment regimen enriched in vitamin D and calcium (i.e., for osteoporosis) or a recent history of sunlight exposure; and (4) an intercurrent condition, or medicinal treatment of an intercurrent condition that increases bone resorption or decreases the glomerular filtration rate (i.e., diuretic therapy for congestive heart failure).

Hypercalciuria almost always precedes the development of overt hypercalcemia in this set of disorders. Patients at risk should be checked for the presence of occult hypercalciuria; since hypercalciuria is presumably caused by increased bone resorption, this is best accomplished by a fasting 2-h urine collection for calcium and creatinine. If the calcium/creatinine ratio (g/g) is not abnormally high (<0.16), then a 24-h urine collection for the fractional calcium excretion rate is necessary to establish hypercalciuria. If screening is to be done only on an annual basis, then the late summer or early autumn, when 25(OH)D levels are usually at their peak, is the best time *(62)*. For patients determined from the appropriate monitoring analyses to be at risk, measures to prevent worsening hypercalciuria and frank hypercalcemia should be instituted. These measures should include (1) the use of UVB-absorbing sunscreens on exposed body parts when the patient anticipates being out of doors for >20–30 min; (2) caution against ingestion of vitamin and food supplements containing ≥400 IU vitamin D; (3) education on the vitamin D and calcium content of foods, vitamin supplements, and medicinal agents like antacids; (4) caution against the regular ingestion of elemental calcium >1000 mg daily; and (5) education regarding the earliest signs of hypercalciuria (i.e., nocturia).

3.3. Treatment of Hypercalciuria and Hypercalcemia

Whether the patient is hypercalciuric or frankly hypercalcemic, there are three general therapeutic goals. First is reduction in the serum concentration of the offending vitamin D metabolite or derivative. In patients suffering from endogenous intoxication with $1,25(OH)_2D$ made by inflammatory cells, reduction in the serum $1,25(OH)_2D$ level can be most reliably achieved by treatment with antiinflammatory doses of glucocorticoid (adult dose of 40 mg prednisone or equivalent per day). Glucocorticoids inhibit two key pathways in macrophage-like cells, phospholipase C *(64)* and the iNOS *(71)*, whose distal products stimulate the 1-hydroxylase. At these doses, glucocorticoids have little inhibitory effect on the renal 1-hydroxylase, so there is no real concern for inducing hypocalcemia with these drugs. Steroid therapy should lower the serum $1,25(OH)_2D$ concentration within a matter of 3–4 d, with a resultant decrease in the filtered load of calcium and urinary calcium excretion rate (provided that the patient's glomerular filtration rate is maintained). In patients who fail glucocorticoid therapy or in whom glucocorticoids are contraindicated, treatment with chloroquine (250 mg twice daily) or hydroxy-chloroquine (up to 400 mg daily) may be effective *(37–39)*. A less desirable therapeutic alternative is the cytochrome P-450 inhibitor ketoconazole *(124)*. It will reduce the serum $1,25(OH)_2D$ concentration *(125–127)*, but the therapeutic margin of safety is narrow; doses of the drug that inhibit the macrophage P-450 system will also inhibit endogenous glucocorticoid and sex steroid production *(128)*.

The second goal of therapy is to limit the actions of the vitamin D derivative at its target tissues, the gut and bone. A reduction in intestinal calcium absorption is best accomplished by elimination of as much calcium as possible from the diet. However, it should be noted that such measures are rarely effective in patients with active, widespread disease. If this is the case, then glucocorticoid administration is also usually required to block vitamin D-mediated calcium absorption and bone resorption. Because of the relative effectiveness of glucocorticoid management of this problem, the use of antiresorptive agents, like calcitonin and the bisphosphonates, is not recommended.

Increasing the urinary calcium excretion to reduce the filtered load of calcium is the third goal of therapy. This can be achieved by maintenance of the glomerular filtration rate and urinary flow rate and by the use of a loop diuretic like furosemide to inhibit calcium reabsorption from the urine. The long-term effects on the patient's skeleton of successfully reducing the serum $1,25(OH)_2D$ concentration and managing hypercalcemia/hypercalciuria are not known; there is preliminary evidence that successful treatment of exogenous vitamin D intoxication may result in a transient increase in bone mineral density *(129)*.

4. LOCAL IMMUNOREGULATORY EFFECTS OF ACTIVE VITAMIN D METABOLITES IN HUMAN DISEASE

4.1. Intracrine Effects on the Macrophage

The possibility that $1,25(OH)_2D_3$ may modulate the immune response was raised by three somewhat divergent lines of evidence. First was the observation that a number of mouse and human cultured cell lines of immune cell origin as well as activated human lymphocytes possess the high-affinity receptor for $1,25(OH)_2D_3$ *(130)*. Second was the observation that mouse and human receptor-containing myeloid leukemia cells could be induced to differentiate when exposed to $1,25(OH)_2D_3$ *(131)*. Third was the finding that pulmonary alveolar macrophages isolated from hypercalcemic patients with sarcoidosis were capable of synthesizing $1,25(OH)_2D_3$ *(18,19)*. Collectively, these data suggested that immunologically active cells were capable of synthesizing the active metabolite of vitamin D_3, whereas other cells of immune origin were responsive to the hormone. Control of the immunologic actions of $1,25(OH)_2D_3$ and control of the calcium-phosphorus regulatory actions of the sterol are dissimilar. In the latter case, regulation is apparently achieved by altering circulating concentrations of $1,25(OH)_2D_3$ through stimulation or inhibition of the renal $25(OH)D_3$-1α-hydroxylase. The immunologic actions of the hormone, on the other hand, are likely to be regulated by changes in the local production of $1,25(OH)_2D_3$, as well as by expression of receptor in "activated," neighboring target cells.

Because the activated tissue macrophage also expresses the VDR *(132,133)*, the same cell that makes the active vitamin D hormone can also serve as a target for the hormone. Indeed, investigators have suggested that $1,25(OH)_2D$ has the potential to interact with the monocyte/macrophage in either an intracrine or autocrine mode *(134–136)* (Fig. 2). For example, incubation with a VDR-saturating concentration of $1,25(OH)_2D$ increases interleukin-1β (IL-1β) expression by one order of magnitude and decreases by three orders of magnitude the concentration of stimulator LPS required to achieve maximal IL-1β gene expression *(136)*. This extraordinary priming effect of the sterol for LPS stimulation on expression of the IL-1 gene can also be observed for another monokine gene product, tumor necrosis factor-α (TNF-α) *(137)*; in the case of TNF, increased responsiveness to LPS is due to induction of the CD14 gene, the high-affinity receptor for LPS, by $1,25(OH)_2D$.

It is now widely held that the actions of active vitamin D metabolites on the macrophage are directed toward stimulation of cellular immune function, including the enhancement of (1) giant cell formation *(138)*, (2) monokine production *(139–141)*, (3) antigen processing *(142)*, and (4) cytotoxic function *(143,145)*. As depicted in Fig. 2, these functional consequences of hormone action indicate that elaboration of $1,25(OH)_2D$ by activated macrophages in human diseases like sarcoidosis is important in modulation of the local cellular immune response to the granuloma-causing antigen, promoting antigen processing, containment, and destruction *(145)*. The system is designed for maximal efficiency in that the hormone interacts with the VDR in the same cell in which the hormone is synthesized, permitting relatively high intracellular and local concentrations of hormone to be achieved without having a generalized, endocrine effect on the host.

4.2. Paracrine Effects on the Lymphocyte

If lymphokine stimulation of macrophage $1,25(OH)_2D$ synthesis were persistent, because of difficulty in macrophage-mediated elimination of the offending antigen, for instance, then one might conceive of a situation in which the lipid-soluble vitamin D hormone escapes the confines of the macrophage (Fig. 2). Once outside the macrophage, 1,25(OH)$_2$D would be free to interact in a paracrine fashion with antigen-activated, VDR-expressing lymphocytes in the local inflammatory microenvironment *(146)*. In contrast to stimulatory effects of the hormone on macrophages, 1,25(OH)$_2$D and most of the nonhypercalcemic analogs of vitamin D will inhibit lymphocyte responsiveness to mitogen or antigen challenge. Interaction of the VDR with its cognate ligand in activated lymphocytes and natural killer cells (1) inhibits cellular proliferation *(147–149)*, (2) generally decreases lymphokine production *(134,150)*, (3) inhibits T-lymphocyte-directed B-cell immunoglobulin synthesis *(151,152)*, and (4) inhibits delayed-type hypersensitivity reactions *(153,154)*. It is postulated *(33,143,155,156)* that this apparent paradox, dampening of lymphocyte activity while stimulating monocyte/macrophage function, is designed to maximize the ability of the host to combat and contain the granuloma-causing antigen while controlling the potentially self-destructive lymphocytic response to that offending antigen. Only at times of heightened immunoreactivity (i.e., extraordinary disease activity) does monokine production escape the confines of the site of inflammation and spill over into the general circulation, causing elevated 1,25(OH)$_2$D concentrations (Fig. 2). If this is truly the case, then the endocrine action of a locally produced active vitamin D metabolite to increase bone resorption and the intestinal calcium absorption are the only signs of the disorder manifest clinically and are probably the exception rather than the rule.

4.3. Accumulation of 1,25(OH)$_2$D at Sites of Inflammation

If 1,25(OH)$_2$D is truly a naturally occurring "cytokine," then one should be able to document accumulation of the metabolite at sites of inflammation and show that the inflammatory cells at this site are under the influence of the locally produced vitamin D metabolite. This was first accomplished by Barnes et al. *(156)*, who determined that the pleural space in nonhypercalcemic/calciuric patients infected with *M. tuberculosis* was one such site of 1,25(OH)$_2$D accumulation. These investigators detected a steep gradient for free 1,25(OH)$_2$D across the visceral pleura in patients with tuberculous effusions but not in patients with nontuberculous effusions. They showed that purified protein derivative-reacting T-lymphocyte clones from these patient expressed the VDR and determined that the stimulated proliferation of these T-cell clones was susceptible to 1,25(OH)$_2$D-mediated inhibition. They also showed that the pleural fluid of these patients contained an IFN-like peptide that stimulated the synthesis of 1,25(OH)$_2$D by macrophages *(157)*. Collectively, these data support the idea put forward by Rook and colleagues *(158)* and recapitulate the model set forth in Fig. 2, that there exists in the pleural microenvironment of patients with active pulmonary tuberculosis a system whereby (1) mycobacterium-activated macrophages are stimulated to make 1,25(OH)$_2$D; (2) this synthetic reaction is supported by proliferating and lymphokine (i.e., IFN)-producing lymphocytes at the local site of inflammation; and (3) the local accumulation of lymphokines, in turn, acts to augment further the local production of 1,25(OH)$_2$D by the macrophage. Investigators *(159)* have viewed this sort of positive feedback effect of IFN on macrophage 1,25(OH)$_2$D production in vivo as an efficient mechanism for dealing with antigens (like mycobacteria, the "sarcoid antigen," or certain viruses) that are obviously difficult for the host to irradicate. Although it less well characterized than the situation in the pleural space of patients with tuberculosis, it should also be noted that the joint space of patients with inflammatory arthritis *(160,161)* and the peritoneal space of patients with peritonitis *(162,163)* are two additional sites of documented vitamin D hormone accumulation.

REFERENCES

1. Sharma OP. Vitamin D, calcium, and sarcoidosis. Chest 1996; 109:535–539.
2. Harrell GT, Fisher S. Blood chemical changes in Boeck's sarcoid. J Clin Invest 1939; 18:687–693.
3. Henneman PH, Dempsey EF, Carrol EJ, Albright F. The causes of hypercalcemia in sarcoid and its treatment with cortisone. J Clin Invest 1956; 35:1229–1242.
4. Taylor RL, Lynch HJ Jr., Wysor WG. Seasonal influence of sunlight on the hypercalcemia of sarcoidosis. Am J Med 1963; 35:67–89.
5. Dent CE, Flynn FV, Nabarro JDN. Hypercalcemia and impairment of renal function in generalized sarcoidosis. Br Med J 1953; 2:808–810.
6. Hendrix JZ. The remission of hypercalcemia and hypercalciuria in systemic sarcoidosis by vitamin D depletion. Clin Res 1963; 11:220–225.
7. Bell NH, Gill JR Jr, Barter FC. Abnormal calcium absorption in sarcoidosis: evidence for increased sensitivity to vitamin D. Am J Med 1964; 36:500–513.
8. Holick MF, Schnoes THK, DeLuca HF, Suda T, Cousins RJ. Isolation and identification of 1,25-dihydroxycholecalciferol. A metabolite of vitamin D active in intestine. J NIH Res 1992; 4:88–96.
9. Bell NH. Vitamin D-endocrine system. J Clin Invest 1985; 76:1–6.
10. Hughes MF, Baylink DJ, Jones PG, Haussler MR. Radioligand receptor assay for 25-hydroxyvitamin D_2/D_3 and $1\alpha,25$-dihydroxyvitamin D_2/D_3. J Clin Invest 1976; 58:61–70.
11. Clemens TL, Hendy GN, Graham RF. A radioimmunoassay for 1,25-dihydroxycholecalciferol. Clin Sci Mol Med 1978; 54:329-332.
12. Gray TK, McAdoo T. Radioimmunoassay for 1,25-dihydroxyvitamin D_3. In: Vitamin D: Basic Research and Its Clinical Application. Norman AW, Schaefer K, von-Herrath D, eds. Berlin: de Gruyter, 1979; 763–767.
13. Bouillon R, DeMoor P, Baggiolini EG, Uskokovic MR. A radioimmunoassay for 1,25-dihydroxycholecalciferol. Clin Chem 1980; 26:562–567.
14. Holick MF. The use and interpretation of assays for vitamin D and its metabolites. J Nutr 1990; 120: 1464–1469.
15. Bell NH, Stern PH, Pantzer E, Sinha TK, DeLuca. Evidence that increased circulating 1,25-dihydroxyvitamin D is the probable cause for abnormal calcium metabolism in sarcoidosis. J Clin Invest 1979; 64:218–225.
16. Papapoulos SE, Clemens TL, Fraher LJ, Lewin IG, Sandler LM, O'Riordan JL. 1,25-Dihydroxycholecalciferol in the pathogenesis of the hypercalcemia of sarcoid. Lancet 1979; 1:627–630.
17. Stern PH, Olazabal J, Bell NH. Evidence for abnormal regulation of circulating 1,25-dihydroxyvitamin D in patients with sarcoidosis. J Clin Invest 1980; 66:852–855.
18. Barbour GL, Coburn JW, Slatopolsky E, et al. Hypercalcemia in an anephric patient with sarcoidosis. N Engl J Med 1981; 305:440–443.
19. Adams JS, Sharma OP, Gacad MA, Singer FR. Metabolism of 25-hydroxyvitamin D_3 by cultured alveolar macrophages in sarcoidosis. J Clin Invest 1983; 72:1856–1860.
20. Adams JS, Singer FR, Gacad MA, et al. Isolation and structural identification of 1,25-dihydroxyvitamin D_3 produced by cultured alveolar macrophages in sarcoidosis. J Clin Endocrinol Metab 1985; 60:960–966.
21. James DG, Neville E, Siltzbach LE, et al. A worldwide review of sarcoidosis. Ann NY Acad Sci 1976; 278:321–334.
22. Studdy PR, Bird R, Neville E, James DG. Biochemical findings in sarcoidosis. J Clin Pathol 1980; 33:528–533.
23. Meyrier A, Valeyre D, Bouillon R, Paillard F, Battesti JP, Georges R. Resorptive versus absorptive hypercalciuria in sarcoidosis. Q J Med 1985; 54:269–281.
24. Rizzato G, Montemurro L, Fraioli P. Bone mineral content in sarcoidosis. Semin Respir Med 1992; 13:411–423.
25. Anderson J, Dent CE, Harper C, Philpot GR. Effect of cortisone on calcium metabolism in sarcoidosis with hypercalcemia. Lancet 1954; 2:720–724.
26. Vergnon GM, Chappard D, Mounier D, et. al. Phosphocalcic metabolism, bone quantitative histomorphometry and clinical activity in 10 cases of sarcoidosis. In: Sarcoidosis and Other Granulomatous Disorders. Grassi C, Rizzato G, Pozzi E, eds. Amsterdam: Elsevier, 1988; 499–502.
27. Fallon MD, Perry HM III, Teitelbaum SL. Skeletal sarcoidosis with osteopenia. Metab Bone Dis Res 1981; 3:171–174.

28. Bell NH. Endocrine complications of sarcoidosis. Endocrinol Metab Clin North Am 1991; 20:645–654.
29. Basile, JN, Leil Y, Shary J, Bell NH. Increased calcium intake does not suppress circulating 1,25-dihydroxyvitamin D in normocalcemic patients with sarcoidosis. J Clin Invest 1993; 91:1396–1398.
30. Sandler LM, Winearls CG, Fraher LJ, Clemens TL, Smith R, O'Riordan JLH. Studies of the hypercalcemia of sarcoidosis. Q J Med 1984; 53:165–180.
31. Papapoulos SE, Clemens TL, Fraher LJ. Dihydroxycholecalciferol in the pathogenesis of the hypercalcemia of sarcoid. Lancet 1979; 1:627–630.
32. Cronin CC, Dinneen SF, O'Mahony MS, Bredin CP, O'Sullivan DJ. Precipitation of hypercalcaemia in sarcoidosis by foreign sun holidays: report of four cases. Postgrad Med J 1990; 66:307–309.
33. Singer FR, Adams JS. Abnormal calcium homeostasis in sarcoidosis. N Engl J Med 1986; 315:755,756.
34. Adams JS, Gacad MA, Anders A, Endres DB, Sharma DP. Biochemical indicators of disordered vitamin D and calcium homeostasis in sarcoidosis. Sarcoidosis 1986; 3:1–6.
35. Insogna KL, Dreyer BE, Mitnich M, Ellison AF, Broadus A. Enhanced production of 1,25-dihydroxyvitamin D in sarcoidosis. J Clin Endocrinol Metab 1988; 66:72–75.
36. Shulman LE, Schoenrich E, Harvey A. The effects of adrenocorticotropic hormone (ACTH) and cortisone on sarcoidosis. Bull John Hopkins Hosp 1952; 91:371–415.
37. O'Leary TJ, Jones G, Yip A, Lohnes D, Cohanim M, Yendt ER. The effects of chloroquine on serum 1,25-dihydroxyvitamin D and calcium metabolism in sarcoidosis. N Engl J Med 1986; 315:727–730.
38. Barre PE, Gascon-Barre M, Meakins JL, Goltzman D. Hydroxychloroquine treatment of hypercalcemia in a patient with sarcoidosis. Am J Med 1987; 82:1259–1262.
39. Adams JS, Diz MM, Sharma OP. Effective reduction in the serum 1,25-dihydroxyvitamin D and calcium concentration in sarcoidosis-associated hypercalcemia with short-course chloroquine therapy. Ann Intern Med 1989; 111:437,438.
40. Adams JS, Gacad MA. Characterization of 1α-hydroxylation of vitamin D_3 sterols by cultured macrophages from patients with sarcoidosis. J Exp Med 1985; 161:755–765.
41. Shany S, Ren S-Y, Arbelle JE, Clemens TL, Adams JS. Subcellular localization of the 25-hydroxyvitamin D_3-1-hydroxylase and partial purification from the chick myelomonocytic cell line HD-11. J Bone Miner Res 1993; 8:269–276.
42. Adams JS, Ren S-Y, Arbelle JE, Horiuchi N, Gray RW, Clemens TL, Shany S. Regulated production and intracrine action of 1,25-dihydroxyvitamin D_3 in chick myelomonocytic cell line HD-11. Endocrinology 1994; 134:2567–2573.
43. Reichel H, Koeffler HP, Norman AW. Synthesis in vitro of 1,25-dihydroxyvitamin D_3 and 24,25-dihydroxyvitamin D_3 by interferon-gamma-stimulated normal human bone marrow and alveolar macrophages. J Biol Chem 1987; 262:10,931–10,937.
44. Reichel H, Koeffler HP, Barbers R, Norman AW. Regulation of 1,25-dihydroxyvitamin D_3 production by cultured alveolar macrophages from normal human donors and patients with pulmonary sarcoidosis. J Clin Endocrinol Metab 1987; 65:1201–1209.
45. Fraser DR. Regulation of the metabolism of vitamin D. Physiol Rev 1980; 60:551–613.
46. Nesbitt T, Drezner MK. Insulin-like growth factor-1 regulation of renal 25-hydroxyvitamin D-1-hydroxylase activity. Endocrinology 1993; 132:133–138.
47. Henry HH. 25(OH)D_3 metabolism in kidney cell cultures: lack of a direct effect of estradiol. Am J Physiol 1981; 240:E119–E124.
48. Adams ND, Garthwaite TL, Gray RW. The interrelationship among prolactin, 1,25-dihydroxyvitamin D and parathyroid hormone in humans. J Clin Endocrinol Metab 1979; 49:628–630.
49. Kumar R, Merimee TJ, Silva P, Epstein FH. The effect of chronic excess or deficiency of growth hormone on plasma 1,25-dihydroxyvitamin D levels in man. In: Vitamin D, Basic Research and Its Clinical Application. Norman AW, Schaefer K, von Herrath D, eds. Berlin: de Gruyter, 1979; 1005–1009.
50. Brixen K, Nielsen HK, Bouillon R, Flyvbjerg A, Mosekilde L. Effects of short-term growth hormone treatment on PTH, calcitriol, thyroid hormones, insulin and glucagon. Acta Endocrinol 1992; 127:331–336.
51. Henry HL. Vitamin D hydroxylases. J Cell Biochem 1992; 49:4–9.
52. Ohyama Y, Noshiro M, Okuda K. Cloning and expression of cDNA encoding 25-hydroxyvitamin D_3-24-hydroxylase. FEBS Lett 1991; 278:195–198.
53. Chen K, Goto H, DeLuca HF. Isolation and expression of human 1,25-dihydroxyvitamin D_3 24-hydroxylase cDNA. J Bone Miner Res 1992; 7:S148.

54. Chen KS, Prahl JM, DeLuca HF. Isolation and expression of human 1,25-dihydroxyvitamin D_3-24-hydroxylase. Proc Natl Acad Sci USA 1993; 90:4543–4547.
55. Ismail R, Elaroussi MA, DeLuca HF. Regulation of chicken kidney vitamin D_3 24-hydroxylase mRNA by 1,25-dihydoxyvitamin D_3 and parathyroid hormone. J Bone Miner Res 1993; 8:S208.
56. Uchida M, Shinki T, Ohyama Y, Noshiro M, Okda K, Suda T. Protein kinase C upregulates $1\alpha,25$-dihydroxyvitamin D_3 induced expression of the 24-hydroxylase gene. J Bone Miner Res 1993; 8:S171.
57. Chen ML, Boltz MA, Armbrecht HJ. Effects of 1,25-dihydroxyvitamin D_3 and phorbol ester on 25-hydroxyvitamin D_3-24-hydroxylase cytochrome P-450 messenger ribonucleic acid levels in primary cultures of rat renal cells. Endocrinology 1993; 132:1782–1788.
58. Tanaka Y, Lorenc RS, DeLuca HF. The role of 1,25-dihydroxyvitamin D_3 and parathyroid hormone in the regulation of chick renal 25-hydroxyvitamin D_3-24-hydroxylase. Arch Biochem Biophys 1975; 171:521–526.
59. Henry H, Luntao EM. Further studies on the regulation of 25-OH-D metabolism in kidney cell cultures. In: Vitamin D; Chemical, Biochemical and Clinical Update. Norman AW, Schaefer K, Grigoleit HG, von-Herrath D, eds. Berlin: de Gruyter, 1985; 505–514.
60. Henry HL. Regulation of the hydroxylation of 25-hydroxyvitamin D_3 in vivo and in primary cultures of chick kidney cells. J Biol Chem 1979; 254:2722–2729.
61. Adams JS, Gacad MA, Singer FR, Sharma OP. Production of 1,25-dihydroxyvitamin D_3 by pulmonary alveolar macrophages from patients with sarcoidosis. NY Acad Sci 1986; 465:587–594.
62. Adams JS. Hypercalcemia and hypercalciuria. Sem Respir Med 1992; 13:402–410.
63. Hunninghake GW. Role of alveolar macrophage- and lung T cell-derived mediators in pulmonary sarcoidosis. Ann NY Acad Sci 1986; 465:82–90.
64. Adams JS, Gacad MA, Diz MM, Nadler JL. A role for endogenous arachidonate metabolites in the regulated expression of the 25-hydroxyvitamin D-1-hydroxylation reaction in cultured alveolar macrophages from patients with sarcoidosis. J Clin Endocrinol Metab 1990; 70:595–600.
65. Celada A, Schreiber RD. Role of protein kinase C and intracellular calcium mobilization in the induction of macrophage tumoricidal activity by interferon-gamma. J Immunol 1986; 137:2373–2379.
66. Wightman PD, Humes JL, Davies P, Bonney RJ. Identification of two phospholipase A2 activities in resident mouse peritoneal macrophages. Biochem J 1981; 195:427.
67. Wightman PD, Dahlgren M, Bonney RS. Protein kinase activation of phospholipase A2 in sonicates of mouse peritoneal macrophages. J Biol Chem 1982; 257:6650.
68. Schumann RR, Leong SR, Flaggs GW, Wright SD, Mathison JC, Tobias PS, Ulevitch RJ. Structure and function of LPS binding protein. Science 1990; 249:1429–1431.
69. Marletta, MA. Nitric oxide synthase: aspects concerning structure and catalysis. Cell 1994; 78:927–930.
70. Nathan C, Xie Q-W. Nitric oxide synthases: roles, tolls, and controls. Cell 1994; 78:915–918.
71. Adams, JS, Ren SY, Arbelle JE, Clemens TL, Shany S. A role for nitric oxide in the regulated expression of the 25-hydroxyvitamin D-1-hydroxylation reaction in the chick myelomonocytic cell line HD-11. Endocrinology 1994; 134:499–502.
72. Adams JS, Ren S-Y, Arbelle J, Shany S, Gacad MA. Coordinate regulation of nitric oxide and 1,25-dihydroxyvitamin D production in the avian myelomonocytic cell line HD-11. Endocrinology 1995; 136:2262–2269.
73. Lowenstein CJ, Glatt CS, Bredt DS, Snyder SH. Cloned and expressed macrophage nitric oxide synthase contrasts with the brain enzyme. Proc Natl Acad Sci USA 1992; 89:6711–6715.
74. Valance P, Collier J. Biology and clinical relevance of nitric oxide. Br Med J 1994; 309:453–457.
75. Leone AM, Palmer RMJ, Knowles RG, Francis PL, Ashton DS, Moncada S. Constitutive and inducible nitric oxide synthases incorporate molecular oxygen into both nitric oxide and citrulline. J Biol Chem 1991; 266:23,790–23,795.
76. Chartrain NA, Geller DA, Koty PP, Sitrin NF, Nussler AK, Hoffman EP, Billiar TR, Hutchinson NI, Mudgett JS. Molecular cloning, structure, and chromosomal localization of the human inducible nitric oxide synthase gene. J Biol Chem 1994; 269:6765–6772.
77. Stamler JS, Singel DJ, Loscalzo J. Biochemistry of nitric oxide and its redox-activated forms. Science 1992; 258:1898–1902.
78. Stamler JS. Redox signaling: nitrosylation and related target interactions of nitric oxide. Cell 1994; 78:931–936.
79. Adams, JS. Extrarenal production and action of active vitamin D metabolites in human lymphoproliferative diseases. In: Vitamin D. Feldman D, Glorieux FH, Pike JW, eds. San Diego: Academic, 1997, pp. 903–922.

80. Adams JS, Ren S-Y. Autoregulation of 1,25-dihydroxyvitamin D synthesis in macrophage mitochondria by nitric oxide. Endocrinology 1996; 137:4514–4517.
81. Griscavage JM, Rogers NE, Sherman MP, Ignarro LJ. Inducible nitric oxide synthase from a rat alveolar macrophage cell line is inhibited by nitric oxide. J Immunol 1993; 151:6329–6337.
82. Khatsenko OG, Gross SS, Rifkind AB, Vane JR. Nitric oxide is a mediator of the decrease in cytochrome P450-dependent metabolism caused by immunostimulants. Proc Natl Acad Sci USA 1993; 90:11,147–11,151.
83. Griscavage JM, Fukuto JM, Komori Y, Ignarro LJ. Nitric oxide inhibits neuronal nitric oxide synthase by interacting with the heme prosthetic group. J Biol Chem 1994; 269:21,644–21,649.
84. Stadler J, Trockfeld J, Schmalix WA, Brill T, Siewert JR, Greim , Doehmer J. Inhibition of cytochromes P450-1A by nitric oxide. Proc Natl Acad Sci USA 1994; 91:3559–3563.
85. Morris SM, Jr, Billiar TR. New insights into the regulation of inducible nitric oxide synthesis. Am J Physiol 1994; 266:E829–E839.
86. Van Voorhis BJ, Dunn MS, Snyder GD, Weiner CP. Nitric oxide: an autocine regulator of human granulosa-luteal cell steroidogenesis. Endocrinology 1994; 135:1799–1806.
87. Hartl FU. Molecular chaperones in cellular protein folding. Nature 1996; 381:571–580.
88. Polla BS, Healy AM, Wojno WC, Krane SM. Hormone $1\alpha,25$-dihydroxyvitamin D_3 modulates heat shock response in monocytes. Am J Physiol 1987; 252:C640–C649.
89. Gacad MA, LeBon TR, Chen H., Arbelle JE, Adams JS. Functional characterization and purification of an intracellular vitamin D binding protein in vitamin D resistant New World primate cells: amino acid sequence homology with proteins in the hsp-70 family. J Biol Chem 1997; 272:8433–8440.
90. Berthold J, Bauer MF, Schneider H-C, Klaus C, Dietmeier K, Neupert W, Brunner M. The MIM complex mediates preprotein translocation across the mitochondrial inner membrane and couples it to the mt-Hsp70/ATP driving system. Cell 1995; 81:1085–1093.
91. Need AG, Phillips PJ. Hypercalcaemia associated with tuberculosis. Br Med J 1980; 280:831.
92. Felsenfeld AJ, Drezner MK, Llach F. Hypercalcemia and elevated calcitriol in a maintenance dialysis patient with tuberculosis. Arch Intern Med 1986; 146:1941–1945.
93. Gkonos PJ, London R, Hendler ED. Hypercalcemia and elevated 1,25-dihydroxyvitamin D levels in a patient with end-stage renal disease and active tuberculosis. N Engl J Med 1984; 311:1683–1685.
94. Epstein S, Stern PH, Bell NH, Dowdeswell I, Turner RT. Evidence for abnormal regulation of circulating $1\alpha,25$-dihydroxyvitamin D in patients with pulmonary tuberculosis and normal calcium metabolism. Calcif Tissue Int 1984; 36:541–544.
95. Bell NH, Shary J, Shaw S, Turner RT. Hypercalcemia associated with increased circulating 1,25-dihydroxyvitamin D in a patient with pulmonary tuberculosis Calcif Tissue Int 1985; 37:588–591.
96. Cadranel J, Garabedian M, Milleron B, Guillozo H, Akoun G, Hance AJ. $1,25(OH)_2D_3$ production by T lymphocytes and alveolar macrophages recovered by lavage from normocalcemic patients with tuberculosis. J Clin Invest 1990; 85:1588–1593.
97. Cadranel JL, Garabedian M, Milleron B, Guillozzo H, Valeyre D, Paillard F, Akoun G, Hance AJ. Vitamin D metabolism by alveolar immune cells in tuberculosis: correlation with calcium metabolism and clinical manifestations. Eur Res J 1994; 7:1103–1110.
98. Isaacs RD, Nicholson GI, Holdaway IM. Miliary tuberculosis with hypercalcaemia and raised vitamin D concentrations. Thorax 1987; 42:555,556.
99. Shai F, Baker RK, Addrizzo JR, Wallach S. Hypercalcemia in mycobacterial infection. J Clin Endocrinol Metab 1972; 34:251–256.
100. Braman SS Goldman AL, Schwarz MI. Steroid-responsive hypercalcemia in disseminated bone tuberculosis. Arch Intern Med 1973; 90:327,328.
101. Hoffman VH, Korzeniowski OM. Leprosy, hypercalcemia, and elevated serum calcitriol levels. Ann Intern Med 1986; 105:890,891.
102. Ryzen E, Rea TH, Singer FR. Hypercalcemia and abnormal 1,25-dihydroxyvitamin D concentrations in leprosy. Am J Med 1988; 84:325–329.
103. Kantarijian HM, Saad MF, Estey EH, Sellin RV, Samaan NA. Hypercalcemia in disseminated candidiasis. Am J Med 1983; 74:72l–724.
104. Walker JV, Baran D, Yakub YN, Freeman RB. Histoplasmosis with hypercalcemia, renal failure, and papillary necrosis. Confusion with sarcoidosis. JAMA 1977; 237:1350–1352.
105. Parker MS, Dokoh S, Woolfenden JM, Buchsbaum HW. Hypercalcemia in coccidioidomycosis. Am J Med 1984; 76:341–343.
106. Ahmed B, Jaspan JB. Case report: hypercalcemia in a patient with AIDS and *Pneumocystis carinii* pneumonia. Am J Med Sci 1993; 306:313–316.

157. Adams JS, Modlin RL, Diz MM, Barnes PF. Potentiation of the macrophage 25-hydroxyvitamin D-1-hydroxylation reaction by human tuberculous pleural effusion fluid. J Clin Endocrinol Metab 1989; 69:457–460.
158. Rook GAW, Taverne J, Leveton C, Steele J. The role of gamma-interferon, vitamin D_3 metabolites and tumour necrosis factor in the pathogenesis of tuberculosis. Immunology 1987; 62:229–234.
159. Rook GAW. The role of vitamin D in tuberculosis. Am Rev Respir Dis 1988; 138:768–770.
160. Mawer EB, Hayes ME, Still PE, Davies M, Lumb GA, Palit J, Holt PJL. Evidence for nonrenal synthesis of 1,25-dihydroxyvitamin D in patients with inflammatory arthritis. J Bone Miner Res 1988; 6:733–739.
161. Hayes ME, Bayley D, Still P, Palit J, Denton J, Freemont AJ, Cooper RG, Mawer EB. Differential metabolism of 25-hydroxyvitamin D_3 by cultured synovial fluid macrophages and fibroblast-like cells from patients with arthritis. Ann Rheum Dis 1992; 51:220–226.
162. Hayes ME, O'Donoghue DJ, Ballardie FW, Mawer EB. Peritonitis induces the synthesis of $1\alpha,25$-dihydroxyvitamin D_3 in macrophages from CAPD patients. FEBS 1987; 220:307–310.
163. Shany S, Rapoport J, Zuili I, Gavriel A, Lavi N, Chaimovitz C. Metabolism of 25-OH-vitamin D_3 by peritoneal macrophages from CAPD patients. Kid Int 1991; 39:1005–1011.

21 Clinical Utility of 1,25-Dihydroxyvitamin D₃ and Its Analogs for the Treatment of Psoriasis and Other Skin Diseases

J. Reichrath and Michael F. Holick

1. INTRODUCTION

Vitamin D is photochemically sythesized by ultraviolet (UV)B action in the skin *(1,2)*. It is now known that the skin itself is a target tissue for the secosteroid hormone 1α,25-dihydroxyvitamin D_3 [1,25$(OH)_2D_3$, calcitriol], the biologically active vitamin D metabolite *(3,4)*. 1,25$(OH)_2D_3$ exerts genomic and nongenomic effects. Nongenomic effects of 1,25$(OH)_2D_3$ and its analogs are related to effects on intracellular calcium *(5,6)*. In keratinocytes and other cell types, calcitriol rapidly increases free cytosolic calcium levels *(5,6)*. Genomic effects of 1,25$(OH)_2D_3$ are mediated via binding to a nuclear receptor protein that is present in target tissues and binds 1,25$(OH)_2D_3$ with high affinity (K_D 10^{-9}–10^{-10} M) and low capacity *(7,8)*. The human vitamin D receptor (VDR) has been cloned *(9)*, and sequence analysis has demonstrated that this protein belongs to the superfamily of *trans*-acting transcriptional regulatory factors, which includes the steroid and thyroid hormone receptors and the retinoic acid receptors *(9)*. Interaction of 1,25$(OH)_2D_3$ with VDR results in the phosphorylation of the receptor complex, which in turn activates the transcription of calcitriol-sensitive target genes, especially genes involved in cellular differentiation and proliferation. Recently, it was shown that VDRs require auxillary factors for sufficient DNA binding *(10)*. These auxillary proteins were identified as the retinoid X receptors (RXR-α, -β,- γ), which were shown to heterodimerize with VDR, thus increasing the transcriptional function and DNA binding to the respective vitamin D response elements (VDRE) in the promotor region of target genes *(10,11)*. In the skin, both VDR (Fig. 1) and RXR-α are expressed in keratinocytes, fibroblasts, Langerhans cells, sebaceous gland cells, endothelial cells, and most cell types related to the skin immune system *(12,13)*. In vitro studies revealed that 1,25$(OH)_2D_3$ is extremely effective in inducing terminal differentiation and inhibiting the proliferation of cultured human keratinocytes in a dose-dependent manner *(14–16)*. Additionally, 1,25$(OH)_2D_3$ acts on many cell types involved in immunologic reactions, including lymphocytes, macrophages, and Langerhans cells *(17,18)*. Data on the effects of 1,25$(OH)_2D_3$ on the melanin pigmentation system are still conflicting, but most studies do not support the possibility that 1,25$(OH)_2D_3$ regulates melanogenesis in human skin *(19)*.

From: *Vitamin D: Physiology, Molecular Biology, and Clinical Applications*
Edited by: M. F. Holick © Humana Press Inc., Totowa, NJ

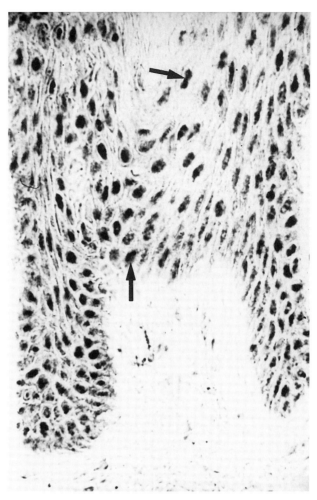

Fig. 1. Immunohistochemical demonstration of 1,25-dihydroxyvitamin D_3 receptors (VDR) in human skin. Notice strong nuclear VDR immunoreactivity in cells of all layers of the viable epidermis (arrows). Labeled avidin-biotin technique using MAb 9A7γ directed against VDR. Original magnification ↔400.

2. PSORIASIS: PATHOGENESIS, IMMUNOLOGY, AND HISTOLOGY

Psoriasis is a chronic dermatosis of unknown etiology characterized by hyperproliferation and inflammation of the skin. The peak age of onset for this psychologically debilitating and disfiguring disease is the second decade, but psoriasis may first appear at any age from infancy to the aged [20]. It is considered a multifactorial disease and has a prevalence of about 1-2% in the United States. Population, family, and twin studies clearly demonstrate a strong and highly complex genetic component leading to the development of psoriatic skin lesions [21]. Multiple genes are probably involved in the pathogenesis of psoriasis. Molecular biology techniques have recently been developed that allow analysis of psoriasis susceptibility genes, but no specific genetic marker has been found so far. Psoriasis has long been known to be associated with certain HLA antigens, particularly HLA-Cw6, although there is no evidence that a psoriasis susceptibility gene

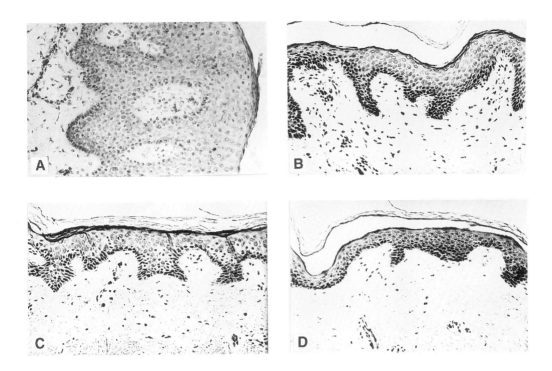

Fig. 2. Histologic demonstration of morphologic changes in lesional psoriatic skin after 6 weeks of topical treatment with calcitriol (15 µg/g, **B**) and calcipotriol (50 µg/g, **C**). (**A**) Lesional psoriatic skin before treatment (note that because of the marked increase in the epidermal thickness the picture is on its side). (**D**) Nonlesional psoriatic skin. Note strong reduction of epidermal thickness after topical treatment with vitamin D analogs. Hematoxylin & eosin staining. Original magnification ↔ 200.

exists at this locus *(22)*. It is still unknown what cell types in human skin are primarily affected by the disease. Recent studies support the hypothesis that epidermal hyperproliferation in psoriasis may be mediated by cells of the immune system, most likely T-lymphocytes *(23)*. Activated T-cells in psoriatic lesions express HLA-DR and the interleukin (IL)-2 receptor (CD25) and secrete specific immune mediators and cytokines, such as IL-2 and interferon-γ *(24–26)*. Thus, psoriasis represents a *Th1 profile disease* (characterized by T-lymphocyte secretion of IL-2, IL-12, and interferon-γ) *(27)*. By contrast, atopic dermatitis represents a *Th2 profile disease* (characterized by T-cell secretion of IL-4, IL-5 and IL-10) *(28)*. The activation signal for the development of psoriatic lesions is still unknown, although there is increasing evidence that superantigens such as the N-terminal component of bacterial M-proteins may be of importance for the initiation of T-cell proliferation in psoriasis *(29)*.

The precise histologic appearance depends on the age of the psoriatic lesion and the site of the biopsy. In general, epidermal hyperplasia, in which the granular layer may be lost and the stratum corneum shows parakeratosis, can be found (Fig. 2). Typical lesions show elongation of the dermal papillae, with a relatively thin epidermis at the top of the papillae. Epidermis may show intercellular edema in suprapapillar compartments, as well as infiltration with T-lymphocytes and neutrophils, which can extend into spongiform pustules of Kogoj or Munro microabscesses *(30)*.

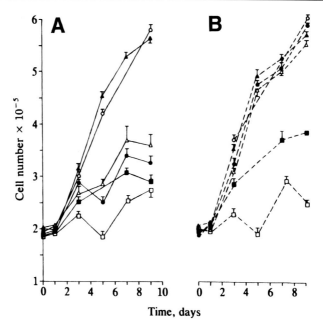

Fig. 3. Human dermal fibroblasts cultured from a skin biopsy from a normal age-matched volunteer (A) or an uninvolved area from a psoratic patient (B) were incubated at day 0 with either EtOH (m) or 1,25(OH)$_2$D$_3$ at one of the following concentrations: 0.0001 (p), 0.01 (r), 1 (●), 10 (n), or 100 (") μ*M*. Each point represents the mean ± SEM of fibroblasts, plated in triplicate. (Reproduced with permission from ref. *31.*)

3. CLINICAL STUDIES OF 1,25(OH)$_2$D$_3$ AND ANALOGS IN PSORIASIS AND OTHER SKIN DISEASES

The use of 1,25(OH)$_2$D$_3$ and its analogs for the treatment of psoriasis resulted from two independent lines of investigation. Since psoriasis is a hyperproliferative skin disorder, it seemed reasonable that the antiproliferative effects of calcitriol could be used for the treatment of this disease. However, before launching clinical trials in 1985, MacLaughlin and associates *(31)* reported the observation that psoriatic fibroblasts are partially resistant to the antiproliferative effects of 1,25(OH)$_2$D$_3$ (Fig. 3). This observation prompted these authors to speculate, that 1,25(OH)$_2$D$_3$ may be effective in the treatment of the hyperproliferative skin disease psoriasis. The other line of investigation resulted from a clinical observation. In 1985, Morimoto and Kumahara *(32)* reported that a patient who was treated orally with 1α-hydroxyvitamin D$_3$ for osteoporosis had a dramatic remission of psoriatic skin lesions.

Morimoto et al. *(33)* reported on a follow-up study demonstrating that almost 80% of 17 patients with psoriasis who were treated orally with 1α-hydroxyvitamin D$_3$ at a dose of 1.0 μg/d for up to 6 month showed clinically significant improvement.

Numerous studies have reported that various vitamin D analogs (Fig. 4), including 1,25(OH)$_2$D$_3$, calcipotriol (calcipotriene), and 1,24-dihydroxyvitamin D$_3$ [1,24(OH)$_2$D$_3$; tacalcitol], are effective and safe in the topical treatment of psoriasis *(34–37)*. Recently it was shown that topical 1,25(OH)$_2$D$_3$ is highly effective and safe in the long-term treatment of psoriasis vulgaris *(35)* (Figs. 5-7). Calcipotriol, the synthetic analog of 1,25(OH)$_2$D$_3$, was similarly applied twice daily topically in amounts of up to 100 g ointment (50 μg

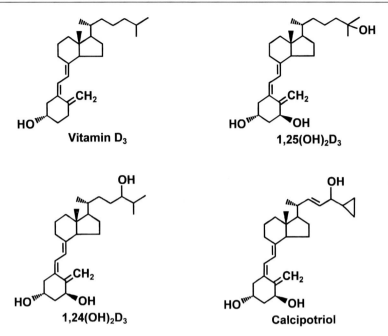

Fig. 4. Structures of vitamin D_3, 1,25-dihydrovitamin D_3 [1,25(OH)$_2$D$_3$], and its analogs 1,24-dihydroxyvitamin D_3 [1,24(OH)$_2$D$_3$] and calcipotriol (calcipotriene, MC-903), which are used for the treatment of psoriasis.

calcipotriol/g ointment) per week; it was shown to be slightly more effective in the topical treatment of psoriasis than betametasone 17-valerate ointment *(39)*. It has been reported that a mild dermatitis can be seen in about 10% of patients treated with calcipotriol (50 µg/g), particularly on the face *(40)*. This side effect has not been reported after topical treatment with 1,25(OH)$_2$D$_3$ *(35)*.

Recently, a long-term (36-month) follow-up study demonstrated the efficacy and safety of oral 1,25(OH)$_2$D$_3$ in the treatment of psoriasis *(38)* (Figs. 8 and 9). Of the 85 patients who received oral 1,25(OH)$_2$D$_3$, 78.0% had some improvement in their disease, and 26.5, 26.3, and 25.3% had complete, moderate, and slight improvement, respectively (Fig. 10). Serum calcium concentrations and 24-h urinary calcium excretion increased by 3.9 and 148.2%, respectively (Fig. 11), but were not outside the normal range. Bone mineral density of these patients remained unchanged. An important consideration for the use of orally administered 1,25(OH)$_2$D$_3$ is the dosing technique. To avoid its effects on enhancing dietary calcium absorption, it is very important to provide calcitriol at night. Perez et al. *(38)* showed that as a result of this dosing technique, doses of 2–4 µg/night are well tolerated by psoriatic patients (Fig. 11).

Patients with psoriasis may need intermittent treatment for the rest of their lives. Vitamin D analogs have been shown not to exhibit tachyphylaxis during treatment of psoriatic lesions and can be continued indefinitely. They are effective and safe for the treatment of skin areas that are usually difficult to treat in psoriatic patients and that respond slowly. Most importantly, the topical active vitamin D therapies do not cause thinning of the skin, as do topical steroids. Additionally, 1,25-dihydrovitamin D_3 and vitamin D analogs are effective in the treatment of psoriatic skin lesions in children and patients with the human immunodeficiency virus (HIV).

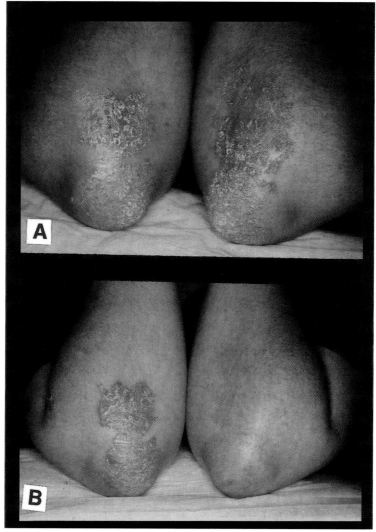

Fig. 5. Topical application of 1,25(OH)$_2$D$_3$ ointment in a double-blind placebo-controlled manner after 2 mo. The left forearm received 1,25(OH)$_2$D$_3$ ointment, and the right forearm received placebo ointment.

3.1. Scalp Psoriasis

Recently, a double-blind, randomized multicenter study demonstrated that calcipotriol solution is effective in the topical treatment of scalp psoriasis *(41)*. Forty-nine patients were treated twice daily over a 4-wk period, 60% of patients on calcipotriol showed clearance or marked improvement vs 17% in the placebo group, and no side effects were reported.

3.2. Nail Psoriasis

Nails in general are very difficult to treat and respond slowly. Nail psoriasis has been reported in up to 50% of patients, and up to now, there has been no consistently effective treatment. Recently, it was shown that calcipotriol ointment is effective in the treatment of nail psoriasis *(42)*. Oral 1,25(OH)$_2$D$_3$ has also been found to be effective in treating this very difficult complication of psoriasis *(38)*.

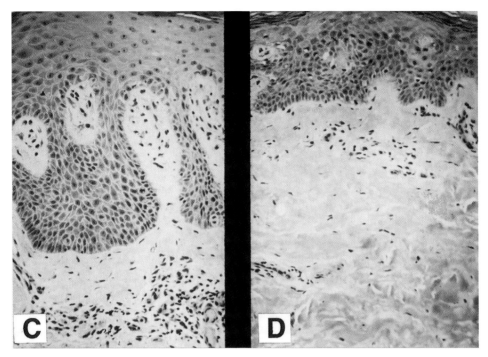

Fig. 5. (Continued). The histology of skin biopsies from both sites showed marked regression in the proliferative epidermis for the lesion treated with 1,25(OH)$_2$D$_3$ ointment. (Reproduced with permission from ref. *35*.)

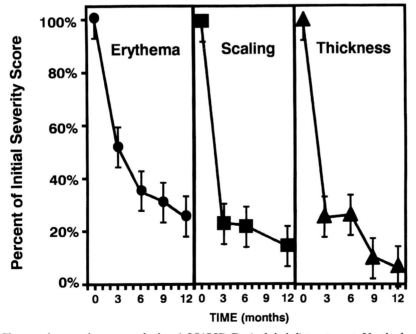

Fig. 6. Changes in severity scores during 1,25(OH)$_2$D$_3$ (calcitriol) treatment. Vertical range bars indicate ± standard error of mean (SEM). The severity scores during treatment with calcitriol showed a statistically significant improvement ($p < 0.001$) in comparison with baseline values. The numbers of subjects studied at 0, 3, 6, 9, and 12 mo were 22, 18, 13, 7, and 6, respectively *(35)*.

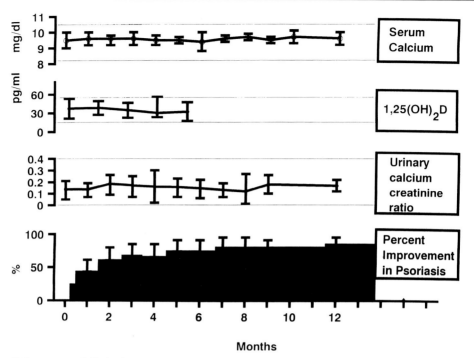

Fig. 7. Sequence of clinical events over time. Changes in blood and urine calcium, as well as serum $1,25(OH)_2D_3$ (calcitriol) concentrations, are shown along with the improvement in psoriasis. (Reproduced with permission from ref. *35*.)

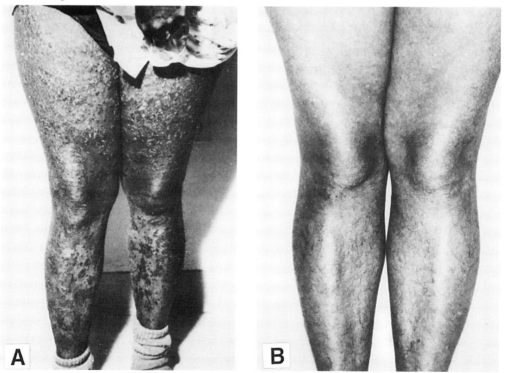

Fig. 8. Views of the legs of a patient with erythrodermal psoriasis before treatment (**A**) and after 3 mo of treatment with oral $1,25(OH)_2D_3$ (**B**). (Reproduced with permission from ref. *66*.)

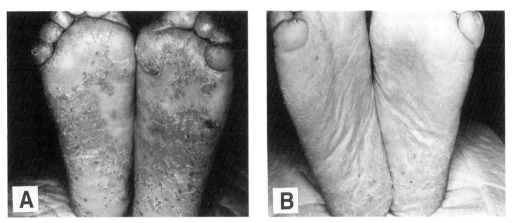

Fig. 9. Oral treatment with 1,25(OH)$_2$D$_3$ for a 64-yr-old male patient with plaque psoriasis on the bottom of his feet. Before (**left**) and 6 mo after oral 1,25(OH)$_2$D$_3$ (1.5 µg) (**right**).

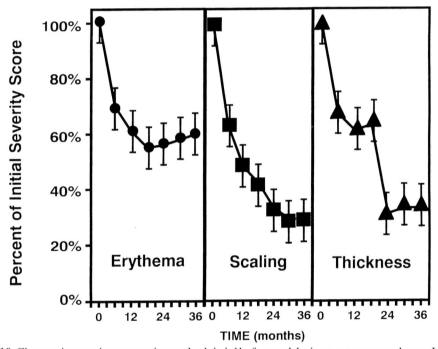

Fig. 10. Changes in severity score using oral calcitriol before and during treatment are shown. Vertical range bars indicate ± standard error of the mean (SEM). The severity scores during treatment with calcitriol were statistically significant ($p < 0.001$ in comparison with baseline values. (Reproduced with permission from ref. *38*.)

3.3. Face and Flexures

Although the use of calcipotriol ointment is not recommended on face and flexures due to irritancy, most patients tolerate vitamin D analogs on these sites. There are no reports of irritation with the use of topical or oral 1,25(OH)$_2$D$_3$ *(35,38)*.

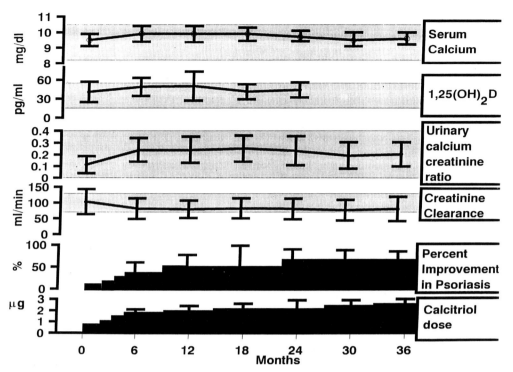

Fig. 11. Sequence of clinical events over time. Changes in blood and urine calcium as well as serum $1,25(OH)_2D_3$ (calcitriol) concentrations are shown at the same time the clinical improvement occurred. (Reproduced with permission from ref. *38.*)

3.4. Skin Lesions in Children

It has been shown that topical application of $1,25(OH)_2D_3$ ointment is an effective, safe and reliable therapy to treat psoriatic skin lesions in children *(43,44)*.

3.5. Psoriatic Lesions in HIV Patients

We recently treated an HIV-positive patient suffering from psoriatic skin lesions with topical and oral $1,25(OH)_2D_3$. The patient responded well, and there was no evidence of enhancement of HIV disease activity or alterations in the number of T-lymphocytes or CD4+ and CD8+ cells.

3.6. Combination of Vitamin D Analogs with Other Therapies

It was recently reported that efficacy of topical treatment with vitamin D analogs in psoriasis can be increased by combination with other therapies, including very low dose oral cyclosporine (2 mg/kg/d), oral acitretin, topical dithranol, topical steroids, and UVB or narrow band UVB phototherapy *(45–49)*. Complete clearing or 90% improvement in Psoriasis Area Severity Index (PASI) was observed in 50% of patients treated with calcipotriol/cyclosporine vs 11.8% in the placebo/cyclosporine group. No difference in side effects was found between the groups.

Addition of calcipotriol ointment to oral application of acitretin (a vitamin A analog) was shown to produce a significantly better treatment response, which was achieved with a lower cumulative dose of acitretin in patients with severe extensive psoriasis vulgaris, compared with the group of patients treated with oral acitretin alone. The number of patients reporting adverse events was similar between the two treatment groups *(47)*. Combined topical treatment with calcipotriol ointment (50 µg/g) and betametasone ointment was recently shown to be slightly more effective and caused less skin irritation than calcipotriol used twice daily *(48)*. Kragballe *(49)* reported that the efficacy of topical calcipotriol treatment in psoriasis can be improved by simultaneous UVB phototherapy. Combination therapy with topical calcipotriol and narrow-band UVB has recently been shown to be very effective for the treatment of psoriatic plaques.

4. TREATMENT OF OTHER SKIN DISORDERS WITH VITAMIN D ANALOGS

Earlier in this century, vitamin D_3 was used in dermatology in huge pharmacologic doses for the treatment of scleroderma, psoriasis, lupus vulgaris, and atopic dermatitis. These first attempts were abandoned because of severe vitamin D intoxication, which caused hypercalcemia, hypercalciuria, and kidney stones, and because other treatments were introduced for the therapy of these diseases.

4.1. Ichthyosis

Recently, a double-blind, bilaterally paired, comparative study showed the effectiveness of topical treatment with calcipotriol ointment on congenital ichthyoses *(50)*. Reduction in scaling and roughness on the calcipotriol-treated side was seen in all patients with lamellar ichthyosis and bullous ichthyotic erythroderma of Brocq. The only patient with Comel-Netherton syndrome showed mild improvement, and the only patient with ichthyosis bullosa of Siemens showed no change in severity on the calcipotriol-treated compared with the vehicle-treated side.

4.2. Scleroderma

Recent findings indicate efficacy of vitamin D analogs for the treatment of scleroderma. Humbert et al. *(51)* reported that oral administration of 1.0–2.5 µg/d 1,25$(OH)_2D_3$ improved skin involvement, probably via inhibition of fibroblast proliferation and dermal collagen deposition.

4.3. Skin Cancer

In vitro studies have demonstrated strong antiproliferative and prodifferentiating effects of vitamin D analogs in many VDR-expressing tumor cell lines, including maligoma, squamous cell carcinoma, and leukemic cells *(18,52)*. In vivo studies supported these results and showed that active vitamin D analogs block proliferation and tumor progression of epithelial tumors in rats *(53)*. Additionally, it was shown that administration of calcitriol reduced the number of lung metastases after implantation of lung carcinoma cells in mice *(54)*. Inhibition of tumor growth of human malignant melanoma and colonic cancer xenografts was also demonstrated in immune-suppressed mice, but only at high doses of calcitriol *(55)*. Little is known regarding the effects of 1,25$(OH)_2D_3$ on the formation of metastases in patients with malignant melanoma or squamous cell carcinoma of the skin.

4.4. Other Skin Diseases

A number of case reports demonstrate positive effects of topical treatment with vitamin D analogs in a variety of skin diseases such as transient acantholytic dermatosis

(Grover's disease), inflammatory linear verrucous epidermal naevus, disseminated superficial actinic porokeratosis, pityriasis rubra pilaris, epidermolytic palmoplantar keratoderma of Vorner, and Sjögren-Larsson syndrome. These promising observations will have to be further evaluated in clinical trials.

5. MECHANISM OF ACTIONS OF 1,25(OH)$_2$D$_3$ ANALOGS IN PSORIASIS

The mechanisms underlying the therapeutic effectiveness of vitamin D analogs in psoriasis are still not completely understood. Results from immunohistochemical and molecular biology studies indicate that the antiproliferative effects of topical 1,25(OH)$_2$D$_3$ on epidermal keratinocytes are more pronounced compared with effects on dermal inflammation. Modulation of various markers of epidermal proliferation (proliferating cell nuclear antigen, Ki-67 antigen) and differentiation (involucrin, transglutaminase K, filaggrin, cytokeratins 10 and 16) in lesional psoriatic skin after topical application of vitamin D analogs was shown *in situ (56)* (Fig. 12). Interestingly, effects of topical treatment with vitamin D analogs on dermal inflammation are less pronounced (CD antigens, cytokines, HLA-DR, etc.). One reason for this observation may be that the bioavailability of this potent hormone in the dermal compartment may be markedly reduced compared with the epidermal compartment *(56)*.

Molecular biology studies have recently demonstrated that clinical improvement in psoriatic lesions treated with 1,25(OH)$_2$D$_3$ correlated with an elevation of VDR mRNA *(57)*. It is known that some patients suffering from psoriasis are resistant to calcitriol treatment. It was demonstrated recently that responders can be distinguished from nonresponders on the molecular level since nonresponders show no elevation of VDR mRNA in skin lesions along with the treatment. These data suggest that the ability of 1,25(OH)$_2$D$_3$ to regulate keratinocyte growth is closely linked to the expression of VDR. The target genes of topical 1,25(OH)$_2$D$_3$ that are responsible for its therapeutic efficacy in psoriasis are still unknown. Major candidates for 1,25(OH)$_2$D$_3$ target genes that are responsible for the 1,25(OH)$_2$D$_3$-induced terminal differentiation in keratinocytes are distinct cell cycle-associated proteins (i.e., the INK4 family), including p21/WAF-1 *(58)*.

6. NEW 1,25(OH)$_2$D$_3$ ANALOGS WITH LESS CALCEMIC ACTIVITY

The use of vitamin D analogs in dermatology and other medical fields was shown to be limited, since serious side effects, mainly on calcium metabolism, may occur at the supraphysiologic doses needed to reach clinical improvement. The evaluation of new vitamin D compounds with strong immunosuppressive, antiproliferative, and differentiating effects but only marginal effects on calcium metabolism introduces new important therapies for the treatment of various skin diseases. The goal of creating new vitamin D analogs with selective biologic activity and no undesirable side effects has not been reached, but recent findings have introduced new and promising concepts.

Calcipotriol (MC 903), a recently developed vitamin D analog with VDR binding properties similar to those of 1,25(OH)$_2$D$_3$ but low affinity for the vitamin D binding protein (DBP), was shown to be effective and safe in the topical treatment of psoriasis *(39)*. In vivo studies in rats showed that the effects of calcipotriol on calcium metabolism are 100–200 times lower than those of 1,25(OH)$_2$D$_3$; in vitro effects on proliferation and differentiation on human keratinocytes are comparable *(59)*. These differential

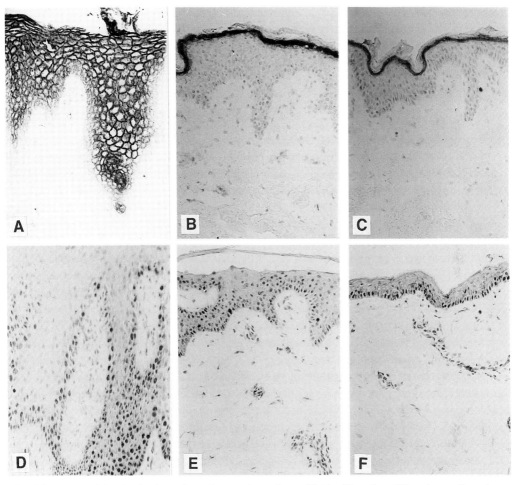

Fig. 12. Immunohistologic detection of transglutaminase K (**A–C**) and proliferating cell nuclear antigen (**D–F**) in lesional psoriatic skin before treatment (**A,D**), lesional psoriatic skin after 6 weeks of topical treatment with calcipotriol (**B,E**), and nonlesional psoriatic skin (**C,F**). Labeled avidin-biotin technique. Original magnification ×160 (**A–C**) and ×400 (**D–F**).

effects are probably caused by the different pharmacokinetic profiles of calcipotriol and $1,25(OH)_2D_3$ (different affinity for DBP). Serum half-life in rats of these vitamin D analogs was shown to be 4 min after treatment with calcipotriol, in contrast to 15 min after treatment with $1,25(OH)_2D_3$ *(59)*. However, most of the calcium studies comparing $1,25(OH)_2D_3$ and calcipotriol were done in vivo, whereas most studies analyzing proliferation or differentiation were done in vitro.

A different approach to creating new vitamin D analogs that are effective in the topical treatment of hyperproliferative or inflammatory skin diseases is the attempt to create new synthetic compounds (with a high degree of dissociation) that are metabolized in the skin and therefore exert few systemic side effects. New analogs of vitamin D, obtained by a combination of 20-methyl modification with biologically interesting artificial side chain subunits *(60)* or 2β-substituted $1,25(OH)_2D_3$ *(61)*, are promising candidates for this approach. Another interesting approach to enhance the concentration of $1,25(OH)_2D_3$ in the skin locally without systemic side effects involves attempts to inhibit specifically the activity of vitamin D-metabolizing enzymes, for example, various hydroxylases [catabolic D_3-OHases, i.e., 24-hydroxylase for $1,25(OH)_2D_3$] that are present in the skin and are responsible for the catabolism of $1,25(OH)_2D_3$ *(62)*. It is known that various pharmacologic active compounds (including other steroidal hormones but also cytochrome P-450 inhibitors such as ketoconazole) specifically inhibit the activity of D_3-OHases in the skin *(63)*. It may be possible to enhance the concentration of endogeneous $1,25(OH)_2D_3$ locally in the skin by topical application of these compounds without obtaining systemic side effects. It can be speculated that the therapeutic effects of various antimycotic compounds including ketoconazole in the treatment of seborrheic dermatitis may at least in part be caused by this mechanism.

It is now known that VDR requires nuclear accessory proteins for efficient binding to VDREs in promoter regions of target genes, thereby inducing VDR-mediated transactivation *(64)*. As a consequence, different vitamin D analogs may have (depending on their chemical structure) different affinities for the various homo- or heterodimers of VDR and nuclear cofactors including RXR-α *(65)*. The synthesis of new vitamin D analogs that activate different vitamin D signaling pathways may lead to the introduction of new therapeutics for the topical or oral treatment of various skin diseases. These new drugs may induce strong effects on target cell proliferation and differentiation in the skin or the immune system, but only marginal effects on calcium metabolism.

Another approach to enhance the therapeutic effects of orally or topically administered $1,25(OH)_2D_3$ may be combination with synergistically acting drugs. The recent discovery of different vitamin D signaling pathways that are determined and regulated by cofactors of VDR including RXR-α and their corresponding ligands suggests that 9-*cis* retinoic acid or all-*trans* retinoic acid may act synergistically with vitamin D analogs in inducing VDR-mediated transactivation and regulating the transcriptional activity of distinct gene networks. Not much is known about the effects of combined application of vitamin D and vitamin A analogs under physiologic or pathophysiologic conditions in vivo. This combination may selectively enhance or block different biologic effects of vitamin D analogs that are mediated by different vitamin D signaling pathways.

In conclusion, it can be speculated that new vitamin D analogs will introduce new alternatives for the treatment of various skin disorders. If the final goal of creating strong antiproliferative and anti-inflammatory vitamin D analogs with only a little calcemic activity is reached, then these new agents may herald a new era in dermatologic therapy, possibly to be compared with the introduction of synthetic corticosteroids or retinoids. These new drugs that may activate selective vitamin D signaling pathways but may exert only a little calcemic activity may also be effective in the systemic treatment of various malignant lymphomas of the skin, including lymphomas, squamous cell carcinoma, and basal cell carcinoma.

ACKNOWLEDGMENT

This work was supported in part by NIH grant M01RR-00533.

REFERENCES

1. Holick MF, MacLaughlin JA, Clark MB, Holick SA, Potts JT, Anderson RR, Blank IH, Parrish JA, Elias P. Photosynthesis of previtamin D_3 in human skin and the physiological consequences. Science 1980; 210:203–205.

2. Holick MF, MacLaughlin JA, Anderson RR, Parrish J. Photochemistry and photobiology of vitamin D. In: Photomedicine. Regan JD, Parrish JA eds. New York: Plenum Press, 1982; 195–218.
3. Holick MF, Smith E, Pincus S. Skin as the site of vitamin D synthesis and target tissue for 1,25-dihydroxyvitamin D_3. Arch Dermatol 1987; 123:1677–1682.
4. Holick MF. Photobiology, physiology and clinical applications for Vitamin D. In: Physiology, Biochemistry and Molecular Biology of the Skin, 2nd ed. Goldsmith LA, ed. New York: Oxford University Press, 1991; 928–956.
5. Bittiner B, Bleehen SS, Mac Neil S. 1α-25-$(OH)_2$Vitamin D_3 increases intracellular calcium in human keratinocytes. Br J Dermatol 1991; 124:12,230–12,235.
6. MacLaughlin JA, Cantley LC, Holick MF. 1,25$(OH)_2D_3$ increases calcium and phosphatidylinositol metabolism in differentiating cultured human keratinocytes. J Nutr Biochem 1990; 1:81–87.
7. Haussler MR. Vitamin D receptors: nature and function. Annu Rev Nutr 1986; 6:527–562.
8. Haussler MR, Mangelsdorf DJ, Komm BS, Terpening CM, Yamaoka K, Allegretto EA, Baker AR, Shine J, McDonnell DP, Hughes M, Weigel NL, O'Malley BW. Molecular biology of the vitamin D hormone. Recent Prog Horm Res 1988; 44:263–305.
9. Baker AR, Mc Donnell DP, Hughes M, Crisp TM, Mangelsdorf DJ, Haussler MR, Pike JW, Shine J, O'Malley BW. Cloning and expression of full-length cDNA encoding human vitamin D receptor. Proc Natl Acad Sci USA 1988; 85:3294–3298.
10. Yu VC, Deisert C, Andersen B, Holloway JM, Devary OV, Näär AM, Kim SY, Boutin JM, Glass CK, Rosenfeld MG. RXRβ: a coregulator that enhances binding of retinoic acid, thyroid hormone and vitamin D receptors to their cognate response elements. Cell 1991; 67:1251–1266.
11. Leid M, Kastner P, Lyons R, Nakshatri H, Saunders M, Zacharewski T, Chen J, Staub A, Garnier J, Mader S, Chambon P. Purification, cloning, and RXR identity of the HeLa cell factor with which RAR or TR heterodimerizes to bind target sequences efficiently. Cell 1992; 68:377–395.
12. Milde P, Hauser U, Simon R, Mall G, Ernst V, Haussler MR, Frosch P, Rauterberg EW. Expression of 1,25-dihydroxyvitamin D_3 receptors in normal and psoriatic skin. J Invest Dermatol 1991; 97: 230–239.
13. Reichrath J, Münssinger T, Kerber A, Rochette-Egly C, Chambon P, Bahmer FA, Baum HP. In situ detection of retinoid-X receptor expression in normal and psoriatic human skin. Br J Dermatol 1995; 133:168–175.
14. Smith EL, Walworth NC, Holick MF. Effect of 1α-25-dihydroxyvitamin D_3 on the morphologic and biochemical differentiation of cultured human epidermal keratinocytes grown under serum-free conditions. J Invest Dermatol 1986; 86:709–714.
15. Hosomi J, Hosoi J, Abe E, Suda T, Kuroki T. Regulation of terminal differentiation of cultured mouse epidermal cells by 1-alpha 25-dihydroxy-vitamin D_3. Endocrinology 1983; 113:1950–1957.
16. Gniadecki R, Serup J. Stimulation of epidermal proliferation in mice with 1 alpha, 25-dihydroxy-vitamin D_3 and receptor-active 20-EPI analogues of 1 alpha, 25-dihydroxyvitamin D_3. Biochem Pharmacol 1995; 49:621–624.
17. Rigby WFC. The immunobiology of vitamin D. Immunol Today 1988; 9:54–58.
18. Texereau M, Viac J. Vitamin D, immune system and skin. Eur J Dermatol 1992; 2:258–264.
19. Ranson M, Posen S, Mason RS. Human melanocytes as a target tissue for hormones: in vitro studies with 1α,25-dihydroxyvitamin D_3, alpha-melanocyte stimulating hormone, and beta-estradiol. J Invest Dermatol 1988; 91:593–598.
20. Christophers E, Henseler T. Psoriasis of early and late onset: characterization of two types of psoriasis vulgaris. J Am Acad Dermatol 1985; 13:450–456.
21. Christophers E, Henseler T. Patient subgroups and the inflammatory pattern in psoriasis. Acta Derm Venereol (Stockh) 1989; 69:88–92.
22. Elder JT, Henseler T, Christophers E, Voorhees JJ, Nair RP. Of genes and antigens: the inheritance of psoriasis. J Invest Dermatol 1994; 103:150S–153S
23. Valdimarsson H, Baker BS, Jonsdittir I, Fry L. Psoriasis: a disease of abnormal keratinocyte proliferation induced by T lymphocytes. Immunol Today 1986; 7:256–259.
24. Lee RE, Gaspari AA, Lotze MT, Chang AE, Rosenberg SA. Interleukin 2 and psoriasis. Arch Dermatol 1988; 124:1811–1815.
25. Barker JN, Jones ML, Mitra RS, Crockett Torab E, Fantone JC, Kunkel SL, Warren JS, Dixit VM, Nickoloff BJ. Modulation of keratinocyte derived interleukin-8 which is chemotactic for neutrophils and T lymphocytes. Am J Pathol 1991; 139:869–876.
26. Gottlieb AB. Immunologic mechanisms in psoriasis. J Invest Dermatol 1990; 95:18S–29S.

27. Schlaak JF, Buslau M, Jochum W, Hermann E, Girndt M, Gallati H, Meyer zum Büschenfelde KH, Fleischer B. T cells involved in psoriasis vulgaris belong to the Th1 subset. J Invest Dermatol 1994; 102:145–149.
28. van Reijsen FC, Druijnzeel-Koomen CAFM, Kalthoff FS, Maggi E, Romagnani S, Westland JKT, Mudde GC. Skin-derived aeroallergen-specific T-cell clones of Th2 phenotype in patients with atopic dermatitis. J Allergy Clin Immunol 1992; 90:184–192.
29. Leung DY, Walsh P, Giorno R, Norris DA. A potential role for superantigens in the pathogenesis of psoriasis. J Invest Dermatol 1993; 100:225–228.
30. Chowaniec O, Jablonska S, Beutner EH, Proniewska M, Jarzabek Chorzelska M, Rzesa G. Earliest clinical and histological changes in psoriasis. Dermatologica 1981; 163:42–51.
31. MacLaughlin JA, Gange W, Taylor D, Smith E, Holick MF. Cultured psoriatic fibroblasts from involved and uninvolved sites have partial but not absolute resistance to the proliferation-inhibition activity of 1,25-dihydroxyvitamin D_3. Proc Natl Acad Sci USA 1985; 82:5409–5412.
32. Morimoto S, Kumahara Y. A patient with psoriasis cured by 1α-hydroxyvitamin D_3. Med J Osaka Univ 1985; 35:3–4, 51–54.
33. Morimoto S, Yochikawa K, Kozuka T, Kitano Y, Imawaka S, Fukuo K, Koh E, Kumahara Y. An open study of vitamin D_3 treatment in psoriasis vulgaris. Br J Dermatol 1986; 115:421–429.
34. Holick MF, Chen ML, Kong XF, Sanan DK. Clinical uses for calciotropic hormones 1,25-dihydroxyvitamin D_3 and parathyroid hormone related peptide in dermatology: a new perspective. J Invest Dermatol (Symp Proc) 1996; 1:1–9.
35. Perez A, Chen TC, Turner A, Raab R, Bhawan J, Poche P, Holick MF. Efficacy and safety of topical calcitriol (1,25-dihydroxyvitamin D_3) for the treatment of psoriasis. Br J Dermatol 1996; 134: 238–246.
36. Kragballe K, Beck HI, Sogaard H. Improvement of psoriasis by topical vitamin D_3 analogue (MC 903) in a double-blind study. Br J Dermatol 1988; 119:223–230.
37. van de Kerkhof PCM, van Bokhoven M, Zultak M, Czarnetzki BM. A double-blind study of topical 1α-25-dihydroxyvitamin D_3 in psoriasis. Br J Dermatol 1989,120:661–664.
38. Perez A, Raab R, Chen TC, Turner A, Holick MF. Safety and efficacy of oral calcitriol (1,25-dihydroxyvitamin D_3) for the treatment of psoriasis. Br J Dermatol 1996; 134:1070–1078.
39. Kragballe K, Gjertsen BT, de Hoop D, Karlsmark T, van de Kerhof PCM, Larko O, Nieboer C, Roed-Petersen J, Strand A, Tikjob B. Double-blind right/left comparison of calcipotriol and betamethasone valerate in treatment of psoriasis vulgaris. Lancet 1991; 337:193–196.
40. Serup J. Calcipotriol irritation: mechanism, diagnosis and clinical implication. Acta Derm Venereol (Stockh) 1994; 186:42S (abstract).
41. Green C, Ganpule M, Harris D, Kavanagh G, Kennedy C, Mallett, R, Rustin M, Downes N. Comparative effects of calcipotriol (MC 903) solution and placebo (vehicle of MC 903) in the treatment of psoriasis of the scalp. Br J Dermatol 1994; 130:483–487.
42. Petrow W. Treatment of a nail psoriasis with calcipotriol. Akt Dermatol 1995; 21:396–400.
43. Saggese G, Federico G, Battini R. Topical application of 1,25 dihydroxyvitamin D_3 (calcitriol) is an effective and reliable therapy to cure skin lesions in psoriatic children. Eur J Pediatr 1993; 152: 389–392.
44. Perez A, Chen TC, Turner A, Holick MF. Pilot study of topical calcitriol (1,25-dihydroxyvitamin D_3) for treating psoriasis in children. Arch Dermatol 1995; 131:961–962.
45. Grossman RM, Thivolet J, Claudy A, Souteyrand P, Guilhou JJ, Thomas P, Amblard P, Belaich S, de Belilovsky C, de la Brassinne M, et al. A novel therapeutic approach to psoriasis with combination calcipotriol ointment and very low-dose cyclosporine: a result of a multicenter placebo-controlled study. J Am Acad Dermatol 1994; 31:68–74.
46. Kerscher M, Volkenandt M, Plewig G, Lehmann P. Combination phototherapy of psoriasis with calcipotriol and narrow band UVB. Lancet 1993; 342:923.
47. Cambazard, van de Kerkhof PCM, Hutchinson PE, and the Calcipotriol Study Group. In: Proceedings of the 3rd International Calcipotriol Symposium, Munich, March 23, 1996.
48. Ortonne JP. Calcipotriol in combination with betamethasone diproprionate. Nouv Dermatol 1994; 13: 736–751.
49. Kragballe K. Combination of topical calcipotriol (MC 903) and UVB radiation for psoriasis vulgaris. Dermatologica 1990; 181:211–214.
50. Lucker GP, van de Kerkhof PC, van Dijk MR, Steijlen PM. Effect of topical calcipotriol on congenital ichthyosis. Br J Dermatol 1994; 131:546–550.

51. Humbert P, Dupond JL, Agache P, Laurent R, Rochefort A, Drobacheff C, de Wazieres B, Aubin F. Treatment of scleroderma with oral 1,25-dihydroxyvitamin D_3: evaluation of skin involvement using non-invasive techniques. Results of an open prospective trial. Acta Derm Venereol (Stockh) 1993; 73:449–451.
52. Koeffler HP, Hirji K, Itri L. 1,25-Dihydroxyvitamin D_3: in vivo and in vitro effects on human preleukemic and leukemic cells. Cancer Treat Rep 1985; 69:1399–1407.
53. Colston KW, Chander SK, Mackay AG, Coombes RC. Effects of synthetic vitamin D analogues on breast cancer cell proliferation in vivo and in vitro. Biochem Pharmacol 1992; 44:693–702.
54. Franceschi RT, Linson CJ, Peter CT, Romano PR. Regulation of cellular adhesion and fibronectin synthesis by $1\alpha,25$-dihydroxyvitamin D_3. J Biol Chem 1987; 262:4165–4171.
55. Eisman JA, Barkla DH, Tutton PJM. Suppression of in vivo growth of human cancer solid tumor xenografts by $1\alpha,25$-dihydroxyvitamin D_3. Cancer Res 1987; 47:21–25.
56. Reichrath J, Müller SM, Kerber A, Baum HP, Bahmer FA. Biological effects of topical calcipotriol (MC 903) treatment in psoriatic skin. J Am Acad Dermatol 1997; 36:19–28.
57. Chen ML, Perez A, Sanan DK, Heinrich G, Chen TC, Holick MF. Induction of vitamin D receptor mRNA expression in psoriatic plaques correlates with clinical response to 1,25-dihydroxyvitamin D_3. J Invest Dermatol 1996; 106:637–641.
58. Missero C, Calautti E, Eckner R, Chin J, Tsai LH, Livingston DM, Dotto GP. Involvement of the cell-cycle inhibitor Cip1/WAF1 and the E1A-associated p300 protein in terminal differentiation. Proc Natl Acad Sci USA 1995; 92:5451–5455.
59. Binderup L, Latini S, Binderup E, Bretting C, Calverley M, Hansen K. 20-Epi-vitamin D_3 analogues: a novel class of potent regulators of cell growth and immune response. Biochem Pharmacol 1991; 42: 1569–1575.
60. Neef G, Kirsch G, Schwarz K, Wiesinger H. Menrad A, Fähnrich M, Thieroff-Eckerdt R, Steinmeyer A. 20-Methyl vitamin D analogues. In: Vitamin D. A Pluripotent Steroid Hormone: Structural Studies, Molecular Endocrinology and Clinical Applications. Norman AW, Bouillon R, Thomasset M, eds. Berlin: Walter de Gruyter, 1994; 97–98.
61. Schönecker B, Reichenbächer M, Gliesing S, Prousa R, Wittmann S, Breiter S, Thieroff-Eckerdt R, Wiesinger H, Haberey M, Scheddin D, Mayer H. 2β-Substituted calcitriols and other A-ring substituted analogues—synthesis and biological results. In: Vitamin D. A Pluripotent Steroid Hormone: Structural Studies, Molecular Endocrinology and Clinical Applications. Norman AW, Bouillon R, Thomasset M, eds. Berlin: Walter de Gruyter, 1994; 99,100.
62. Schuster I, Herzig G, Vorisek G. Steroidal hormones as modulators of vitamin D metabolism in human keratinocytes. In: Vitamin D. A Pluripotent Steroid Hormone: Structural Studies, Molecular Endocrinology and Clinical Applications. Norman AW, Bouillon R, Thomasset M eds. Berlin: Walter de Gruyter, 1994;184–185.
63. Zhao J, Marcelis S, Tan BK, Verstuaf A, Boillon R. Potentialisation of vitamin D (analogues) by cytochrome P-450 enzyme inhibitors is analog- and cell-type specific. In: Vitamin D. A Pluripotent Steroid Hormone: Structural Studies, Molecular Endocrinology and Clinical Applications. Norman, AW, Bouillon R, Thomasset M, eds. Berlin: Walter de Gruyter, 1994; 186,187.
64. Carlberg C, Bendik I, Wyss A, Meier E, Sturzenbecker LJ, Grippo JF, Hunziker W. Two nuclear signalling pathways for vitamin D. Nature 1993; 361:657–660.
65. Schräder M, Müller KM, Becker-Andre M, Carlberg C. Response element selectivity for heterodimerization of vitmain D receptors with retinoic acid and retinoid X receptors. J Mol Endocrinol 1994; 12:327–339.
66. Smith EL, Pincus SH, Donovan L, Holick MF. A novel approach for the evaluation and treatment of psoriasis. J Am Acad Dermatol 1988; 19:516–528.

22 Epidemiology of Cancer Risk and Vitamin D

Cedric F. Garland, Frank C. Garland, and Edward D. Gorham

1. INTRODUCTION

Even though great differences exist in incidence and mortality rates of colon cancer *(1)* and breast cancer *(2,3)* according to latitude of residence, the etiologies of these cancers remain largely unexplained. Two other cancers, ovarian *(4)* and prostate *(5–7)*, are also inversely related to latitude. Profound changes in incidence and mortality rates from colon and breast cancer also occur in immigrants from low-risk to high-risk areas. The rise in incidence rates occurs in the immigrants themselves for colon cancer *(8)*, but mainly in their daughters for breast cancer *(9)*.

The defining example of a disease due to vitamin D deficiency is rickets *(10,11)*. Although rickets was recognized as a disease in the medical literature for nearly 500 years *(12,13)*, it was not until 1890 that Theodore Palm *(10)* showed that it was associated with sunlight deficiency. Palm indicated variations in prevalence rates of rickets on a map. The map revealed that latitude was the major correlate of the disease, with some contribution of urbanization and industrialization. Rickets occurred mostly at latitudes of 37° or greater, with particularly high prevalence in urban areas such as London, Manchester, the industrialized British midlands, and the Netherlands. It was one of the earliest examples of a noninfectious disease that had a highly specific etiology. Ultraviolet (UV) deficiency in the dark, polluted urban canyons of northern cities was identified as the cause and was separated from other suspected factors such as poverty, heredity, and overall malnutrition *(11)*. Later investigators discovered that UV was needed for synthesis of the compound that prevented rickets, which was later isolated as vitamin D *(14,15)*.

The possible role of UV in risk of colon cancer was not considered until other explanations were evaluated *(16–18)*. Since these alternatives have been explored in epidemiologic studies, it has become clear that differences in dietary intake of fiber or fat cannot account for the strong association between latitude and these cancers. Other explanations are necessary.

Since colon *(1)* and breast cancer *(2,3)* mortality rates vary according to UV-absorbing sulfate air pollution (acid haze) *(19)* as well as latitude, UV deficiency may be the common factor in a chain of causation. If so, the analogy with rickets will be close, since death rates from rickets in England rose with the increasing use of high-sulfur-content coal from Newcastle during the 17th century *(19)*.

From: *Vitamin D: Physiology, Molecular Biology, and Clinical Applications*
Edited by: M. F. Holick © Humana Press Inc., Totowa, NJ

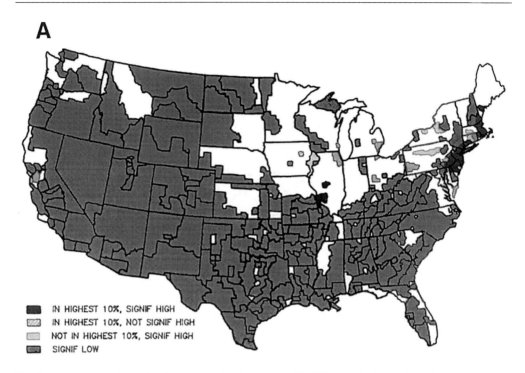

Fig. 1. Annual age-adjusted colon cancer death rates per 100,000 population, selected areas, women, 1986–1990. Category includes cancer of the rectum. Data are from the World Health Organization *(64)* and National Cancer Institute *(65)*. Rates for countries were directly age-adjusted to the world standard population. Rates for states were directly age-adjusted to the 1970 US standard population and normalized to the world standard population to allow comparisons. Data for Mexico were for 1983–1983; data for Guatemala were for 1984. Belg, Belgium; CT, Connecticut; Eng, England and Wales; Ger, Germany; MA, Massachusetts, USA; NH, New Hampshire, USA; NJ, New Jersey, USA; RI, Rhode Island, USA.

2. COLON CANCER

2.2. Latitude

Mortality rates of colon cancer in the United States are highest at northern latitudes, where the fall, winter, and spring UVB flux is far weaker than at lower latitudes *(1)* (Fig. 1). Death rates are also high in areas with high atmospheric concentrations of the acid haze air pollutants that inhibit transmission of solar UVB *(3,19)*. In the United States, age-adjusted mortality rates in the heavily polluted and less sunny industrial northeastern states are about one-third higher than in sunnier states, such as Hawaii, Arizona, and New Mexico. Age-adjusted annual mortality rates from colon cancer rise by approx 3 in 100,000 population/$10°$ of latitude within the United States (Fig. 1).

2.3. Studies of Immigrants

Studies of immigrants have helped to isolate environmental from genetic factors. Migration from a high latitude (New York) to a sunny latitude (Florida), even during adulthood, is associated with a decline in the risk of colon cancer *(20)*. Residence at a sunny latitude during childhood and adolescence confers reduction in risk that persists throughout life. For example, immigrants from Mexico and Puerto Rico to the northern

Chapter 22 / Epidemiology of Cancer Risk and Vitamin D

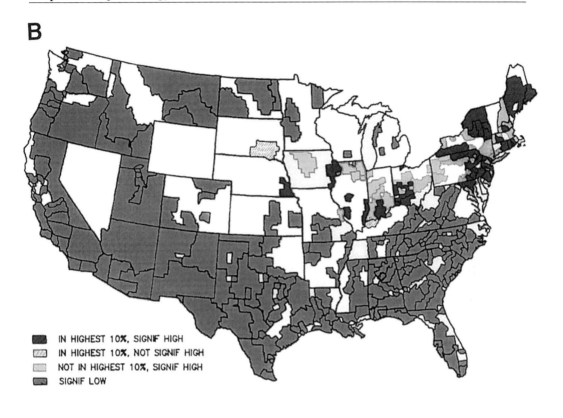

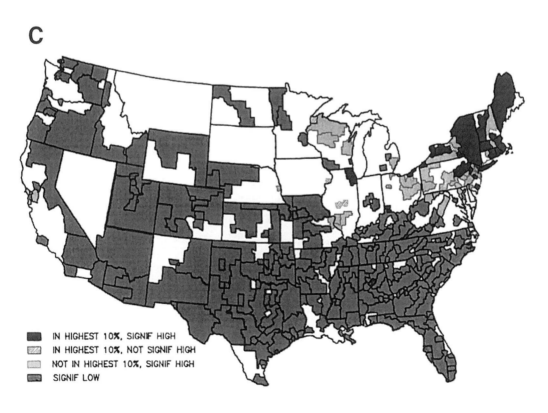

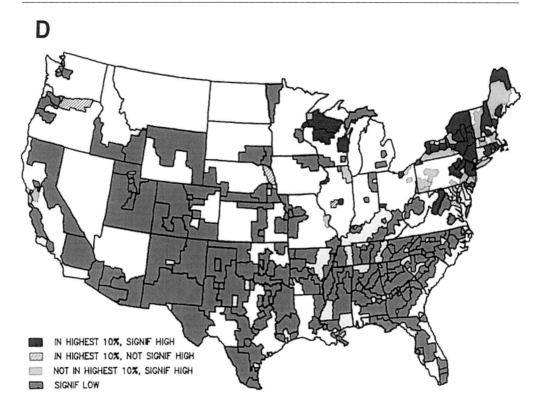

United States have age-specific colon cancer death rates that are about half those of persons born in the northern United States (21). The advantage of originating in a sunny location is usually lost or greatly attenuated after a generation has passed in the new area of residence.

2.4. Observational Studies

The association between dietary intake of vitamin D and incidence of colon cancer was reported in a cohort study of 1954 men followed for 19 yr (22). Participants who consumed ≥150 IU/d of vitamin D at baseline had about half the incidence rate of colon cancer as consumers of less.

A 12-yr prospective study of 89,448 female registered nurses based on a food-frequency questionnaire included 501 incident cases of colorectal cancer (396 colon and 105 rectal cancers). Relative risks and 95% confidence intervals (CIs) were calculated using the lowest quintile of intake as a reference. The relative risk for the highest versus the lowest category of intake of total vitamin D was 0.42 (95% CI 0.19–0.91). When analyses were limited to those who reported no change in intake of foods rich in vitamin D, the relative risk for colorectal cancer was 0.33 (95% CI 0.16–0.70). Incidence was reduced by two-thirds in association with dietary vitamin D intake (23).

A 6-yr prospective study of 47,935 professional men found that the incidence rate of colon cancer in those in the highest quintile of vitamin D intake was about half that of those in the lowest quintile (24).

2.5. Circulating Vitamin D Metabolites and Risk of Colon Cancer

Exposure to UVB causes a transient increase in circulating vitamin D, followed by a more persistent rise in the principal circulating metabolite, 25-hydroxyvitamin D *(25–27)*. The persistence of 25-hydroxyvitamin D makes it the optimal marker of vitamin D status. A nested case–control study was conducted using subjects from the Johns Hopkins Operation Clue cohort. This cohort consisted of 25,620 health adult residents of Washington County, MD, who provided samples of serum in 1974–1975. Serum samples were thawed for all cases of colon cancer and for two controls per case and matched for age, race, sex, county of residence, and date of serum collection. Sera were analyzed blindly for 25-hydroxyvitamin D. Individuals whose 25-hydroxy vitamin D levels were >20 ng/mL had one-third the risk of colon cancer compared with those with lower concentrations ($p < 0.05$) *(28)*.

Since the association was detected within the first decade of follow-up, the principal role of vitamin D in colon cancer may be during the promotional or decoupling phase *(see below)* of carcinogenesis. In this cohort, for example, most of the association of serum 25-hydroxyvitamin D with incidence appeared during the first decade of follow-up *(29)*.

2.6. Potentially Confounding Factors

Factors other than sunlight and vitamin D are also believed to contribute to risk of colon cancer. Three factors have been examined in detail: genetic predisposition, dietary fiber, and dietary fat.

Genetic predisposition has long been known to increase risk of colon cancer, specifically in patients who have Gardner syndrome/familial polyposis *(30)*. These well-known genetic disorders are associated with colon polyps, and the mutations involved have been described *(31,32)*. There are also pedigrees in which there is a high degree of aggregation of nonpolyposis colon cancer, and a genetic basis for a nonpolyposis cancer family syndrome has been identified *(33)*.

Despite the increased risk of colon cancer in individuals with known or suspected genetic syndromes, colon cancer has relative mild familial aggregation compared with diseases of known genetic etiology. Eighty to 90% of colon cancer occurs in people with no first-degree relative (parent, sibling, or child) with the disease.

Disorders that do not occur considerably more often in close relatives than in the population are usually of environmental cause. Even when incidence of a disease is greater than expected in close relatives, shared diet and environment within the family need to be considered.

Dietary factors have long been thought to influence risk of colon cancer. The most prominent is the dietary fiber hypothesis that grew out of observations of low rates of colon cancer in Africa *(18)*. The traditional African diet consists mainly of fibrous carbohydrates, such as roots and tubers, suggesting that fiber could be a factor in the low risk of colon cancer in Africans. Some epidemiologic studies have reported a moderate benefit of fiber, but others have been mixed, and some have shown no benefit *(34)* or a mild adverse effect *(35)* on risk of colon cancer.

An alternative explanation for high rates of colon cancer in northern parts of the United States is that populations living in northern states consumed more fat than populations living at latitudes closer to the equator *(17)*. This hypothesis was consistent with studies in rodents exposed to dietary fat and strong chemical carcinogens *(16)*. However intake of dietary fat does not vary by region in the United States *(1)*. In one of the longest cohort

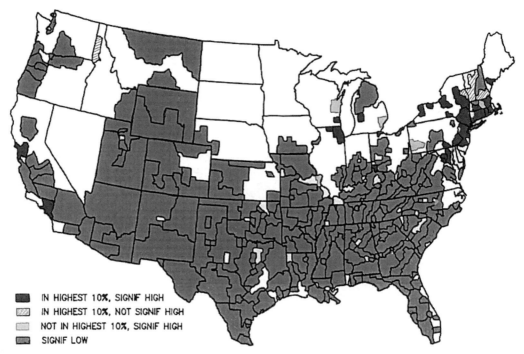

Fig. 2. Annual age-adjusted breast cancer death rate per 100,000 population, selected areas, women, 1986–1990. Sources: World Health Organization *(64)* and National Cancer Institute *(65)*. Rates for countries were directly age-adjusted to the world standard population. Rates for states were directly age-adjusted to the 1970 US standard population and normalized to the world standard population to allow comparisons. Data for Mexico were for 1983–1983; data for Guatemala were for 1984. Bel, Belgium; CT, Connecticut, USA; Eng, England and Wales; Ger, Germany; Ir, Ireland; MA, Massachusetts, USA; N, Netherlands; NJ, New Jersey, USA; R, Rhode Island, USA. Some points are too closely clustered to label clearly.

studies to date, baseline dietary intake of fat was identical (42% of energy) in participants who developed colon cancer and those who did not during a 19-yr period of follow-up *(22)*.

3. BREAST CANCER

3.1. Latitude

Age-adjusted breast cancer mortality rates, like those of colon cancer, follow a latitude gradient in the United States (Fig. 2) *(2)*. Like colon cancer, death rates from breast cancer in North America are highest in the northeastern and New England states and Canada, where sunlight intensity and UVB levels are relatively low during most of the year. In these areas, age-adjusted mortality rates from breast cancer tend to be about 40% higher than in the sunniest state in the United States, Hawaii, and considerably higher than in the sunny states of the southwest (Arizona and New Mexico).

As with colon cancer, climatologic factors and air pollution are also associated with breast cancer mortality rates. Places at high latitudes (generally above approx 37°) with high levels of air pollution from acid haze (mainly ammonium sulfate) aerosols are at the highest risk of breast cancer *(19)*. The need to cover the skin with clothing to protect against cold may also contribute to reduction in synthesis of vitamin D.

Acid haze concentrations in the atmosphere are associated with higher death rates from breast cancer in polluted eastern Canadian provinces than in provinces with clean air. In eastern Canada and the northeastern United States, soft coal with high sulfur content is burned extensively for electricity generation, heating, and smelting, resulting in release of large quantities of the combustion product sulfur dioxide. Sulfur dioxide is rapidly chemically transformed to a persistent sulfate aerosol that acts as a filter layer capable of scattering and absorbing in the atmosphere the UVB photons that have penetrated the stratospheric ozone layers *(19)*.

The association of latitude with breast cancer death rates is dependent on age, with the strongest association during the postmenopausal period (generally, age 50 yr and older). The slopes of breast cancer mortality rate by latitude were obtained for four age groups in Italy *(36)*. The slope of age-specific breast cancer mortality with latitude is nearly flat at ages 25–44 yr, but rises to six deaths per degree of latitude at ages 65–79 yr.

As with other types of cancer, incidence rates of breast cancer are available for only a few areas in the world that maintain full-coverage population-based cancer registries. The largest geographic area with complete registry coverage includes the 10 republics of the Soviet commonwealth (the former USSR). Age-adjusted incidence rates of breast cancer in the northern republics are approximately twice those in the southern, with intermediate rates at intermediate latitudes *(37)*.

Unlike mortality rates, age-adjusted annual breast cancer incidence rates that are available for 10 areas of the United States do not reveal a readily discernible association with latitude *(38)*. This suggests that the association of high latitude with high mortality rates of breast cancer is mainly due to higher case-fatality rates at higher latitudes.

4. OVARIAN CANCER

Although less prominent than the associations for colon and breast cancer, age-adjusted mortality rates of ovarian cancer also follow a latitude gradient within the United States, with the strongest effect present for deaths during the perimenopausal years (45–54 yr) *(4)*.

5. PROSTATE CANCER

Incidence rates of prostate cancer follow a gradient with UV light intensity within the United States *(5)*. Mortality rates in whites are also higher in northern counties than in southern, with rates in the range of 26 in 100,000 in the most northern areas compared with 18–19 in the most southern *(6,7)* (Fig. 3).

Prostate cancer is rare in sub-Saharan Africa but common in blacks living in the United States *(38)*. Two nested case–control studies have found no association between prediagnostic levels of 25-hydroxyvitamin D and incidence of prostate cancer *(39,40)*. All the findings combined suggest that the favorable association of UV levels with mortality rates *(6,7)* is the result of a reduction in case-fatality rates of prostate cancer rather an effect on incidence rates.

6. LATITUDE AND VITAMIN D STATUS

Latitude is the principal determinant of the UVB flux and is a useful proxy indicator of the probable vitamin D status of a population (Tables 1 and 2). Latitude has a particularly

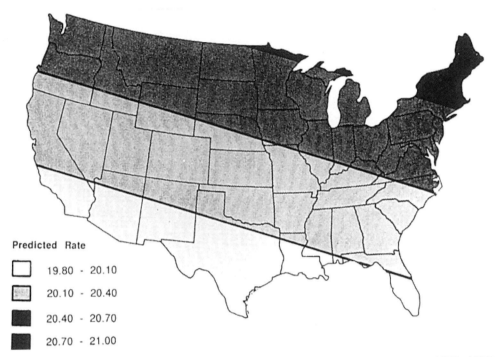

Fig. 3. Age-adjusted prostate cancer mortality rates per 100,000 population, whites, 1973–1987 (smoothed) (Data from refs. 6 and 7.)

strong effect on UVB, since this portion of the solar spectrum is far more subject than visible light or infrared energy to attenuation while penetrating the atmosphere. Apart from thickness of the stratospheric ozone layers, the maximum daily solar elevation relative to the horizon is of critical importance in determining UVB flux. This angle determines the length of the slant path of photons through the atmosphere. It is along this path that UVB photons are absorbed and scattered; the longer the path, the greater the loss.

Circulating levels of 25-hydroxyvitamin D are considerably higher in populations living at lower latitudes (such as Florida, 26°N) compared with those at higher latitudes (such as Finland, 61°N) *(41)*. The mean springtime concentration of 25-hydroxy vitamin D in healthy women is 36 ng/mL in Florida but only 15 ng/mL in Finland *(41)*. The median winter value in normal adults in Middlesex, England, historically in the rickets belt, is 10 ng/mL, which is the upper end of the range in which rickets may develop in children *(42)*. Circulating 25-hydroxyvitamin D concentrations were similarly low in children in Middlesex. These values contrast markedly with a median 25-hydroxyvitamin D concentration of 45 ng/mL in healthy adults in San Diego, California (C. Garland, unpublished data, 1996).

6.1. Survival Rates

Currently the only uniform source for cancer survival rates over a large geographic range is the Surveillance Epidemiology and End Results (SEER) registry of the National Cancer Institute *(38)*. Nine cancer registries in the United States report survival rates to SEER. The most recent data available from SEER are described below.

Table 1
Annual Age-Adjusted Death Rates from Colon Cancer per 100,000 Population by Latitude of Residence for Women in Selected Areas, 1986–1990[a]

Country	Latitude (°)	Death rate per 100,000 population
Northern Ireland	54	16.4
Ireland	53	16.6
England and Wales	52	15.3
Netherlands	52	14.7
Germany	51	16.5
Belgium	50	15.5
Austria	47	15.2
Switzerland	47	12.2
France	46	11.2
Canada	45	13.5
New Hampshire, USA	44	11.5
New York, USA	43	12.4
Connecticut, USA	42	11.5
Rhode Island, USA	42	12.2
Massachusetts, USA	42	12.1
Italy	42	10.5
New Zealand	41	19.7
New Jersey, USA	40	12.9
Spain	40	7.8
Greece	39	5.2
Japan	36	9.3
New Mexico, USA	34	9.1
Arizona, USA	34	8.8
Australia	33	15.8
Israel	31	11.8
Chile	30	6.1
Florida, USA	28	9.9
Mexico[b]	23	2.7
Hawaii, USA	20	8.5
Guatemala[c]	15	0.5

[a] Category includes cancer of the rectum. Data for countries were from the World Health Organization (64) and refer to rates directly adjusted using a world standard population. Data for states within the United States were for 1986–1990, directly age-adjusted to 1970 US Standard Population (65) and normalized to the world standard population to allow comparison with other areas.
[b] Latest data available for Mexico were for 1982–1983.
[c] Latest data available for Guatemala were for 1984.

6.2. Colon Cancer

Hawaii is at the lowest latitude of any SEER cancer registry (18° N), and the 5-yr relative case-fatality rate from colon cancer in whites is lowest in Hawaii (32%) of the nine cancer registries reporting survival rates to SEER (overall mean for all registries, 38%) (38). Hawaii also has the lowest case-fatality rate from colon cancer for all races combined. No confounding factor has been identified that could account for the difference.

Table 2
Annual Age-Adjusted Death Rates from Breast Cancer per 100,000 Population by Latitude of Residence for Women in Selected Areas, 1986–1990[a]

Country	Latitude (°)	Death rate per 100,000 population
Northern Ireland	54	26.9
Ireland	53	25.7
England and Wales	52	29.0
Netherlands	52	25.8
Germany	51	21.9
Belgium	50	25.6
Austria	47	22.0
Switzerland	47	24.9
France	46	19.0
Canada	45	23.5
New Hampshire, USA	44	25.0
New York, USA	43	25.6
Connecticut, USA	42	23.6
Rhode Island, USA	42	25.7
Massachusetts, USA	42	25.0
Italy	42	20.4
New Zealand	41	25.0
New Jersey, USA	40	25.8
Spain	40	15.0
Greece	39	15.1
Japan	36	5.8
New Mexico, USA	34	19.4
Arizona, USA	34	20.0
Australia	33	20.5
Israel	31	22.5
Chile	30	12.7
Florida, USA	28	20.9
Mexico[b]	23	6.3
Hawaii, USA	20	15.0
Guatemala[c]	15	2.3

[a]Category includes cancer of the rectum. Data for countries were from the World Health Organization (64) and refer to rates directly adjusted using a world standard population. Data for states within the United States were for 1986–1990, directly age-adjusted to 1970 US Standard Population (65) and normalized to the world standard population to allow comparison with other areas.
[b]Latest data available for Mexico were for 1982–1983.
[c]Latest data available for Guatemala were for 1984.

6.3. Breast Cancer

The 5-yr relative case-fatality rate from breast cancer in whites is lowest in Hawaii (9%) of all cancer registries currently reporting survival rates to SEER (overall mean 16%) (38). Hawaii also has the lowest case-fatality rate from breast cancer for all races combined. Stated in another way, the risk of dying from breast cancer within 5 yr of diagnosis is slightly more than half as high in Hawaii on the average as in the registry areas located in the continental Unites States. As with colon cancer, no confounding factor has been identified that could account for the improved survival.

6.4. Ovarian Cancer

As with colon and breast cancer, 5-yr case-fatality rates in whites are lower in Hawaii (50%) than for any other registry in the United States (mean for all 56%). This difference is mainly caused by better survival in postmenopausal cases in Hawaii *(38)*. A similar pattern is present in other racial groups.

6.5. Prostate Cancer

As with the other cancers described in this chapter, 5-yr case-fatality rates for prostate cancer in whites are lower in Hawaii (7%) than in any other registry area (mean for all 12%) *(38)*. The risk of a white man dying from prostate cancer within 5 yr of diagnosis is slightly more than half as high in Hawaii as in the continental United States. Other racial groups have case-fatality rates that are similar to those found elsewhere in the United States. The latter may be due to delays in detection associated with lower socioeconomic status of nonwhite Hawaiians.

7. ROLE OF AIR POLLUTION

Some types of air pollution reduce the transmission of UVB. The greatest filtering of UVB by air pollution occurs for sulfate particles and sulfuric acid aerosols. These aerosols have been termed acid haze *(19)*. Acid haze is the precursor of acid precipitation *(19)*. Acid haze aerosols consist of ammonium sulfate crystals in a range of sizes and sulfuric acid droplets, both having a median diameter of 300 nm *(43)*. These particulates and droplets absorb and scatter UVB energy *(44–46)*. Sulfuric acid droplets that result from the oxidation of sulfur dioxide in the presence of water may remain aloft in the atmosphere for several days. When ammonia is present in the atmosphere, the sulfuric acid reacts with it to form ammonium sulfate, which may remain in the atmosphere for 1–2 wk in the absence of precipitation.

Scattering of solar radiation depends on particle size, and maximal scatter occurs when the diameter of the particle is identical to the wavelength of the energy *(47)*. The effect of sulfate aerosols therefore is greater in the UVB range than in other parts of the UV spectrum. Due to their median 300-nm diameter, acid haze particles scatter far more light than the same mass per unit volume of large particles, and scatter far more UV away from the earth's surface (backscatter) than large particles (Fig. 4) *(47)*.

Almost all solar UVB is absorbed by stratospheric ozone, except in polar ozone discontinuities *(48,49)*. Only a small proportion of UVB penetrates the ozone layer under normal circumstances. If pollution is present, some of the photons that penetrate to the troposphere are absorbed by sulfur dioxide in very polluted environments, since this gaseous pollutant has a UV absorption spectrum nearly identical to that of ozone *(50)*.

These data have important epidemiologic implications, since absorption by the atmosphere of UVB photons is much greater in the eastern United States than in other parts of the country, and is much higher in much of England and northern Europe than in surrounding regions *(51)*. The source of the air pollution in the northeastern United States is the combustion of high-sulfur coal in the Ohio and Tennessee valleys, which is

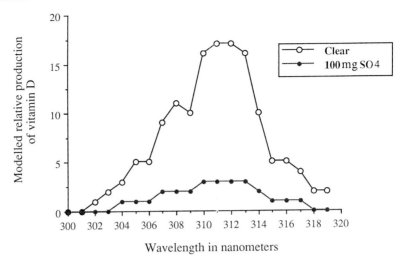

Fig. 4. Predicted influence of acid haze sulfate air pollution on synthesis of vitamin D, according to UV transmission model taking into account latitude, season, time of day, atmospheric ozone, and molecular scattering. The area under the two curves corresponds to vitamin D synthesis under clear conditions and in the presence of acid haze. The data were obtained by modeling loss of energy from 300 to 310 nm using backscattering and absorption coefficients for acid haze particles in polluted air compared with transmission through clear air, taking into account latitude, time of day, and Rayleigh scatter, and assuming a 1000-foot boundary layer. The modeled photon flux was convoluted with the action spectrum for cutaneous synthesis of vitamin D to produce the points plotted in the curves *(66)*. The difference in area under the curves represents the quantity of vitamin D not synthesized under polluted conditions. The model assumed a relatively nonabsorptive clean aerosol; if the aerosol were less clean and more absorptive, the reduction in vitamin D synthesis could be much greater.

transported to the northeast by winds. The effects of air pollution are accentuated by moisture and formation of clouds; moisture from arctic masses dominates the climate of much of the northeastern United States, New England, and the British Isles during the winter months, also reducing transmission of solar UV *(49)*. Finally, a measurable (although considerably smaller) proportion of loss of solar UV energy penetrating the stratosphere occurs through Rayleigh scatter, which increases as the reciprocal of the fourth power of the wavelength. This form of scatter is about 10 times greater for terrestrial UVB (wavelength 295–315 nm) than for visible light (median wavelength 550 nm) *(52)*. A considerable proportion of Rayleigh scattering is in the forward direction, but some energy is reflected away from the earth's surface and the optical path that the UVB photon must follow is increased at the low elevations of the sun above the horizon characteristic of winters at high latitudes.

8. IMPORTANCE OF 25-HYDROXYVITAMIN D IN EPIDEMIOLOGIC UV STUDIES

Different levels of UV exposure result in different levels of 25-hydroxyvitamin D in the serum. Except at the extremes, though, serum levels of 1,25-dihydroxyvitamin D are not affected.

Therefore the epidemiologic associations of UV with the cancers described in this chapter are mainly related to variation among individuals or populations in the concentration of 25-hydroxyvitamin D.

It may be that 25-hydroxyvitamin D acts as an analog of 1,25-dihydroxyvitamin, producing some of the same responses as 1,25-dihydroxyvitamin D. The concentration of 25-hydroxyvitamin D must be markedly higher than that of 1,25-dihydroxyvitamin D to produce the same effect. In human serum, and apparently in the extra cellular fluid compartment in general, the concentration of 25-hydroxyvitamin D is typically approx 900 times greater than that of 1,25-dihydroxyvitamin D, allowing 25-hydroxyvitamin D to have an influence despite its being less efficacious per molecule. For these reasons, epidemiologic studies have focused on 25-hydroxyvitamin D as the main metabolite of interest.

9. ROLE OF VITAMIN D IN COLON AND BREAST CANCER: MODELING THE POSSIBLE CHAIN OF CAUSATION

The causal path of action of sunlight in the four types of cancer described in this chapter is still uncertain, but might occur as follows: (1) solar UVB stimulates synthesis of vitamin D precursors in the skin *(25,26)*; (2) vitamin D is translocated from the skin to the serum and converted to 25-hydroxyvitamin D in the liver *(25)*; and (3) 25-hydroxyvitamin D circulates to the tissues and enters the cells *(53)*. These steps are all well documented. The next steps involve mutual adhesion of epithelial cells.

Low intracellular calcium concentration reduces mutual cellular adhesion *(54,55)*. The low calcium level in tumor cells is unique and is due to decreased calcium uptake by these cells *(56)*. To continue the steps above, (4) vitamin D metabolites help to maintain differentiation of the epithelium by increasing the expression of intercellular junctions, e.g., in the breast *(57)*; (5) intercellular junctions bind the epithelial cells tightly to each other; (6) the high degree of binding between the differentiated cells tends to inhibit proliferation (by contact inhibition) *(58)*; (7) a similar growth inhibition exists when there are junctional contacts between normal and transformed cells *(58–61)*; (8) the binding also tends to immobilize the cells into a single plane (in the colon) or a few layers in the other tissues affected by vitamin D-induced differentiation, limiting movement to a relatively coordinated flow of cells atop the basement membrane; and (9) the resulting reduced mobility reduces the chances for meaningful competition among cells for local resources (glucose, oxygen), reducing the opportunity for natural selection of aggressive cells within the epithelium.

None of these steps is deterministic, but all may be needed for expression of malignancy. The data on survival suggest that vitamin D may also play some role in metastasis, since death from these cancers occurs mainly as a result of distant metastases (e.g., to liver, bone, or brain).

Decreased mutual adhesiveness is tumors associated with invasiveness into surrounding tissues *(55)*. Well-differentiated cells expressing normal intercellular junctions of the degree associated with stimulation of vitamin D metabolites tend to clump, at least in tissue culture. Such behavior may make budding off of daughter cells less likely for most cells. Under the influence of intercellular binding, daughter cells may remain adherent to the precursor cells and less likely to detach from the primary clump and enter the lymphatics, reducing the potential for metastasis.

Malignant cells typically have lost most junctional structures and have achieved a degree of autonomy from the surrounding epithelium. Even once metastases have occurred, the metastatic cells in contact with normal cells may have their growth inhibited by junctional contact with neighboring heterogeneous cells, increasing the likelihood that the metastases will remain latent and therefore asymptomatic *(61)*. For this reason, it has been suggested that an adjunct to other treatment of metastatic tumors would be to find strategies to promote junctional linkages between metastatic cells and surrounding normal cells *(61)*. The effect would not apply to all malignant cells, as some cancer cells are communication competent, and others are not *(62)*.

The net effects of the potential influence of vitamin D metabolites on the epithelium may be to maintain the integrity of the tumor mass sufficiently to reduce the evolution and continued selection of aggressive tumor cells. This could theoretically result in a less aggressive tumor with lower potential for metastasis.

The prodifferentiational effect of 25-hydroxyvitamin D could be due to *in situ* synthesis of the more potent metabolite, 1,25-dihydroxyvitamin D. Although the calcitropic effect of 25-hydroxy vitamin D is only 1/30th–1/1000 that of 25-hydroxyvitamin D, the circulating concentration of 25-hydroxyvitamin D, as described previously, is approx 900 times that of 1,25-dihydroxyvitamin D. By mass action, 25-hydroxyvitamin D might exert an effect at least as great as 1,25-dihydroxyvitamin D even if 1,25-hydroxyvitamin D is more effective per molecule *(53)*.

The foregoing is a theoretical framework for the possible influence of vitamin D on carcinogenesis. Since it is a model, many aspects remain to be tested. It is provided to serve as a means of organizing thinking about the possible effects of vitamin D metabolites on incidence and case-fatality rates of colon, breast, ovarian, and prostate cancers.

The value of the theoretical model is that it can help with placement of existing research into a context. The model differs from the classical model in that it incorporates decoupling, or loss of intercellular junctions, as an important feature, and emphasizes the role of vitamin D metabolites in regulation of this phenomenon.

10. SUMMARY

Intensity of solar UV exposure and intake of vitamin D from diet are associated with decreased mortality rates from colon and breast cancer and, to a lesser degree, from ovarian and prostate cancer. The circulating concentration associated with significantly low rates of colon cancer is approx 30 ng/mL. Higher (but physiologic and clearly nontoxic) serum levels of 25-hydroxyvitamin D may be associated with lower mortality rates from breast cancer, but this possibility has not been tested, and further research clearly is needed.

Levels of 25-hydroxyvitamin D of 50 or more ng/mL are common in populations regularly exposed to sunlight. Such levels can be achieved by daily exposure to sunlight for 15–20 min/d at noon, although the efficacy of the exposure depends greatly on latitude, season, climate, and air pollution. No exposures to sunlight during the winter in the northeastern United States are sufficient for synthesis of vitamin D in the winter, however *(63)*, and longer exposures may be required in the elderly. For individuals who cannot tolerate regular moderate sun exposure because of photosensitivity or inability to form pigment, vitamin D is also available from such sources as fatty fish or supplements.

REFERENCES

1. Garland C, Garland F. Do sunlight and vitamin D reduce the likelihood of colon cancer? Int J Epidemiol 1980; 9:227–231.
2. Garland F, Garland C, Gorham E, Young J Jr. Geographic variation in breast cancer mortality in the United States: a hypothesis involving exposure to solar radiation. Prev Med 1990; 19:614–622.

3. Gorham E, Garland F, Garland C. Sunlight and breast cancer incidence in the USSR. Int J Epidemiol 1990; 19:820–824.
4. Lefkowitz E, Garland C. Sunlight, vitamin D, and ovarian cancer mortality rates in U.S. women. Int J Epidemiol 1994; 23:1133–1136.
5. Hanchette C, Schwartz G. Geographic patterns of prostate cancer mortality: evidence for a protective effect of ultraviolet radiation. Cancer 1992; 70:2681–2689.
6. Kafadar X. Geographic trends in prostate cancer mortality. Ann Epidemiol 1997; 7:35–45.
7. Pommerenke F, Srivastava S. State cancer control map and data program targeting cancer at the local level. In: Cancer Prevention and Control. Greenwald P, Kramer B, Weed D, eds. New York: Marcel Dekker, 1995; 771–775.
8. Haenszel W, Kurihara M. Studies of Japanese migrants. J Natl Cancer Inst 1968; 40:43–68.
9. Buell P. Changing incidence of breast cancer in Japanese-American women. J Natl Cancer Inst 1973; 51:1479–1483.
10. Palm T. The geographic distribution of rickets. Practitioner 1890.
11. Park E. The etiology of rickets. Physiol Rev 1924.
12. Glisson F. De Rachitide, sive, Morbo Puerili: Qui Vulgo The Rickets Dicitur Tactatus, Editio Secunda, Priori Adcuratior Lorge, & Emendatior (Treatise on Rickets, 2nd ed.). London: Laurentii Sadler, 1660.
13. Whistler D. Disputatio Medica Inauguralis, de Morbo Puerili Anglorum, quem Patrio Idiomate Indigenae Vocant The Rickets. (University of Leiden Doctoral Dissertation). Oxford: Alexander Cooke, 1645 ca.
14. Hess A, Weinstock M, Hellman F. The antirachitic value of irradiated phytosterol and cholesterol. J Biol Chem 1925; 63:305–308.
15. Steenbock H, Black A. Fat soluble vitamins XVII. The induction of growth-promoting and calcifying properties of a ration by exposure to ultraviolet light. J Biol Chem 1924; 61:405–422.
16. Tannenbaum A. The genesis and growth of tumors: effects of a high-fat diet. Cancer Res 1942; 2:468.
17. Carroll K, Khor H. Effects of level and type of dietary fat on incidence of mammary tumors in female Sprague-Dawley rats by 7,12-dimethyl(a)anthracene. Lipids 1971; 6:415–420.
18. Cleave TL. The Saccharine Disease: Conditions Caused by the Taking of Refined Carbohydrates, such as Sugar and White Flour. Bristol, England: J Wright, 1974.
19. Gorham E, Garland C, Garland F. Acid haze air pollution and breast and colon cancer in 20 Canadian cities. Can J Public Health 1989; 80:96–100.
20. Ziegler R. Epidemiologic patterns of colorectal cancer. In: Important advances in oncology, 1986. DeVita VT HS, Rosenberg SA, eds. Philadelphia: JP Lippincott, 1986; 209–232.
21. Mallin K, Anderson K. Cancer mortality in Illinois Mexican and Puerto Rican immigrants, 1979–1984. Int J Cancer 1988; 41:670–676.
22. Garland C, Shekelle R, Barrett-Connor E. Dietary vitamin D and calcium and risk of colorectal cancer: a 19-year prospective study in men. Lancet 1985; 1:307–309.
23. Martinez ME, Giovannucci EL, Colditz GA, Stampfer MJ, Hunter DJ, Speizer FE, Wing A, Willett WC. Calcium, vitamin D, and the occurrence of colorectal cancer among women. J Natl Cancer Inst 1996; 88:1375–1382.
24. Kearney J, Giovanucci E, Rimm E, et al. Calcium, vitamin D, and dairy foods and the occurrence of colon cancer in men. Am J Epidemiol 1996; 143:907–917.
25. Holick M, MacLaughlin J, Clark M, et al. Photosynthesis of previtamin D_3 in human skin and the physiological consequences. Science 1980; 210:203–205.
26. Holick M. The cutaneous photosynthesis of previtamin D_3: a unique photoendocrine system. J Invest Dermatol 1981; 76:51–58.
27. Holick M. Photosynthesis of vitamin D in the skin: effect of environment and life-style variables. Fed Proc 1987; 46:1876–1882.
28. Garland C, Comstock G, Garland F, Helsing K, Shaw E, Gorham E. Serum 25-hydroxyvitamin D and colon cancer: eight-year prospective study. Lancet 1989; 2:1176–1178.
29. Braun MM, Helzlsouer KJ, Hollis BW, Comstock GW. Colon cancer and serum vitamin D metabolite levels 10-17 years prior to diagnosis. Am J Epidemiol 1995; 142:608–611.
30. Rustgi A. Hereditary gastrointestinal polyposis and non polyposis syndromes. N Engl J Med 1994; 331:1694–1702.
31. Bisgaard M, Fenger K, Bulow S, Niebuhr E, Mohr J. Familial adenomatous polyposis (FAP): frequency, penetrance, and mutation. Hum Mutat 1994; 3:121–125.

32. Maher E, Barton D, Slatter R, et al. Evaluation of molecular genetic diagnosis in the management of familial adenomatous polyposis coli: a population based study. J Med Genet 1993; 30:675–678.
33. Konishi M, Kikuchi-Yanoshita R, Tanaka K, et al. Molecular nature of colon tumors in hereditary nonpolyposis colon cancer, familial polyposis, and sporadic colon cancer. Gastroenterology 1996; 111:307–317.
34. Potter J, Slattery M, Bostick R, Gapstur S. Colon cancer: a review of the epidemiology. Epidemiol Rev 1993; 15:499–545.
35. Potter JD. Dietary fiber, vegetables and cancer. J Nutr 1988; 118:1591,1592.
36. Gonzalez D. Breast Cancer Death Rates in Italy. Independent Study Project Thesis. San Diego: University of California, San Diego, School of Medicine, 1996.
37. Gorham E, Garland F, Garland C. Sunlight and breast cancer incidence in the USSR. Int J Epidemiol 1990; 19:820–824.
38. Kosary C, et al. SEER Cancer Statistics Review, 1973–1992. Publication No. 96-2798. Bethesda: National Cancer Institute, 1996; 131,168.
39. Braun M, Helzlhauer K, Hollis B, Comstock G. Prostate cancer and prediagnostic levels of serum vitamin D metabolites. Cancer Causes Control 1995; 6:235–239.
40. Gann PH, Ma J, Hennekens CH, Hollis BW, Haddad JG, Stampfer MJ. Circulating vitamin D metabolites in relation to subsequent development of prostate cancer. Cancer Epidemiol Biomarkers Prev 1996; 5:121–126.
41. Punnonen R, Gillespy M, Hahl M, et al. Serum 25-OHD, vitamin A and vitamin E concentrations in healthy Finnish and Floridian Women. Int J Vit Nutr Res 1988; 58:37–39.
42. Preece M, O'Riordan J, Lawson D, Kodicek E. A competitive protein-binding assay for 25-hydroxycholecalciferol and 25-hydroxyergocalciferol in serum. Clin Chim Acta 1974; 54:235–242.
43. Sandberg JS LD, DeMandel RE, Siu W. Sulfate and nitrate particulates as related to SO2 and NOx gases and emissions. J Air Poll Control Assoc 1976; 26:559–564.
44. Waggoner A, Vanderpool A, Charlson R, et al. Sulfate light scattering as an index of the role of sulfur in tropospheric optics. Nature 1976; 261:120–122.
45. Waggoner A, Weiss RE, Ahlquist N, Covert D, Will S, Charlson R. Optical characteristics of atmospheric aerosols. Atmospheric Environ 1981; 15:1891–1909.
46. Leaderer BP Tanner R, Lioy PJ, Stolwijk JAJ. Seasonal variations in light scattering in the New York region and their relation to sources. Atmospheric Environ 1981; 15:2407–2420.
47. Lavery TF, .Hidy GM, Baskett RL, Mueller PK. The formation and regional accumulation of sulfate concentrations in the northeastern United States. In: Proceedings of a Conference on Environmental and Climatic Impact of Coal Utilization (Williamsburg VA, April 17–19, 1979). Singh JJ, Deepak A, eds. New York: Academic Press, 1980; 625–647.
48. McCartney EJ. Absorption and Emission by Atmospheric Gases: The Physical Processes. New York: Wiley, 1983.
49. Whitten RC, Prasad SS, eds. Ozone in the Free Atmosphere. New York: Van Nostrand Reinhold, 1985; 130,131.
50. Zerefos C, Mantis H, Bais A, Ziomas I, Zoumakis N. Solar ultraviolet absorption by sulfur dioxide in Thessaloniki, Greece. Atmosphere-Ocean 1986; 24:292–300.
51. Flowers E, McCormick R, Kurfis K. J Appl Meteorol 1969; 8:955–964.
52. Middleton W. Vision Through the Atmosphere. Toronto: University of Toronto Press, 1952; 6.
53. Puzas J, Brand J. *In vitro* uptake of vitamin D metabolites: culture conditions determine cell uptake. Calcif Tissue Int 1985; 37:474–477.
54. Zeidman L. Chemical factors in the mutual adhesiveness of epithelial cells. Cancer Res 1947; 7: 386–389.
55. DeLong R, Coman D, Zeidman I. The significance of low calcium and high potassium content in neoplastic tissue. Cancer 1950; 18:718–721.
56. Lansing A, Rosenthal T, Kamen M. Calcium ion exchange in some normal tissues and in epidermal carcinogenesis. Arch Biochem 1948; 19:177–183.
57. Frappart L FN, Lefebre MF, Bremond A, Vauzelle JL, Saez S. In vitro study of the effects of 1,25 dihydroxyvitamin D_3 on the morphology of human breast cancer cell line BT.20. Differentiation 1989; 40:63–69.
58. Borek C, Sachs L. The difference in contact inhibition of cell replication between normal cells and cells transformed by different carcinogens. Proc Natl Acad Sci USA 1966; 56:1705–1711.

59. Stoker M. Regulation of growth and orientation in orientation in hamster cells transformed by polyoma virus. Virology 1964; 24:165–174.
60. Stoker M. Transfer of growth inhibition between normal and virus-transformed cells: autoradiographic studies using marked cells. J Cell Sci 1967; 2:293–304.
61. Mehta P, Bertram J, Loewenstain W. Growth inhibition of transformed cells correlates with their junctional communication with normal cells. Cell 1986; 44:187–196.
62. Lowenstein W. Junctional intercellular communication and the control of growth. Biochim Biophys Acta 1979; 560:1–65.
63. Webb A, Pilbeam C, Hanofin N, Holick M. An evaluation of the relative contributions of exposure to sunlight and of diet to the circulating concentrations of 25-hydroxyvitamin D in an elderly nursing home population in Boston. Am J Clin Nutr 1990; 51:1075–1081.
64. World Health Organization. Statistical Bulletin. Geneva: World Health Organization, 1995.
65. US National Cancer Institute. Surveillance, Epidemiology, and End Results (SEER) Cancer Statistics Report. Bethesda, MD: National Cancer Institute, 1993; VI–19.
66. MacLaughlin J, Anderson R, Holick M. Spectral character of sunlight modulates photosynthesis of previtamin D_3 and its photoisomers in human skin. Science 1982; 216:1001–1003.

23 Chemotherapeutic and Chemopreventive Actions of Vitamin D_3 Metabolites and Analogs

Thomas A. Brasitus and Marc Bissonnette

1. INTRODUCTION

Over the past 15 years, a number of exciting developments have occurred in the vitamin D field, including the discovery of new target tissues, elucidation of novel mechanisms of action, and the synthesis of analogs of vitamin D_3, which, in general, are more potent and selective in their actions than naturally occurring active metabolites of vitamin D_3 *(1)*. These developments have in turn generated great interest in the potential use of vitamin D-related compounds to treat and/or prevent a number of pathologic conditions, including malignant transformation in a variety of organs.

During the past decade, receptors for $1\alpha,25$-dihydroxyvitamin D_3 [$1,25(OH)_2D_3$], the major active metabolite of vitamin D_3, were found to exist in many normal and transformed cells, in addition to those found in established target organs, i.e., intestine, bone, and kidney *(2,3)*. These findings suggested that $1,25(OH)_2D_3$, and/or other metabolites of vitamin D_3, might have important biologic actions besides those related to mineral metabolism. Through binding to the high-affinity vitamin D receptor (VDR), and subsequent association of this liganded-VDR complex to unique promoter sequences within the genome, $1,25(OH)_2D_3$ has been shown to alter the transcription of a large number of genes involved in calcium and phosphorus regulation, vitamin D metabolism, DNA replication, and cellular differentiation *(2,3)*.

$1,25(OH)_2D_3$, and/or other metabolites/analogs of vitamin D_3, generally at high concentrations, have also been shown to inhibit the cellular proliferation (and concomitantly induce the differentiation), of both normal and malignant cell lines *(3,4)*. One of the earliest demonstrations of these effects of $1,25(OH)_2D_3$ was in human myeloid leukemia cells *(5)*. The in vitro addition of $1,25(OH)_2D_3$ suppressed cell growth and induced these cells to undergo differentiation *(5)*. Similar growth-inhibiting and differentiation-inducing effects have subsequently been demonstrated in vitro and in vivo in many other normal and malignant cell lines and/or tissues *(see below)*. Several lines of evidence suggested that these effects were mediated by ligand-activated VDR *(6–10)*, but recent studies have indicated that this may not always be the case *(11,12)*. The potential chemotherapeutic

From: *Vitamin D: Physiology, Molecular Biology, and Clinical Applications*
Edited by: M. F. Holick © Humana Press Inc., Totowa, NJ

and chemopreventive actions of $1,25(OH)_2D_3$, and its analogs, are currently the subject of intense study, and many of the findings related to the use of these analogs for these purposes are summarized in this review.

As recently discussed by Bikle *(1)* and Anzano et al. *(13)*, the practical use of these secosteroids for the treatment and/or prevention of various cancers will undoubtedly depend on the development of synthetic analogs of vitamin D_3 that, even at the high doses required for efficacy, remain cell or organ selective, without inducing hypercalcemia and hypercalciuria. In this regard, a number of analogs of $1,25(OH)_2D_3$ have recently been synthesized that are markedly less calcemic or calciuric, yet are more potent as inhibitors of cellular proliferation and/or inducers of differentiation than $1,25(OH)_2D_3$ in a variety of cell types. Several of these analogs, which possess lower risk/benefit ratios, have also shown cell or organ selectivity *(14)* and appear promising with respect to the treatment and/or prevention of various malignancies.

In the present chapter, we review the data derived from epidemiologic, cell culture, experimental animal, and human studies, which indicate that $1,25(OH)_2D_3$, or other metabolites/analogs of vitamin D_3, may be useful in the treatment and/or prevention of a number of malignancies. We then examine several possible mechanisms of action responsible for their chemotherapeutic and chemopreventive effects. Finally, we discuss new and potentially exciting clinical strategies aimed at the treatment or prevention of these malignancies, using these synthetic analogs of $1,25(OH)_2D_3$, as well as adjunctive therapies utilizing other agents combined with these secosteroids.

2. EPIDEMIOLOGIC STUDIES

Several epidemiologic studies have addressed whether vitamin D, derived from the diet and/or from exposure to ultraviolet irradiation from the sun, are associated with a reduced risk of developing several types of malignancies, particularly cancers of the colon, breast, and prostate. To date, most of these studies have focused on the potential protective role of this vitamin and its metabolites in reducing the risk of colon cancer. Evidence from a variety of sources suggests that increases in vitamin D from the diet and/or from solar radiation are associated with a decreased risk of this malignancy, but this has not been a universal finding. Observations supporting this contention include the fact that colon cancer deaths in the United States are generally highest in places where populations are exposed to the least amount of sunlight, i.e., urban locations and rural areas of high latitudes *(15–17)*. Similarly, geographic patterns of colon cancer around the world indicate that countries with the lowest risk of colon cancer are generally situated within 20° of the equator *(18)*, albeit with a few notable exceptions, such as Japan. In addition, the Garlands and their associates *(19,20)* have also shown in 19- and 8-yr prospective studies that dietary intake of 3.75 µg vitamin D/d, which moderately elevated serum concentrations of 25-hydroxyvitamin D_3, was associated with significant reductions in the incidence of colon cancer. The results of the recently published Iowa Women's Health Study *(21)* also noted a trend between the dietary intake of vitamin D and a reduction in the risk of colon cancer, in keeping with a possible inverse relationship between vitamin D and the incidence of this malignancy.

In contrast to these observations, however, vitamin D from supplements was only slightly, but not significantly, found to have an inverse relationship with the occurrence of colorectal adenomas, precursors of cancer in this organ, in women and to have no inverse relationship in men *(22)*. Moreover, Braun et al. *(23)* in a recent case-control study of the Washington County cohort previously studied by Garland et al. *(20)*, but of larger size and longer duration between serum collections and the diagnosis of colon cancer (10–17 vs 8 yr), failed to confirm that serum levels of vitamin D metabolites affected the subsequent risk for the development of colon cancer. Taken together, these observations would suggest that vitamin D, derived from the diet and from sunlight, may be inversely associated with the risk of developing colon cancer, but that such an inverse association is relatively modest.

The epidemiologic data to date would also support a possible inverse relationship between vitamin D and the risk of breast and prostate cancers *(17,24)*. However, it should be emphasized that, in contrast to the large number of epidemiologic studies on vitamin D and colon cancer, relatively few studies have been published on breast and prostate cancer. Such studies have suggested that an inverse correlation may exist between annual exposure to sunlight and risk of breast cancer in Canada *(17)* and the United States *(25)*. Schwartz and Hulka *(24)* have also recently hypothesized that vitamin D deficiency may increase the risk of prostate cancer, based on the finding that mortality rates of this cancer in the United States were inversely proportional to ultraviolet irradiation. Additional studies with respect to the possible relationship(s) between vitamin D and these cancers will therefore be of interest.

3. CELL CULTURE STUDIES

A number of human malignant cell lines have been shown to possess the intracellular VDR, as well as to undergo growth inhibition and/or differentiation by in vitro treatment with $1,25(OH)_2D_3$ and/or other metabolites/analogs of vitamin D_3. These include HL-60 leukemic cells *(5)* and cell lines derived from human breast cancers *(26–29)*, prostate cancers *(30)*, malignant melanomas *(27)*, colon cancers *(31–34)*, osteogenic sarcomas *(10)*, and pancreatic cancers *(11)*, as well as other types of cancer *(4,35)*. Several generalizations can be made concerning the effects of these secosteroids on malignant cell lines. To inhibit cellular proliferation and/or induce differentiation, pharmacologic concentrations of $1,25(OH)_2D_3$ are frequently necessary *(4)*. In fact, in many malignant cell types, in vitro treatment with physiologic doses of $1,25(OH)_2D_3$ may actually induce proliferation, rather than cause growth arrest *(27)*. The mechanism(s) responsible for this secosteroid-induced biphasic growth response is currently under active investigation.

As noted earlier, several newly available synthetic analogs of vitamin D_3 are more potent in their effects on cellular proliferation and/or differentiation than their parent compounds, including $1,25(OH)_2D_3$ *(1,8,11,14,36–42)* (Fig. 1). Moreover, in general, these analogs have markedly reduced effects on intestinal calcium absorption and bone mineral reabsorption compared with $1,25(OH)_2D_3$ *(1,14,38)*, thereby accounting, at least in part, for their failure to induce hypercalcemia when chronically fed to experimental animals *(13,43)*. The exact mechanisms responsible for these effects are yet to be determined, but they may be related to their decreased affinity for binding to the serum vitamin D binding protein (DBP) compared with $1,25(OH)_2D_3$ *(40,41)*. This, in turn, might be expected to lead to increases in the free (active) concentrations of these analogs, as well as to alterations in their metabolism and duration of action compared with $1,25(OH)_2D_3$ *(40,41)*, which may contribute to the differential effects of these secosteroids. As recently reported by Cheskis et al. *(44)*, these analogs may also differ from $1,25(OH)_2D_3$ in their ability to induce VDR-retinoid X receptor (RXR) heterodimerization, DNA binding, and transactivation. Regardless of their mechanisms of action, several synthetic analogs,

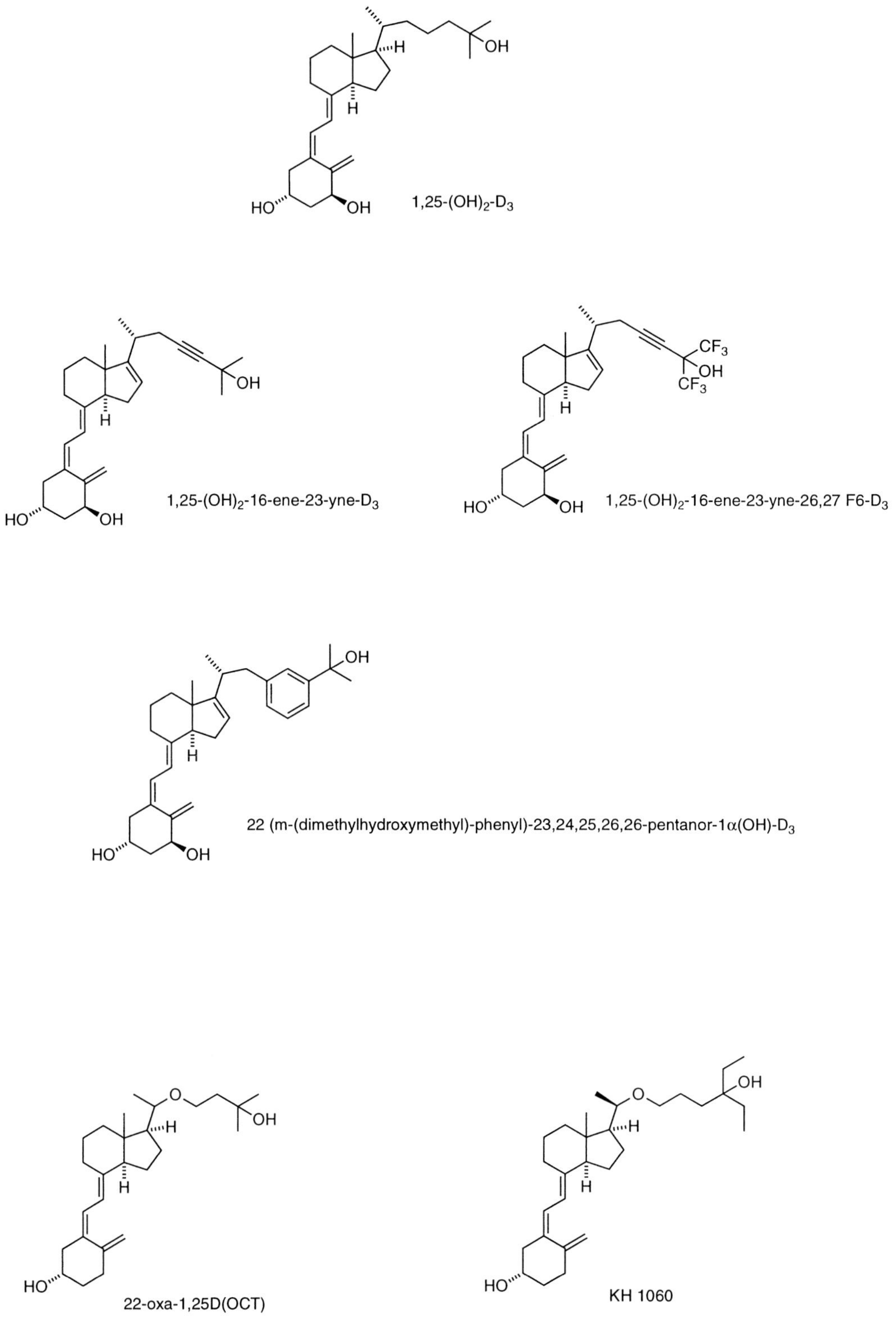

Fig. 1. Structure of 1,25-dihydroxyvitamin D$_3$ and several analogs with potential chemopreventive actions.

whose structure–function relationships have recently been reviewed *(1,14,38,39,42)*, have shown promise as chemopreventive and/or chemotherapeutic agents in several experimental animal models of tumorigenesis *(see below)*.

As exemplified by studies performed in HL-60 cells *(see below)*, $1,25(OH)_2D_3$, and the aforementioned synthetic analogs of vitamin D_3, cause a wide array of biochemical events, which have been implicated in the actions of these secosteroids on cellular proliferation and/or differentiation. Based on these observations, the mechanisms involved in these and other cellular processes such as apoptosis *(see below)*, which are induced by these secosteroids and which may contribute to their anticarcinogenic actions, appear to be highly complex. These effects and their potential consequences, are discussed in more detail *(see* Potential Mechanisms of Action*)*.

As noted earlier, the VDR is expressed in a large number of malignant cell lines. Moreover, several lines of evidence have previously indicated that the cellular actions of $1,25(OH)_2D_3$ and its synthetic analogs are mediated via the VDR. Recent studies, however, have suggested that this hypothesis may not always hold. In this regard, Kawa et al. *(11)* reported that $1,25(OH)_2D_3$ and 22-oxa-calcitriol, a synthetic analog of this secosteroid, markedly inhibited the proliferation of three of nine pancreatic cell lines, yet one of the six nonresponsive cell lines had the second highest VDR content, with no abnormalities in its primary structure. Based on their findings, these investigators concluded that the secosteroid-induced inhibition of the growth of these pancreatic cell lines could not be predicted by the content or mutational status of their VDR *(11)*. In addition, Bhatia et al. *(12)*, demonstrated that the monocytic differentiation of NB4 acute promyelocytic leukemic cells in response to $1,25(OH)_2D_3$ was independent of binding to the VDR. These recent findings suggest that, at least in some malignant cell lines, the cellular actions of $1,25(OH)_2D_3$, and/or its synthetic analogs, may, in fact, not require binding to the VDR. It bears emphasis, however, that gene control by these secosteroids is highly complex. Recent studies, for example, have shown that transcriptional regulation of many genes by these vitamin D_3-related compounds occurs not only by their binding to the VDR, but also via heterodimerization of the VDR with the RXR *(see below)*, as well as by their ability to induce the phosphorylation of serine residues on the VDR *(45)*. Thus, further studies along these lines will be of interest.

Finally, recent studies in HL-60 and breast cancer cell lines have suggested that, in addition to their effects on proliferation and differentiation, $1,25(OH)_2D_3$ and/or its synthetic analogs may also induce these cells to undergo apoptosis *(46,47)*, one form of programmed cell death, as well as inhibit the invasive potential of breast cancer cells in vitro *(48)*. In preliminary studies in CaCo-2 cells, a human colon cancer cell line, our laboratory has also observed that $1,25(OH)_2D_3$, as well as 1,25-dihydroxy-16-ene-23-yne-26,27-hexafluorocholecalciferol (Fig. 1), induced these cells to undergo apoptosis *(49)*. Based on these findings, it would appear that these secosteroids have a number of actions on important cellular processes mediated by VDR-dependent, and perhaps VDR-independent mechanisms, that may contribute to their potential anticarcinogenic and chemotherapeutic actions *(see below)*.

4. EXPERIMENTAL ANIMAL STUDIES

Although epidemiologic and in vitro cell culture studies have suggested that vitamin D_3 metabolites/analogs may have chemotherapeutic and/or chemopreventive actions, to date the strongest data in support of this contention have been derived from the results of in vivo experimental animal studies. The vast majority of these studies, conducted in various experimental models of carcinogenesis, have indicated that several metabolites/ analogs of vitamin D_3 were capable of significantly reducing the incidence of tumors and/or the tumor burden of animals in these models. In this regard, utilizing the 7,12-dimethylbenz(a)-anthracene (DMBA) model of mammary tumorigenesis, Saez et al. *(29)* demonstrated that weekly ip injections of $1,25(OH)_2D_3$ significantly inhibited the growth of mammary tumors. Jacobson et al. *(50)*, using this same cancer model, also demonstrated that by decreasing the calcium and vitamin D content of a diet high in fat in animals administered DMBA, the incidence of mammary tumors were markedly increased beyond that anticipated by the known promotional effects of the high fat content of the diet alone. More recently, Anzano et al. *(13)*, in basic agreement with these previous studies, demonstrated that dietary supplementation with 1,25-dihydroxy-16-ene-23-yne-26,27-hexafluorocholecalciferol alone, or in combination with tamoxifen, significantly inhibited the development of mammary tumors in rats treated with *N*-nitroso-*N*-methylurea. Moreover, chronic dietary administration of this analog was well tolerated by the animals in these experiments and not associated with the development of hypercalcemia during the duration of these studies.

$1,25(OH)_2D_3$ has also been shown to prolong the survival time of mice inoculated with M1 murine myeloid leukemic cells *(51)*, as well as inhibit tumor formation in a TPA-promoted skin cancer model *(52)*. In contrast to other studies, however, administration of this secosteroid has actually been shown to increase, not decrease, the yield of tumors in the DMBA-induced skin cancer model *(53)*. The reasons for these disparate effects of $1,25(OH)_2D_3$ in these latter studies on skin cancer remain unclear and warrant further study.

To date, most animal studies dealing with the potential chemopreventive effects of vitamin D_3 metabolites/analogs have utilized various experimental models of colonic carcinogenesis. In this regard, Kawaura et al. *(54,55)* demonstrated that the intragastric administration of 1α-hydroxyvitamin D_3 not only suppressed colonic tumors in rats induced by intrarectal (i.r.) instillation of *N*-methyl-*N*-nitrosourea, but also inhibited the promotion of tumors in these animals caused by lithocholic acid.

In keeping with these latter observations, Pence and Buddingh *(56)* also demonstrated that in rats fed a high-fat (20%) diet, which promoted the development of 1,2-dimethylhydrazine (DMH)-induced colonic tumors, but not in rats fed a low-fat (5%) diet, supplemental dietary vitamin D_3 significantly reduced the incidence of colonic tumors. Comer et al. *(57)* also failed to detect a protective effect of supplemental dietary vitamin D_3 in rats fed a low-fat (5%) diet and administered a single sc dose of DMH (200 mg/kg body wt). Taken together, these observations suggested that dietary vitamin D_3, as well as intragastrically administered $1\alpha(OH)D_3$, might serve to inhibit the promotion of colonic tumors, i.e., serve as antipromoters.

In further support of this contention, Otoshi et al. *(58)*, in classical postinitiation (promotional) experiments, in which 22-oxa-calcitriol, a synthetic analog of $1,25(OH)_2D_3$, was administered ip to rats several weeks after they were initially treated with DMH, demonstrated that this secosteroid significantly reduced the formation of carcinogen-induced colonic aberrant crypt foci, considered to be a putative preneoplastic lesion in this organ *(58)*. In addition, Sitrin et al. *(59)* demonstrated that although supplemental dietary calcium fed to rats administered DMH failed to influence the incidence of colonic tumors induced by this carcinogen, it did significantly decrease the number of tumors/tumor-bearing rat and the size of those tumors. Moreover, this dietary regimen also totally inhibited the development of K-*ras* mutations in these tumors *(60)*. Interestingly, concomitant vitamin D deficiency abolished the aforementioned anticarcinogenic and antimutagenic actions of supplemented dietary calcium in the promotional phase *(59,60)*.

In contrast to these findings, however, Belleli et al. *(61)*, reported that $1,25(OH)_2D_3$, when injected prior to, but not after the administration of DMH, significantly reduced the incidence of colonic tumors. In basic agreement with the findings of Belleli et al. *(61)*, our laboratory has also recently reported that dietary supplementation with a synthetic vitamin D_3 analog, $1\alpha,25$- dihydroxy-16-ene-23-yne-26,27-hexafluorocholecalciferol (Fig. 1), when fed to animals prior to, but not after, administration of the colonic procarcinogen azoxymethane (AOM), significantly reduced the incidence of colonic adenomas, and totally abolished the development of adenocarcinomas *(43)*. Moreover, in agreement with the findings of Anzano et al. *(13) (see above)*, during the approx 10 months of these studies, the animals fed this analog did not develop hypercalcemia *(43)*. These observations, together with those of Bellelli et al. *(61)*, would in general be in keeping with a hypothesis that $1,25(OH)_2D_3$ and this fluorinated analog of $1,25(OH)_2D_3$, exhibit their chemopreventive actions in the initiation stage, rather than in the promotional stage(s) of the multistage process of colonic malignant transformation, as previously suggested by the findings of Kawaura et al. *(54,55)*, Pence and Buddingh, and other investigators *(56,58,59)*. It is therefore unclear at what stage(s) in this multistage colonic process these secosteroids exert their anticarcinogenic effects. The differences observed in these studies may reflect important experimental design differences, including choice of carcinogen and secosteroid, as well as their routes of administration and dosing intervals.

5. HUMAN STUDIES

Although $1,25(OH)_2D_3$ and other metabolites/analogs of vitamin D_3 are currently being used to treat a number of pathologic conditions *(1,38)*, relatively few authors have studied the potential chemotherapeutic and/or chemopreventive effects of these secosteroids in humans. In this regard, $1,25(OH)_2D_3$ has been shown by Thomas et al. *(62)* to inhibit the cellular proliferation of cultured rectal explants from both normal patients as well as patients with familial adenomatous polyposis. Given the potential importance of hyperproliferation in colorectal malignant transformation, these findings would suggest, albeit indirectly, that these secosteroids, by reducing proliferation, might prove useful in preventing the development of colorectal cancer, as has been suggested for supplemental dietary calcium *(59,62)*.

Several studies have also addressed whether the VDR exists in human colorectal cancers and may contribute to the development of these malignancies *(63–67)*. The results of these studies can be summarized as follows: (1) the VDR is expressed in at least one-third of these cancers; (2) Scatchard analysis showed a single class of specific high-affinity, low-capacity binding sites in these tumors; (3) VDR expression in these malignancies is significantly lower compared with its expression in either paired normal-appearing mucosa from these patients or colonic mucosa of patients without colon cancer; (4) the frequency of detection of the VDR appears to depend on the location of the cancers, i.e., right colon > left colon > rectum; and (5) cancers with the VDR may be histologically more differentiated, although this latter point remains controversial *(68)*.

In addition, Kane et al. *(68)* have more recently demonstrated that RXRs also exist in human colon cancer. These receptors are potentially very important in the actions of

1,25(OH)$_2$D$_3$ and other vitamin D$_3$-related compounds, since they can heterodimerize with the VDR and thereby influence their effects *(42,44,68)*. The expression of these receptors, however, like the VDR, was found to be reduced in most (75%) of these tumors. Neither VDR nor RXR expression in this small study (12 patients) was correlated with site, grade of differentiation, or Dukes staging of these malignancies. Taken together, the results of these studies provide a rational basis for the treatment of colorectal malignancies with these secosteroids. If, however, as indicated by most studies, the VDR is present in a minority of tumors, this approach may potentially prove useful only in a subset of patients who harbor this malignancy and possess the VDR, if these effects are mediated by this receptor.

Finally, it should be noted that Bower et al. *(69)* have reported the results of a limited trial with topical calcipotriol in patients with advanced breast cancer. In this trial, 14 of 19 patients completed the 6-wk study period. Three patients showed a 50% reduction in tumor size, and another showed a slight response, when treated with 100 µg of this vitamin D$_3$ analog daily. Two patients became hypercalcemic during treatment, requiring withdrawal from the study. Interestingly, in all four of the patients who responded to the topical administration of calcipotriol, the tumors contained the VDR, as assessed by immunohistochemistry. This latter finding would again suggest that a subset of breast cancers (like colon cancers) that possess the VDR might be expected to benefit from treatment with these secosteroids. This clinically important issue, however, remains unclear and will require further study.

6. POTENTIAL MECHANISMS IN VITAMIN D$_3$ METABOLITE/ANALOG ACTION

As noted earlier, considerable evidence now exists that the chemopreventive and/or chemotherapeutic effects of 1,25(OH)$_2$D$_3$ and other vitamin D$_3$ metabolites/analogs, at least in part, are likely to involve their actions on such fundamental cellular processes as proliferation, differentiation, and perhaps apoptosis. Based on the current evidence from a variety of sources, several biochemical events induced by these secosteroids that may potentially underlie their aforementioned actions on each of these processes are discussed below.

As noted in the Cell Culture Studies section, 1,25(OH)$_2$D$_3$, as well as several synthetic analogs of this secosteroid, have been shown to inhibit the proliferation of a variety of malignant cell types. Moreover, many of these vitamin D$_3$-related compounds have been shown to induce the arrest of HL-60 and other cells in the G$_1$ stage of the cell cycle *(11,70–73)*, as well as possibly at the G$_2$/M interphase *(70)*. As summarized in Fig. 2, the cell cycle is composed of four distinct stages, G$_1$, S (DNA synthesis), G$_2$, and M (mitosis). Cell cycle progression is governed by sequential formation, activation, and subsequent inactivation of a series of cyclin–cyclin-dependent kinase (Cdk) complexes *(74,75)*. The major cyclin–Cdk complexes are cyclin D–Cdk4 and –Cdk6, cyclin E–Cdk2, cyclin A–Cdk2, and cyclin B–Cdc2, acting in G$_1$ and G$_1$/S, in S, and in G$_2$/M, respectively (Fig. 2). The formation of each of these complexes depends on cell cycle-regulated expression of cyclins, which assemble with preexisting Cdks. Cyclin–Cdks are also regulated positively and negatively by *phosphorylation (74,75)*. Cell cycle progression in most cells is *negatively* regulated by the retinoblastoma susceptibility gene product, pRb, and members of the Cdk inhibitor family, including among others p21^{waf1} (also known as p21^{cip1}) and p27^{kip1}, as well as the tumor suppression gene product p53. Orderly progression

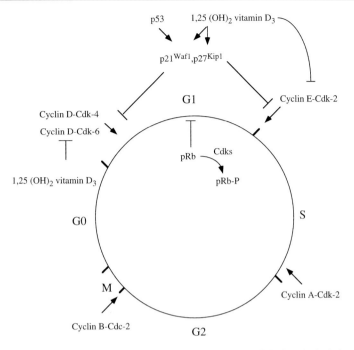

Fig. 2. Positive and negative regulators of the cell cycle. Cyclin D–Cdk4 and –Cdk6 control early G_1 progression, and cyclin E–Cdk2 activation is required for S (synthesis) phase entry. The cyclin–Cdk inhibitors p21^{waf1} and p27^{kip1} can arrest cells in G_1. The retinoblastoma protein, pRb, is phosphorylated by Cdk2, Cdk6, and Cdk4, thereby limiting its ability to inhibit cell cycling. Progression in S phase is regulated by cyclin A–Cdk2; cyclin B–Cdc2 controls entry into the M phase (mitosis). Potential sites of action of 1,25(OH)$_2$ vitamin D$_3$ are shown. In addition to this secosteroid inducing a G_1 arrest, inhibition of G2/M progression has also been suggested *(70)*.

through the cell cycle requires activation of specific cyclins–Cdks at appropriate times. Each of their activations is regulated by feedback mechanisms preventing premature entry of cells into the next phase of the cycle prior to the completion of prerequisite biochemical events. For example, checkpoints in G_1 and G_2/M monitor the readiness of cells to initiate DNA replication and the completion of DNA synthesis/repair, respectively. Once activated, G_1 cyclin–Cdks phosphorylate targets that include pRb and related proteins, thereby releasing and activating transcription factors such as E2F and other nuclear regulators needed for the expression of genes involved in DNA replication *(74,75)*.

In HL-60 cells, for example, 1,25(OH)$_2$D$_3$ has recently been shown to induce the growth arrest of these cells in the G_1 stage of their cell cycle, at least partly, by inducing an increase in the expression of two members of the Cdk inhibitor family, p27^{kip1} and p21^{waf1} *(72)*. Increased p27^{kip1}, in turn, may be responsible for the decrease in cyclin D–Cdk6 activity observed in these cells *(76)*. In addition, in HL-60 cells this secosteroid reduced cyclin E levels that probably contributed to the decreased cyclin E–Cdk2 activity previously reported *(76)*. Moreover, in MCF-7 breast cancer cells, taxol, an agent that induced their G_1 arrest, did so, at least in part, by inducing the expression of p21^{waf1}, via a Raf-1-dependent pathway *(77)*. This suggests a mechanism, whereby 1,25(OH)$_2$D$_3$ may cause G_1 arrest, since Raf-1 has recently been shown to be activated in hepatic Ito cells by this secosteroid via a protein kinase C (PKC)-dependent mechanism *(78)*. It is, therefore, possible that activation of Raf-1 by 1,25(OH)$_2$D$_3$ may lead to the arrest of HL-60 and other cells in their G_1 stage, via an increase in the expression of $p21^{waf1}$ and/or $p27^{kip1}$. If proved to be true, this hypothesis would be very interesting in that Raf-1 has been shown to be activated by PKC-α *(79)*, and by ceramide, via the stimulation of a neutral sphingomyelinase *(80)*. Both of these latter signal transduction elements are stimulated by $1,25(OH)_2D_3$, for example, in CaCo-2 and HL-60 cells, respectively *(81,82)*.

Recent studies by Frey et al. *(83)* have also implicated PKC-α, one of the vitamin D-responsive isoforms of PKC in CaCo-2 cells *(82)* as well as in other cells *(84)*, in the hypophosphorylation of pRb as well as in the increased expression of $p21^{waf1}$. Another PKC isoform, PKC-β_{II}, which has been shown to be responsive to $1,25(OH)_2D_3$ in rat colonocytes *(85)*, has also recently been implicated in the regulation of the cell cycle in human erythroleukemia cells *(86,87)*. In addition, as noted earlier, our laboratory has demonstrated that the vitamin D status of rats can influence the mutational status of the K-*ras* proto-oncogene in the DMH model of colon cancer *(60)*. Given the importance of the K-*ras* mutations in the dysregulation of cell growth in these tumors, these effects may be important in the antiproliferative actions of vitamin D_3 metabolites. Moreover, we have also shown that feeding animals supplemental dietary 1,25-dihydroxy-16-ene-23-yne-26,27-hexafluorocholecalciferol not only markedly reduced the number of AOM-induced colonic tumors and totally abolished the incidence of colonic adenocarcinomas *(43)*, but also preserved the expression of one isoform of PKC in these tumors, PKC-ζ (zeta) *(88)*. These findings are of interest in that two other structurally unrelated chemopreventive agents in this experimental model, ursodeoxycholic acid *(89)*, a secondary bile acid, and piroxicam *(90)*, a nonsteroidal anti-inflammatory drug, have also recently been shown to preserve the expression of PKC-ζ selectively in adenomas harvested from these animals compared with adenomas and carcinomas from carcinogen-treated rats fed control diets. Since in other cells, PKC-ζ has been shown to be intimately involved in the regulation of their proliferation and differentiation *(91–96)*, it would appear that the preservation of the expression of this PKC isoform in adenomas by this fluorinated analog of $1,25(OH)_2D_3$ and these other agents, at least in this model, may prevent the progression of adenomas to adenocarcinomas. Future studies should therefore prove interesting in determining the mechanisms involved in the antiproliferative effects of $1,25(OH)_2D_3$ and other metabolites/analogs of vitamin D_3.

As discussed previously, $1,25(OH)_2D_3$, and particularly the recently developed synthetic analogs of this secosteroid, in addition to their antiproliferative effects, induce the differentiation of a large number of malignant cell types. Most studies that have attempted to elucidate the biochemical mechanisms involved in the differentiating actions of these secosteroids have been performed in HL-60 cells. The following biochemical effects have been implicated in the ability of these secosteroids to induce HL-60 cells to differentiate into monocytes/macrophages: (1) increases in intracellular Ca^{2+} *(97,98)*; (2) increases in intracellular pH *(99)*; (3) increases in ceramide, via activation of a neutral sphingomyelinase *(81)*; (4) alterations in the expression of several oncogenes including c-*fos* and c-*myc* *(100,101)*; and (5) increases in the expression of PKC-α and -β *(102–104)*. In addition, an increase in the expression of $p21^{waf1}$ has also been implicated in the differentiation of many cell types *(105)*. In U937 myelomonocytic cells, $1,25(OH)_2D_3$ has been shown to activate the $p21^{waf1}$ promoter, via a VDR-dependent, p53-independent pathway, and concomitantly induce their differentiation *(106)*. How one or more of these biochemical changes, however, ultimately leads to the induction of differentiation of these cells is still unclear and deserving of further study.

Regardless of the exact biochemical events necessary for these secosteroids to induce the differentiation of HL-60 cells, considerable evidence does exist that this effect is mediated via the VDR in these cells *(see above)*. As noted earlier, however, recent studies in other cell types, including NB4 acute promyelocytic leukemia cells *(12)*, have indicated that $1,25(OH)_2D_3$ may induce their differentiation via a pathway(s) independent of the VDR. These latter observations suggest that cell-specific exceptions exist to the generalization concerning the prerequisite role for the VDR in these secosteroid-induced effects on differentiation.

Recent studies in HL-60 *(46)* and breast cancer cells *(47)* have suggested that, in addition to their effects on proliferation and differentiation, $1,25(OH)_2D_3$, and one or more of its synthetic analogs, may also induce these malignant cells to undergo apoptosis. In agreement with these prior findings, in preliminary studies we have observed that $1,25(OH)_2D_3$ and 1,25-dihydroxy-16-ene-23-yne-26,27-hexfluorocholecalciferol also induced CaCo-2 cells to undergo apoptosis *(49)*. It is therefore likely that, at least in some cell types, the anticarcinogenic actions of these secosteroids may involve the induction of this form of programmed cell death.

In this regard, however, it should be emphasized that the data on induction of apoptosis by these secosteroids, at least in HL-60 cells, are controversial. Xu et al. *(107)* demonstrated that while $1,25(OH)_2D_3$ actually protected these cells from undergoing apoptosis, this effect was accompanied by a downregulation in the expression of bcl-2. This latter finding was surprising, since increases, not decreases, in the expression of bcl-2 have been shown to inhibit apoptosis in many cell types *(46)*. More recently, Elstner et al. *(46)*, demonstrated that a 20-epi-vitamin D_3 analog, KH-1060, only modestly induced the apoptosis of HL-60 cells when used alone, but when concomitantly administered with 9-cis-retinoic acid, it markedly increased the apoptosis of these cells. Moreover, this combination significantly decreased the expression of bcl-2, and increased the ratio of bax to bcl-2 in these cells. This accentuated effect of KH-1060 on the apoptotic process, in the presence of 9-cis-retinoic acid, presumably involved the heterodimerization of the VDR with the RXR in these cells *(44,46,108,109)*. Further studies to determine whether the pro-apoptotic effects of these secosteroids can be detected more widely in other malignant cell lines, or are relatively cell specific, will therefore be of interest. In addition, investigations to elucidate the biochemical actions involved in this phenomenon are likely to provide important insights into the chemopreventive and/or chemotherapeutic mechanisms of these agents.

7. FUTURE APPROACHES TO THE TREATMENT AND PREVENTION OF MALIGNANCIES

As noted in this chapter, several newly synthesized analogs of vitamin D_3 show great promise with respect to the treatment and/or prevention of a number of human malignancies. They offer several potential advantages compared with $1,25(OH)_2D_3$ and other metabolites of vitamin D_3 for these purposes. These analogs, for example, appear to be more active in their effects on cellular growth and differentiation, and in addition their calcemic activities are often orders of magnitude lower than $1,25(OH)_2D_3$ *(1,42)*. In addition, based on studies *(14,42)* to date, it would appear that newer analogs may be developed that are organ specific and perhaps eventually even tumor specific. These features strongly suggest that, compared with $1,25(OH)_2D_3$, these analogs should have a decreased risk/benefit ratio with respect to the treatment and/or prevention of a variety of human malignancies.

It is worth emphasizing, however, that, as with many other chemotherapeutic agents, single-agent treatment with synthetic secosteroids alone may prove to be only minimally effective and/or of limited duration of action. Since combination chemotherapy has proved superior to monotherapy, it is highly likely that in the near future, these secosteroids might be used in combination with other classes of antitumor agents. As recently reviewed in detail *(110)*, based on their combined efficacies in a variety of in vitro and in vivo experimental systems, these other agents might include retinoids, particularly those that bind to the RXR, cytokines, such as interferon-α and -γ, and other more conventional chemotherapeutic drugs.

It is also important to point out that these secosteroids, alone or in combination with one or more of the aforementioned agents, might also prove useful in treating and/or preventing the development of humorally mediated hypercalcemia associated with malignancy. This latter complication of many tumors appears to be caused, at least in part, by a parathyroid hormone (PTH)-related protein (PTHrP) *(111)*. In recent studies a synthetic analog of $1,25(OH)_2D_3$, EB 1089, was shown to prevent the development of hypercalcemia in Fischer rats implanted with a Leydig cell tumor by inhibiting PTHrP production; it also significantly prolonged the survival time of these tumor-bearing animals *(112)*.

Based on all these considerations, we are entering an exciting era with respect to the use of synthetic analogs of vitamin D_3, alone or in combination with other agents, for the prevention or treatment of a variety of malignancies, as well as cancer-associated complications such as hypercalcemia. If, as expected, clinical trials demonstrate that these agents are efficacious, we anticipate that these secosteroids may soon provide important additions to our armamentarium against a number of cancers, particularly those of the colon, breast, and prostate.

REFERENCES

1. Bikle DD. Vitamin D: New actions, new analogs, new therapeutic potential. Endocrinol Rev 1992; 13:765–784.
2. Minghetti PP, Norman AW. $1,25(OH)_2$-vitamin D_3 receptors: gene regulation and genetic circuitry. FASEB J 1988; 2:3043–3053.
3. Walters MR. Newly identified actions of the vitamin D endocrine system. Endocr Rev 1992; 13: 719–763.
4. Pols HA, Birkenhager JC, Foekens JA, van Leeuwen JPM. Vitamin D: a modulator of cell proliferation and differentiation. J Steroid Biochem Mol Biol 1990; 37:873–876.
5. Miyaura C, Abe E, Kuribayashi T, Tanaka H, Konno K, Nishii Y, Suda T. 1 α,25- Dihydroxyvitamin D_3 induces differentiation of human myeloid leukemia cells. Biochem Biophys Res Commun 1981; 102:937–943.
6. Inaba M, Okuno S, Koyama H, Nishizawa Y, Morii H. Dibutyryl cAMP enhances the effect of 1,25-dihydroxyvitamin D_3 on a human promyelocytic leukemia cell, HL-60, at both the receptor and the postreceptor steps. Arch Biochem Biophys 1992; 293:181–186.
7. Kuribayashi T, Tanaka H, Abe E, Suda T. Functional defect of variant clones of a human myeloid leukemia cell line (HL-60) resistant to 1α, 25-dihydroxyvitamin D_3. Endocrinology 1983; 113: 1992–1998.
8. Ostrem VK, Lau WF, Lee SH, Perlman K, Prahl J, Schnoes HK, DeLuca HF, Ikekawa N. Induction of monocytic differentiation of HL-60 cells by 1,25-dihydroxyvitamin D analogs. J Biol Chem 1987; 262:14,164–14,171.
9. Feldman J, Federico MHH, Sonohara S, Katayama MLH, Koike MAA, Roela RA, Da Silva MRP, Brentani MM. Vitamin D_3 binding activity during leukemic cell differentiation. Leuk Res 1993; 17: 97–101.
10. Dokoh S, Donaldson CA, Haussler MR. Influence of 1,25-dihydroxyvitamin D_3 on cultured osteogenic sarcoma cells: correlation with the 1,25-dihydroxyvitamin D_3 receptor. Cancer Res 1984; 44: 2103–2109.

11. Kawa S, Yoshizawa K, Tokoo M, Imai H, Oguchi H, Kiyosawa K, Homma T, Nikaido T, Furihata K. Inhibitory effect of 22-oxa-1,25-dihydroxyvitamin D_3 on the proliferation of pancreatic cancer cell lines. Gastroenterology 1996; 110:1605–1613.
12. Bhatia M, Kirkland JB, Meckling-Gill KA. Monocytic differentiation of acute promyelocytic leukemia cells in response to 1,25-dihydroxyvitamin D_3 is independent of nuclear receptor binding. J Biol Chem 1995; 270:15,962–15,965.
13. Anzano MA, Smith JM, Uskokovic MR, Peer CW, Mullen LT, Letterio JJ, Welsh MC, Shrader MW, Logsdon DL, Driver CL, Brown CC, Roberts AB, Sporn MB. 1 α,25-Dihydroxy-16-ene-23-yne-26,27-hexafluorocholecalciferol (R024-5531), a new deltanoid (vitamin D analogue) for prevention of breast cancer in the rat. Cancer Res 1994; 54:1653–1656.
14. Norman AW, Sergeev IN, Bishop JE, Okamura WH. Selective biological response by target organs (intestine, kidney, and bone) to 1,25-dihydroxyvitamin D_3 and two analogues. Cancer Res 1993; 53: 3935–3942.
15. Garland CF, Garland FC. Do sunlight and vitamin D reduce the likelihood of colon cancer? Int J Epidemiol 1980; 9:227–231.
16. Emerson JC, Weiss NS. Colorectal cancer and solar radiation. Cancer Causes Control 1992; 3:95–99.
17. Gorham ED, Garland CF, Garland FC. Acid haze air pollution and breast and colon cancer mortality in 20 Canadian cities. Can J Public Health 1989; 80:96–100.
18. Muir C, Waterhouse J, Mack T. Cancer Incidence in Five Continents, vol 5. IARC Scientific Publication No. 88. Lyon: International Agency for Research on Cancer, 1987.
19. Garland C, Shekelle RB, Barrett-Connor E, Criqui MH, Rossof AH, Paul O. Dietary vitamin D and calcium and risk of colorectal cancer: a 19 year prospective study in men. Lancet 1985; 1:307–309.
20. Garland CF, Comstock GW, Garland FC, Helsing KJ, Shaw EK, Gorham ED. Serum 25-hydroxyvitamin D and colon cancer: eight-year prospective study. Lancet 1989; 2:1176–1178.
21. Bostick RM, Potter JD, Sellers TA, McKenzie DR, Kushi LH, Folsom AR. Relation of calcium, vitamin D, and dairy food intake to incidence of colon cancer among older women. Am J Epidemiol 1993; 137:1302–1317.
22. Kampman E, Giovannucci E, van't Veer P, Rimm E, Stampfer MJ, Colditz GA, Kok FJ, Willett WC. Calcium, vitamin D, dairy foods, and the occurrence of colorectal adenomas among men and women in two prospective studies. Am J Epidemiol 1994; 139:16–29.
23. Braun MM, Helzlsouer KJ, Hollis BW, Comstock GW. Colon cancer and serum vitamin D metabolite levels 10–17 years prior to diagnosis. Am J Epidemiol 1995; 142:608–611.
24. Schwartz GG, Hulka BS. Is vitamin D deficiency a risk factor for prostate cancer? Anticancer Res 1990; 10:1307–1312.
25. Garland FC, Garland CF, Gorham ED, Young JF. Geographic variation in breast cancer mortality in the United States: a hypothesis involving exposure to solar radiation. Prev Med 1990; 19:614–622.
26. Elstner E, Linker-Israeli M, Said J, Umiel T, de Vos S, Shintaku IP, Heber D, Binderup L, Uskokovic M, Koeffler HP. 20-Epi-vitamin D_3 analogues: a novel class of potent inhibitors of proliferation and inducers of differentiation of human breast cancer cell lines. Cancer Res 1995; 55:2822–2830.
27. Frampton RJ, Osmond SA, Eisman JA. Inhibition of human cancer cell growth by 1,25-dihydroxyvitamin D_3 metabolites. Cancer Res 1983; 43:4443–4447.
28. Koga M, Eisman JA, Sutherland RL. Regulation of epidermal growth factor receptor levels by 1,25-dihydroxyvitamin D_3 in human breast cancer cells. Cancer Res 1988; 48:2734–2739.
29. Saez S, Falette N, Guillot C, Meggouh F, Lefebvre M-F, Crepin M. 1,25$(OH)_2D_3$ modulation of mammary tumor cell growth in vitro and in vivo. Breast Cancer Res Treat 1993; 27:69–81.
30. Skowronski RJ, Peehl DM, Feldman D. Vitamin D and prostate cancer: 1,25 dihydroxyvitamin D_3 receptors and actions in human prostate cancer cell lines. Endocrinology 1993; 132:1952–1960.
31. Halline AG, Davidson NO, Skarosi SF, Sitrin MD, Tietze C, Alpers DH, Brasitus TA. Effects of 1,25-dihydroxyvitamin D_3 on proliferation and differentiation of CaCo-2 cells. Endocrinology 1994; 134:710–1717.
32. Cross HS, Hulla W, Tong W-M, Peterlik M. Growth regulation of human colon adenocarcinoma-derived cells by calcium, vitamin D and epidermal growth factor. J Nutr 1995; 125:2004S–2008S.
33. Wargovich MJ, Lointier PH. Calcium and vitamin D modulate mouse colon epithelial proliferation and growth characteristics of a human colon tumor cell line. Can J Physiol Pharm 1987; 65:472–477.
34. Lointier P, Wargovich MJ, Saez S, Levin B, Wildrick DM, Boman BM. The role of vitamin D_3 in the proliferation of a human colon cancer cell line in vitro. Anticancer Res 1987; 7:817–821.
35. Niendorf A, Arps H, Dietel M. Effect of 1,25-dihydroxyvitamin D_3 on human cancer cells in vitro. J Steroid Biochem 1987; 27:825–828.

36. Masuda S, Byford V, Kremer R, Makin HLJ, Kubodera N, Nishii Y, Okazaki A, Okano T, Kobayashi T, Jones G. In vitro metabolism of the vitamin D analog, 22-oxacalcitriol, using cultured osteosarcoma, hepatoma, and keratinocyte cell lines. J Biol Chem 1996; 271:8700–8708.
37. Zhou JY, Norman AW, Chen D-L, Sun G-W, Uskokovic M, Koeffler HP. 1,25-Dihydroxy-16-ene-23-yne-vitamin D_3 prolongs survival time of leukemic mice. Proc Natl Acad Sci USA 1990; 87: 3929–3932.
38. Pols HA, Birkenhager JC, van Leeuwen JPTM. Vitamin D analogues: from molecule to clinical application. Clin Endocrinol 1994; 40:285–292.
39. Bouillon R, Allewaert K, van Leeuwen JPTM, Tan BK, Xiang DZ, De Clercq P, Vandewalle M, Pols HA, Bos MP, Van Baelen H, Birkenhager JC. Structure function analysis of vitamin D analogs with C-ring modifications. J Biol Chem 1992; 267:3044–3051.
40. Dilworth FJ, Calverley MJ, Makin HL, Jones G. Increased biological activity of 20-epi-1,25-dihydroxyvitamin D_3 is due to reduced catabolism and altered protein binding. Biochem Pharm 1994; 47:987–993.
41. Bouillon R, Allewaert K, Xiang DZ, Tan BK, van Baelen H. Vitamin D analogs with low affinity for the vitamin D binding protein: enhanced *in vitro* and decreased *in vivo* activity. J Bone Miner Res 1991; 6:1051–1057.
42. Brown AJ, Dusso A, Slatopolsky E. Selective vitamin D analogs and their therapeutic applications. Semin Nephrol 1994; 14:156–174.
43. Wali RK, Bissonnette M, Khare S, Hart J, Sitrin MD, Brasitus TA. 1α,25-Dihydroxy-16-ene-23-yne-26,27-hexafluorocholecalciferol, a noncalcemic analogue of 1 α,25-dihydroxyvitamin D_3, inhibits azoxymethane-induced colonic tumorigenesis. Cancer Res 1995; 55:3050–3054.
44. Cheskis B, Lemon BD, Uskokovic M, Lomedico PT, Freedman LP. Vitamin D_3-retinoid X receptor dimerization, DNA binding, and transactivation are differentially affected by analogs of 1,25-dihydroxyvitamin D_3. Mol Endocrinol 1995; 9:1814–1824.
45. Haussler MR, Jurutka PW, Hsieh J-C, Thompson PD, Selznick SH, Haussler CK, Whitfield GK. Receptor mediated genomic actions of $1,25(OH)_2D_3$: modulation by phosphorylation. In: Vitamin D. Norman AW, Bouillon R, Thomasset M, eds. New York: Walter de Gruyter, 1994; 209–216.
46. Elstner E, Linker-Israeli M, Umiel T, Le J, Grillier I, Said J, Shintaku IP, Krajewski S, Reed JC, Binderup L, Koeffler HP. Combination of a potent 20-epi-vitamin D_3 analogue (KH 1060) with 9-cis-retinoic acid irreversibly inhibits clonal growth, decreases bcl-2 expression, and induces apoptosis in HL-60 leukemic cells. Cancer Res 1996; 56:3570–3576.
47. Vandewalle B, Hornez L, Wattez N, Revillion F, Lefebvre J. Vitamin-D_3 derivatives and breast-tumor cell growth: effect on intracellular calcium and apoptosis. Int J Cancer 1995; 61:806–811.
48. Hansen CM, Frandsen TL, Brunner N, Binderup L. 1α,25-Dihydroxyvitamin D_3 inhibits the invasive potential of human breast cancer cells *in vitro*. Clin Exp Metast 1994; 12:195–202.
49. Skarosi S, Abraham C, Bissonnette M, Scaglione-Sewell B, Sitrin MD, Brasitus TA. 1,25-Dihydroxyvitamin D_3 stimulates apoptosis in CaCo-2 cells. Gastroenterology 1997; 112:A608.
50. Jacobson EA, James KA, Newmark HL, Carroll KK. Effects of dietary fat, calcium, and vitamin D on growth and mammary tumorigenesis induced by 7,12- dimethylbenz(a)-anthracene in female Sprague-Dawley rats. Cancer Res 1989; 49:6300–6303.
51. Abe E, Miyaura C, Sakagami H, Takenda M, Konno K, Yamazaki T, Yoshiki S, Suda T. Differentiation of mouse myeloid leukemia cells induced by 1α,25-dihydroxyvitamin D_3. Proc Natl Acad Sci USA 1981; 78:4990–4994.
52. Wood AW, Chang RL, Huang MT, Uskokovic M, Conney AH. 1 α, 25-Dihydroxyvitamin D_3 inhibits phorbol ester-dependent chemical carcinogenesis in mouse skin. Biochem Biophys Res Commun 1983; 116:605–611.
53. Wood AW, Chang RL, Huang MT, Baggiolini E, Partridge JJ, Uskokovic M, Conney AH. Stimulatory effect of 1α, 25-dihydroxyvitamin D_3 on the formation of skin tumors in mice treated chronically with 7,12-dimethylbenz[a]anthracene. Biochem Biophys Res Commun 1985; 130:924–931.
54. Kawaura A, Tanida N, Sawada K, Oda M, Shimoyama T. Supplemental administration of 1α-hydroxyvitamin D_3 inhibits promotion by intrarectal instillation of lithocholic acid in N-methyl-N-nitrosourea-induced colonic tumorigenesis in rats. Carcinogenesis 1989; 10:647–649.
55. Kawaura A, Takahashi A, Tanida N, Oda M, Sawada K, Sawada Y, Maekawa S, Shimoyama T. 1α-hydroxyvitamin D_3 suppresses colonic tumorigenesis induced by repetitive intrarectal injection of N-methyl-N-nitrosourea in rats. Cancer Lett 1990; 55:149–152.
56. Pence BC, Buddingh F. Inhibition of dietary fat-promoted colon carcinogenesis in rats by supplemental calcium or vitamin D_3. Carcinogenesis 1988; 9:187–190.

57. Comer PF, Clark TD, Glauert HP. Effect of dietary vitamin D_3 (cholecalciferol) on colon carcinogenesis induced by 1,2-dimethylhydrazine in male Fischer 344 rats. Nutr Cancer 1993; 19:113–124.
58. Otoshi T, Iwata H, Kitano M, Nishizawa Y, Morii H, Yano Y, Otani S, Fukushima S. Inhibition of intestinal tumor development in rat multi-organ carcinogenesis and aberrant crypt foci in rat colon carcinogenesis by 22-oxa-calcitriol, a synthetic analogue of 1α, 25-dihydroxyvitamin D_3. Carcinogenesis 1995; 16:2091–2097.
59. Sitrin MD, Halline AG, Abrahams C, Brasitus TA. Dietary calcium and vitamin D modulate 1,2-dimethylhydrazine-induced colonic carcinogenesis in the rat. Cancer Res 1991; 51:5608–5613.
60. Llor X, Jacoby RF, Teng BB, Davidson NO, Sitrin MD, Brasitus TA. K-ras mutations in 1,2-dimethylhydrazine-induced colonic tumors: effects of supplemental dietary calcium and vitamin D deficiency. Cancer Res 1991; 51:4305–4309.
61. Belleli A, Shany S, Levy J, Guberman R, Lamprecht SA. A protective role of 1,25-dihydroxyvitamin D_3 in chemically induced rat colon carcinogenesis. Carcinogenesis 1992; 13:2293–2298.
62. Thomas MG, Tebbutt S, Williamson RCN. Vitamin D and its metabolites inhibit cell proliferation in human rectal mucosa and a colon cancer cell line. Gut 1992; 33:1660–1663.
63. Lointier P, Meggouh F, Dechelotte P, Pezet D, Ferrier C, Chipponi J, Saez S. 1,25-Dihydroxyvitamin D_3 receptors and human colon adenocarcinoma. Br J Surg 1991; 78:435–439.
64. Meggouh F, Lointier P, Pezet D, Saez S. Evidence of 1,25-dihydroxyvitamin D_3-receptors in human digestive mucosa and carcinoma tissue biopsies taken at different levels of the digestive tract in 152 patients. J Steroid Biochem 1990; 36:143–147.
65. Lointier P, Meggouh F, Pezet D, Dapoigny M, Dieng PND, Saez S, Chipponi J. Specific receptors for 1,25-dihydroxyvitamin D_3 (1,25-DR) and human colorectal carcinogenesis. Anticancer Res 1989; 9:1921–1924.
66. Meggouh F, Lointier P, Saez S. Sex steroid and 1,25-dihydroxyvitamin D_3 receptors in human colorectal adenocarcinoma and normal mucosa. Cancer Res 1991; 51:1227–1233.
67. Sandgren M, Danforth L, Plasse TF, DeLuca HF. 1,25-Dihydroxyvitamin D_3 receptors in human carcinomas: a pilot study. Cancer Res 1991; 51:2021–2024.
68. Kane KF, Langman MJ, Williams GR. 1,25-Dihydroxyvitamin D_3 and retinoid X receptor expression in human colorectal neoplasms. Gut 1995; 36:255–258.
69. Bower M, Colston KW, Stein RC, Hedley A, Gazet J-C, Ford HT, Combes RC. Topical calcipotriol treatment in advanced breast cancer. Lancet 1991; 337:701,702.
70. Godyn JJ, Xu H, Zhang F, Kolla S, Studzinski GP. A dual block to cell cycle progression in HL60 cells exposed to analogues of vitamin D_3. Cell Prolif 1994; 27:37–46.
71. Rigby WF, Noelle RJ, Krause K, Fanger MW. The effects of 1,25-dihydroxyvitamin D_3 on human T lymphocyte activation and proliferation: a cell cycle analysis. J Immunol 1985; 135:2279–2286.
72. Wang QM, Jones JB, Studzinski GP. Cyclin-dependent kinase inhibitor p27 as a mediator of the G1-S phase block induced by 1,25-dihydroxyvitamin D_3 in HL60 cells. Cancer Res 1996; 56:264–267.
73. Mangelsdorf DJ, Koeffler HP, Donaldson CA, Pike JW, Haussler MR. 1,25- Dihydroxyvitamin D_3-induced differentiation in a human promyelocytic leukemia cell line (HL-60): receptor-mediated maturation to macrophage-like cells. J Cell Biol 1984; 98:391–398.
74. Toyoshima H, Hunter T. p27, a novel inhibitor of G1 cyclin-Cdk protein kinase activity, is related to p21. Cell 1994; 78:67–74.
75. Ohtsubo M, Roberts JM. Cyclin-dependent regulation of G1 in mammalian fibroblasts. Science 1993; 259:1908–1912.
76. Wang QM, Luo X, Studzinski GP. Cyclin-dependent kinase 6 is the principal target of p27/Kip1 regulation of the G_1-phase traverse in 1,25-dihydroxyvitamin D_3-treated HL60 cells. Cancer Res 1997; 57:2851–2855.
77. Blagosklonny MV, Schulte TW, Nguyen P, Mimnaugh EG, Trepel J, Neckers L. Taxol induction of p21WAF1 and p53 requires c-raf-1. Cancer Res 1995; 55:4623–4626.
78. Beno DWA, Brady LM, Bissonnette M, Davis BH. Protein kinase C and mitogen-activated protein kinase are required for 1,25-dihydroxyvitamin D_3-stimulated *egr* induction. J Biol Chem 1995; 270: 3642–3647.
79. Kolch W, Heldecker G, Kochs G, Hummel R, Vahidi H, Mischak H, Finkenzeller G, Marmé D, Rapp UR. Protein kinase C α activates RAF-1 by direct phosphorylation. Nature (Lond) 1993; 364: 249–252.
80. Belka C, Wiegmann K, Adam D, Holland R, Neuloh M, Herrmann F, Kronke M, Brach MA. Tumor necrosis factor (TNF)-α activates c-raf-1 kinase via the p55 TNF receptor engaging neutral sphingomyelinase. EMBO J 1995; 14:1156–1165.

81. Okazaki T, Bell RM, Hannun YA. Sphingomyelin turnover induced by vitamin D_3 in HL-60 cells. Role in cell differentiation. J Biol Chem 1989; 264:19,076–19,080.
82. Bissonnette M, Tien X-Y, Niedziela SM, Hartmann SC, Frawley BP, Jr., Roy HK, Sitrin MD, Perlman RL, Brasitus TA. 1,25(OH)2 vitamin D_3 activates PKC-α in CaCo-2 cells: a mechanism to limit secosteroid-induced rise in [Ca^{2+}]. Am J Physiol 1994; 30:G465–G475.
83. Frey MR, Zhao X, Evans SS, Black JD. Activation of protein kinase C (PKC) isozymes inhibits cell cycle progression, modulates phosphorylation of the retinoblastomas (Rb) protein, and induces expression of $p21^{waf1/cip1}$ in intestinal epithelial cells. Proc Am Assoc Cancer Res 1996; 37:51.
84. Bissonnette M, Wali RK, Hartmann SC, Niedziela SM, Roy HK, Tien X-Y, Sitrin MD, Brasitus TA. 1,25-Dihydroxyvitamin D_3 and 12-O-tetradecanol phorbol 13-acetate cause differential activation of Ca^{2+}-dependent and Ca^{2+}-independent isoforms of protein kinase C in rat colonocytes. J Clin Invest 1995; 95:2215–2221.
85. Wali RK, Bissonnette M, Starvarkos J, Sitrin MD, Brasitus TA. $1,25(OH)_2D_3$ causes persistent activation of PKC-β_{II} in rat colonocytes and increases the baso-lateral membrane association of this isoform. Gastroenterology 1996; 110:A1131.
86. Murray NR, Baumgardner GP, Burns DJ, Fields AP. Protein kinase C isotypes in human erythroleukemia (K562) cell proliferation and differentiation. Evidence that β_{II} protein kinase C is required for proliferation. J Biol Chem 1993; 268:15,847–15,853.
87. Goss VL, Hocevar BA, Thompson LJ, Stratton CA, Burns DJ, Fields AP. Identification of nuclear β_{II} protein kinase C as a mitotic lamin kinase. J Biol Chem 1994; 269:19,074–19,080.
88. Wali R, Bissonnette M, Khare S, Aquino B, Niedziela S, Sitrin M, Brasitus T. Protein kinase C isoforms in the chemopreventive effects of a novel vitamin D_3 analogue in rat colonic tumorigenesis. Gastroenterology 1996; 111:118–126.
89. Wali RK, Frawley BP Jr, Hartmann S, Roy HK, Khare S, Scaglione-Sewell BA, Earnest DL, Sitrin MD, Brasitus TA, Bissonnette M. Mechanism of action of chemoprotective ursodeoxycholate in the azoxymethane model of rat colonic carcinogenesis: potential roles of protein kinase C-α, -β_{II}, and -ζ. Cancer Res 1995; 55:5257–5264.
90. Roy HK, Bissonnette M, Frawley BP, Jr., Wali RK, Niedziela SM, Earnest D, Brasitus TA. Selective preservation of protein kinase C-ζ in the chemoprevention of azoxymethane-induced colonic tumors by piroxicam. FEBS Lett 1995; 366:143–145.
91. Dominguez I, Diaz-Meco MT, Municio MM, Berra E, DE Herreros AG, Cornet ME, Sanz L, Moscat J. Evidence for a role of protein kinase C ζ subspecies in maturation of *Xenopus laevis* oocytes. Mol Cell Biol 1992; 12:3776–3783.
92. Nakanishi H, Brewer KA, Exton JH. Activation of the ζ isoform of protein kinase C by phosphatidylinositol 3,4,5-trisphosphate. J Biol Chem 1993; 268:13–16.
93. Diaz-Meco MT, Lozano J, Municio MM, Berra E, Frutos S, Sanz L, Moscat J. Evidence for the *in vitro* and *in vivo* interaction of Ras with protein kinase C ζ. J Biol Chem 1994; 269:31,706–13,710.
94. Diaz-Meco MT, Berra E, Municio MM, Sanz L, Lozano J, Dominguez I, Diaz-Golpe V, Lain de Lera MT, Alcami J, Paya CV, Arenzana-Seisdedos F, Virelizier J-L, Moscat J. A dominant negative protein kinase C ζ subspecies blocks NF-κB activation. Mol Cell Biol 1993; 13:4770–4775.
95. Ways DK, Posekany K, deVente J, Garris T, Chen J, Hooker J, Qin W, Cook P, Fletcher D, Parker P. Overexpression of protein kinase C-ζ stimulates leukemic cell differentiation. Cell Growth Differ 1994; 5:1195–1203.
96. Wooten MW, Zhou G, Seibenhener ML, Coleman ES. A role for ζ protein kinase C in nerve growth factor-induced differentiation of PC12 cells. Cell Growth Differ 1994; 5:395–403.
97. Levy R, Nathan I, Chaimovitz C, Shany S. The involvement of calcium ions in the effect of 1,25-dihydroxyvitamin D_3 on HL-60 cells. In: Current Advances in Skeletogenesis. Hurvits S, Sela J, eds. Jerusalem: Heiliger, 1987; 240–249.
98. Levy R, Nathan I, Barnea E, Chaimovitz C, Shany S. The involvement of calcium ions in the effect of 1,25-dihydroxyvitamin D_3 on HL-60 cells. Exp Hematatol 1988; 16:290–294.
99. Hazav P, Shany S, Moran A, Levy R. Involvement of intracellular pH elevation in the effect of 1,25-dihydroxyvitamin D_3 on HL-60 cells. Cancer Res 1989; 49:72–75.
100. Brelvi ZS, Studzinski GP. Inhibition of DNA synthesis by an inducer of differentiation of leukemic cells, $1\alpha,25$ dihydroxyvitamin D_3, precedes down regulation of the *c-myc* gene. J Cell Physiol 1986; 128:171–179.
101. Brelvi ZS, Christakos S, Studzinski GP. Expression of monocyte-specific oncogenes *c-fos* and *c-fms* in HL60 cells treated with vitamin D_3 analogs correlates with inhibition of DNA synthesis and reduced calmodulin concentration. Lab Invest 1986; 55:269–275.

102. Obeid LM, Okazaki T, Karolak LA, Hannun YA. Transcriptional regulation of protein kinase C by 1,25-dihydroxyvitamin D_3 in HL-60 cells. J Biol Chem 1990; 265:2370–2374.
103. Simpson RU, Hsu T, Begley DA, Mitchell BS, Alizadeh BN. Transcriptional regulation of the c-myc protooncogene by 1,25-dihydroxyvitamin D_3 in HL-60 promyelocytic leukemia cells. J Biol Chem 1987; 262:4104–4108.
104. Solomon DH, O'Driscoll K, Sosne G, Weinstein IB, Cayre YE. 1α,25-dihydroxyvitamin D_3-induced regulation of protein kinase C gene expression during HL-60 cell differentiation. Cell Growth Differ 1991; 2:187–194.
105. Gartel AL, Serfas MS, Tyner AL. p21-negative regulation of the cell cycle. Proc Soc Exp Biol Med 1996; 213:138–149.
106. Liu M, Lee M-H, Cohen M, Bommakanti M, Freedman LP. Transcriptional activation of the Cdk inhibitor p21 by vitamin D_3 leads to the induced differentiation of the myelomonocytic cell line U937. Genes Dev 1996; 10:142–153.
107. Xu H-M, Tepper CG, Jones JB, Fernandez CE, Studzinski GP. 1,25-Dihydroxyvitamin D_3 protects HL60 cells against apoptosis but down-regulates the expression of the bcl-2 gene. Exp Cell Res 1993; 209:367–374.
108. Oberg F, Botling J, Nilsson K. Functional antagonism between vitamin D_3 and retinoic acid in the regulation of CD14 and CD23 expression during monocytic differentiation of U-937 cells. J Immunol 1993; 150:3487–3495.
109. Brown G, Bunce CM, Rowlands DC, Williams GR. All-trans retinoic acid and 1α,25- dihydroxyvitamin D_3 co-operate to promote differentiation of the human promyeloid leukemia cell line HL60 to monocytes. Leukemia 1994; 8:806–815.
110. Bollag W. Experimental basis of cancer combination chemotherapy with retinoids, cytokines, 1,25-dihydroxyvitamin D_3, and analogs. J Cell Biochem 1994; 56:427–435.
111. Rosol TJ, Nagode LA, Couto CG, Hammer AS, Chew DJ, Peterson JL, Ayl RD, Steinmeyer CL, Capen CC. Parathyroid hormone (PTH)-related protein, PTH, and 1,25-dihydroxyvitamin D in dogs with cancer-associated hypercalcemia. Endocrinology 1992; 131:1157–1164.
112. Haq M, Kremer R, Goltzman D, Rabbani SA. A vitamin D analogue (EB1089) inhibits parathyroid hormone-related peptide production and prevents the development of malignancy-associated hypercalcemia in vivo. J Clin Invest 1993; 91:2416–2422.

24 Vitamin D and Breast Cancer

*Johannes P. T. M. van Leeuwen,
Trudy Vink-van Wijngaarden,
and Huibert A. P. Pols*

1. INTRODUCTION

During the last 15 years, it has become evident that the biologically most active form of vitamin D_3, 1,25-dihydroxyvitamin D_3 [$1,25(OH)_2D_3$], exerts effects on a variety of tissues apparently unrelated to calcium homeostasis. $1,25(OH)_2D_3$ has been shown to promote cellular differentiation and inhibit proliferation of hematopoietic cells, cancer cells, and keratinocytes. In addition, studies with animal models for cancer have shown that $1,25(OH)_2D_3$ administration can prolong the survival of leukemic mice and suppress the growth of tumors of different origin, including breast, colon, skin, and lung *(1,2)*.

These newly discovered properties suggested a possible role of the hormone in the treatment of cancer. However, a major drawback for a clinical application is that high doses are needed. These doses produce serum levels of $1,25(OH)_2D_3$ far above the physiologic level, which may lead to hypercalcemia. To circumvent this problem, many investigators have tried to change the $1,25(OH)_2D_3$ molecule so that it retains its antiproliferative and differentiation-inducing activity, but has a reduced effect on calcium and bone metabolism. This strategy has resulted in new synthetic vitamin D_3 analogs with clinical potential *(3–5)*.

Breast cancer is the most frequent cause of cancer death in women in the western world. Many studies have tried to identify the causal factors responsible for the uncontrolled growth of the tumor cells. A variety of biochemical and genetic changes have been identified in breast carcinomas and have been found to be related to breast cancer growth. However, especially because of the heterogeneity of the disease on the clinical, biologic, and genetic levels, the exact mechanism of breast cancer development and progression is still unclear.

This chapter focuses on the possible role of $1,25(OH)_2D_3$ and analogs for the treatment of breast cancer. It gives an overview of the antiproliferative action of $1,25(OH)_2D_3$ and $1,25(OH)_2D_3$ analogs on breast cancer cells in culture and in breast cancer animal models. Possible combination therapies with established endocrine and cytotoxic agents are discussed, and attention is given to the mechanism of the antiproliferative action of vitamin D_3 compounds.

From: *Vitamin D: Physiology, Molecular Biology, and Clinical Applications*
Edited by: M. F. Holick © Humana Press Inc., Totowa, NJ

2. SUNLIGHT AND BREAST CANCER INCIDENCE

Descriptive epidemiologic studies have indicated that vitamin D might play a protective role against breast cancer. Incidence and mortality of breast cancer vary considerably worldwide and reveal a geographic pattern. The lowest rates of breast cancer generally occur in countries close to the equator. As the latitude increases, reported breast cancer incidence and mortality rates also increase. Studies in the United States and the former Soviet Union have shown a negative correlation between available sunlight and breast cancer death rates. Sunlight exposure is a measure of vitamin D produced in the skin. Furthermore, it is known that the primary source of vitamin D for adults in the United States is casual exposure to sunlight. Therefore, it has been hypothesized that vitamin D formed in the skin may reduce the risk of breast cancer *(6–8)*.

3. THE VITAMIN D RECEPTOR IN BREAST CANCER

The presence of the vitamin D receptor (VDR) in breast cancer was first demonstrated in the human breast cancer cell line MCF-7 *(9)*. Later studies have extended this finding to other breast cancer cell lines and to surgically obtained normal breast and breast tumor tissue *(10)*. The VDR is present in approx 80% of human breast tumor specimens. The presence of the VDR is not correlated to the presence of other steroid hormone receptors (estrogen receptor, progesterone receptor) *(11–14)*. No relationship was observed between VDR status and clinical indices (age, menopausal status, T-stage, histology, lymph node involvement) and overall survival *(12–15)*. However, two studies reported that the receptor status correlated positively with disease-free interval *(14,15)*.

Regulation of the VDR number may affect the cellular responsiveness to $1,25(OH)_2D_3$, the active form of vitamin D_3. In a number of different systems regulation of the VDR by $1,25(OH)_2D_3$ itself (homologous upregulation), and by hormones and growth factors has been demonstrated *(2,16–21)*. In breast cancer, homologous upregulation of the VDR has been observed in MCF-7 cells *(22)*. Heterologous upregulation by serum, growth factors [epidermal growth factor (EGF), insulin, insulin-like growth factor-1 (IGF-I)], and estradiol was noticed in MCF-7 and T47-D cells *(23,24)*, whereas in another study no effect of estradiol or tamoxifen was observed *(25)*.

For cancer cells the presence of a functional VDR is a prerequisite for a growth-regulatory response, and a relationship between VDR level and growth inhibition has been suggested for breast cancer cells *(26,27)*, but also for other cancer cells like colon carcinoma *(28)*, prostate cancer *(29,30)*, and osteosarcoma *(31)*. However, the presence of the VDR is not always coupled to a growth-inhibitory response of $1,25(OH)_2D_3$. Results with breast cancer cells *(32,33)* and several other cell types *(34–38)* demonstrated a lack of growth inhibition by $1,25(OH)_2D_3$ whereas the VDR is present. In the resistant MCF-7 clone, this defect is not located at a very common site in the growth-inhibitory pathway of the cell because with the antiestrogen tamoxifen the growth could still be inhibited *(33)*. So far the underlying mechanism(s) of the VDR-independent resistance to growth inhibition is unknown. That these growth-resistant cells contain a functional VDR is demonstrated by the fact that some of these cells can still be induced to differentiate *(see below)*. Moreover, as demonstrated by gel-shift analyses, the VDR from the growth-resistant MCF-7 clone is still capable of binding to a consensus vitamin D response element *(33)*. Together, these data indicate that it is probably a specific defect in the growth-inhibitory pathway of 1,25(OH)$_2$D$_3$.

4. VITAMIN D EFFECTS ON CULTURED BREAST CANCER CELLS

The first studies on the effect of 1,25(OH)$_2$D$_3$ on breast cancer cells showed a biphasic growth response of the estrogen receptor-positive T47-D human breast tumor cell line. At low concentrations (10^{-11} M), a stimulation of cell growth was observed, whereas at higher concentrations (10^{-8} M), an inhibition was observed *(10,39,40)*. The presence of a biphasic effect and the extent of the growth inhibition was shown to be dependent on the culture conditions, i.e., the concentration and charcoal treatment of fetal calf serum added to the culture medium *(10,40,41)*. The growth-inhibitory effect of 1,25(OH)$_2$D$_3$ was confirmed in other breast tumor cell lines and shown to be independent of the estrogen receptor status *(41,42)*. In the breast cancer cells shown to be growth inhibited by 1,25(OH)$_2$D$_3$ or one of its analogs, this was the result of a cell cycle block in the G$_0$/G$_1$ phase. 1,25(OH)$_2$D$_3$ and 1,25(OH)$_2$D$_3$ analogs caused an increase of the number of cells in the G$_0$/G$_1$ and occasionally in the G$_2$ phase together with a decrease of the number of cells in the S phase *(32,41,43–46)*. Apoptosis (programmed cell death), which is related to the cell cycle, is an orderly and characteristic sequence of biochemical and structural changes resulting in the death of the cell *(47)*. Induction of apoptosis can be a way in which 1,25(OH)$_2$D$_3$ inhibits tumor cell growth. For several cancer cell types, 1,25(OH)$_2$D$_3$ has been shown to induce apoptosis. Only for the MCF-7 breast cancer cell line has induction of apoptosis been described *(32,46,48,49)*. A central role for apoptosis in the action of 1,25(OH)$_2$D$_3$ is uncertain because growth inhibition of several breast cancer cells appeared to be independent of apoptosis *(32)*. Also, MCF-7 cells, growth inhibited by 1,25(OH)$_2$D$_3$, could, after removal of 1,25(OH)$_2$D$_3$, again be growth stimulated *(50)*. Possibly in these latter cases induction of differentiation is more prominent. Treatment of breast cancer cells with 1,25(OH)$_2$D$_3$ resulted in morphologic changes, which may resemble a more differentiated status of the cells *(40,51,52)*. Induction of differentiation was recently supported by data on increase in cellular lipid content *(32)*. Elstner et al. *(32)* reported that 1,25(OH)$_2$D$_3$ and several of its analogs increased the number of lipid-positive MCF-7, BT 474, MDA-MB-231, SK-BR-3, BT 20, and MDA-MB-436 breast cancer cells. In particular, their observation that 1,25(OH)$_2$D$_3$ did not inhibit the growth of BT 20 and MDA-MB-436 cells suggests that induction of differentiation and growth inhibition are two independent processes. The antiproliferative effects of various synthetic vitamin D$_3$ analogs on breast cancer cells are summarized in Table 1. These analogs have been shown to be more potent than 1,25(OH)$_2$D$_3$ in the growth inhibition of several cancer cell types, whereas their in vivo calcemic activity was similar or even reduced compared with 1,25(OH)$_2$D$_3$.

5. IN VIVO TUMOR SUPPRESSION BY VITAMIN D

Several animal models for breast cancer are being used. Mammary tumors can be induced in rats by oral administration of the carcinogens *N*-nitroso-*N*-methylurea (NMU) or 7,12-dimethylbenz[a]anthracene (DMBA). NMU- and DMBA-induced tumors form a model for hormone-dependent tumors, as they contain considerable amounts of estrogen receptors and regress on ovariectomy and antiestrogen treatment *(59,60)*. The presence

Table 1
Inhibition of Breast Cancer Cell Growth by Vitamin D_3 Analogs[a]

Vitamin D_3 analog	Breast cancer cell line	Relative potency compared with $1,25(OH)_2D_3$	Ref.
OCT	MCF-7	approx. 10; 0.25	25,42
	ZR-75-1	approx. 10; 0.5	25,42
	T47-D	approx. 10	42
	MDA-MB-231	approx. 10	42
	BT-20	approx. 10	42
EB1089	MCF-7	50,67	25,53,54
	ZR-75-1	29	25
KH1060	MCF-7	100, 1400	25,32
	ZR-75-1	100	25
	T47-D	800	56
	BT-474	100	32
	MDA-MB-231	100,000	32
	SK-BR-3	10	32
	BT-20	N.I.	32
	MDA-MB-436	N.I.	32
MC903	MCF-7	1	57
CB966	MCF-7	2	25
	ZR-75-1	1	25
MC1288	MCF-7	90	32
	BT-474	10	32
	MDA-MB-231	500	32
	SK-BR-3	8	32
	BT-20	N.I.	32
	MDA-MB-436	N.I.	32
Ro24-2637	MCF-7	1	32
	BT-474	6	32
	MDA-MB-231	N.I.	32
	SK-BR-3	0.08	32
	BT-20	N.I.	32
	MDA-MB-436	N.I.	32
Ro23-7553	MCF-7	2	32
	BT-474	2	32
	MDA-MB-231	N.I.	32
	SK-BR-3	3	32
	BT-20	N.I.	32
	MDA-MB-436	N.I.	32
Ro24-5531	MCF-7	10 - 100	55
	T47-D	10 - 100	55

[a]Based on the EC_{50} values for the growth inhibition, the relative potency compared with $1,25(OH)_2D_3$ (=1) was calculated. N.I., not inhibited. MCF-7, ZR-75-1, T47-D, BT-474 cells are estrogen receptor positive; MDA-MD-231, BT-20, MDA-MB-436, and SK-BR-3 cells are estrogen receptor negative. For $1,25(OH)_2D_3$, MC903, OCT, CB966, KH1060, EB1089, Ro24-5531 see Fig. 1; Ro23-7553, $1,25(OH)_2$-16-ene-23-yne-D_3; Ro24-2637, $1,25(OH)_2$-16-ene-D_3; MC1288, 20-epi-$1,25(OH)_2D_3$. Ro25-5317 [$1,25(OH)_2$-16,23Z-diene-26,27-hexafluoro-D_3] and Ro24-5583 [1α-fluoro-25(OH)-16-ene-23-yne-26,27-hexafluoro-D_3] have been tested in MCF-7 cells, but no EC_{50}s were presented (58).

of the VDR has been demonstrated in NMU tumors *(61)*. In the nude mice model, immune-deficient athymic mice are inoculated subcutaneously with human breast cancer cells derived from continuous cell cultures. Alternatively, transplantable breast tumors can be propagated in vivo in the nude mice model. Both estrogen-dependent and estrogen-independent tumors can be studied in the nude mouse model *(62)*.

Oral or intraperitoneal administration of $1,25(OH)_2D_3$ or 1α-hydroxyvitamin D_3, which is rapidly converted to $1,25(OH)_2D_3$ in the liver, resulted in an inhibition of the growth of NMU- *(15,57)* and DMBA-induced rat mammary tumors *(63,64)*. By contrast, Noguchi et al. *(65)* did not find an effect of $1,25(OH)_2D_3$ on the incidence and growth of DMBA-induced rat mammary tumors. In the latter study, $1,25(OH)_2D_3$ was applied on shaved dorsal skin at doses that produced mild hypercalcemia. To achieve tumor suppression, high doses of $1,25(OH)_2D_3$ or 1α-hydroxyvitamin D_3 were needed. These high doses, approx 0.5 µg/kg BW, resulted in the development of hypercalcemia and subsequent weight loss. In an attempt to overcome this problem, synthetic vitamin D_3 analogs with low in vivo calcemic activity have been developed. So far, only a few analogs have been evaluated in vivo for their potential use in the treatment of breast cancer. The modifications in the $1,25(OH)_2D_3$ structure of the analogs so far tested in in vivo breast cancer studies (*see* Section 5.) are shown in Fig. 1. The in vivo data are summarized in Table 2 and show that with some analogs an inhibition of tumor growth can indeed be achieved, without raising serum calcium levels.

To date, only two clinical studies on the effect of vitamin D_3 analogs on cancer growth in humans have been reported. Topical application of calcipotriol (MC903) in a small group of patients with locally advanced or cutaneous metastatic breast cancer showed a reduction of the size of treated lesions in 4 of 14 patients *(70)*, whereas another study could not confirm this observation *(71)*. In a phase I trial the analog EB1089 is being examined in advanced breast cancer, but no detailed analyses have been published.

6. COMBINATION THERAPIES WITH VITAMIN D: IN VITRO AND IN VIVO LABORATORY RESULTS

The data obtained with $1,25(OH)_2D_3$ and synthetic vitamin D_3 analogs offer promise for the use of vitamin D in endocrine treatment of estrogen receptor-positive and -negative breast cancer. Single-agent treatment with a low calcemic vitamin D_3 analog could provide a new endocrine therapy; however, combination therapy with other tumor-effective drugs may provide an even more beneficial effect. Hitherto several in vitro and in vivo studies have focussed on possible future combination therapies with $1,25(OH)_2D_3$ and $1,25(OH)_2D_3$ analogs.

Presently, the antiestrogen tamoxifen is the most widely used endocrine agent. Tamoxifen is effective in prolonging both disease-free interval and overall survival. About one-third of all patients with breast cancer will respond to tamoxifen therapy. Patients with mammary carcinomas that are both estrogen receptor and progesterone receptor positive display response rates of approx 70% compared with response rates of <20% in patients with mammary carcinomas that are estrogen receptor and progesterone receptor negative *(72–74)*. A major problem of tamoxifen therapy is that in case of response, the tumor almost inevitably progresses to a tamoxifen-resistant state during prolonged therapy. Furthermore, long-term tamoxifen therapy has been linked to an increased risk of endometrial cancer. Therefore, despite the efficacy of tamoxifen for breast cancer, alternative additional endocrine therapies are needed.

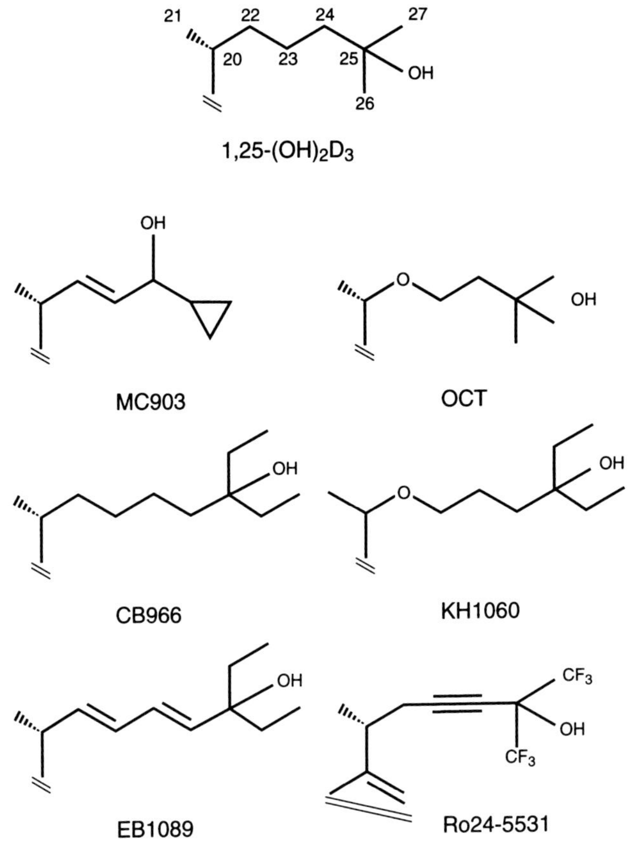

Fig. 1. Side-chain analogs of 1,25(OH)$_2$D$_3$. MC903 is also referred to as calcipotriol, OCT as 22-oxa-calcitriol, and Ro24-5531 as 1α,25-dihydroxy-16-ene-23-yne-26,27-hexafluoro-vitamin D$_3$. Note that Ro24-5531 has also a modification of the D-ring. MC903, CB966, EB1089, KH1060, and MC1288 (Table 1) are compounds from Leo Pharmaceutical Products (Denmark); OCT is a compound of Chugai Pharmaceutical Co., (Japan); and Ro24-5531 and other Ro analogs (Table 1) are compounds of Hoffmann-LaRoche (USA).

Several in vitro and in vivo studies have focused on possible future combination therapies with vitamin D$_3$ compounds. Abe-Hashimoto et al. *(68)* observed a synergistic antiproliferative effect of submaximum dosages of 22-oxa-calcitriol and tamoxifen in MCF-7 and ZR-75-1 breast cancer cells. In addition, a synergistic action of 22-oxa-calcitriol and tamoxifen was observed in vivo in athymic mice implanted with MCF-7 cells. Vink et al. *(50)* have observed that combined treatment with 1,25(OH)$_2$D$_3$ and tamoxifen resulted in a stronger growth inhibition of MCF-7 and ZR-75-1 cells than treatment with either compound alone. With a number of vitamin D$_3$ analogs (CB966, EB1089, KH1060, and 22-oxa-calcitriol), a similar effect was observed *(25)*. In combination with tamoxifen the cells were more sensitive to the antiproliferative action of 1,25(OH)$_2$D$_3$ and the analogs, i.e., the EC$_{50}$s of the vitamin D$_3$ compounds in the presence of tamoxifen were lower than those in the absence of tamoxifen. The magnitude of the shift was the dependent on the vitamin D$_3$ compound tested and varied from 2- to 4000-fold *(25)*.

Anzano et al. *(55)* have studied a possible role of 1,25-dihydroxy-16-ene-23-yne-26, 27F6-vitamin D$_3$ (Ro24-5531) in combination with low doses of tamoxifen for the

Table 2
Antitumor Effects of Vitamin D_3 Analogs in Animal Models for Breast Cancer[a]

Vitamin D_3 analog	Model	Effective dose (administration)	Antitumor effect	Calcemic effect	Ref.
OCT	Athymic mice implanted with ER-neg. MX-1 tumor	0.01–1 µg/kg BW (oral and intratumor)	Tumor suppression	No rise in serum Ca	42,68
OCT	DMBA-induced rat mammary tumor	0.1, 1 µg/kg BW (i.m.)	Tumor suppression	No rise in serum Ca	66
OCT	Athymic mice implanted with ER-pos. MCF-7 cells	0.001–1 µg/kg BW (oral)	Tumor suppression	No rise in serum Ca	68
MC903	NMU-induced rat mammary tumor	50 µg/kg BW (i.p.)	Tumor suppression	Small rise in serum Ca	57
EB1089	NMU-induced rat mammary tumor	0.5 µg/kg BW (oral)	Tumor suppression	No rise in serum Ca	67,69
CB966	NMU-induced rat mammary tumor	1 µg/kg BW (oral)	Tumor suppression	Small rise in serum Ca	69
Ro24-5531	NMU-induced rat mammary tumor	1.2 and 2.5 nmol/kg diet	Decreased tumor incidence	No rise in serum Ca	55

[a] The chemical structures of the vitamin D_3 analogs are shown in Fig. 1.

prevention of breast cancer. They have observed that Ro24-5531 significantly enhanced the ability of tamoxifen to reduce the total tumor burden of rats treated with the carcinogen NMU.

Also, combinations with endocrine therapies other than antiestrogens have been studied. In a study of Iino et al. *(63)*, 1α-hydroxyvitamin D_3 (0.5 µg/kg BW) was as effective as the progestin medroxyprogesterone acetate (MPA) (1 mg/kg BW) in the DMBA-induced rat mammary tumor model. Although a combination therapy did not result in an enhanced antitumor effect, MPA reduced the weight loss induced by 1α-hydroxyvitamin D_3. Vitamin A derivatives like fenretinide are currently being tested in clinical trials as a preventive agent against recurrence of breast cancer, and animal studies point to a potential use of these compounds as therapeutic agents for breast cancer *(75)*. A combination of retinoic acid and $1,25(OH)_2D_3$ was studied by Koga et al. *(76)*, and they observed a synergistic growth inhibition of T47-D breast cancer cells. Also, a synergistic growth inhibition of MCF-7 cells by $1,25(OH)_2D_3$ and tumor necrosis factor has been described *(77)*. Hassan et al. *(78)* showed with MCF-7 cells that in the presence of granulocyte-macrophage colony-stimulating factor lower concentrations of $1,25(OH)_2D_3$ could be used to achieve a similar antiproliferative effect.

Furthermore, combinations with vitamin D_3 compounds and cytotoxic drugs have been studied. Using athymic mice implanted with estrogen receptor-negative MX-1 breast tumors, Abe et al. *(42)* have shown a more beneficial antiproliferative response by combined treatment with 22-oxa-calcitriol and Adriamycin than by these agents alone. Iino et al. *(63)* have tested the effect of combined treatment with 1α-hydroxyvitamin D_3 and 5-fluorouracil in the DMBA-induced rat mammary tumor model. They measured a similar tumor-suppressive effect of 5-fluorouracil and 1α-hydroxyvitamin D_3 after 4 wk. However, the combination therapy did not result in an enhanced antitumor effect. Cho et al. *(79)* observed that low concentrations of carboplatin and cisplatin interacted synergistically with $1,25(OH)_2D_3$ to inhibit MCF-7 cell growth.

The data on combinations of $1,25(OH)_2D_3$ and $1,25(OH)_2D_3$ analogs with various other anticancer compounds are promising and justify further analyses. For example, the development of effective combination therapies may result in better response rates and, lower dosages, thereby reducing the risk of negative side effects.

7. MECHANISMS FOR THE ANTIPROLIFERATIVE ACTION OF VITAMIN D ON BREAST CANCER CELLS

7.1. Modulation of (Onco)gene Expression

Oncogenes affect cellular growth and differentiation through their protein products, which belong to various categories of the cell signaling machinery, such as growth factors, growth factor receptors, guanosine triphosphate binding proteins, and nuclear proteins involved in transcription regulation *(80,81)*. Oncogene expression at an inappropriate location or time during cell cycle or maturation, overexpression of oncogenes, or mutations in oncogenes cause the transformation of cells in culture and induction of tumors in animals *(80–82)*. A second class of cancer-related genes, the tumor suppressor genes, normally prevents tumor growth. Mutations or deletions in these genes cause their functional inactivation, which in turn contributes to cellular transformation *(83,84)*.

A number of oncogenes and tumor suppressor genes have been implicated in the development of breast cancer *(85–91)*. It has been proposed that oncogenes (and tumor suppressor genes) play a role in the tumor-suppressive activity of vitamin D_3 compounds. The role of oncogenes in the cell-regulating activity of $1,25(OH)_2D_3$ has been investigated most extensively in the leukemic cell line HL60. In these cells $1,25(OH)_2D_3$ induced a progressive downregulation of *myc* and a stimulation of *fos* and *fms*, which was initiated after approximately 4 h of incubation and preceded the differentiation into monocyte-like cells and loss of proliferation capacity *(92,93)*. In addition, $1,25(OH)_2D_3$ modulated the expression of *myc*, *fos*, and *jun* in other cell types than HL60 cells *(36,94–97)*. However, very little data are available on the regulation of oncogenes by $1,25(OH)_2D_3$ in breast cancer cells. A decreased c-*myc* expression *(53,98)* and a transient induction of *fos* expression *(53)* by $1,25(OH)_2D_3$ and EB1089 treatment of MCF-7 cells have been reported. However, another study did not observe an inhibition of *myc* mRNA expression by $1,25(OH)_2D_3$ and analogs in MCF-7 and ZR-75-1 cells, either under basal growth conditions or during growth stimulation *(25)*. In this situation $1,25(OH)_2D_3$ did inhibit the growth of these cells; thus it is questionable whether regulation of c-*myc* expression plays a crucial role in their growth-inhibitory action. Also, $1,25(OH)_2D_3$ has been found to have no effect on induction of *fos* mRNA expression in MCF-7 cells *(99)*.

Two recent studies demonstrated in MCF-7 cells a stimulation of tumor suppressor protein p53 expression by $1,25(OH)_2D_3$ and the analogs KH1060 and EB1089 *(32,48, 99a)*. Considering the growth-inhibitory effect of $1,25(OH)_2D_3$ and analogs, this is an interesting observation. p53 is postulated to be involved in preventing progression of the cell cycle from G_1 to S *(100)* and, as described above (*see* Section 4.), $1,25(OH)_2D_3$ and analogs have been shown to block the cell cycle of breast cancer cells in the G_0/G_1 phase. However, at present the precise role of p53 in the growth-inhibitory action of $1,25(OH)_2D_3$ and analogs needs to be established because Elstner et al. *(32)* also reported growth inhibition independent of $1,25(OH)_2D_3$- or analogs-induced changes in p53. In line with the effects on p53 is the stimulation observed of the cyclin-dependent kinase inhibitor p21$^{WAF1, CIP1}$ *(99a,100a,100b)*. The tumor suppressor gene retinoblastoma is also interesting in relation to the cell cycle. The retinoblastoma gene product can be in either a phosphorylated or a dephosphorylated state. In the phosphorylated form it can activate several transcription factors and cause transition to the S phase and DNA synthesis. In several cancer cells and also in a breast cancer cell line, 1,25(OH)$_2$D$_3$ caused dephosphorylation of the retinoblastoma gene product, which is related to growth inhibition *(27,46,100b)*. Besides p21 and p53, effects of the vitamin D analog EB1089 on other regulatory components (cyclins, cyclin-dependent-kinases) of the cell cycle have also been noted *(100c)*.

In regard to gene expression involved in the 1,25(OH)$_2$D$_3$/analogs inhibition of breast cancer growth and induction of apoptosis, the *bcl-2* gene is a candidate. The *bcl-2* gene product has been shown to inhibit apoptosis *(101)*. 1,25(OH)$_2$D$_3$, KH1060, and EB1089 decrease *bcl-2* expression in several breast cancer cell lines (MCF-7, BT-474, MDA-MB-231) *(32,48,101a)*. However, only in MCF-7 cells was this change in *bcl-2* expression accompanied by apoptosis, whereas the growth inhibition of other breast cancer cell lines (BT-474, MDA-MB-231, SK-BR-3) was independent of apoptosis *(32)*.

At present it is clear that 1,25(OH)$_2$D$_3$ and 1,25(OH)$_2$D$_3$ analogs have effects on the expression of various oncogenes and tumor suppressor genes. The data so far are, however, not conclusive with regard to which genes are crucial in growth inhibition by 1,25(OH)$_2$D$_3$, possibly because their postulated role is often complex, e.g., *c-myc* expression has been related to induction of apoptosis as well as cell cycle progression.

7.2. Interaction with Steroid Hormones and Polypeptide Growth Factors

Breast cancer cell growth can be regulated by steroid hormones, acting via nuclear steroid hormone receptors, and by polypeptide growth factors, acting via membrane receptors. Estradiol is the most important steroidal growth stimulator of breast cancer cell proliferation. Studies of James et al. *(54)* and Demirpence et al. *(102)* pointed to an interaction of 1,25(OH)$_2$D$_3$ with estradiol. 1,25(OH)$_2$D$_3$ and EB1089 suppressed the mitogenic effect of estradiol in MCF-7 cells. In other studies, an interaction with estradiol was less clear. A very small inhibition of estradiol-induced growth by 1,25(OH)$_2$D$_3$ and analogs in MCF-7 cells, but a potent inhibition in ZR-75-1 cells was shown *(25,50)*. James et al. *(54)* have suggested that the suppression of estradiol-induced growth involved modulation of estrogen receptor expression. They have observed that EB1089, and to a lesser extent 1,25(OH)$_2$D$_3$, decreased the estrogen receptor levels of MCF-7 cells. However, Vink et al. *(25)* did not observe modulation of estrogen receptor levels in either MCF-7 or ZR-75-1 cells. Demirpence et al. *(102)* have observed an interaction of 1,25(OH)$_2$D$_3$ and estradiol on the transcriptional level. They have shown that 1,25(OH)$_2$D$_3$ inhibited estradiol-induced gene expression of an endogenous gene (pS2), as well as various exogenous transfected genes in MCF-7 cells.

Steroid hormones may act directly on the tumor cell to stimulate growth, or indirectly via regulation of growth factor production and growth factor receptors. Breast cancer cells secrete a number of growth factors including transforming growth factor-β (TGF-β), IGF-1, IGF-2, and platelet-derived growth factor, which may act by autocrine loops, when the cells possess the adequate receptors. Alternatively, growth factors are derived from the circulation or are produced by stromal cells in the tumor and act in a paracrine manner on breast cancer cells *(103–106)*.

There are indications that $1,25(OH)_2D_3$ can interfere with paracrine or autocrine acting growth factors in breast cancer. Stimulation of breast cancer cell proliferation by coculture with fibroblasts is inhibited by $1,25(OH)_2D_3$ *(107)*. A good example of a specific factor is TGF-β, a negative growth factor produced by breast cancer cell lines *(108–110)*. TGF-β can inhibit the ability of cells to enter the S phase when it is present during the G_1 phase. $1,25(OH)_2D_3$ and several analogs stimulated the expression of TGF-β mRNA and secretion of active and latent TGF-β1 by a breast cancer cell line *(111)*. In addition, $1,25(OH)_2D_3$ growth inhibition of MCF-7 cells is inhibited by a TGF-β neutralizing antibody *(46)*. Comparable data were obtained with other cancer cell types *(112)*. Therefore, TGF-β is a likely candidate to play a role in $1,25(OH)_2D_3$-induced growth inhibition.

Interactions with other growth factors have also been described. We have observed that $1,25(OH)_2D_3$ and analogs reversed the mitogenic effect of exogenous insulin and IGF-I in MCF-7 cells *(99)* and EGF in ZR/HERc cells *(25)*. An antigrowth factor activity of vitamin D_3 compounds could involve the regulation of growth factor receptor numbers. It has been shown that the number of EGF binding sites was decreased by $1,25(OH)_2D_3$ treatment in some breast cancer cell lines (MCF-7, T47-D) *(113)*, whereas in other cell lines (MDA-MB-231, BT-20), EGF binding was increased by $1,25(OH)_2D_3$ *(64,113,114)*. Comparably, $1,25(OH)_2D_3$ and the analog EB1089 inhibit insulin and IGF-stimulated breast cancer cell growth, while this is accompanied by an increase in their respective receptor numbers *(99)*. As $1,25(OH)_2D_3$ reduced the growth of all these cell lines, the relation between receptor regulation and growth inhibition by $1,25(OH)_2D_3$ is not clear. It can be speculated that it is a feedback mechanism of the cells to growth inhibition. This would be in line with the observation that MCF-7 cells can be restimulated to proliferate after removal of $1,25(OH)_2D_3$ *(50)*. Xie et al. *(114a)* also showed growth inhibition of IGF-1-stimulated growth of breast cancer cells; however, they observed a concomitant decrease in IGF-1 binding. Within the IGF-1 system specifically stimulatory effects on IGF binding protein 5 were observed *(114b,* our unpublished observations), which may contribute to the growth-inhibitory effect of vitamin D, as it has been shown to have a growth-inhibitory action on breast cancer cells *(114c)*.

Together, $1,25(OH)_2D_3$ affects breast cancer cell growth by influencing these complex growth-regulatory systems of steroid hormones and growth factors. $1,25(OH)_2D_3$ may inhibit the secretion of stimulatory growth factors or stimulate the secretion of negative growth factors. Furthermore, $1,25(OH)_2D_3$ may regulate the number of growth factor or steroid receptors. In addition, $1,25(OH)_2D_3$ may interfere with the action of steroid hormones on a nuclear level, or with the intracellular signaling pathways of membrane-bound growth factors.

7.3. The Role of Calcium

Calcium (Ca^{2+}) is an important intracellular regulatory molecule and acts as an intracellular messenger for extracellular signaling molecules. Because of the well-known effects of $1,25(OH)_2D_3$ on calcium translocation in the intestine, it has been suggested that the antiproliferative and differentiation-inducing effects of $1,25(OH)_2D_3$ are mediated via the regulation of calcium-related processes.

It has been demonstrated that $1,25(OH)_2D_3$ increases calcium uptake in a number of different cell types. This phenomenon can either be a rapid mobilization of calcium, probably via a nongenomic mechanism, or a slow calcium influx, which may proceed via a genomic mechanism (e.g., via the production of vitamin D-dependent calbindins)

(2,115). An elevation of intracellular calcium appears to participate in monocytic differentiation of leukemic cells *(116,117)*. VandeWalle et al. *(49)* demonstrated that 1,25(OH)$_2$D$_3$ and the analog 1,25(OH)$_2$-16-ene-13-yne-D$_3$ deplete the intracellular calcium stores of MCF-7 breast cancer cells, which appeared to be paralleled by the induction of apoptosis.

The antiproliferative action of 1,25(OH)$_2$D$_3$ on malignant cells was shown to be dependent on the extracellular calcium concentration. The growth inhibition by 1,25(OH)$_2$D$_3$ was enhanced by a low extracellular calcium concentration in breast cancer cells *(118)*, leukemic cells *(119,120)*, and colon cancer cells *(121)*. In a more general way an inverse association between calcium and vitamin D intake and development of breast cancer has been described *(122,123)*.

Finally, research on the cellular mechanisms involved in growth inhibition of breast cancer cells by 1,25(OH)$_2$D$_3$ and its analogs has resulted in several possible candidates. However, hitherto for every candidate both positive and negative data have been available, and therefore no distinct growth-inhibitory mechanism can be put forward. This is probably because of the heterogeneity of the breast cancer cell lines but also to the heterogeneity within a cell lines (clones, subclones) used in the various laboratories, which each may represent specific phenotypes of breast cancer. On basis of the data that 1,25(OH)$_2$D$_3$ and analogs inhibit breast cancer cell growth induced by various stimuli, it can be concluded that they interfere at a common site, distal in the growth-stimulatory pathways used by the various stimuli. Alternatively, 1,25(OH)$_2$D$_3$ may exert its growth-inhibitory effects indirectly, via regulation of growth factor production or growth factor receptor expression.

8. VITAMIN D AND ANGIOGENESIS, INVASION, AND METASTASIS

The mechanisms of the antiproliferative action of vitamin D$_3$ compounds discussed in the previous section are studied in cell culture systems. For the tumor-suppressive activity of vitamin D$_3$ compounds in vivo, additional aspects may be involved. Vitamin D has been suggested to play a role in the processes of angiogenesis, invasion, and metastasis.

Angiogenesis, the formation of new capillary blood vessels, is an essential requirement for the growth of solid tumors. Inhibition of angiogenesis may contribute to the tumor-suppressive activity of vitamin D compounds because antiangiogenic effects of 1,25(OH)$_2$D$_3$ and 22-oxa-calcitriol have been shown using different experimental model systems *(124–126)*. In other tumor models antiangiogenic effects of 1,25(OH)$_2$D$_3$ have been described *(127–129)*.

Invasion and metastasis of tumor cells are the primary causes for the fatal outcome of cancer diseases. A recent report of Mork Hansen et al. *(130)* indicated that 1,25(OH)$_2$D$_3$ may be effective in reducing the invasiveness of breast cancer cells. They have shown that 1,25(OH)$_2$D$_3$ inhibited the invasion and migration of a metastatic human breast cancer cell line (MDA-MB-231) using the Boyden chamber invasion assay.

A fact to be considered in relation to metastasis is that bone is the most frequent site of metastasis of advanced breast cancer *(131)*. There are some indications from clinical studies that bone metastases develop preferentially in areas with high bone turnover *(132,133)*. By contrast, agents that inhibit bone resorption have been reported to reduce the incidence of skeletal metastasis *(134–136)*. As 1,25(OH)$_2$D$_3$ is an important stimulator of bone resorption and consequently of bone turnover, treatment with 1,25(OH)$_2$D$_3$ or vitamin D$_3$ analogs for breast cancer might increase the risk of skeletal metastases. So far, no in vivo studies have been reported on this subject with respect to breast cancer. However, Krempien *(137)* has reported in abstract form that following intraarterial injection with Walker 256 tumor cells, rats treated with 1,25(OH)$_2$D$_3$ developed significantly more bone metastases than untreated controls.

In relation to this latter aspect, we have shown that tamoxifen, which exerts positive estrogenic effects on bone *(138)*, considerably suppressed the bone resorption induced by 1,25(OH)$_2$D$_3$, EB1089, and KH1060, using an in vitro bone resorption assay *(139)*. These data suggest that in a vitamin D–tamoxifen combination therapy for breast cancer *(see above)*, tamoxifen may offer protection against the bone resorption induced by vitamin D$_3$ compounds and thereby decrease the risk of bone metastases.

The data on agiogenesis and metastasis show that these two processes might be involved in the inhibitory effects of vitamin D$_3$ on breast cancer but also more generally in other types of cancer.

9. CONCLUDING REMARKS

The first reports on the antiproliferative action of 1,25(OH)$_2$D$_3$ on breast cancer cells in vitro, 10–15 years ago, have since been confirmed in vivo in different animal models for breast cancer. It appears that the problem of the strong calcemic activity of 1,25(OH)$_2$D$_3$ can be circumvented by the use of new synthetic vitamin D$_3$ analogs, since it has been shown that a number of these analogs have more potent antiproliferative effects on breast cancer cells in vitro, and suppress breast cancer growth in vivo without marked calcemic effects. However, it should be noted that apart from calcemic side effects, other negative side effects may arise, in particular immunosuppressive effects and an increased risk of bone metastases. The development of new vitamin D$_3$ analogs continues, and in the future analogs with even stronger antiproliferative action and better selectivity may become available.

The mechanism of the antiproliferative action of vitamin D remains largely unclear. There are some indications that vitamin D interferes with the action of estrogen and polypeptide growth factors that influence breast cancer growth. More research is needed to define the roles of oncogenes, tumor suppressor genes, and intracellular calcium in the antiproliferative action of vitamin D. Irrespective of the lack of precise knowledge on the mechanism, it can be concluded that vitamin D and analogs act on a general site in the growth-stimulatory pathways because most breast cancer cells stimulated by a wide variety of stimuli are growth inhibited. These vitamin D and analogs have a potential broad therapeutic spectrum.

A major advantage of an endocrine therapy with vitamin D may be that vitamin D suppresses breast tumor growth independent of the presence of the estrogen receptor. Most established endocrine therapies are based on antiestrogenic action, and for estrogen receptor-negative tumors therapeutic choices are limited. Vitamin D treatment could theoretically be beneficial for a large group of patients, since the VDR is expressed in about 80% of human breast cancers. Another promising aspect of vitamin D treatment might be its combination with other established endocrine (tamoxifen) or cytotoxic agents. Finally, epidemiologic studies and laboratory results have suggested a role for vitamin D in the prevention of breast cancer. In the next few years, clinical studies are needed to confirm that vitamin D$_3$ analogs, either alone or in combination with other antitumor agents, can provide an effective treatment for breast cancer.

REFERENCES

1. Reichel H, Koeffler HP, Norman AW. The role of the vitamin D endocrine system in health and disease. N Engl J Med 1989; 320:980–991.
2. Walters MR. Newly identified actions of the vitamin D endocrine system. Endocr Rev 1992; 13: 719–764.
3. Bikle DD. Clinical counterpoint: vitamin D: new actions, new analogs, new therapeutic potential. Endocr Rev 1992; 13:765–784.
4. Pols HAP, Birkenhäger JC, van Leeuwen JPTM. Vitamin D analogues: from molecule to clinical application. Clin Endocrinol 1994; 40:285–291.
5. Bouillon R, Okamura WH, Norman AW. Structure-function relationships in the vitamin D endocrine system. Endocr Rev 1995; 16:200–257.
6. Garland FC, Garland CF, Gorham ED, Young JF. Geographic variation in breast cancer mortality in the United States: a hypothesis involving exposure to solar radiation. Prev Med 1990; 19:614–622.
7. Gorham ED, Garland FC, Garland CF. Sunlight and breast cancer incidence in the USSR. Int J Epidemiol 1990; 19:820–824.
8. Ainsleigh HG. Beneficial effects of sun exposure on cancer mortality. Prev Med 1993; 22:132–140.
9. Eisman JA, Martin TJ, MacIntyre I, Moseley JM. 1,25-Dihydroxyvitamin D receptor in breast cancer cells. Lancet 1979; 2:1335,1336.
10. Eisman JA. 1,25-Dihydroxyvitamin D3 receptor and role of 1,25-$(OH)_2D_3$ in human cancer cells. In: Vitamin D Metabolism: Basic and Clinical Aspects. Kumar R, ed. The Hague: Martinus Nijhoff, 1984; 365–382.
11. Eisman JA, Suva LJ, Sher E, Pearce PJ, Funder JW, Martin TJ. Frequency of 1,25-dihydroxyvitamin D_3 receptor in human breast cancer. Cancer Res 1981; 41:5121–5124.
12. Freake HC, Abeyasekera G, Iwasaki J, Marcocci C, MacIntyre I, McClelland RA, Skilton RA, Easton DF, Coombes RC. Measurement of 1,25-dihydroxyvitamin D_3 receptors in breast cancer and their relationship to biochemical and clinical indices. Cancer Res 1984; 44:1677–1681.
13. Berger U, Wilson P, McClelland RA, Colston K, Haussler MR, Pike JW, Coombes RC. Immunocytochemical detection of 1,25-dihydroxyvitamin D_3 receptor in breast cancer. Cancer Res 1987; 47: 6793–6799.
14. Berger U, McClelland RA, Wilson P, Greene GL, Haussler MR, Pike JW, Colston K, Easton D, Coombes RC. Immunocytochemical determination of estrogen receptor, progesterone receptor, and 1,25-dihydroxyvitamin D_3 receptor in breast cancer and relationship to prognosis. Cancer Res 1991; 51:239–244.
15. Colston KW, Berger U, Coombes RC. Possible role for vitamin D in controlling breast cancer cell proliferation. Lancet 1989; 1:188–191.
16. van Leeuwen JPTM, Birkenhäger JC, Buurman CJ, Schilte JP, Pols HAP. Functional involvement of calcium in the homologous up-regulation of the 1,25-dihydroxyvitamin D_3 receptor in osteoblast-like cells. FEBS Lett 1990; 270:165–167.
17. van Leeuwen JPTM, Pols HAP, Schilte JP, Visser TJ, Birkenhäger JC. Modulation by epidermal growth factor of the basal 1,25-$(OH)_2D_3$ receptor level and the heterologous up-regulation of the 1,25-$(OH)_2D_3$ receptor in clonal osteoblast-like cells. Calcif Tissue Int 1991; 49:35–42.
18. Pike JW. Vitamin D_3 receptors: structure and function in transcription. Annu Rev Nutr 1991; 11: 189–216.
19. van Leeuwen JPTM, Birkenhäger JC, Vink-van Wijngaarden T, van den Bemd GJCM, Pols HAP. Regulation of 1,25-dihydroxyvitamin D_3 receptor gene expression by parathyroid hormone and cAMP agonists. Biochem Biophys Res Commun 1992; 185:881–886.
20. Darwish H, De Luca HF. Vitamin D-regulated gene expression. Crit Rev Eukaryot Gene Expr 1993; 3:89–116.
21. Staal A, Birkenhäger JC, Pols HAP, Buurman CJ, Vink-van Wijngaarden T, Kleinekoort WMC, van den Bemd GJCM, van Leeuwen JPTM. Transforming growth factor β-induced dissociation between vitamin D receptor level and 1,25-dihydroxyvitamin D3 action in osteoblast-like cells. Bone Miner 1994; 26:27–42.
22. Costa EM, Hirst MA, Feldman D. Regulation of 1,25-dihydroxyvitamin D_3 receptors by vitamin D analogues in cultured mammalian cells. Endocrinology 1985; 117:2203–2210.
23. Krishnan AV, Feldman D. Stimulation of 1,25-dihydroxyvitamin D_3 receptor gene expression in cultured cells by serum and growth factors. J Bone Miner Res 1991; 6:1099–1107.

24. Escaleira MTF, Sonohara S, Brentani MM. Sex steroids induced up-regulation of 1,25-$(OH)_2$ vitamin D_3 receptors in T 47D breast cancer cells. J Steroid Biochem Mol Biol 1993; 45:257–263.
25. Vink-van Wijngaarden T, Pols HAP, Buurman CJ, van den Bemd GJCM, Dorssers LCJ, Birkenhäger JC, van Leeuwen JPTM. Inhibition of breast cancer cell growth by combined treatment with vitamin D_3 analogs and tamoxifen. Cancer Res 1994; 54:5711–5717.
26. Buras RR, Schumaker LM, Davoodi F, Brenner RV, Shabahang M, Nauta RJ, Evans SRT. Vitamin D receptors in breast cancer cells. Breast Cancer Res Treatment 1994; 31:191–202.
27. Fan FS, Yu WC. 1,25-Dihydroxyvitamin D3 suppresses cell growth, DNA synthesis, and phosphorylation of retinoblastoma protein in a breast cancer cell line. Cancer Invest 1995; 13:280–286.
28. Shabahang M, Buras RR, Davoodi F, Schumaker LM, Nauta RJ, Evans SRT. 1,25-Dihydroxyvitamin D_3 receptor as a marker of human colon carcinoma cell line differentiation and growth inhibition. Cancer Res 1993; 53:3712–3718.
29. Miller GJ, Stapleton GE, Hedlund TE, Moffatt KA. Vitamin D receptor expression, 24-hydroxylase activity, and inhibition of growth by $1\alpha,25$-dihydroxyvitamin D_3 in seven prostatic carcinoma cell lines. Clin Cancer Res 1995; 1:997–1003.
30. Hedlund TE, Moffatt KA, Miller GJ. Stable expression of the nuclear vitamin D receptor in the human prostatic carcinoma cell line JCA-1: evidence that the antiproliferative effects of $1\alpha,25$-dihydroxyvitamin D_3 are mediated exclusively through the genomic signaling pathway. Endocrinology 1996; 137:1554–1561.
31. Dokoh S, Donaldson CA, Haussler MR. Infuence of 1,25-dihydroxyvitamin D_3 on cultured osteogenic sarcoma cells: correlation with the 1,25-dihydroxyvitamin D_3 receptor. Cancer Res 1984; 44:2103–2109.
32. Elstner E, Linker-Israeli M, Said J, Umiel T, de Vos S, Shintaku PI, Heber D, Binderup L, Uskokovic M, Koeffler HP. 20-epi-Vitamin D_3 analogues: a novel class of potent inhibitors of proliferation and inducers of differeniation of human breast cancer cells. Cancer Res 1995; 55:2822–2830.
33. Narvaez CJ, Vanweelden K, Byrne I, Welsh J. Characterization of a vitamin D-resistant MCF-7 cell line. Endocrinology 1996; 137:400–409.
34. Trydal T, Lillehaug JR, Aksnes L, Aarskog D. Regulation of cell growth, c-myc mRNA, and 1,25-$(OH)_2$-vitamin D_3 receptor in C3H/10T1/2 mouse embryo fibroblasts by calcipotriol and 1,25-$(OH)_2$-vitamin D_3. Acta Endocrinol 1992; 126:75–79.
35. Xu H-M, Kolla SS, Goldenberg NA, Studzinski GP. Resistance to 1,25-dihydroxyvitamin D_3 of a deoxycytidine kinase deficient variant of human leukemia HL-60 cells. Exp Cell Res 1992; 203:244–250.
36. Sebag M, Henderson J, Rhim J, Kremer R. Relative resistance to 1,25-dihydroxyvitamin D_3 in a keratinocyte model of tumor progression. J Biol Chem 1992; 267:12,162–12,167.
37. Taoka T, Collins ED, Irino S, Norman AW. 1,25-$(OH)_2$-vitamin D_3 mediated changes in mRNA for c-myc and 1,25-$(OH)_2D_3$ receptor in HL-60 cells and related subclones. Mol Cell Endocrinol 1993; 95:51–57.
38. Lasky SR, Posner MR, Iwata K, Santos-Moore A, Yen A, Samuel V, Clark J, Maizel AL. Characterization of a vitamin D_3-resistant human chronic myelogenous leukemia cell line. Blood 1994; 84:4283–4294.
39. Freake HC, Marcocci C, Iwasaki J, MacIntyre I. 1,25-Dihydroxyvitamin D_3 specifically binds to a human breast cancer cell line (T47D) and stimulates growth. Biochem Biophys Res Commun 1981; 101:1131–1138.
40. Frampton RJ, Omond SA, Eisman JA. Inhibition of human cancer cell growth by 1,25-dihydroxyvitamin D_3 metabolites. Cancer Res 1983; 43:4443–4447.
41. Chouvet C, Vicard E, Devonec M, Saez S. 1,25-Dihydroxyvitamin D_3 inhibitory effect on the growth of two human breast cancer cell lines (MCF-7, BT-20). J Steroid Biochem 1986; 24:373–376.
42. Abe J, Nakano T, Nishii Y, Matsumoto T, Ogata E, Ikeda K. A novel vitamin D_3 analog, 22-oxa-1,25-dihydroxyvitamin D_3, inhibits the growth of human breast cancer in vitro and in vivo without causing hypercalcemia. Endocrinology 1991; 129:832–837.
43. Eisman JA, Koga M, Sutherland RL, Barkla DH, Tutton PJM. 1,25-Dihydroxyvitamin D_3 and the regulation of human cancer cell replication. Proc Soc Exp Med 1989; 191:221–226.
44. Eisman JA, Sutherland RL, McMenemy ML, Fragonas JC, Musgrove EA, Pang GYN. Effects of 1,25-dihydroxyvitamin D_3 on cell cycle kinetics of T47D human breast cancer cells. J Cell Physiol 1989; 138:611–616.

45. Pols HAP, Birkenhäger JC, Foekens JA, van Leeuwen JPTM. Vitamin D: a modulator of cell proliferation and differentiation. J Steroid Biochem Mol Biol 1990; 6:873–876.
46. Simboli-Campbell M, Welsh J. 1,25-Dihydroxyvitamin D_3: coordinate regulator of active cell death and proliferation in MCF-7 breast cancer cells. In: Apoptosis in Hormone Dependent Cancers. Tenniswood M, Michna H, eds. Berlin: Springer-Verlag, 1995; 181–200.
47. Kerr JFR, Wyllie AG, Currie AR. Apoptosis: a basic biological phenomenon with wide ranging implications in tissue kinetics. Br J Cancer 1972; 26:239–257.
48. James SY, Mackay AG, Colston KW. Vitamin D derivatives in combination with 9-cis-retinoic acid promote active cell death in breast cancer cells. J Mol Endocrinol 1995; 14:391–394.
49. VandeWalle B, Hornez L, Wattez N, Revillion F, Lefebvre J. Vitamin-D_3 derivatives and breast-tumor cell growth: effect on intracellular calcium and apoptosis. Int J Cancer 1995; 61:806–811.
50. Vink-van Wijngaarden T, Pols HAP, Buurman CJ, Birkenhäger JC, van Leeuwen JPTM. Combined effects of 1,25-dihydroxyvitamin D_3 and tamoxifen on the growth of MCF-7 and ZR-75-1 human breast cancer cells. Breast Cancer Res Treat 1993; 29:161–168.
51. Gross M, Bollman Kost S, Ennis B, Stumpf W, Kumar R. Effect of 1,25-dihydroxyvitamin D_3 on mouse mammary tumor (GR) cells: evidence for receptors, cellular uptake, inhibition of growth and alteration in morphology at physiologic concentrations of hormone. J Bone Miner Res 1986; 1:457–467.
52. Frappart L, Falette N, Lefebvre MF, Bremond A, Vauzelle JL, Saez S. In vitro study of effects of 1,25-dihydroxyvitamin D_3 on the morphology of human breast cancer cell line BT.20. Differentiation 1989; 40:63–69.
53. Mathiasen IS, Colston KW, Binderup L. EB1089, a novel vitamin D analogue, has strong antiproliferative and differentiation inducing effects on cancer cells. J Steroid Biochem Mol Biol 1993; 46:365–371.
54. James SY, Mackay AG, Colston KW. Effects of a new synthetic vitamin D analogue, EB1089, on the oestrogen-responsive growth of human breast cancer cells. J Endocrinol 1994; 141:555–563.
55. Anzano MA, Smith JM, Uskokovic MR, Peer CW, Mullen LT, Letterio JJ, Welsh MC, Shrader MW, Logsdon DL, Driver CL, Brown CC, Roberts AB, Sporn MB. 1α,25-dihydroxy-16-ene-23-yne-26,27-hexafluorocholecalciferol (Ro24-5531), a new deltanoid (vitamin D analogue) for prevention of breast cancer in the rat. Cancer Res 1994; 54:1653–1656.
56. Binderup L, Latini S, Binderup E, Bretting C, Calverley M, Hansen K. 20-Epi-vitamin D_3 analogues: a novel class of potent regulators of cell growth and immune responses. Biochem Pharmacol 1991; 42:1569–1575.
57. Colston KW, Chander SK, Mackay AG, Coombes RC. Effects of synthetic vitamin D analogues on breast cancer cell proliferation in vivo and in vitro. Biochem Pharmacol 1992; 44:693–702.
58. Brenner RV, Shabahang M, Schumaker LM, Nauta RJ, Uskokovic MR, Evans SR, Buras RR. The antiproliferative effect of vitamin D analogs on MCF-7 human breast cancer cells. Cancer Lett 1995; 92:77–82.
59. Furr BJA, Nicholson RI. Use of analogues of luteinizing hormone-releasing hormone for treatment of cancer. J Reprod Fertil 1982; 64:529–539.
60. Wilkinson JR, Williams JC, Singh D, Goss PE, Easton D, Coombes RC. Response of nitrosomethylurea-induced rat mammary tumors to endocrine therapy and comparison with clinical response. Cancer Res 1986; 46:4862–4865.
61. Colston K, Wilkinson JR, Coombes RC. 1,25-Dihydroxyvitamin D_3 binding in estrogen responsive rat breast tumor. Endocrinology 1986; 119:397–403.
62. Osborne CK, Hobbs K, Clark GM. Effect of estrogens and antiestrogens on growth of human breast cancer in athymic nude mice. Cancer Res 1985; 45:584–590.
63. Iino Y, Yoshida M, Sugamata N, Maemura M, Ohwada S, Yokoe T, Ishikita T, Horiuchi R, Morishita Y. 1α-Hydroxyvitamin D_3, hypercalcemia, and growth suppression of 7,12-dimethylbenz[a]anthracene-induced rat mammary tumors. Breast Cancer Res Treat 1992; 22:133–140.
64. Saez S, Falette N, Guillot C, Meggouh F, Lefebvre MF, Crepin M. 1,25(OH)$_2$D$_3$ Modulation of mammary tumor cell growth in vitro and in vivo. Breast Cancer Res Treat 1993; 27:69–81.
65. Noguchi S, Tahara H, Miyauchi K, Koyama H. Influence of 1α,25-dihydroxyvitamin D_3 on the development and steroid hormone receptor contents of DMBA-induced rat mammary tumors. Oncology 1989; 46:273–276.
66. Oikawa T, Yoshida Y, Shimamura M, Ashino-Fuse H, Iwaguchi T, Tominaga T. Antitumor effect of 22-oxa-1α,25-dihydroxyvitamin D_3, a potent angiogenesis inhibitor, on rat mammary tumors induced by 7,12-dimethylbenz[a]anthracene. Anticancer Drugs 1991; 2:475–480.

67. Colston KW, Mackay AG, James SY, Binderup L. EB1089: a new vitamin D analogue that inhibits the growth of breast cancer cells in vivo and in vitro. Biochem Pharmacol 1992; 44:2273–2280.
68. Abe-Hashimoto J, Kikuchi T, Matsumoto T, Nishii Y, Ogata E, Ikeda K. Antitumor effect of 22-oxa-calcitriol, a noncalcemic analogue of calcitriol, in athymic mice implanted with human breast carcinoma and its synergism with tamoxifen. Cancer Res 1993; 53:2534–2537.
69. Colston KW, Mackay AG, Chandler S, Binderup L, Coombes RC. Novel vitamin D analogues suppress tumour growth in vivo. In: Vitamin D: Gene Regulation, Structure-Function Analysis and Clinical Application. Proceedings of the Eghth Workshop on Vitamin D. Norman AW, Bouillon R, Thomasset M, eds. Berlin: Walter de Gruyter, 1991; 465,466.
70. Bower M, Colston KW, Stein RC, Hedley A, Gazet JC, Ford HT, Coombes RC. Topical calcipotriol treatment in advanced breast cancer. Lancet 1991; 337:701,702.
71. O'Brien MER, Talbot D, Maclennan K, Smith IE. Inefficacy of calcipotriol in skin metastases from breast cancer. Lancet 1993; 342:994.
72. Muss HB. Endocrine therapy for advanced breast cancer: a review. Breast Cancer Res Treat 1992; 21:15–26.
73. Santen RJ, Manni A, Harvey H, Redmond C. Endocrine treatment of breast cancer in women. Endocr Rev 1990; 11:221–265.
74. Litherland S, Jackson IM. Antioestrogens in the management of hormone-dependent cancer. Cancer Treat Rev 1988; 15:183–194.
75. Costa A. Breast cancer chemoprevention. Eur J Cancer 1993; 29A:589–592.
76. Koga M, Sutherland RL. Retinoic acts synergistically with 1,25-dihydroxyvitamin D_3 or anti-oestrogen to inhibit T-47D human breast cancer cell proliferation. J Steroid Biochem Mol Biol 1991; 39:455–460.
77. Rocker D, Ravid A, Liberman UA, Garach-Jehoshua O, Koren R. 1,25-Dihydroxyvitamin D_3 potentiates the cytotoxic effect of TNF on human breast cancer cells. Mol Cell Endocrinol 1994; 106:157–162.
78. Hassan HT, Eliopoulos A, Maurer HR, Spandidos DA. Recombinant human GM-CSF enhances the anti-proliferative activity of vitamin D in MCF-7 human breast cancer clonogenic cells. Eur J Cancer 1992; 28A:1588,1589.
79. Cho YL, Christensen C, Saunders DE, Lawrence WD, Deppe G, Malviya VK, Malone JM. Combined effects of 1,25-dihydroxyvitamin D_3 and platinum drugs on the growth of MCF-7 cells. Cancer Res 1991; 51:2848–2853.
80. Druker BJ, Mamon HJ, Roberts TM. Oncogenes, growth factors, and signal transduction. N Engl J Med 1989; 321:1383–1391.
81. Studzinski GP. Oncogenes, growth, and the cell cycle: an overview. Cell Tissue Kinet 1989; 22:405–424.
82. Travali S, Koniecki J, Petralia S, Baserga R. Oncogenes in growth and development. FASEB J 1990; 4:3209–3214.
83. Weinberg RA. Tumor suppressor genes. Science 1991; 254:1138–1145.
84. Hollingsworth RE, Lee WH. Tumor suppressor genes: new prospects for cancer research. J Natl Cancer Inst 1991; 83:91–96.
85. Cattoretti G, Rilke F, Andreola S, D'Amato L, Domenico D. p53 in breast cancer. Int J Cancer 1988; 41:178–183.
86. Devilee P, Cornelisse CJ. Genetics of human breast cancer. Cancer Surv 1990; 9:605–630.
87. Groner B, Hynes NE. Mutations in human breast cancer cells: dominantly-acting oncogenes and tumor suppressor genes suggest strategies for targeted interference. Int J Cancer 1990; 5:40–46.
88. Mackay J, Thompson AM, Coles C, Steel CM. Molecular lesions in breast cancer. Int J Cancer 1990; 5:47–50.
89. Van de Vijver MJ, Nusse R. The molecular biology of breast cancer. Biochim Biophys Acta 1991; 1072:33–50.
90. Groner B. Oncogene expression in mammary epithelial cells. J Cell Biochem 1992; 49:128–136.
91. Moll UM, Riou G, Levine AJ. Two distinct mechanisms alter p53 in breast cancer: mutation and nuclear exclusion. Proc Natl Acad Sci USA 1992; 89:7262–7266.
92. Reitsma PH, Rothberg PG, Astrin SM, Trial J, Bar-Shavit Z, Hall A, Teitelbaum SL, Kahn AJ. Regulation of myc gene expression in HL-60 leukaemia cells by a vitamin D metabolite. Nature 1983; 306:492–494.

25 Anitcancer Activity of Vitamin D Analogs

Milan R. Uskoković, Candace S. Johnson, Donald L. Trump, and Robert H. Getzenberg

1. INTRODUCTION

From its discovery in 1920 to the late 1960s, vitamin D was considered to function as a nutritional cofactor for biologic processes related to calcium metabolism. Discovery and chemical characterization of the vitamin D plasma metabolites 25-hydroxy D_3 [25(OH)D_3] and 1,25-dihydroxy D_3 [1,25(OH)$_2D_3$] led to recognition of the true significance of vitamin D_3 as a precursor of a hormone, biosynthesis of which is tightly regulated, as in the case of other steroid hormones (Scheme 1). During the last 10 years, significant progress has been made at defining the molecular mechanism of 1,25(OH)$_2D_3$ activity in various cellular systems. 1,25(OH)$_2D_3$ as a hormone plays an essential role in a host of cellular processes involved in calcium-phosphate homeostasis, inhibition of cell growth, and induction of cell differentiation. It is now well established that this hormonally active form of vitamin D generates biologic responses by two distinct mechanistic pathways, via regulation of gene transcriptions mediated by 1,25(OH)$_2D_3$-specific nuclear receptors [the vitamin D receptor (VDR)], and a nongenomic mode initiated at cellular membranes. The genomic effects of 1,25(OH)$_2D_3$ are especially well portrayed by its multifaceted role in bone-forming processes, regulation of parathyroid hormone release, induction of differentiation of myeloid leukemia cells, and inhibition of proliferation of a large number of cancer cell lines, such as breast, prostate, and colon. As therapeutic agents, 1,25(OH)$_2D_3$ and some of its analogs are being used in the treatment of renal osteodystrophy, secondary hyperparathyroidism, osteoporosis, psoriasis, and scleroderma. These compounds also have the potential to be used in the treatment of leukemia and solid tumors based on the in vivo results discussed in this chapter. The immunosuppressive activity of 1,25(OH)$_2D_3$ and its analogs is defined by their downregulation of interleukin-12 (IL-12), interferon-γ (INF-γ), and IL-2 release, and it is also supported by disease suppression in in vivo models for experimental autoimmune encephalomyelitis, diabetes and lupus *(1)*.

Colon cancer, breast cancer, and cancer of the prostate are leading causes of death among cancer patients in the United States. Despite significant advances in treatment using surgery, radiotherapy, and chemotherapy, the cure rate for these diseases has remained low. New strategies are being intensively investigated, among them the use of

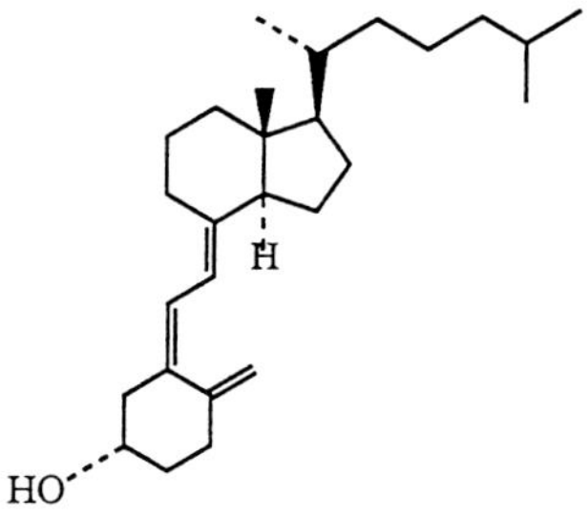

Scheme 1. Chemical structure of 25-hydroxy D_3 and 1,25-dihydroxy D_3.

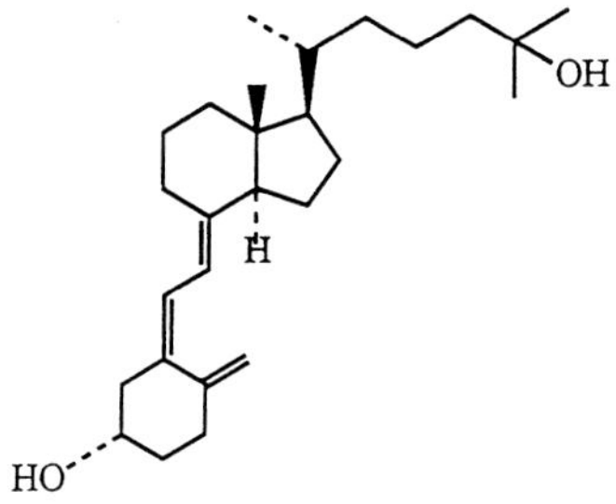

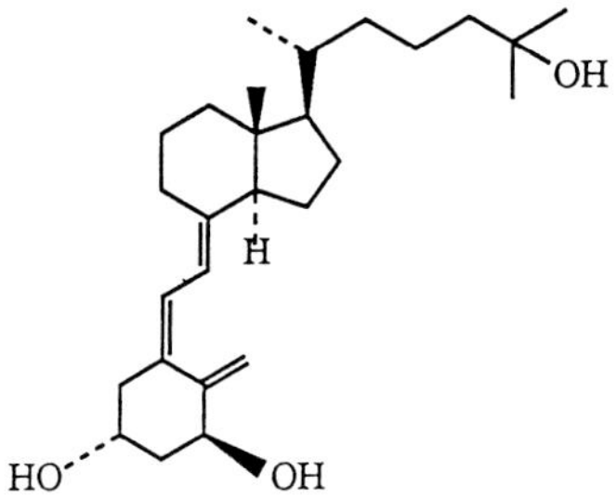

retinoids and vitamin D compounds, because cell-differentiating and antiproliferative activities has been observed in numerous cancer cell lines in vitro *(1)*.

Data have accumulated from a number of different sources in support of the possibility that vitamin D_3, or its metabolites, may play a preventive role in the development of colon cancer. Several epidemiologic studies have suggested that vitamin D_3, derived from the diet or from its biosynthesis in skin due to sunlight exposure, may decrease the risk of colon cancer in humans. Mortality rates from colon cancer in the United States are highest in populations exposed to the least amounts of natural sunlight. In a 1985 study, an association of dietary vitamin D and calcium with 19-yr risk of colorectal cancer was examined in 1954 men who had completed detailed, 28-d dietary histories in the period 1957–1959. Risk of colorectal cancer was inversely correlated with dietary vitamin D and calcium *(2)*. An 8-yr prospective study was completed in 1989 that correlates the major circulating vitamin D metabolite, 25-hydroxy D_3, to colon cancer. Blood samples taken in 1974 from 25,620 volunteers were used to investigate the relationship of serum 25-hydroxy D_3 with subsequent risk of getting colon cancer. Thirty-four cases of colon cancer diagnosed between August 1975 and January 1983 were matched to 67 controls by age, race, and sex. The risk of colon cancer was reduced by 75% in people with 27–32 ng/mL of serum 25-hydroxy D_3, and by 80% with a 33–41 ng/mL level. The risk of getting colon cancer decreased threefold in people with serum 25-hydroxy D_3 concentration of ≥ 20 ng/ml *(3)*.

2. VITAMIN D PREVENTS CARCINOGEN-INDUCED TUMOR FORMATION

Dietary factors have been extensively studied and correlated with incidence of different cancers, especially colon cancer. High intake of dietary fat is implicated as a cause of colorectal tumors. A high fat diet, rich in polyunsaturated fat, appears to act as a

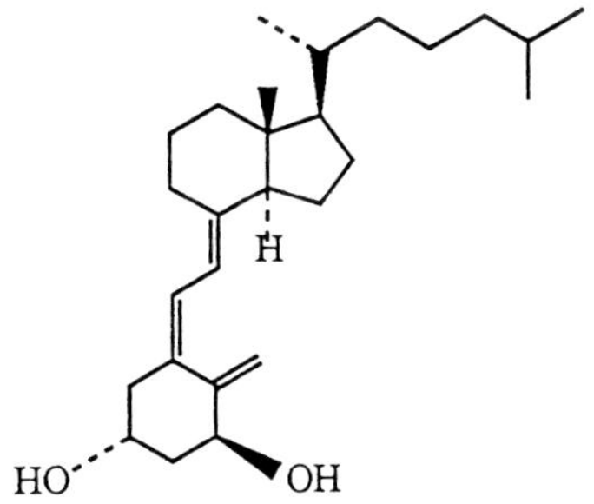

Scheme 2. Chemical structure of 1α-hydroxy D_3.

promoter rather than initiator in colon carcinogenesis by expanding cell populations previously exposed to initiating carcinogens. The colon lipids increase free unesterified fatty acids, which in the ionized form are known to be cytotoxic to the colonic epithelium. This damage causes a regenerative cell proliferation. The administration to rodents of the potent procarcinogen 1,2-dimethylhydrazine results in development of colon cancer. The murine model of colon carcinogenesis closely resembles human colon neoplasia in pathologic features. In the Pence and Buddingh *(4)* study of fat-promoted colon carcinogenesis in male F344 male rats, the tumors were induced with 1,2-dimethylhydrazine and promoted with 20% corn oil fat diet. Animals on this high-fat diet had increased tumor incidence compared with the group on a low-fat diet (86% vs 53%). Supplemental calcium and vitamin D counteract this increase by reducing the incidence below that of the low-fat diet. The result of this investigation suggests that supplemental calcium and vitamin D in the prevention of colon cancer are only effective in the cases of high-fat diet *(4)*.

Bile acids also play an important role in colonic carcinogenesis. A high-fat diet increases the secretion of bile acids, which act as a colonic tumor promoter. The effect of 1α-hydroxyvitamin D_3 on promotion of N-methyl-N-nitrosourea-induced colonic tumorogenesis by intrarectal instillation of lithocholic acid was investigated in female F344 rats by Kawaura et al. *(5)*. In the liver 1α-hydroxy D_3 (Scheme 2) is converted to 1,25(OH)$_2$D$_3$, thus acting as a prodrug for the natural hormone. Coadministration of 1α-hydroxy D_3 caused a 55% reduction in tumor formation. A significant increase in serum calcium, however, was observed in 1α-hydroxy D_3-treated animals.

Numerous aspects of 1,2-dimethylhydrazine-induced rat colonic carcinogenesis in respect to biologic expression of 1,25(OH)$_2$D$_3$ in colonic cells were analyzed by Belleli et al. *(6)*. Charles River rats received five injections of 18 mg dimethylhydrazine, each 1 wk apart. At the end of this treatment the rat colonic mucosal crypts were dilated, often bifurcated, and exhibited an increase in mitotic features in lower, middle, and upper crypt zones. Concurrently with these histologic examinations, the time-course of changes in ornithine decarboxylase (ODC) activity was portrayed. ODC is the rate-limiting enzyme in the polyamine biosynthetic pathway, which plays a key role in normal and neoplastic cell proliferation. Colonic mucosal ODC exhibited a biphasic activity response to dimethylhydrazine administration. The first peak was observed 48 h after the first dimethylhydrazine injection and the second 4 wk after. In control age- and weight- matched animals, ODC activity was low at all periods examined. Simultaneous administration of 1,25(OH)$_2$D$_3$ with dimethylhydrazine (400 ng/rat) did not alter basal colonic ODC activity. By contrast, five doses of 1,25(OH)$_2$D$_3$, each delivered 1 wk apart prior to initiation of dimethylhydrazine treatment, resulted in abrogation of the early ODC activity peak, but not in the later ODC peak response.

The administration of $1,25(OH)_2D_3$ caused a five- to eightfold increase of $1,25(OH)_2D_3$ serum level and the calcium serum levels from 9 to 11.5 mg/dL without any hypercalcemic symptoms. Dimethylhydrazine appears to prevent $1,25(OH)_2D_3$-induced hypercalcemia. Specific $1,25(OH)_2D_3$ receptors (VDR) were identified in extracted rat colonic fraction. In the control group, the maximal binding capacity was 42.4 ± 3.7 fmol/mg protein. In the colonic fraction of the dimethylhydrazine-treated group, a marked decrease in the maximal binding capacity of $1,25(OH)_2D_3$ receptors was indicated (21.9 ± 4.9 fmol/mg protein).

Following dimethylhydrazine administration, tumor incidence and distribution in rat treatment groups were assessed along the whole colon length 10 wk after termination of dimethylhydrazine treatment. Only colonic adenocarcinomas were found. Adenocarcinomas were present in all groups, irrespective of $1,25(OH)_2D_3$ administration schedules. In the rat group that received $1,25(OH)_2D_3$ prior to dimethylhydrazine administration, tumor number decreased by 50%. No tumor suppression was observed in the rat groups treated with $1,25(OH)_2D_3$ simultaneously or after dimethylhydrazine administration. The findings of this study suggest that the colon-specific carcinogen 1,2-dimethylhydrazine interferes with the physiologic expression of $1,25(OH)_2D_3$ in the colon. When the active metabolite of vitamin D_3 was administered prior to carcinogenic challenge, a significant reduction in tumor incidence was observed.

In a very similar study, Sprague-Dawley rats were randomized into two groups, a control receiving vehicle and an experimental group receiving 0.1 µg of $1,25(OH)_2D_3$. One week after initiation of treatment, 1,2-dimethylhydrazine was administered to all rats subcutaneously on a weekly basis at a dose of 20 mg/kg, for 20 consecutive weeks; treatments with $1,25(OH)_2D_3$ were continued for 30 wk. Animals were maintained on a low calcium diet, and the serum calcium for all animals remained in the normal range throughout the study. All tumors were examined histologically. In the control group, 12 malignant colonic tumors were identified in 11 of 24 rats (46%). This compared with five tumors in 4 of 22 animals (18%) in the $1,25(OH)_2D_3$-treated group. The average tumor volume in the former group was 1901 mm^3 (range 523–19,654). In the latter group, the average tumor volume was 689 mm^3 (range 33–4178). In the control group, six rats (25%) showed lymph node involvement and three (12.5%) had distant metastases to the pancreas and liver, whereas none of the $1,25(OH)_2D_3$-treated animals showed any lymph node or distant metastases (M. Shalahang et al., unpublished results).

Although these studies indicate that vitamin D_3, 1α-$(OH)D_3$, and $1,25(OH)_2D_3$ may prevent colon cancer, there is considerable concern about their potential toxicity, particularly with respect to elevation of serum calcium levels and soft tissue calcification. A number of $1,25(OH)_2D_3$ analogs have recently been synthesized, which are markedly less calcemic, yet are more potent inhibitors of cellular proliferation and inducers of differentiation than $1,25(OH)_2D_3$.

In the Hoffmann-La Roche Laboratory the following structural modifications of $1,25(OH)_2D_3$ molecule have incrementally improved the desired pharmacological profile:

Introduction of 16-ene double bond
Introduction of 23E, 23Z, or 23-triple bond
Substitution of 26- and 27-hydrogens with fluorine
Removal of the 10-methylene group

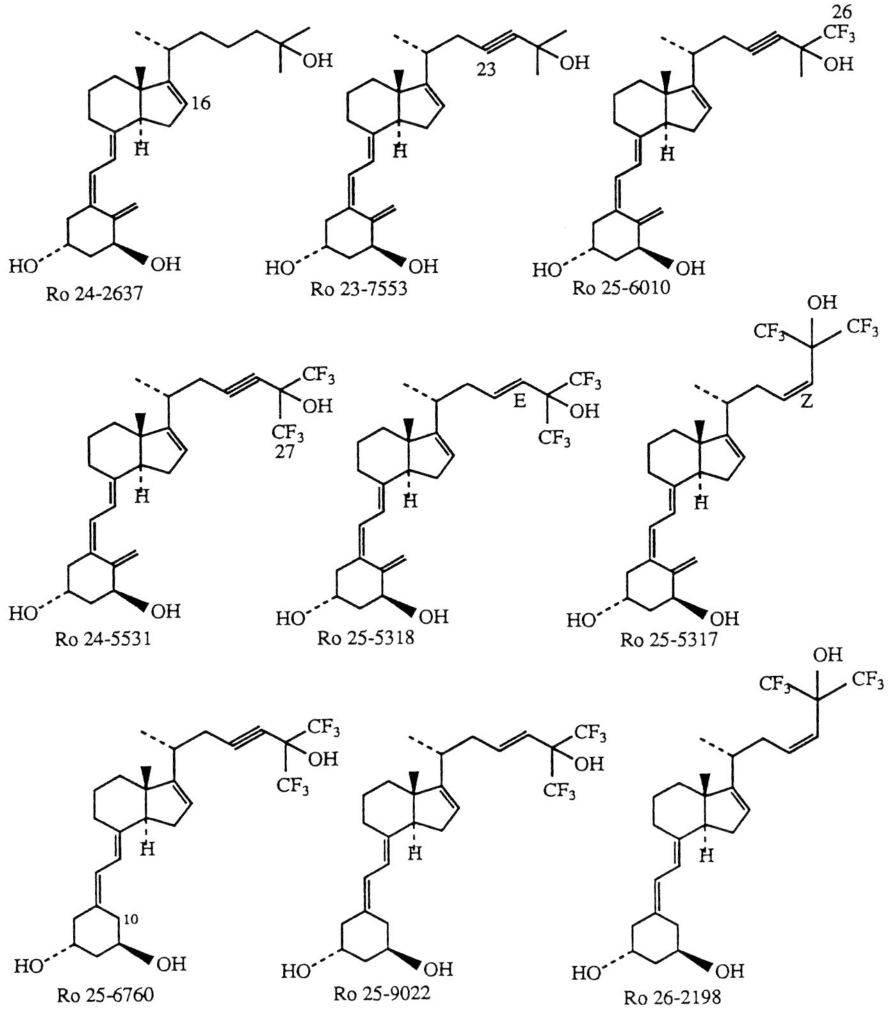

Scheme 3. 1,25(OH)$_2$D$_3$ analogs developed at Hoffmann-La Roche.

These structural modifications have produced several analogs, shown in Scheme 3, that exhibit reduced capacity for causing hypercalcemia in vivo despite a substantial increase in their antiproliferative and cell-differentiating properties in various leukemia and solid tumor cell lines *(6a)*.

In the Leo Pharmaceutical Laboratory, other types of 1,25(OH)$_2$D$_3$ side chain alterations have been examined (Scheme 4). The following modifications have produced powerful vitamin D analogs:

Epimerization at C$_{20}$
Substitution of C$_{22}$ with oxygen
Elongation of the side chain
Introduction of 22E-double bond
Introduction of 22E, 24E conjugated double bonds
Substitution of 25-methyl with ethyl groups
Connecting C$_{26}$ and C$_{27}$ into cyclopropyl ring
Moving the C$_{25}$-hydroxy group to C$_{24}$

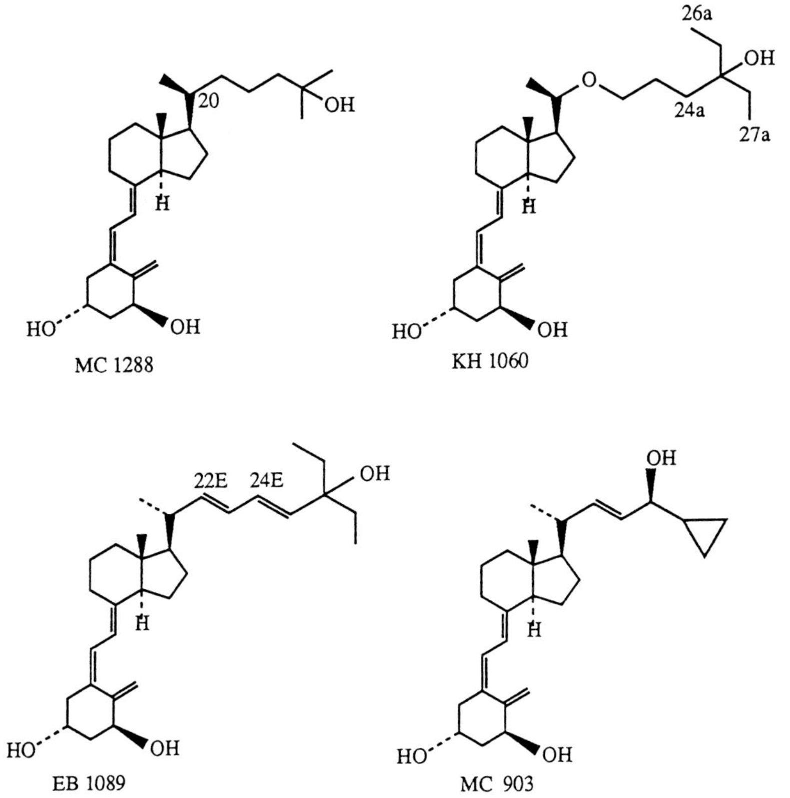

Scheme 4. 1,25(OH)$_2$D$_3$ analogs developed at Leo Pharmaceutical Laboratory.

The vitamin D analog 1,25-dihydroxy-16-ene-23-yne-26,27-hexafluoro D$_3$, Ro 24-5531 (Scheme 3), has attracted significant attention because of its low capacity to cause hypercalcemia and its very high activities in growth inhibition of various cancer cell lines. Brasitus' group *(7)* has studied this analog as a dietary supplement in rats with azoxymethane-induced colon carcinoma. The control and the first treatment group of rats were kept on a standard diet for first 2 wk, and then given sc injections of azoxymethane 15 mg/kg once a week for 2 wk. The second treatment group was kept on a standard diet supplemented with Ro 24-5531 (2.5 nmol/kg feed) for the first 2 wk, and then given azoxymethane 15 mg/kg once a week for 2 wk. The control group was kept on a standard diet for all 34 wk of the study. The first treatment group was switched after the initial 5 wk to the standard diet supplemented with Ro 24-5531 for the rest of the 34 wk of study. The second treatment group, after the first 5 wk on the standard diet supplemented with Ro 24-5531, was switched to standard diet alone. After 34 wk of study, colons of animals were examined macroscopically and microscopically for presence and types of tumors. In comparison with the control group, in which 50% of the animals developed colon tumors, the first treatment group experienced a 40% reduction in tumor incidence, and the second treatment group 70% reduction. In the control group, 40% of the observed tumors were malignant adenocarcinomas, whereas the Ro 24-5531-treated groups experienced only benign adenomas.

Similarly to colon cancer, data have accumulated that support the application of vitamin D metabolites and analogs in the prevention and treatment of breast cancer.

A specific 1,25(OH)$_2$D$_3$ nuclear receptor (VDR) is present in the established estrogen receptor-positive breast cell lines MCF-7, T-47D, and ZR-75, and 1,25(OH)$_2$D$_3$ inhibits proliferation of these cells in vitro. It has also been established that >80% of breast tumors contain VDR, the status of which correlates positively with time between first tumor detection and relapse in breast cancer patients. Patients with VDR-negative tumors relapse significantly earlier than those with receptor positive tumors *(8)*.

The effects of 1α-hydroxyD$_3$ (Scheme 2) in vivo on the growth of primary rat mammary tumors induced with nitrosomethylurea was assessed by Colston et al. *(8)*. It has been previously established that nitrosomethylurea-induced mammary tumors have VDR and that they are potentially responsive to 1,25(OH)$_2$D$_3$. In this study tumor-bearing rats received 0.1 µg of 1α-hydroxy D$_3$ or vehicle three times weekly and were maintained on a 0.1% calcium diet for the whole treatment period of 28 d. In comparison with the control group, 1α-hydroxy D$_3$ completely prevented tumor progression. A small but significant rise in serum calcium concentration was observed. All 12 treated and 11 control animals survived the study period. Similar treatment with 0.25 µg of 1α-hydroxy D$_3$ caused serious hypercalcemia, with the death of 25% of animals involved.

Using the same nitrosomethylurea-induced breast cancer model, Colston et al. *(9)* have investigated the effects of 1,25(OH)$_2$D$_3$ and the analogs MC 903 and EB 1089 (Scheme 4). Treatment with 1,25(OH)$_2$D$_3$ at 0.25, 0.5, and 1.25 µg/kg resulted in minimal or no antitumor effects, but significant hypercalcemia. With 50 µg/kg of MC 903, inhibition of tumor growth was observed. In the case of EB 1089 at 0.5, 1.0, and 2.5 µg/kg, all three doses resulted in a significant antitumor effect, with hypercalcemia and weight loss in rats treated with 1.0 and 2.5 µg/kg doses.

In the same Sprague-Dawley rat breast cancer model, Anzano et al. *(10)* have used the vitamin D analog Ro 24-5531 (Scheme 3), 1,25-dihydroxy-16-ene-23-yne-26,27-hexafluoro vitamin D$_3$, for inhibition of mammary carcinogenesis induced by *N*-nitroso-*N*-methylurea. In vitro, Ro 24-5531 was 10–100 times more potent than the parent 1,25(OH)$_2$D$_3$ for inhibition of proliferation of human MCF-7 and T-47D breast cancer cell lines, as well as primary cultures of cells from two patients with acute myelogenous leukemia. In vivo, rats were first treated with a single dose of either 15 or 50 mg/kg *N*-nitroso-*N*-methylurea and then fed Ro 24-5531 (2.5 or 1.25 nmol/kg of diet) for 5–7 mo. Ro 24-5531 significantly extended tumor latency and lessened tumor incidence as well as tumor number in rats treated with the lower dose of *N*-nitroso-*N*-methylurea. In comparison with control, the tumor incidence decreased from 22 tumors/40 animals to 12/40. The number of tumors decreased from 30 to 15. In rats treated with the higher dose of carcinogen, Ro 24-5531 was fed in combination with tamoxifen; in these experiments, Ro 24-5531 significantly enhanced the ability of tamoxifen to reduce total tumor burden, as well as to increase the probability that an animal would be tumor free at the end of the experiment.

Chugai in Japan have also contributed significantly to the elite list of vitamin D analogs, with their 22-oxa-1,25-dihydroxy D$_3$ derivative OCT (Scheme 5). Several elaborate studies, both in vitro and in vivo, have been completed with this analog, the results of which have significantly enhanced the potential of vitamin D analogs as anticancer agents. OCT inhibited the proliferation of both estrogen receptor-positive MCF-7, T-47D, and ZR-75 and estrogen receptor-negative MDA-MB-231 and BT-20 breast cancer cells in a time- and dose-dependent manner. The antiproliferative effect was observed with concentrations as low as 10^{-11} *M* of OCT, which was approximately one order of magnitude more potent than 1,25(OH)$_2$D$_3$ *(11)*.

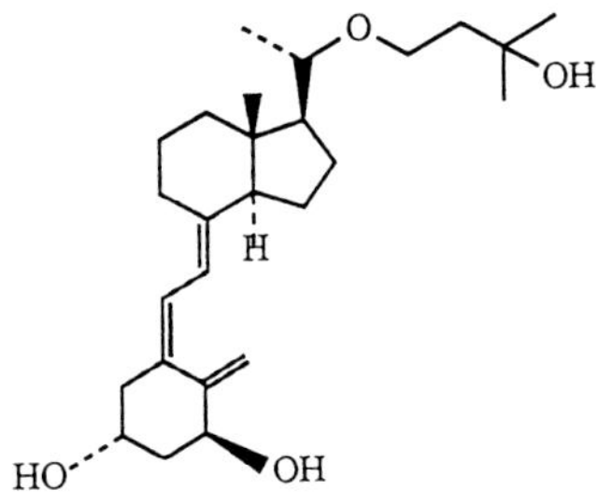

Scheme 5. The 22-oxa-1,25dihydroxy D_3 derivative OCT developed by Chugai.

Inhibition of intestinal tumor development and aberrant crypt foci in rat colon carcinogenesis by the 22-oxa analog OCT was investigated in great detail *(12)*. In the first experiment 6-wk-old F344 rats were given a cocktail of carcinogens: at the commencement of the experiment a single ip administration of 100 mg/kg *N*-diethylnitrosamine, four ip administration of 20 mg/kg *N*-methyl-*N*-nitrosourea at d 2, 5, 8, and 11, four sc injections of 40 mg/kg *NN*-dimethylhydrazine (DMH) at d 14, 17, 20, and 23, 0.05% of *N*-butyl-*N*-(4-hydroxybutyl)nitrosamine in the drinking water for 2 wk, and 0.1% di-*N*-propylnitrosamine in the drinking water during wk 3 and 4. Two wk after the administration of carcinogens was completed, a group of rats was given 30 µg/kg OCT three times a week. The second group was given 3 µg/kg. The third group received vehicle. The total period of experimental observation was 30 wk. Histopathologic examination of the small intestines in the group treated with 30 µg/kg OCT show adenomas in 8% of animals and adenocarcinomas in none; in the group treated with 3 µg/kg OCT, 12% of animals exhibited adenomas, another 12% exhibited adenocarcinomas; in the control group, 5% had adenomas and 18% adenocarcinomas. In the large intestine, the group treated with 30 µg/ OCT, had 12% adenomas and another 12% adenocarcinomas; in the group treated with 3 µg/kg OCT, 16% had adenomas and 28% adenocarcinomas; in the control group, 14% had adenomas and 36% adenocarcinomas.

In the second experiment, 9-wk-old rats were injected sc 20 with mg/kg of DMH, four times at 1-week intervals. Half of the animals were then treated with ip injections of 30 µg/kg OCT six times a week; the other half received vehicle. The experiment lasted for 12 wk. The results showed that the OCT-treated animals exhibited only half or less of aberrant crypt foci in comparison with the control group. Although all animals in both experiments survived without significant changes in body weights, the animals treated with OCT had significant elevations in serum calcium and phosphorus concentrations. No significant elevation of VDR content in the rat colonic epithelium was observed.

Our review thus far has described the significant effects of $1,25(OH)_2D_3$ and its analogs on carcinogen-induced tumor formations in various in vivo models. Both prevention of tumor formation and suppression of existing tumors growth were observed. However, inhibition of carcinogenesis is mechanistically different from inhibition of tumor cell proliferation or direct tumor effects in established tumor growth. Models to examine the latter involve syngeneic or xenograft implantation of tumor cells that either grow and metastasize to other organ sites or grow only where implanted with no metastasis. Whether these models mimic the human situation remains to be determined; however, each model offers a somewhat different perspective with regard to the therapeutic potential of vitamin D analogs.

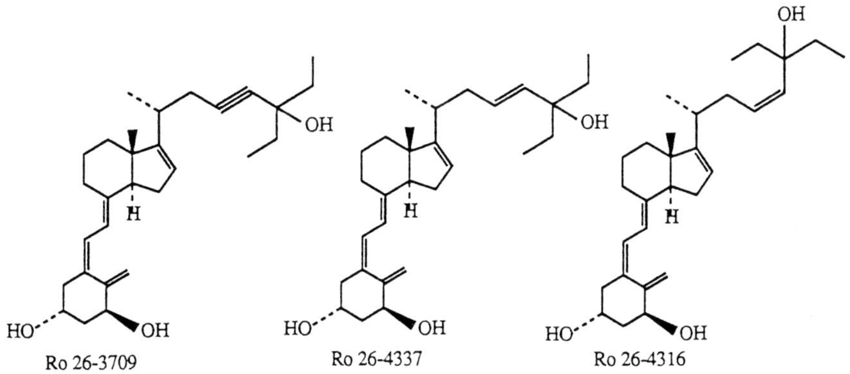

Scheme 6. 1,25(OH)$_2$D$_3$ analogs Ro 36-3709, Ro 26-4337, and Ro 26-4316.

3. SYNGENEIC MODELS

A syngeneic Dunning rat prostatic adenocarcinoma model system was recently used for evaluation of 1,25(OH)$_2$D$_3$ and the 16-ene-23-yne-hexafluoro-19-nor analog Ro 25-6760 (Scheme 3). Epidemiologic data indicate that many risk factors for clinically significant prostate cancer can be correlated with low serum levels of 1,25(OH)$_2$D$_3$. A recent study has focused investigation of the 16-ene vitamin D analogs on three prostate cancer cell lines: LNCap obtained from a lymph node metastasis of a patient with hormonally refractory prostate cancer, PC-3 derived from a primary adenocarcinoma of the prostate, and DU-145 from a prostate cancer metastatic to the brain. LNCap cells have wild-type p53, and the other two cell types have mutant p53 genes. LNCap cells are androgen responsive, and PC-3 and DU-145 are not. These three prostate cancer cell lines have been demonstrated to have functional 1,25(OH)$_2$D$_3$ nuclear receptors (VDR) *(13)*.

The natural hormone 1,25(OH)$_2$D$_3$ (Ro 21-5535) exhibited weak clonal growth-inhibitory activity in PC-3 cell line, but no ED$_{50}$ was achieved in LNCap or DU-145 cells. The 16-ene-23-yne analog Ro 23-7553 was active in LNCap cells but inactive in the other two cell lines. Replacement of two C$_{25}$-methyl groups in Ro 23-7553 with ethyl groups produced the analog Ro 26-3709 (Scheme 6), which retained LNCap activity, but it was also highly active in the PC-3 cell line. The two corresponding 23-double bond epimers Ro 26-4337 and Ro 26-4316 showed a very similar activity profile. Activity in all three cell lines was achieved when the side-chain end methyl hydrogens of Ro 23-7553 were substituted with fluorine to give the analog Ro 24-5531 (Scheme 3). In this case the corresponding 23-double bond epimers Ro 25-5318 and Ro 25-5317 were also active, Ro 25-5318 being the most active in the notoriously resistant DU-145 cell line. The highest specificity for inhibition of prostate cancer cell clonal growth was observed in the 19-nor series, when the 19-nor modification of ring A was combined with a 16-double bond, 23E and Z double bonds, or a 23-triple bond and 26,27-hexafluoro substitution (Scheme 3). Compounds Ro 25-6760, Ro 25-9022, and Ro 26-2198 were fully active in LNCap, PC-3 and DU-145 cell lines at low nonamolar and subnonamolar ED$_{50}$s (Table 1).

In preparation for an in vivo study of 1,25(OH)$_2$D$_3$ and its analogs in the syngeneic Dunning rat prostate adenocarcinoma model, in vitro growth inhibition of the moderately metastatic R3327-AT-2 (AT-2) and the highly metastatic Mat-lylu (MILL) Dunning cell lines were studied by Getzenberg et al. *(14)*. The MILL cell line is an aggressive tumor that when injected subcutaneously quickly metastasizes to regional lymph nodes and

Table 1
Inhibition of Clonal Growth of Prostate Cancer Cell Lines (ED$_{50}$ nM)[a]

Analog	LNCap	PC-3	DU-145
1,25-(OH)$_2$D$_3$, Ro 21-5535	na	100.0	na
1,25-(OH)$_2$-16-ene-23-yne D$_3$, Ro 23-7553	2.5	na	na
1,25-(OH)$_2$-16-ene-23-yne-26,27-bishomoD$_3$, Ro 26-3709	8.0	0.5	na
1,25-(OH)$_2$-16,23E-diene-26,27-bishomoD$_3$, Ro 26-4337	15.0	5.0	na
1,25-(OH)$_2$-16,23Z-diene-26,27-bishomoD$_3$, Ro 26-4316	17.0	1.5	na
1,25-(OH)$_2$-16-ene-23-yne-26,27-F$_6$D$_3$, Ro 24-5531	15.0	35.0	100.0
1,25-(OH)$_2$-16,23E-diene-26,27-F$_6$D$_3$, Ro 25-5318			2.5
1,25-(OH)$_2$-16,23Z-diene-26,27-F$_6$D$_3$, Ro 25-5317	1.5	20.0	100.0
1,25-(OH)$_2$-16-ene-23-yne-19-nor-26,27-F$_6$D$_3$, Ro 25-6760	6.0	4.5	20.0
1,25-(OH)$_2$-16,23E-diene-19-nor-26,27-F$_6$D$_3$, Ro 25-9022	0.025	0.75	1.0
1,25-(OH)$_2$-16,23Z-diene-19-nor-26,27-F$_6$D$_3$, Ro 26-2198	0.075	0.5	1.0

[a]na, not active (compound does not reach ED$_{50}$).

lung. Both AT-2 and MILL cell lines growth were similarly inhibited by 1,25(OH)$_2$D$_3$, with an IC$_{50}$ of 20 µM. To study the action of 1,25(OH)$_2$D$_3$ on cell cycle, MILL cells in culture were treated with 10 µM 1,25(OH)$_2$D$_3$ for 48 h and analyzed by flow cytometry. The analysis revealed a G$_0$/G$_1$ phase block effect when compared with untreated cells, with a significant increase in the G$_0$/G$_1$ peak. In the control group, 50% of cells were in G$_0$/G$_1$ phase and 40% in S phase; in the 1,25(OH)$_2$D$_3$-treated group, 74% of cells were in G$_0$/G$_1$ and 22.5% in S phase.

To examine in vivo efficacy, the ability of 1,25(OH)$_2$D$_3$ and of the analog 1,25-dihydroxy-16-ene-23-yne-26,27-hexafluoro-19-nor D$_3$ (Ro 25-6760; Scheme 3) has been tested to alter the growth and metastasis of MLL implanted tumors. MLL cells (5 × 10^4) were implanted subcutaneously into the flank of Copenhagen rats on the same day that treatment was initiated with either 1,25(OH)$_2$D$_3$ at 1 µg, or analog Ro 25-6760 at 1 or 5 µg per rat three times a week for 3 wk. Treatment of animals with 1 µg 1,25(OH)$_2$D$_3$ and 5 µg Ro 25-6760 resulted in almost complete inhibition in the increase of tumor volume and significant reduction in the number and size of lung metastases in comparison with a control group. These antitumor effects, however, were accompanied by a significant increase in serum calcium in both cases, from control of 9.3 to 12.3 mg/dL, with 30–34% weight loss. A similar treatment with the analog 1,25-dihydroxy-16-ene-23-yne D$_3$, Ro 23-7553, at 20 µg/rat, showed no antitumor activity in MLL cell-induced tumors of Copenhagen rats. As shown previously, unlike 1,25(OH)$_2$D$_3$ and Ro 25-6760, the analog Ro 23-7553 was inactive in growth inhibition of the PC-3 cell line in culture. The structural specificity of vitamin D analogs plays a significant role in various cancer models, in vitro and in vivo. Ro 23-7553 is very potent in inhibiting proliferation and inducing differentiation of myeloid leukemic cells and squamous carcinoma in vitro, as well as a leukemic model in vivo, as will be discussed shortly.

A variety of syngeneic murine tumor models exist for leukemia and solid tumors. Each model has its strengths and weaknesses, and although claims cannot be made these studies can be directly translated into the clinical setting, useful and applicable information can be obtained. Another important feature of these models is that they have intact immune systems and are therefore not compromised by immunologic defects.

Zhou, et al. (15) have explored the therapeutic potential of Ro 23-7553 (Scheme 3) by developing and using the following three leukemia models:

1. Injection of 2.5×10^5 myeloid leukemic cells (WEHI 3BD$^+$) into syngeneic BALB/c mice resulted in leukemic death of all diluent-injected mice by d 26. Mice that received the same number of leukemic cells and also received 1,25(OH)$_2$D$_3$ (0.1 µg qod, ip) had a nearly identical survival curve. Those that received the leukemic cells and the analog Ro 23-7553 (1.6 µg qod, ip) had a significantly ($p = 0.003$) longer survival, with the last mouse dying of leukemia on d 50.
2. Injection of 50% fewer leukemic cells (1×10^5 cells) into syngeneic BALB/c mice resulted in 86% dead of leukemia at 51 d. Experimental mice that received the same number of leukemic cells and Ro 23-7553 (0.8 µg qod) had a significantly ($p = 0.0006$) longer survival than controls; only 53% of the mice were dead by d 100.
3. After injection of 1.5×10^4 leukemic cells, 13% of syngeneic BALB/c mice were free of disease at d 180. By contrast, 43% of mice that received leukemic cells and Ro 23-7553 (1.6 µg qod) were still free of disease at d 180. Thus the 1,25-(OH)$_2$-16-ene-23-yne D$_3$ analog Ro 23-7553 significantly increased the survival of mice who had myeloid leukemia.

Using the murine SCCVII/SF squamous cell carcinoma (SCC) model system, McElwain et al. (16) examined the in vitro and in vivo effects of 1,25(OH)$_2$D$_3$ and the analog Ro 23-7553. Both compounds were equally effective in growth inhibition of SCC cells in vitro at similar concentrations. To determine whether in vitro growth inhibition translates into a significant effect in vivo, C3H/HeJ mice with SCCVII/SF established tumors (d 9–14 after subcutaneous inoculation) were treated intraperitoneally three times a week with either 1,25(OH)$_2$D$_3$ or Ro 23-7553 and monitored for tumor growth. In a dose-dependent manner, both vitamin D compounds significantly inhibited tumor cell growth. At higher doses of 1,25(OH)$_2$D$_3$ (0.5 µg/dose), animals experienced a significant increase in serum calcium. By contrast, Ro 23-7553 could be administered at doses up to 10–15 µg/dose without any hypercalcemia. The effect of 1,25(OH)$_2$D$_3$ and Ro 23-7553 on tumor initiation and early tumor growth has also been examined in a preventive model in which treatments were initiated at the same time as tumor inoculation. In this model, a significant inhibition of tumor induction was seen with both compounds, with hypercalcemia, however, in animals treated with 1,25(OH)$_2$D$_3$.

Using this same model system, studies have been performed to examine the effect of glucocorticoids on 1,25(OH)$_2$D$_3$-mediated antitumor activity (17). Glucocorticoids have been shown to upregulate the expression of the VDR. To examine the effect of glucocorticoids on 1,25(OH)$_2$D$_3$-mediated antiproliferative activity in vivo, 1,25(OH)$_2$D$_3$ in combination with dexamethasone has been administered for an effect on established tumors as well as on tumor initiation. In both treatment regimens, dexamethasone significantly potentiated the antiproliferative effect of 1,25(OH)$_2$D$_3$. In addition, serum calcium levels were significantly lower in groups of animals treated with dexamethasone plus 1,25(OH)$_2$D$_3$ compared with 1,25(OH)$_2$D$_3$-treated animals alone. To determine whether dexamethasone mediates its effects through action on binding to the VDR, the effect of dexamethasone on VDR ligand binding in SCC cell lines as well as in tumor and other tissues isolated from dexamethasone-treated tumor-bearing mice was examined. Using the cytosol ligand binding assay, 1,25(OH)$_2$D$_3$ binding was significantly increased in vitro after treatment with dexamethasone for 24 h in SCC cells. Similarly, when tumor-bearing animals were treated with either 3 or 7 d of dexamethasone, cytosol preparations from tumor showed significantly greater 1,25(OH)$_2$D$_3$ binding compared with control (no treatment). Enhancement of 1,25(OH)$_2$D$_3$ binding was not observed in cytosol preparations isolated from the muscle, bone, esophagus, blood, and skin of the same animals treated or untreated with dexamethasone for 7 d. By contrast, in intestinal mucosa a significant decrease in VDR ligand binding after treatment for 7 d with dexamethasone compared with control was observed. Therefore, these studies demonstrate that dexamethasone may specifically enhance the antiproliferative effect of $1,25(OH)_2D_3$ through upregulation of $1,25(OH)_2D_3$ receptor binding in cancer tissue.

$1,25(OH)_2D_3$ has a significant effect on the cell cycle. In HL-60 cells, a human myelomonocytic leukemia cell line, $1,25(OH)_2D_3$ has been shown to arrest cells in G_1 phase, the process mediated through the cyclin-dependent kinase inhibitor p27 *(18–20)*. Studies in the murine SCC model have demonstrated that >90% of SCC cells are in G_1/G_0 phase, with an accompanying decrease of cells in S phase after 48 h of treatment with $1,25(OH)_2D_3$ or Ro 23-7553 in vitro. In this study, the ability to arrest cells in G_1/G_0 phase was exploited by combining Ro 23-7553 with the cytotoxic agent cisplatin, in the murine SCC model system. Cisplatin was shown to have an enhanced cytotoxic effect on cells in G_1 phase just prior to DNA replication, the same phase in which Ro 23-7553 has been shown to arrest cells *(21)*. Tumor-bearing animals (11 d post implant) were treated intraperitoneally for 3 d with 0.5 µg/kg/d of Ro 23-7553 and on the third day they received cisplatin. Pretreatment with Ro 23-7553 significantly enhanced cisplatin-mediated antitumor activity, even at low doses of cisplatin, as measured by a significant enhancement of tumor clonogenic cell kill and a decrease in fractional tumor volume with an accompanying increase in tumor regrowth delay. These studies suggest that $1,25(OH)_2D_3$ and analogs act to sensitize the tumor cells to the cytotoxic effects of cisplatin through effects on the cell cycle. Another advantage of this approach centers on the fact that higher doses of $1,25(OH)_2D_3$ can be administered before cytotoxic drugs without hypercalcemia to achieve the optimum antitumor effect.

4. HUMAN XENOGRAFT MURINE MODELS

To examine the effect of $1,25(OH)_2D_3$ analogs on human tumor cells in vivo, investigators have utilized xenograft models in which human tumors grow and proliferate in nude or immunocompromised mice. Not all fresh human tumor isolates or established human tumor cell lines grow in nude mice, and thus tumor cells must be selected for in vivo growth. Cultures of human cancer cell lines were grown to near confluence and injected subcutaneously (10^5–10^6 cells/site) into each flank of the immunosuppressed mice. Successful "takes" and the production of solid tumors were first achieved with three cell lines in the original study by Eisman et al. *(22)*. These cell lines were two malignant melanoma cell lines COLO 239F and RPMI 7932, and one colonic cancer cell line COLO 206F. RPMI 7932 was VDR negative, and COLO 239F and COLO 206F had detectable VDR levels. Xenografts obtained in the first group of immunosuppressed mice were removed, minced, and passaged further in immunosuppressed mice to establish stable xenografted tumors for the subsequent growth studies. After xenografts had grown to a diameter of 8–10 mm, equal numbers of male and female mice were randomly assigned to treatments and control groups with each group containing 10–14 xenografts. All animals were kept on a low-calcium diet; the treatment group received 0.1 µg of $1,25(OH)_2D_3$, and the control received vehicle for 45 d of the study. For the COLO 206F colon carcinoma xenografts, volume doubling time in control mice was 7 d. The growth of these xenografts was markedly inhibited by treatment with $1,25(OH)_2D_3$ over the entire treatment period. At the end of the study, the relative tumor volume decreased from the pretreatment size and was <0.5% of that in the control mice. The melanoma line RPMI

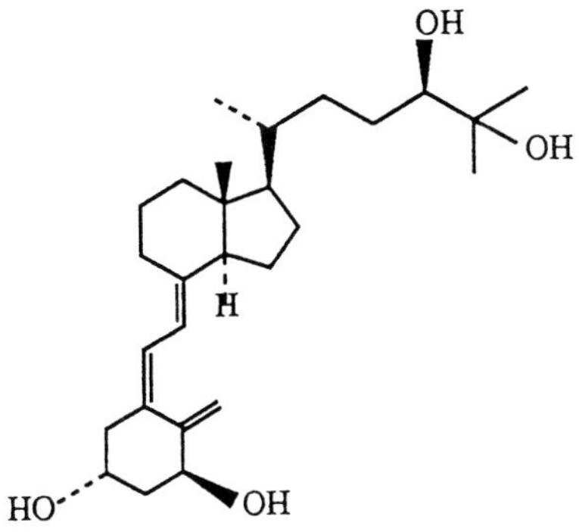

Scheme 7. The analog DD-003.

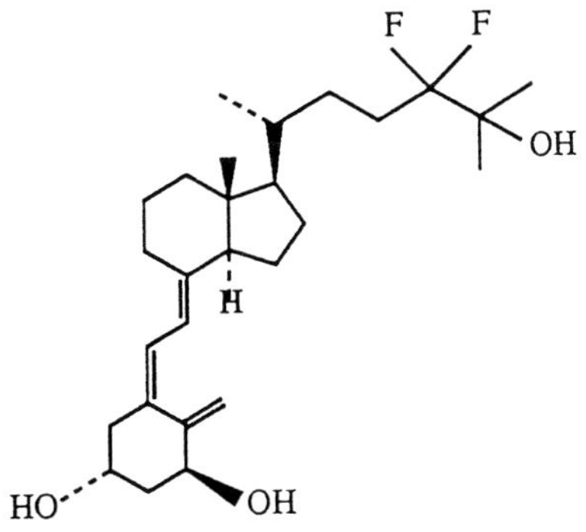

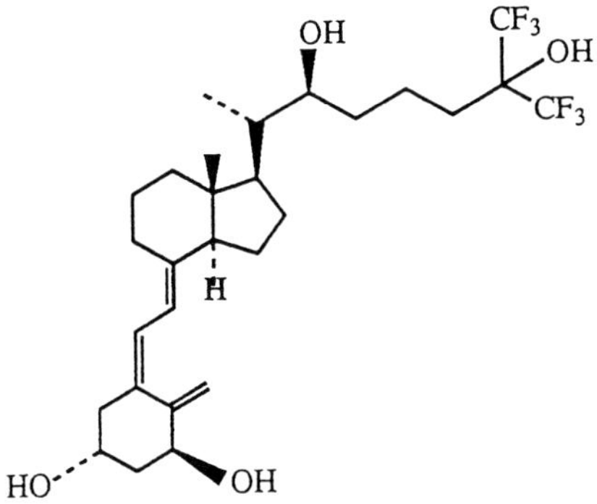

7932 xenografts had a volume doubling time of 8 d. The growth of these xenografts was unaffected by the treatment. The other malignant melanoma xenograft line, COLO 239F, also had a volume doubling time of 8 d. However, whereas the control and the treated tumors of these xenografts initially grew at similar rates, from d 12 onward the treated group grew significantly more slowly ($p < 0.05$) than control animals; the volume doubling time for control was 8 d compared with 27 d in the treated group for the entire time of the experiment. In this case, no regression from the pretreatment tumor volume was observed. In a similar accompanying study with 1,24R,25-trihydroxy D_3 (Ro 21-7729) or 24-difluoro-1,25-dihydroxy D_3 (Ro 22-9343), no antitumor effect was achieved.

The analog 1,22S,25-trihydroxy-24-homo-26,27-hexafluoro D_3 (DD-003; Scheme 7) was 10 times more active than 1,25-dihydroxy D_3 in growth inhibition of HT-29 human colonic adenocarcinoma cell line (23). To examine the antitumor activity of DD-003 in vivo, a fibrin clot of HT-29 cells was constructed with fibrinogen and thrombin and implanted under the renal capsule of immunodeficient mice. Starting 7 d after implantation, mice were given 3 µg/kg of DD-003 or vehicle intraperitoneally every other day for 2 wk. The HT-29 tumor grew rapidly in control mice, with mitosis, massive tumor angiogenesis, and invasion into normal kidney tissue. Tumors in DD-003-treated mice were smaller, with less invasion. Administration of DD-003 inhibited growth of HT-29 tumors by 63%. No sign of hypercalcemic effects was observed. DD-003 inhibited growth of HT-29 tumors in a dose-dependent manner over the range of 0.1–10 µg/kg. When DD-003 was withdrawn after 2 wk of treatment, tumor growth resumed. These results are highly encouraging since the subrenal capsule assay correlates well with clinical response.

In a xenograft model for breast cancer, the estrogen receptor (ER) negative human breast cancer cell line MX-1 has been shown to be growth inhibitory when animals are treated with the analog 22-oxa-1,25-dihydroxy-vitamin D_3 (OCT) (Scheme 5) *(11)*. This analog did not induce hypercalcemia and significantly inhibited the proliferation of human breast cancer cell lines. In vivo, OCT is effective at delaying the growth of the MX-1 tumor when administered by intratumor injection three times a week. Interestingly, OCT was also combined in such a treatment regime with Adriamycin, a commonly used anticancer agent for breast cancer. Treatment with OCT alone resulted in a greater antitumor effect than Adriamycin alone; however, in combination, the antitumor effects of these two agents was additive. In these studies both intratumor and oral routes of administration were utilized, with no difference observed in antitumor effect or in the lack of observed hypercalcemia.

In a xenograft model for cancer of prostate, Skowronski, Peehl, and Feldman (unpublished data) inoculated nude mice with either LNCap or PC-3 cells. Animals received intraperitoneal injections of 100 ng $1,25(OH)_2D_3$ or vehicle on alternate days starting 2 d prior to cell inoculation and continuing for 6 wk after inoculation. In the LNCap xenografts, tumor weights in vehicle treated mice ($n = 9$) were 364 ± 33 mg, whereas tumors in $1,25(OH)_2D_3$-treated mice ($n = 9$) were 260 ± 24 mg, 30% smaller. In the PC-3 xenografts, tumors in control mice ($n = 8$) were 284 ± 38 mg, whereas tumors in $1,25(OH)_2D_3$ treatment group ($n = 8$) were 169 ± 30 mg, 40% smaller. Schwartz et al. *(24)* treated immunodeficient BALB/c nude mice harboring PC-3 xenografts with the 1,25-dihydroxy-16-ene-23-yne analog Ro 23-7553, but no sustained decrease in tumor weights in comparison with control was observed. Ro 23-7553 was also inactive in a syngeneic model for cancer of the prostate, as previously discussed.

5. SUMMARY

The hormonally active form of vitamin D_3, $1,25(OH)_2D_3$, has demonstrated significant antiproliferative or antitumor effects in a wide variety of cancer model systems. The potential therapeutic use of 1,25-dihydroxy D_3 as an ananticancer agent is, however, limited because of dose-limiting hypercalcemia. A number of analogs obtained by chemical modifications of the 1,25-dihydroxy D_3 side chain exhibit increased activities both in in vitro and in vivo antitumor models, but with decreased ability to cause hypercalcemia and thus offer a promise of usefulness in the therapy of leukemia and solid tumors. A number of these analogs have potent effects in the inhibition of rat carcinogenesis and metastasis as well as growth effects in human xenograft models. Alternative means may be utilized to diminish or decrease 1,25-dihydroxy D_3-mediated hypercalcemia such as use of glucocorticoids, or the use of 1,25-dihydroxy D_3 in combination with chemotherapeutic agents. It is hoped that the information gained by these preclinical studies in animal model systems can be exploited and applied toward the design of a more effective clinical approach.

REFERENCES

1. Bouillon R, Okamura WH, Norman AW, Structure-function relationships in the vitamin D endocrine system. Endocr Rev 1995; 16:200–257.
2. Garland FC, Barrett-Connor E, Rossof AH, Shekelle RB, Criqui MH, Oglesby P. Dietary vitamin D and calcium and risk of colorectal cancer: a 19-year prospective study in men. Lancet 1985; XX:307.
3. Garland CF, Garland FC, Shaw EK, Comstock GW, Helsing KF, Gorham ED. Serum 25-hydroxy-vitamin D and colon cancer: eight-year prospective study. Lancet 1989; XX:1176.

4. Pence BC, Buddingh X. Inhibition of dietary fat-promoted colon carcinogenesis in rats by supplemental calcium or vitamin D. Carcinogenesis. 1988; 9:187–190.
5. Kawaura A, Tanida N, Sawada K, Oda M, Shimoyama T. Supplemental Administration of 1α-hydroxy-vitamin D_3 inhibits promotion by intrarectal installation of lithocholic acid in N-methyl-N-nitrosourea-induced colonic tumorigenesis in rats. Carcinogenesis 1989; 10:647–649.
6. Belleli A, Shany S, Levy F, Guberman R, Lamprecht SA. A protective role of 1,25-dihydroxy D_3 in chemically induced rat colon carcinogenesis. Carcinogenesis 1992; 13:2293–2298.
6a. Uskokovic MR, Studzinski GP, Reddy SG. The 16-ene vitamin D analogs. Vitamin D 1997; 1045–1069.
7. Wali RK, Bissonnette M, Khare S, Hart F, Sitrin MD, Brasitus TA. 1α,25-Dihydroxy-16-ene-23-yne-26,27-hexafluorocholecalciferol, a noncalcemic analogue of 1α,25-dihydroxyvitamin D_3, inhibits azoxymethane-induced colonic tumorigenesis. Cancer Res 1995; 55:3050–3054.
8. Colston KW, Berger U, Coombes RC. Possible role for vitamin D in controlling breast cancer cell proliferation. Lancet 1989; XX:188–191.
9. Colston KW, MacKay AG, James SY, Binderup L, Chander S, Coombes RC. EB1089: A new vitamin D analogue that inhibits the growth of breast cancer cells *in vivo* and *in vitro*. Biochem Pharmacol 1992; 44:2273–2280.
10. Anzano MA, Smith FM, Uskokovic MR, Peer CW, Mullen LT, Letterio FF, Welsh MC, Shrader MW, Logsdon DL, Driver CL, Brown CC, Roberts AB, Sporn MB. 1α,25-Dihydroxy-16-ene-23-yne-26,27-hexafluoro-cholecalciferol, a new deltanoid (vitamin D analogue) for prevention of breast cancer in rat. Cancer Res 1994; 54:1653–1656.
11. Abe J, Nakano T, Nishii Y, Matsumoto T, Ogata E, Ikeda K. A novel vitamin D_3 analog, 22-oxa-1,25-dihydroxyvitamin D_3, inhibits the growth of human breast cancer *in vitro* and *in vivo* without causing hypercalcemia. Endocrinology 1991; 129:832–837.
12. Otoshi T, Iwata H, Kitano M, Nishizawa Y, Morii H, Yano Y, Otani S, Fukushima S. Inhibition of intestinal tumor development in rat multiorgan carcinogenesis and aberrant crypt focci in rat colon carcinogenesis by 22-oxa-calcitriol, a synthetic analogue of 1α,25-dihydroxyvitamin D_3. Carcinogenesis 1995; 16:2091–2097.
13. Campbell MJ, Elstner E, Holden S, Uskokovic M, Koeffler HP. 19-Norhexafluoride analogs of vitamin D_3 induce a cell cycle arrest-mediated differentiation of prostate cells: LNCaP, PC-3 and DV-145. Unpublished results.
14. Getzenberg RH, Light BW, Lapco PE, Konety BR, Nangia AK, Acierno JS, Dhir R, Shurin Z, Day RS, Trump DL, Johnson CS. Vitamin D inhibition of prostate adenocarcinoma growth and metastasis in the dunning rat prostate model system. Unpublished results.
15. Zhou JY, Norman AW, Chen DL, Sun GW, Uskokovic M. 1,25-Dihydroxy-16-ene-23-yne-vitamin D_3 prolongs survival time of leukemic mice. Proc Natl Acad Sci USA 1990; 87:3929–3932.
16. McElwain MC, Dettelbach MA, Modzelewski RA, Russell DM, Uskokovic MR, Smith DC, Trump DL, Johnson CS. Anti-proliferative effects *in vitro* and *in vivo* of 1,25-dihydroxy vitamin D_3, and a vitamin D_3 analogue in squamous cell carcinoma model system. Mol Cell Differen 1995; 3:31–50.
17. Yu WD, McElwain MC, Modzelewski RA, Russell DM, Smith DC, Trump DL, Johnson CS. Dexamethasone enhancement of 1,25-dihydroxy-vitamin D_3-mediated anti-proliferative activity through effects on binding to the vitamin D receptor. Unpublished results.
18. Studzinski GP, Bhandal AK, Brelvi ZS. Cell Cycle Sensitivity of HL-60 cells to the differentiation-inducing effects of 1α,25-dihydroxyvitamin D_3. Cancer Res 1985; 45:3898–3905.
19. Godyn JJ, Xu H, Zhang F, Kolla S, Studzinski GP. A dual block to cell cycle progression in HL60 cells exposed to analogues of vitamin D_3. Cell Prolif 1994; 27:37–46.
20. Wang GM, Jones JB, Studzinski GP. Cyclin-dependent kinase inhibitor p27 as a mediator of the G_1S phase block induced by 1,25-dihydroxyvitamin D_3 in HL60 Cells. Cancer Res 1996; 56:264–267.
21. Light BW, McElwain MC, Russell DM, Trump DL, Johnson CS. Potentiation of cisplatin antitumor activity using the vitamin D analogue, 1,25-dihydroxy-16-ene-23-yne-cholecalciferol in a murine squamous cell carcinoma model system. Unpublished results.
22. Eisman JA, Barkla DH, Tutton PJM. Suppression of *in vivo* growth of human cancer solid tumor xenografts by 1,25-dihydroxyvitamin D_3. Cancer Res 1987; 47:21–25.
23. Tanaka Y, Wu AYS, Ikekawa N, Isekik, Kawai M, Kobayashi Y. Inhibition of HT-29 human colon cancer growth under the renal capsule of severe combined immunodeficient mice by an analogue of 1,25-dihydroxyvitamin D_3, DD-003. Cancer Res 1994; 54:5148–5153.
24. Schwartz GG, Hill CC, Oeler TA, Becich MF, Bahnson RR. 1,25-Dihydroxy-16-ene-23-yne-vitamin D_3 and prostate cancer cell proliferation *in vivo*. Urology 1995; 46:365–369.

INDEX

A

Acid haze concentration, 385
   breast cancer, 381
Acitretin
   psoriasis, 266
Actin binding
   DBP, 104
Actin-DBP, 104
Adolescents
   recommended dietary intake, 8
   vitamin D deficiency, 277
Adriamycin, 417
Adults
   bone health, 287–300
African diet, 379
Age, 10
   breast cancer, 381
Aging, 6, 29, 30
   bone loss, 296, 297
1α-hydroxycholecalciferol, 296
1α-hydroxyD$_3$
   breast cancer, 437
1α-hydroxylase, 95, 310
1α-hydroxylase enzyme, 59
1α-hydroxylase regulation, 50
1α-hydroxylase vitamin D metabolites, 311
1α-hydroxylation, 64, 65
1α-hydroxylation
   metabolism, 57–59
1α-hydroxyvitamin D
   vitamin D-dependent rickets type I, 309
1α-hydroxyvitamin D$_3$
   breast cancer, 415
1α-hydroxyvitamin D$_3$
   breast cancer, 417
   colon tumors, 398
   5-fluorouracil, 417
Air pollution
   cancer incidence, 385, 386
Alaska, 7, 8
Albumin, 103, 104
Alkaline phosphatase
   serum levels, 310

Alopecia
   HVDRR, 328–330
Alopecia totalis, 311
Alphacalcidol, 296
American Medical Association Council on Foods and Nutrition recommendations, 19
Analog XII, 157
Analog XIV, 158
AOAC chick assays, 243, 244
1α,25(OD)$_2$D$_3$, 147
1α(OH)D$_2$
   activation, 64–67
1α(OH)D$_3$
   activation, 64–67
1α,25(OH)$_2$D$_3$
   catabolism, 62
   membrane binding, 197
   nongenomic effects
   intestine, 200, 201
   osteoblasts, 196–199
1α-OH group, 152
Arg274Leu mutant VDR
   HVDRR, 328
Asia, 29
Association of Official Agricultural Chemists chick assays, 243, 244
Atopic dermatitis, 359
aVDR, 111, 113
Avian VDR, 111, 113

B

Basic multicellular unit, 288
BB genotype, 293
3B-BODIPY-calcitriol, 133, 134
B-cell neoplasms, 345–347
1β-epimer, 198, 199
Berylliosis, 345
Betametasone 17-valerate ointment
   psoriasis, 361, 367
Bile acids, colon cancer and , 433
Bioassays
   1,25-dihydroxyvitamin D assay methodology, 256

Bipartite NLS
Black skin, 27, 32
BMD, 292, 293
BMU, 288
Bone
    vitamin D, 175–189
Bone acquisition, 293, 294
Bone formation, 177, 178
    gene expression
        vitamin D regulation, 178–181
Bone gla protein, *see* Osteocalcin
Bone health
    adults and elderly, 287–300
Bone homeostasis, 330
Bone loss
    age-associated, 296, 297
Bone matrix organization
    vitamin D, 175–177
Bone mineral density, 292, 293
Bone remodeling
    1,25-dihydroxyvitamin D_3, 289–291
    pathophysiologic states, 289
    vitamin D homeostasis, 287–291
Bone resorption, 177, 178, 290, 339
Bone turnover, 279
Bowed legs, 3
BP-calcitriol, 133–136
Breast cancer, 398–400, 411–422, 437
    acid haze concentration, 381
    age, 381
    angiogenesis, 421
    calcium, 420, 421
    combination therapies, 415–418
    invasion, 421
    in vivo tumor suppression, 413–415
    latitude, 380, 381
    metastasis, 421, 422
    polypeptide growth factors, 419, 420
    risk, 395
    SEER registry, 384, 385
    steroid hormones, 419, 420
    sunlight, 412
    VDR, 412, 413
    vitamin D antiproliferative action, 418–421
    vitamin D role, 387, 388
    xenograft model, 444
Breast cancer cells
    cultured, 413
Breast milk, 6
Brush border membrane, 164
*Bsm*I, 293

C

C5a cochemotaxis
    DBP, 105
Calbindin-D9k
    renal calcium transport, 168
Calbindin-D28k
    renal calcium transport, 168
Calcipotriene, *see* Calcipotriol
Calcipotriol, 213, 222
    breast cancer, 400
    catabolism, 67, 68
    face, 365
    flexures, 365
    ichthyosis, 367
    metabolites, 73, 74
    nail psoriasis, 362
    pharmacokinetics, 67
    psoriasis, 368, 369
    scalp psoriasis, 362
    side effects, 361
Calcitriol, 38, 139, 219–222, 295
    secondary hyperparathyroidism, 231
    skin cancer, 367
Calcitroic acid, 41, 62
Calcium
    breast cancer, 420, 421
    HVDRR, 329
    $1,25(OD)_2D_3$, 164–166
    PTH gene expression regulation, 223–227
    transcellular movement, 164, 165
    vitamin D, 1, 2
    vitamin D deficiency rickets, 282
    vitamin D-dependent rickets type II, 314
Calcium absorption
    duodenum, 291
Calcium balance
    sarcoidosis, 338, 339
Calcium carbonate, 231
Calcium homeostasis
    mechanism, 95
    vitamin D, 40
Calcium hydroxyapatite epiphyseals, 5
Calmodulin, 130
Calreticulin, 130, 135, 225
cAMP, 279
cAMP-responsive element, *see* CRE
Canada, 6–8, 19
Cancer
    1,25-dihydroxyvitamin D_3, 212
    treatment and prevention
        future, 403, 404

Cancer cells
    1,25-dihydroxyvitamin D_3, 207, 208
Cancer risk
    vitamin D
        cell culture studies, 395–397
        epidemiological studies, 394, 395
        epidemiology, 375–388
        experimental animal studies, 397–399
        human studies, 399, 400
Carcinogen-induced tumor formation
    prevention, 432–438
Carpopedal spasms, 275
CAT gene, 219
Caucasian skin, 26
C-22=C-23 double bond, 73, 74
Celiac disease, 299
Cell-associated DBP, 104, 105
Cereal, 19
Cerebrotendinous xanthomatosis, 86–88
cGMP
    VDR, 138–140
Chemotherapy-induced alopecia, 213
Chick bioassays, 243, 244
Children
    recommended dietary intake, 7, 8
    vitamin D deficiency, 277
    vitamin D deficiency prevention, 282
Chinese infants, 7
Chloroquine, 339, 347
Cholecalciferol, 294
Cholestanetriol 27-hydroxylase, 88
Chromatin structure, 187
Chromosome 12q13-14, 319
Chromosome 12q14, 308
Chronic renal failure, 222
Chvostek's sign, 275
Chylomicron, 103, 104
Clothing, 30, 31
c-myc, 418
Cod liver oil, 18
Collagen type I, 176
COLO 206F, 442–445
COLO 239F, 442–445
Colon cancer, 398–400, 402, 436, 442, 443
    bile acids, 433
    circulating vitamin D metabolites, 379
    dietary factors, 379, 380
    dietary fat, 432, 433
    genetic predisposition, 379
    immigrants, 376–378
    latitude, 376

observational studies, 378
risk, 394, 395
SEER registry, 383
vitamin D role, 387, 388
Colonic adenocarcinoma, 434
Colorectal adenomas, 394, 395
Colorimetric assay, 245
Comel-Netherton syndrome, 367
Competitive protein binding assays, 245, 248, 249
Costochondral junctions, 277
Cow's milk, 6
Coxa vara, 276, 298
C-24 oxidation pathway, 62
CPBA, 245, 248, 249
CRE, 219
C-terminal domain, 115–118
    phosphorylation, 118
    transactivation, 117
CTX, 86–88
Cultured breast cancer cells, 413
Cyclic guanosine monophosphate, see cGMP
Cyclodextrins, 25
Cyclosporine
    psoriasis, 266
CYP24, 62, 77
CYP27, 57–59, 77, 86
CYP24cDNA, 90, 91
CYP24cDNA expression
    bacterial cells, 91
CYP24 knockout animals, 62
CYP24mRNA, 62
Cys 288, 154
Cysteine residues, 114
Cytokines
    sarcoidosis, 342, 343
Cytoreceptor assay
    1,25-dihydroxyvitamin D assay
        methodology, 256, 257

D

DBD, see DNA-binding domain
DBP, 57
    actin binding, 104
    C5a cochemotaxis, 105
    cell-associated, 104, 105
    clinical significance, 101–105
    clinical studies, 104
    gelsolin, 104
    human serum concentrations, 101–103

osteoclastogenesis, 105
plasma actin scavenger system, 104
sterol transport, 101–104
three-dimensional structure
   crystallographic approach, 158, 159
   vitamin D_3, 149, 150
vitamin D metabolites, 101
vitamin D_3 metabolites, 149, 150
DBP amino acid mutations
   ligand-binding, 153, 154
DBP gene knockouts, 77
DBP ligand-binding domain
   covalent modification
      affinity and photoaffinity labeling, 154–158
7-dehydrocholesterol, 3, 6
Demineralization, 298
Dexamethasone, 441
7-DHC, 20, 21, 40
Diet, 3–13, 40, 41, 347
Dietary 1,25-dihydroxy-16-ene-23-yne26,27-hexafluorocholecalciferol, 402
Dietary factors
   colon cancer, 379, 380
Dietary fat, 379
   colon cancer, 432, 433
Dietary vitamin D_3
   colon tumors, 398
Dihydrotachysterol
   activation, 64
   structure, 64
1,25-dihydroxy-16-ene-23-yne-26,27-hexafluorocholecalciferol
   breast cancer, 398, 437
   colon cancer, 436
   tamoxifen, 416, 417
1,25-dihydroxyvitamin D
   DBP, 101
   extrarenal production, 337–349
   PTH gene expression regulation, 219–223
1,25-dihydroxyvitamin D_3
   bone remodeling, 289–291
   breast cancer, 398, 411–422
   cancer, 207, 208, 212
   cell culture studies, 395–397
   colon cancer, 399, 433, 434
   historical perspective, 207
   immunoregulatory effects, 348, 349
   intestinal calcium absorption, 163–166
   intestinal phosphorus absorption, 166
   intestine, 291

   kidney, 169, 170
   noncalcemic actions, 207–214
   production and catabolism regulation, 41–45
   prostate, 210
   renal calcium transport, 166–170
   secondary hyperparathyroidism, 230, 231
   skin, 208–210, 291, 292
   skin and hair proliferative disorders, 213
   skin cancer, 398
   smooth muscle, 210, 211
   transcellular intestinal calcium absorption, 164–166
24,25-dihydroxyvitamin D_3, 89, 90
1,25-dihydroxyvitamin D_3 analogs
   colon cancer, 434–436
   psoriasis, 368–370
24,25-dihydroxyvitamin D_3 assay, 253–255
1,25-dihydroxyvitamin D assay
   methodology, 255–262
   bioassays, 256
   cytoreceptor assay, 256, 257
   GLC/MS assays, 257
   radioimmunoassays, 258–260, 262
   radioreceptor assays, 257, 258, 260–262
1,25-dihydroxyvitamin D_3-26,23-lactone assay, 263, 264
Direct repeat 3, 45
Distal growth plates, 277
Dithranol
   psoriasis, 266
DNA-binding domain, 318, 319
DNA-binding domain mutations
   HVDRR, 322–325
DR3, 45
Dunning rat prostatic adenocarcinoma model, 439, 440
Duodenum
   calcium absorption, 291
Dynein, 137

E

EB 1089, 72, 73, 418, 419, 437
Elderly
   bone health, 287–300
   bone loss, 296, 297
   recommended dietary intake, 11
Endogenous molybdate factor, 140
Endogenous vitamin D intoxication
   diagnosis, 346
   management, 346, 347

screening and prevention, 346, 347
  treatment, 347
Endoplasmic reticulum, 135, 136
Epiphyseal plates, 4
ER, 114
ERE, 119
Ergocalciferol, *see* Vitamin D_2
Ergosterol, 19
Escherichia coli, 91
Eskimos, 6
Estradiol, 419
Estrogen receptor, 114
Europe, 6, 17, 19, 29
Extrinsic vitamin D depletion, 298, 299

F

Face
  calcipotriol, 365
Failure to thrive, 276
Familial adenomatous polyposis, 399
F6-1α,25(OH)$_2$D$_3$, 72
Fenretinide, 417
Fibroblasts, 321, 322
Fish, 6
Flexures
  calcipotriol, 365
Flucocorticoids, 441
5-fluorouracil, 417
Fms, 418
Food and Drug Administration guidelines, 19
Foods, 6
Fos, 418
Free hormone hypothesis, 102, 103
Free sterol hypothesis, 102, 103
French Canadian population, 308
Furosemide, 347

G

Gal 4, 117
Gardner syndrome/familial polyposis, 379
Gas/liquid chromatography, 245, 249, 250
Gc2 isomorph, 154
Gelsolin
  DBP, 104
Gelsolin-DBP, 104
Gene expression
  vitamin D regulation
    osteoblasts bone formation, 178–181
Gene transcription
  VDR phosphorylation, 50

GH, 43
GLC, 245, 249, 250
GLC/MS assays
  1,25-dihydroxyvitamin D assay methodology, 257
Glucocorticoid receptor, 114
Glucocorticoids, 339, 347
GR, 114
Granuloma-forming disease, 344, 345
GRE, 119
Great Britain, 29
Grover's disease, 368
Growth factors
  breast cancer, 419, 420
Growth hormone, 43
GR translocation, 130
Guanylate cylcase, 139, 140

H

Harrison's groove, 276
Hereditary vitamin D-resistant rickets, *see* HVDRR
High-pressure liquid chromatography, 245–247, 249–251
Hinge, 319
Hip fracture, 296
His 305
HL-60, 395–397, 401–403, 418, 442
HLA-Cw6, 348–359
Hodgkin's lymphoma, 345–347
HPLC, 245–247, 249–251
hsp90, 137
HT-29, 443, 444
Human xenograft murine models, 442–445
HVDRR, *see also* Vitamin D-dependent rickets type II
  cellular basis, 321, 322
  clinical features, 317, 318
  DNA-binding domain mutations, 322–325
  etiology, 318
  ligand binding domain mutations, 325–328
  molecular basis, 322–328
  prenatal diagnosis, 331
  spontaneous improvement, 331
  therapy, 328, 329
  VDR receptor defects, 317–331
Hydroxyapatite, 179
Hydroxyapatite crystals, 175, 176
Hydroxy-chloroqine, 347

Hydroxylase, 77
1-hydroxylase, 43
24-hydroxylase, 95, 341
25-hydroxylase, 86–88
   gene structure anomaly, 86–88
   unique function, 86
25-hydroxylase deficiency, 314
24-hydroxylase regulation, 50
24-hydroxylated vitamin D metabolism, 44
1-hydroxylation
   abnormal, 299, 300
24-hydroxylation, 59, 60
25-hydroxylation
   impaired, 299
   metabolism, 57–59
26-hydroxylation, 60, 61
1,25-hydroxyvitamin D
   vitamin D-dependent rickets type I, 309
24-hydroxyvitamin D
   DBP, 101
25-hydroxyvitamin D
   epidemiologic UV studies, 386
   vitamin D-dependent rickets type I, 308
25-hydroxyvitamin D_3 1α-hydroxylase, 89
25-hydroxyvitamin D_3 24-hydroxylase, 89–91
   cDNA cloning, 90
   COS cells expression, 90, 91
   gene structure, 91
   purification, 90
   regulation, 92–95
      kidney, 92
      mechanism, 94
      Northern blot analysis, 92–94
Hypercalcemia, 43
Hypercalciuria, 345, 347
Hyperparathyroidism, 4
Hyperphosphatemia, 43
Hypocalcemia, 43, 64, 278

I

Ichthyosis
   calcipotriol, 367
IGF-I, 43, 420
Ileum
   phosphate absorption, 291
Immigrants
   colon cancer, 376–378
Immune system
   1,25-dihydroxyvitamin D_3, 211, 212
Indian skin, 27
Inducible nitric oxide synthase, 343, 344
Infant food enrichment, 283
Infants
   recommended dietary intake, 6
   vitamin D deficiency, 275–277
INF receptor, 343
iNOS, 343, 344
Inositol triphosphate, 196
Institute of Medicine recommendations, 7–13
Insulin-like growth factor-1, 43
Intestinal calcium absorption
   1,25-dihydroxyvitamin D_3, 163–166
Intestinal calcium absorption assay, 244
Intestinal calcium binding proteins
   stimulation, 244
Intestinal cell
   calcium extrusion, 165, 166
Intestinal tumor, 439
Intestine
   1α,25(OH)$_2$D$_3$
      nongenomic effects, 200, 201
   1,25-dihydroxyvitamin D_3, 291
Intrinsic vitamin D depletion, 299
Intron polymorphisms, 293
Iowa Women's Health Study, 394
IP3, 196

J

Jejunum
   phosphate absorption, 291
Jun, 418

K

Ketoconazole, 370
KH1060, 418
   metabolism, 69–72
Kidney
   1,25-dihydroxyvitamin D_3, 292
   1,25(OH)$_2$D$_3$, 169, 170
Kidney microsomal 1α-hydroxylase, 89
Kidney mitochondrial enzyme, 89
Knock knees, 3
K-ras proto-oncogene, 402
Kyphosis, 276, 298

L

Lactation
   recommended dietary intake, 11, 12
26,23-lactone formation, 60, 61
Latitude
   breast cancer, 380, 381
   colon cancer, 376
   vitamin D, 381–385

LBD, 115, 116
Leukemia, 440, 441
Ligand binding domain, 115, 116, 318, 319
Ligand binding domain mutations
    HVDRR, 325–328
Ligand-induced transcription factor, 45
Line test, 243
Lipid-membrane environment, 33
Lipopolysaccharide, 343
    sarcoidosis, 342, 343
Liver, 6
Liver microsomal 1α-hydroxylase, 89
Liver microsomal enzyme, 86
Liver mitochondrial 1α-hydroxylase, 89
Liver mitochondrial enzyme, 86
Loop of Henle, 167
Low-pressure liquid chromatography, 249
LPLC, 249
LPS, 343
Lymphocytes
    paracrine effects, 348, 349
Lymphokine, 349

M

Macrophage
    paracrine effects, 349
Macrophage 1-hydroxylase, 340–344
Malignant lymphoproliferative disorders, 346
Margarine, 6, 19
Maternal osteomalacia, 274
Maternal vitamin D deficiency, 274
MCF-7, 412, 416, 418, 419
MD 903, 437
Medroxyprogesterone acetate, 417
MediSun, 30
Melanin pigmentation, 26, 27
Membrane vitamin D receptor
    covalent labeling, 158
Metaphyseoepiphyseal zones, 277, 278
Microtubule
    VDR, 136–138
Milk, 6
    vitamin D fortification, 19, 20
Missense mutations, 324
Mitochondrial 25-hydroxylase, 86
Molybdate, 137, 138
MPA, 417
Multiple assays, 264
Multivitamins, 6
myc, 418
Mycobacterium tuberculosis, 344, 345

Myocardial smooth muscle
    1,25-dihydroxyvitamin D_3, 210, 211

N

NADPH-cytochrome P450 reductase, 86
Nail psoriasis
    calcipotriol, 362
Narrow band UVB phototherapy
    psoriasis, 266
nCaRE, 224, 225
N-CoR, 46, 47, 122
Negative calcium regulatory element, *see* nCaRE
Neonatal hypocalcemia, 274
Neonates
    vitamin D deficiency, 274
        atypical pattern, 281
New Zealand, 295
Nitric oxide, 343, 344
NMU tumors, 415
Non-Hodgkin's lymphoma, 345–347
Nonpolyposis cancer family syndrome, 379
Nonpolyposis colon cancer, 379
Norwegian infants, 7
NPC, 129
N-telopeptide, 296
N-terminal DNA binding domain, 113–118
    localization and structure, 113, 114
    VDR-DNA contacts, 114, 115
Nuclear localization sequences, 129
Nuclear matrix, 187
Nuclear pore complex, 129
Nuclear receptor co-repressor, 46, 47
nVDR, 195, 196

O

OC box I, 181, 182
OC box II, 181, 182
Ocean-dwelling animals, 1
OCT, 68–70, 397, 398, 416, 417
$1,25(OH)_2D_3$, 40
    calcium, 164–166
$25(OH)D_3$
    metabolism, 57–61
$1,25(OH)_2D_3$ (Ro 21-5535), 439, 440
$1,25(OH)_2D_3$, *see* 1,25-dihydroxyvitamin D_3
$24,25(OH)_2D_3$, *see* 24,25-dihydroxyvitamin D_3
$25(OH)D$, 3
    catabolism, 299
$25,26(OH)_2D_3$, 60, 61
$25(OH)D$ assay, 247–253

Scalp psoriasis
  calcipotriol, 362
SCCVII/SF squamous cell carcinoma model, 441
Scleroderma
  vitamin D analogs, 367
Scoliosis, 276
Seasons, 6, 27, 28
Secondary hyperparathyroidism
  1,25-dihydroxyvitamin D_3, 230, 231
  PTH cell proliferation, 225–227
  renal disease, 230, 231
Secosteroids
  cancer risk, 393–404
SEER registry, 382–385
  ovarian cancer, 385
Serum calcium levels, 42, 43
Serum phosphorus levels, 42, 43
Sialoprotein, 176
Silicone-induced granulomata, 345
Skeletal mineralization, 5, 40
Skeletal smooth muscle
  1,25-dihydroxyvitamin D_3, 210, 211
Skin, 2, 3, 5, 6
  1,25-dihydroxyvitamin D_3, 208–210, 291, 292
  previtamin D_3, 20–23
Skin cancer, 398
  calcitriol, 367
Skin pigmentation, 5
SMART, 122
SRC-1, 46, 122
Steroid receptor coactivator-1, 46
Sterol transport, 101–104
  secondary carriers, 103, 104
Submariners, 8
Sunlight, 5, 6
  breast cancer, 412
Sunscreen, 5, 30, 31
Suntanning lamp, 30
Surveillance Epidemiology and End Results registry, 382–385
Syngeneic cancer models, 430–442

T

Tamoxifen, 415–417
TATA binding protein, 121
TATA motif, 181, 182
T-box, 114
TBP, 121
Terrestrial vertebrates, 3

TFIIB, 46, 121, 122, 184
TGF-B, 420
Th1 profile disease, 359
Th2 profile disease, 359
Thyroid replacement receptor, 119
Tierra del Fuego, 8
TR, 119
Transcaltachia, 166, 200
Transcription factor IIB, see TFIIB
Transient acantholytic dermatosis, 367, 368
TRE, 119
Trihydroxyvitamin D_3 metabolites assay, 263, 264
Trousseau's sign, 275
Trp 145, 154
Tuberculosis, 344, 345
Type II osteoporosis, 296, 297
Type I osteoporosis, 292–296

U

United States, 6, 17, 19, 29
Unsaturated analogs, 72–74
UV, 26, 27
  epidemiologic studies
    25-hydroxyvitamin D, 386
UVB
  psoriasis, 266

V

VDDR I, see vitamin D-dependent rickets type I
VDDR II, see vitamin D-dependent rickets type II
v7-DHC
  subcellular localization, 24
VDR, 77
  breast cancer, 412, 413
  cGMP accumulation, 138–140
  colorectal cancer, 399, 400
  C-terminal domain, 115
  1,25-dihydroxyvitamin D action, 318
  distribution and translocation
    living cells, 133–135
  docking sites
    endoplasmic reticulum, 135, 136
  domain structure, 318, 319
  malignant cell lines, 397
  microtubule interaction, 136–138
  molecular biology, 109–123
  N-terminal DNA binding domain, 113–115

1,25(OH)$_2$D$_3$ ligand, 121
  RXR interaction, 46
  subcellular distribution, 131–133
  VDRE, 118–120
  vitamin D$_3$, 149, 150
  vitamin D$_3$ metabolites, 149, 150
VDR amino acid mutations
  ligand-binding, 153, 154
VDR-associated fibers, 136
VDRcDNA, 45
VDR cDNA
  characteristics, 112, 113
  isolation, 111, 112
  sequence comparisons, 112, 113
VDR cDNA sequences, 111–113
VDR complex
  formation, 183, 184
VDR DNA sequences
  cloning and characterization, 110–113
VDRE, 45, 47, 94, 109, 223
  nucleotide sequence, 184
  25(OH)D$_3$-24-hydroxylase gene
    promoter, 49
  protein-DNA, 182–197
  protein-protein, 182–197
VDR functional domains
  molecular analysis, 113–118
VDR function defects, 133
VDR gene
  chromosomal localization, 110, 111
  genetic mutations, 39
  mutations, 324, 325
  organization, 111
VDR gene mutations, 133
VDR gene organization, 319–321
VDR ligand-binding domain
  covalent modification
    affinity and photoaffinity labeling,
      154–158
VDR ligand-binding pocket studies, 78
VDR mRNA transcript, 319–321
VDR phosphorylation, 47
  gene transcription, 50
VDR polymorphisms, 293
VDR receptor defects
  HVDRR, 317–331
VDR-RXR heterodimer, 76, 110, 116, 117,
  120, 121
VDR-RXR heterodimeric complex, 184
VDR-RXR-VDRE complex, 76
VDR transcriptional control, 118–122

VDR translocation, 129–141
v-erb-A superfamily, 131
Vertebral fracture, 294–296
Vesicular transport model, 166
Vitamin D
  bone, 175–189
  calciotropic role, 109
  calcium connection, 1, 2
  calcium homeostasis, 40
  cancer risk
    cell culture studies, 395–397
    epidemiological studies, 394, 395
    experimental animal studies, 397–399
    human studies, 399, 400
  DBP, 101
  evolution, 1–3
  extrarenal production, 337–349
  functional metabolism, 39–45
  genetic defects, 39
  latitude, 381–385
  metabolism, 3
  molecular biology, 45–50
  nongenomic rapid effects, 195–202
  phosphorus homeostasis, 40
  photobiology, 17–33
  recommended dietary allowance, 3–6
  recommended dietary intake, 6–13
    birth to 6 months, 6, 7
    lactation, 12
    pregnancy, 11, 12
    tolerable upper levels, 12
    6 to 12 months, 7
    1 to 8 years, 7, 8
    9 to 18 years, 8
    19 to 50 years, 8, 9
    51 to 70 years, 9, 10
    71 years and older, 10, 11
  skin photosynthesis, 2, 3
  sources, 5, 6, 40, 41
Vitamin D$_2$, 19, 20
  activation, 62, 63
Vitamin D$_3$
binding proteins, 148, 149
DBP, 149, 150
  metabolism, 57–61
  photodegradation, 32
  photoisomers, 32
  translocation from skin to circulation, 25, 26
Vitamin D analogs
  anticancer activity, 431–444
  calcemic vs noncalcemic, 77, 78

colon cancer, 434–436
combination
    psoriasis, 366, 367
cyclopropane ring-containing, 67, 68
HVDRR, 328
mechanism of action implications, 77, 78
metabolism and catabolism, 62–74
metabolism-resistant, 72–74
metabolism-sensitive, 67–72
oxa-group containing, 68–72
scleroderma, 367

Vitamin D_3 analogs
    action mechanisms, 400–403
    chemotherapeutic actions, 393–404

Vitamin D assays
    clinical utility, 239–265
    methodology, 242–247

Vitamin D binding protein, *see* DBP

Vitamin D deficiency
    biochemical signs, 278–282
    clinical features, 274–277
        children and adolescents, 277
        infants, 275–277
        neonates, 274
    consequences, 305
    neonates
        atypical pattern, 281
    prevention, 282
    radiologic signs, 277, 278
    redefinition, 12, 13
    subclinical, 282

Vitamin D deficiency rickets, 282
    treatment, 282

Vitamin D-dependent calcium binding proteins
    renal calcium transport, 168

Vitamin D-dependent rickets type I, 39, 307–311
    clinical presentation, 308–311
    management, 308–311
    vitamin D, 308

Vitamin D-dependent rickets type II, 39, 311–314, *see also* HVDRR
    calcium therapy, 314
    clinical presentation, 311–314
    management, 311–314
    pregnancy, 315
    vitamin D metabolites, 311–314

Vitamin D endocrine system, 149, 159

Vitamin D fortification
    milk, 19, 20

Vitamin D homeostasis
    bone remodeling, 287–291
    sarcoidosis, 339–346

Vitamin D intoxication, 6
    endogenous, 346, 347

Vitamin D ligands, 102, 103

Vitamin D metabolism
    inherited defects, 307–314
    osteomalacia, 298–300

Vitamin D metabolites
    DBP, 101
    immunoregulatory effects, 348, 349

Vitamin D_2 metabolites, 72

Vitamin D_3 metabolites
    action mechanisms, 400–403
    binding
        structural requirements, 150–152
    binding proteins, 148, 149
    chemotherapeutic actions, 393–404
    VDR, 149, 150

Vitamin D molecule
    novel synthetic modifications, 78

Vitamin D/PTH/calcitonin endocrine system, 41, 42

Vitamin D receptor, *see* VDR

Vitamin D-regulated genes, 47–50

Vitamin D regulation
    gene expression
        osteoblasts bone formation, 178–181
    gene transcription
        nuclear structure influences, 187–189

Vitamin D response elements, *see* VDRE

Vitamin D signaling pathways, 77

Vitamin D signal transduction cascade, 45

Vitamin D signal transduction system, 77

Vitamin D supplementation, 283

Vitamin D synthesis
    percutaneous
        regulation, 291, 292

W

Women
    recommended dietary intake, 7–13
        postmenopausal, 10
        premenopausal, 8

Y

Yeast two-hybrid system, 50
YY-1, 184

Z

Zinc fingers, 319
ZR-75-1, 416, 419